Dielectric Materials and Applications

For a complete listing of the *Artech House Microwave Library*,
turn to the back of this book

Dielectric Materials and Applications

Arthur von Hippel, editor

Artech House
Boston • London

Library of Congress Cataloging-in-Publication Data
Von Hippel, Arthur R. (Arthur Robert), 1898-
Dielectric Materials and Applications / Arthur von Hippel, editor.
Originally published: Cambridge : Technology Press of MIT, 1954
Includes bibliographical references and index.
ISBN 0-89006-805-4
1. Dielectrics I. Title
QC585.V6   1995   94-25138
537.–dc20   CIP

British Library Cataloguing in Publication Data
Hippel, Arthur von
Dielectric Materials and Applications. New ed
I. Title
537.24

ISBN 1-58053-123-7

© 1995 Arthur von Hippel

All rights reserved. Printed and bound in the United States of America. No part of this book may be reproduced or utilized in any form or by any means, electronic or mechanical, including photocopying, recording, or by any information storage and retrieval system, without permission in writing from the publisher.

International Standard Book Number: 1-58053-123-7
Library of Congress Catalog Card Number: 94-25138

10 9 8 7 6 5 4 3 2 1

To the memory of

KARL TAYLOR COMPTON,

**beloved friend and counselor of
scientists and engineers**

CONTRIBUTORS

Fred H. Behrens
MAJOR, U. S. AIR FORCE, WRIGHT AIR DEVELOPMENT CENTER, DAYTON, OHIO

Leo J. Berberich
MANAGER, LIAISON ENGINEERING DEPT., WESTINGHOUSE ELECTRIC CORP., EAST PITTSBURGH, PA.

Francis Bitter
PROFESSOR OF PHYSICS, M.I.T.

John T. Blake
DIRECTOR OF RESEARCH, SIMPLEX WIRE AND CABLE CO., CAMBRIDGE, MASS.

Dudley A. Buck
STAFF MEMBER OF DIGITAL COMPUTER LABORATORY, M.I.T.

Warren F. Busse
POLYCHEMICALS DEPARTMENT, EXPERIMENTAL STATION, E. I. DU PONT DE NEMOURS AND CO., WILMINGTON, DEL.

Thomas D. Callinan
NAVAL RESEARCH LABORATORIES, WASHINGTON, D. C.

David J. Epstein
STAFF MEMBER OF LABORATORY FOR INSULATION RESEARCH, M.I.T.

Robert F. Field
FORMERLY OF GENERAL RADIO CO., CAMBRIDGE, MASS.

Clarence W. Hewlett
CHIEF, PHYSICS SECTION, THOMSON LABORATORY, GENERAL ELECTRIC CO., LYNN, MASS.

Louis Kahn
DIRECTOR OF RESEARCH, AEROVOX CORPORATION, NEW BEDFORD, MASS.

Charles P. Lascaro
SIGNAL CORPS, SQUIER SIGNAL LABORATORY, FORT MONMOUTH, N. J.

William N. Papian
STAFF MEMBER OF DIGITAL COMPUTER LABORATORY, M.I.T.

John D. Piper
HEAD OF CHEMICAL DIVISION, ENGINEERING LABORATORY AND RESEARCH DEPT., DETROIT EDISON CO., DETROIT, MICH.

Samuel J. Rosch
MANAGER, INSULATED PRODUCTS DEVELOPMENT, ANACONDA WIRE AND CABLE CO., HASTINGS-ON-HUDSON 6, NEW YORK

Malcolm W. P. Strandberg
ASSOCIATE PROFESSOR OF PHYSICS AND STAFF MEMBER OF RESEARCH, LABORATORY OF ELECTRONICS, M.I.T.

Hans Thurnauer
DIRECTOR OF RESEARCH, AMERICAN LAVA CORP., CHATTANOOGA, TENN.

John G. Trump
PROFESSOR OF ELECTRICAL ENGINEERING, AND DIRECTOR OF HIGH VOLTAGE RESEARCH LABORATORY, M.I.T.

Karl S. Van Dyke
PROFESSOR OF PHYSICS, WESLEYAN UNIVERSITY, MIDDLETOWN, CONN.

Arthur von Hippel
PROFESSOR OF ELECTROPHYSICS AND DIRECTOR OF LABORATORY FOR INSULATION RESEARCH, M.I.T.

Arthur J. Warner
TECHNICAL DIRECTOR, CHEMICAL AND PHYSICAL LABORATORIES, FEDERAL TELECOMMUNICATION LABORATORIES, NUTLEY, N. J.

William B. Westphal
STAFF MEMBER OF LABORATORY FOR INSULATION RESEARCH, M.I.T.

PREFACE TO THE SECOND EDITION

As a practicing engineer, I have used in my technical work the two-volume von Hippel set as an indispensable reference for dielectric properties of materials. When I realized that they were out of print, I sent copies of selected pages to Artech House for evaluation with the recommendation that both volumes be republished for the benefit of other interested engineers. Several months later, Artech House notified me that a decision had been made to republish, that the rights to republish had been secured from the retired author, and that, as implausible as it may seem, neither the previous publishers nor the author have any remaining copies of this set, as would be required for the republication process. Therefore, I supplied the original editions to Artech House for reproduction. I would like to thank the entire staff at Artech House for their professionalism in the handling of this republication.

Huntington Beach, California ALEXANDER S. LABOUNSKY
August 1994 Principal engineer/scientist
McDonnell Douglas Aerospace

PREFACE TO THE FIRST EDITION

Many foresee that science and industry are building a Tower of Babel and that this undertaking will be halted as in Biblical times: the laborers, more and more specialized, will finally cease to understand each other. The editor, for one, does not share this gloomy conviction. On the contrary, as our knowledge grows, old boundaries vanish and the view expands to broader horizons. However, people accustomed to boundaries in certain places tend still to respect them after their actual disappearance. To make them feel at home with their new neighbors is a driving motive of this book and its companion volume. In *Dielectrics and Waves* we have tried to bring together physicist, chemist, and electrical engineer; in *Dielectric Materials and Applications* we hope to establish alliances between research worker, development engineer, manufacturer, field engineer, and actual user of "nonmetals."

It is frequently maintained that a true scientist should not think of applications, but new insight must be tested in order to reveal loopholes in our knowledge. Pasteur, in the domestication of microbial life, went from the laboratory to the sheep farm, wine cellar, and brewery, and returned, invigorated by success (the sheep survived, the wine improved) and humiliated by failure (the French beer is still terrible). His continuous reshaping of ideas through practical applications was the necessary prerequisite for the development of the germ theory of disease, for an understanding of infection and its attenuation by imparted immunity. We are no Pasteurs, but our ideas can stand some testing.

Preface

This conviction that scientists, engineers, manufacturers and users of dielectrics should learn to speak each other's language and appreciate mutual problems, failures, and advances was the driving impulse which led to the Summer Session Course of the Laboratory for Insulation Research at the Massachusetts Institute of Technology in September, 1952. The lectures, given by experts from science, industry, Government, and our own group, and revised and edited in book form, are herewith presented as the outcome of this experiment. We hope the book conveys not only useful knowledge but also some of the eagerness with which we talked and listened. The content will show that we did not try to stay always on Main Street but peered into darker side alleys whenever the spirit moved us. Thus problems are brought into the limelight which could stimulate much fruitful new activity.

The editor wants to thank all contributors for their patient understanding of his task and their unfailing co-operation. To make a book out of disconnected contributions requires much cutting and rewriting, for which the editor needs a "blank check of confidence."

Our task has been lightened by the invaluable help of Dr. L. G. Wesson and Miss Aina Sils in readying the manuscript for the printer and of our excellent draftsman, J. J. Maccarone, who provided a major part of the finished illustrations.

A vote of gratitude is also in order for the support of the Laboratory for Insulation Research by the ONR, the Army Signal Corps, and the Air Force (ONR Contracts N5ori-07801 and N5ori-07858). Without their sponsorship of our research program and full understanding and support of our aims, we would not have been able to undertake this task.

Cambridge, Massachusetts
July 1954

A. VON HIPPEL

CONTENTS

Survey	Arthur von Hippel	1
I · Theory	Arthur von Hippel	3
A · *Macroscopic Properties of Dielectrics*		3
1 · Complex Permittivity and Permeability		3
2 · Polarization and Magnetization		5
3 · Description of Dielectrics by Various Sets of Parameters		9
4 · Reflection and Refraction of Electromagnetic Waves on Boundaries; Measurement of Dielectrics by Standing Waves		13
B · *Molecular Properties of Dielectrics*		18
1 · Molecular Mechanisms of Polarization		18
2 · Polarization and Atomic Structure		21
3 · Structure and Dielectric Response of Molecules		30
4 · Relaxation Polarization in Liquids and Solids		36
5 · Piezoelectricity and Ferroelectricity		40
II · Dielectric Measuring Techniques		47
A · *Permittivity*		47
1 · Lumped Circuits	Robert F. Field	47
2 · Distributed Circuits	William B. Westphal	63
B · *Permeability*	David J. Epstein	122
C · *Microwave Spectroscopy*	Malcolm W. P. Strandberg	134
D · *Magnetic Resonance*	Francis Bitter	139
III · Dielectric Materials and Their Applications		147
A · *Dielectric Materials*		147
1 · Insulation Strength of High-Pressure Gases and of Vacuum . . .	John G. Trump	147
2 · Liquid Dielectrics	John D. Piper	156
3 · Plastics as Dielectrics	Warren F. Busse	168
4 · Ceramics	Hans Thurnauer	179

Contents

- B · *Dielectrics in Equipments* ... 189
  - 1 · Dielectrics in Power and Distribubution Equipment ... Leo J. Berberich ... 189
  - 2 · Dielectrics in Electronic Equipments ... Arthur J. Warner ... 211
  - 3 · Dielectrics in Capacitors ... Louis Kahn ... 221
  - 4 · Rubber and Plastics in Cables ... John T. Blake ... 225
  - 5 · Problems of the Cable Engineer ... Samuel J. Rosch ... 233
- C · *Dielectric Materials as Devices* ... 242
  - 1 · Rectifiers ... Clarence W. Hewlett ... 242
  - 2 · Piezoelectric Transducers and Resonators ... Karl S. Van Dyke ... 250
  - 3 · Magnetic and Dielectric Amplifiers ... Dudley A. Buck ... 261
  - 4 · Memory Devices ... William N. Papian ... 274

IV · **Dielectric Requirements of the Armed Services** ... 283

- A · *The Air Force* ... Fred H. Behrens ... 283
- B · *The Army* ... Charles P. Lascaro ... 285
- C · *The Navy* ... Thomas D. Callinan ... 288

V · **Tables of Dielectric Materials** ... Laboratory for Insulation Research Massachusetts Institute of Technology ... 291

- Introduction ... 291
- Contents ... 292
- Dielectric Parameters ... 294
- Measurements and Accuracy ... 300
- Tabulated Dielectric Data ... 301
- Data at Fixed Frequencies f as a Function of Temperature ... 371
- Company Index ... 426
- Materials Index ... 429

Index ... 435

Survey

By ARTHUR VON HIPPEL

Dielectrics, in the sense of this book, are not a narrow class of so-called insulators, but the broad expanse of *nonmetals* considered from the standpoint of their interaction with electric, magnetic, or electromagnetic fields. Thus we are concerned with gases as well as with liquids and solids, and with the storage of electric and magnetic energy as well as with its dissipation.

Taking this point of view, we may summarize the dielectric properties of matter in the three terms *polarization, magnetization,* and *conduction*. How these phenomena can be described macroscopically and interpreted from the standpoint of molecular theory, how they can be measured, what the properties of present-day materials are, how dielectrics are applied in equipments or used as devices, these are the questions under consideration. Semiconductors form a part of this dielectrics domain, but they will be treated only in passing, since many special conferences in recent time have been devoted to their discussion.

In introducing the scientific background (I, "Theory"), we must first develop a concise language that describes phenomenologically what we observe when subjecting a material to electric or to magnetic fields (IA, "Macroscopic Properties of Dielectrics"). The concepts *complex permittivity* ϵ^* and *complex permeability* μ^*, defined by a lumped-circuit approach (IA, Sec. 1) and incorporated into the field theory through the vectors *polarization* and *magnetization* (IA, Sec. 2), provide a generally useful description for sinusoidal fields. Various other sets of parameters, such as the complex propagation factor and impedance of the electrical engineer or the complex index of refraction which serve the physicist in the optical spectral range, can be reduced to ϵ^* and μ^* in short order (IA, Sec. 3). This macroscopic description is concerned not only with the free volume of dielectrics but also with boundary phenomena; in fact, the reflection and refraction of waves on boundaries provide the essential means for measuring the interaction between fields and matter. Hence, a short discussion of these boundary effects, especially for standing waves, is presented as an introduction for the measurement of dielectrics in the microwave range (IA, Sec. 4).

The field vectors, polarization and magnetization, can be interpreted as the electric and the magnetic dipole moment per unit volume of the dielectric. By visualizing that these moments express the additive action of a multitude of elementary moments, a straightforward entry may be made from the macroscopic into the molecular world (IB, "Molecular Properties of Dielectrics"). Various molecular mechanisms of polarization have to be distinguished (IB, Sec. 1): the formation of induced moments by the displacement of electrons or nuclei (electronic and atomic polarization), moments resulting from the orientation of permanent dipoles (orientation polarization), and finally field distortions caused by traveling charge carriers (space-charge or interfacial polarization). By applying these considerations to atoms and molecules in the gaseous state, the basic facts of spectroscopy emerge and the structures of atoms and molecules become apparent (IB, Secs. 2 and 3). These facts lead simultaneously to a break with classical physics and to a substitution of quantum mechanics for Newtonian mechanics in the description of molecular events.

The dielectric response of gases expresses itself in resonance spectra; when gases condense to the various structures of liquids and solids, the electrical frequency range, previously empty, becomes populated with various types of relaxation spectra (IB, Sec. 4). The major part of these spectra is caused by the hindered rotation of permanent dipole moments. In crystals such moments are usually built into the structure, and their freedom of individual orientation is thus lost completely. However, under favorable conditions they can still act concertedly, as the phenomena *piezoelectricity* and *ferroelectricity* testify (IB, Sec. 5).

After this introduction has established a common background of language and ideas, various methods are discussed for measuring ϵ^* and μ^* in the range from direct current to ca. 3×10^{10} cycles per second (Part II, "Dielectric Measuring Techniques"). The standard subdivision into lumped circuits (IIA, Sec. 1) and distributed circuits (IIA, Sec. 2) is retained to give a comprehensive survey of permittivity measurements, from charging and discharging currents to bridge and resonance techniques and finally to the various waveguide techniques developed during the last ten years. The techniques for the determination of the magnetic permeability follow in part parallel lines, but the nonlinear character of ferromagnetics requires additional high field-strength measurements and a careful analysis of the parameters actually under investigation (IIB). A corresponding situation arises in permittivity measurements on ferroelectrics.

For readers faced with specific measuring problems, enough information has been added in fine print and in the form of appendices with charts, tables, and references, to guide not only in the proper selection of the method but also in its actual operation.

In addition to the more or less classical techniques, two resonance methods are presented which have gained fundamental importance since the Second World War: microwave spectroscopy (IIC) and magnetic resonance (IID). In both cases quantized transitions between well-defined energy states are measured with the tools and in the frequency range of the electrical engineer. Extremely accurate information can thus be obtained on the electric and magnetic moments and other structure parameters of nuclei, atoms, and molecules and, by magnetic resonance, on the electric and magnetic fields in which nuclei and electrons find themselves in liquids and solids. The powerful impact of these new techniques is just beginning to be felt in many problems of solid-state research.

This general survey on the theory of dielectrics and the measurement of their characteristics is followed by a discussion of dielectric materials and their applications. What the properties and present limitations of materials (IIIA) are, how they are being used in equipments (IIIB), and how they can serve as devices (IIIC) are the developing themes of this third part of the book. Appropriately, it begins with the gaseous state, where the breakdown strength of compressed gases and of vacuum is of foremost technical interest (IIIA, Sec. 1). Then the spotlight shifts to liquids (IIIA, Sec. 2), plastics (IIIA, Sec. 3), and ceramics (IIIA, Sec. 4), and the main emphasis turns to dielectric constant and loss, mechanical properties, structure, stability, and considerations of manufacture.

Two main fields of application may be distinguished: dielectrics in power and distribution equipment (IIIB, Sec. 1) and in electronics equipment (IIIB, Sec. 2). However, there is really no strict boundary, and the two subsequent sections, "Dielectrics in Capacitors" (IIIB, Sec. 3) and "Rubber and Plastics in Cables" (IIIB, Sec. 4), switch freely between the power and communications field. All four contributions return frequently to problems discussed in previous sections with a new emphasis on questions of practical performance. The last word in this discussion (IIIB, Sec. 5) is properly left to a cable engineer who knows from bitter experience that, not the office and laboratory, but deserts and sewers are the final proving ground for his product.

Turning to dielectric materials as devices, we concentrate on four applications of urgent present-day interest, which simultaneously show dielectrics at work under very different circumstances. Rectifiers (IIIC, Sec. 1) introduce dielectrics as semiconductors; piezoelectric transducers and resonators (IIIC, Sec. 2) center on the electromechanical response; magnetic and dielectric amplifiers (IIIC, Sec. 3) make use of the nonlinear characteristics of ferromagnetics and ferroelectrics; and memory devices (IIIC, Sec. 4) are based on the retentivity of these materials.

A host of problems remains to be solved in this unified field of nonmetals, and a number of them are especially important to the Armed Services. Their spokesmen, in this book, have appraised the dielectric requirements for the Air Force (IV, A), the Army (IV, B), and the Navy (IV, C).

The book closes with quantitative data on dielectrics, the "Tables of Dielectric Materials" (Part V). Here are summarized the measurements of the Laboratory for Insulation Research on the complex permittivity and permeability of more than 600 dielectrics, for a frequency range of 10^2 to 2.5×10^{10} cycles per second, and for temperatures reaching up to 500°C.†

† Part V is reproduced without change from the *Tables of Dielectric Materials*, Vol. IV, Laboratory for Insulation Research, Massachusetts Institute of Technology (Tech. Rep. 57), sponsored by the Office of Naval Research, the Army Signal Corps, and the Air Force under ONR Contracts N5ori-07801 and N5ori-07858.

I · THEORY

By ARTHUR VON HIPPEL

A · Macroscopic Properties of Dielectrics
1 · Complex Permittivity and Permeability

A capacitor, connected to a sinusoidal voltage source

$$\mathcal{V} = \mathcal{V}_0 e^{j\omega t} \quad \dagger \quad (1.1)$$

of the *angular frequency*

$$\omega = 2\pi\nu \quad (1.2)$$

stores, when vacuum is its dielectric, a charge

$$Q = C_0 \mathcal{V}, \quad (1.3)$$

and draws a *charging current*

$$I_c = \frac{dQ}{dt} = j\omega C_0 \mathcal{V} \quad (1.4)$$

leading the voltage by a temporal phase angle of 90° (Fig. 1.1). C_0 is the *vacuum* (or geometrical) *capacitance* of the condenser.

When filled with some substance, the condenser increases its capacitance to

$$C = C_0 \frac{\epsilon'}{\epsilon_0} = C_0 \kappa', \quad (1.5)$$

where ϵ' and ϵ_0 designate the real *permittivities* or *dielec-*

---

† Throughout this book we will use complex quantities in treating periodic phenomena and represent them in the complex plane. Here the *x*-axis corresponds to the *axis of reals* and the *y*-axis to the *axis of imaginaries*. The factor $j = \sqrt{-1}$ in front of a real quantity signifies an imaginary component oriented in the $+y$-axis or $+j$-axis direction. A complex quantity $z = x + jy$ plotted in the complex plane corresponds in polar co-ordinates to a radius vector $\rho = \sqrt{x^2 + y^2}$ inclined by an angle $\theta = \tan^{-1}(y/x)$ towards the real axis: $z = \rho e^{j\theta}$.

The complex function $\mathcal{V} = \mathcal{V}_0 e^{j\omega t} = \mathcal{V}_0(\cos \omega t + j \sin \omega t)$ consequently can be plotted in the complex plane as a radius vector of length $\mathcal{V}_0$, the voltage amplitude, making an angle of ωt radians with the axis of reals. As long as the voltage and current vectors rotate at the same angular velocity of ω radians per second, we can forget this rotation in discussing their relative positions in the complex plane.

We return from the complex functions to actual currents and voltages by taking the real or the imaginary part: Re ($\mathcal{V}$) = $\mathcal{V}_0$

tric constants of the dielectric and of vacuum, respectively, and their ratio κ' the *relative dielectric constant* of the material. Simultaneously, there may appear, in

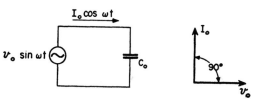

Fig. 1.1. Current-voltage relation in ideal capacitor.

addition to the charging current component I_c, a *loss current* component

$$I_l = G\mathcal{V} \quad (1.6)$$

in phase with the voltage; G represents the conductance of the dielectric. The total current traversing the condenser,

$$I = I_c + I_l = (j\omega C + G)\mathcal{V}, \quad (1.7)$$

is inclined by a *power factor angle* $\theta < 90°$ against the applied voltage $\mathcal{V}$, that is, by a loss angle δ against the $+j$-axis (Fig. 1.2).

It would be premature to conclude that the dielectric material corresponds in its electrical behavior to a capacitor paralleled by a resistor (RC circuit) (Fig. 1.3).

cos ωt, Im ($\mathcal{V}$) = $\mathcal{V}_0 \sin \omega t$. In dealing with products of complex functions it has to be kept in mind that the product of the real parts of two complex quantities A_1 and A_2 is not equal to the real part of their product, but

$$\text{Re}(A_1)\text{Re}(A_2) = \tfrac{1}{4}(A_1 + \tilde{A}_1)(A_2 + \tilde{A}_2).$$

The symbol $\tilde{A}_1$ signifies the conjugate of A_1; for example, if $A_1 = (x + jy)e^{j\omega t}$, then $\tilde{A}_1 = (x - jy)e^{-j\omega t}$.

The time average of a periodic function A is $\bar{A} = \frac{1}{T}\int_0^T A\, dt$, where T is the period of the function. If A_1 and A_2 are such functions, the product of the averages of their real parts is $\overline{\text{Re}(A_1)\text{Re}(A_2)} = \tfrac{1}{2}\text{Re}(A_1\tilde{A}_2)$.

The frequency response of this circuit, which can be expressed by the ratio of loss current to charging current, that is, the *dissipation factor D* or *loss tangent* tan δ as

$$D \equiv \tan \delta = \frac{I_l}{I_c} = \frac{1}{\omega RC}, \quad (1.8)$$

may not at all agree with that actually observed because the conductance term need not stem from a migration

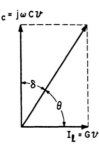

Fig. 1.2. Capacitor containing dielectric with loss.

of charge carriers, but can represent any other energy-consuming process. It has therefore become customary to refer to the existence of a loss current in addition to

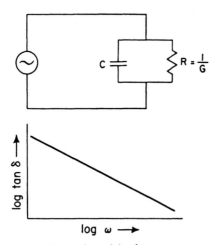

Fig. 1.3. RC circuit and its frequency response.

a charging current noncommittally by the introduction of a *complex permittivity*

$$\epsilon^* = \epsilon' - j\epsilon''. \quad (1.9)$$

The total current I of Eq. 1.7 may thus be rewritten

$$I = (j\omega\epsilon' + \omega\epsilon'')\frac{C_0}{\epsilon_0}\mathcal{V} = j\omega C_0 \kappa^* \mathcal{V}, \quad (1.10)$$

where

$$\kappa^* \equiv \frac{\epsilon^*}{\epsilon_0} = \kappa' - j\kappa'' \quad (1.11)$$

is the *complex relative permittivity* of the material, and

ϵ'' and κ'' are the *loss factor* and *relative loss factor*, respectively. The loss tangent becomes

$$\tan \delta = \frac{\epsilon''}{\epsilon'} = \frac{\kappa''}{\kappa'}. \quad (1.12)$$

Since a parallel-plate condenser of the area A and the plate separation d, fringing effects neglected, has the vacuum capacitance

$$C_0 = \frac{A}{d}\epsilon_0, \quad (1.13)$$

the current density J traversing a condenser under the applied field strength

$$E = \mathcal{V}/d \quad (1.14)$$

becomes, according to Eq. 1.10,

$$J = (j\omega\epsilon' + \omega\epsilon'')E = \epsilon^* \frac{dE}{dt} \quad (1.15)$$

(Fig. 1.4). The product of angular frequency and loss

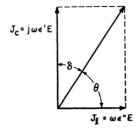

Fig. 1.4. Charging and loss current density.

factor is equivalent to a *dielectric conductivity*

$$\sigma = \omega\epsilon''. \quad (1.16)$$

This dielectric conductivity sums over all dissipative effects and may represent as well an actual conductivity caused by migrating charge carriers as refer to an energy loss associated with a frequency dependence (dispersion) of ϵ', for example, to the friction accompanying the orientation of dipoles.

If the dielectric material is transferred from the electric field of the capacitor into the magnetic field of a coil, the voltage $\mathcal{V}$ drives through the coil a magnetization current I_m according to Faraday's inductance law $\left(\mathcal{V} = L\frac{dI}{dt}\right)$ as

$$I_m = \frac{\mathcal{V}}{j\omega L_0 \frac{\mu'}{\mu_0}} = -j\frac{\mathcal{V}}{\omega L_0 \kappa_m'}. \quad (1.17)$$

L represents the *inductance* and L_0 the *vacuum* (or *geometrical*) *inductance* of the coil. This magnetization current lags behind the applied voltage by 90° (Fig. 1.5). The *permeabilities* μ' and μ_0 designate the magnetization of the material and of vacuum, respectively, and their ratio

$$\kappa_m' \equiv \mu'/\mu_0 \quad (1.18)$$

the *relative permeability* of the material in which the magnetic field of the coil resides.

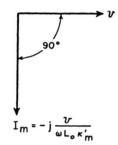

Fig. 1.5. Current-voltage relation in ideal inductor.

Because of the resistance R of the coil windings, an ohmic current component $\mathcal{V}/R$ exists. In addition, there may appear, in phase with $\mathcal{V}$, a magnetic loss current I_l caused by energy dissipation during the magnetization cycle. We shall allow for this magnetic loss by introducing a *complex permeability*

$$\mu^* = \mu' - j\mu'' \quad (1.19)$$

and a *complex relative permeability*

$$\kappa_m^* = \frac{\mu^*}{\mu_0} = \kappa_m' - j\kappa_m'' \quad (1.20)$$

in complete analogy to the electric case. Thus we obtain the total magnetization current

$$I = I_m + I_l = \frac{\mathcal{V}}{j\omega L_0 \kappa_m^*} = -\frac{j\mathcal{V}(\mu' + j\mu'')}{\omega \dfrac{L_0}{\mu_0}(\mu'^2 + \mu''^2)}. \quad (1.21)$$

According to these lumped circuit considerations the macroscopic electric and magnetic behavior of a dielectric material in sinusoidal fields is determined by the two complex parameters ϵ^* and μ^*.[1] We have now to incorporate these parameters systematically into the concepts of the field theory.

[1] The real and imaginary part (ϵ' and ϵ'' or μ' and μ'', respectively) of these complex parameters are not entirely independent of each other. When one of them is given over the whole frequency spectrum, the other one is prescribed. See, for example, A. von Hippel, *Dielectrics and Waves*, Appendix AI, 1.

2 · Polarization and Magnetization

A dielectric material increases the storage capacity of a condenser by neutralizing charges at the electrode surfaces which otherwise would contribute to the external field. Faraday[1] was the first to recognize this phenomenon of *dielectric polarization*. We may visualize it as the action of dipole chains which form under the influence of the applied field and bind countercharges with their free ends on the metal surfaces (Fig. 2.1).

By writing the voltage of the capacitor according to Eqs. 1.3 and 1.5 as

$$\mathcal{V} = \frac{Q}{\kappa'} \cdot \frac{1}{C_0}, \quad (2.1)$$

we may interpret this equation as stating that only a fraction of the *total charge* Q, the *free charge* Q/κ', contributes to the voltage whereas the remainder, the *bound charge* $Q\left(1 - \dfrac{1}{\kappa'}\right)$, is neutralized by the polarization of the dielectric.

To obtain a clearer conception of the charge distribution and its effect in space, we represent charge densities

[1] M. Faraday, *Phil. Trans.*, 1837–1838.

by field vectors. The *total* (or *true*) *charge* Q concentrated in the capacitor is distributed over the surface

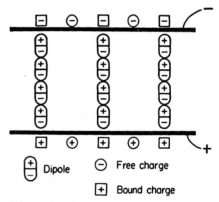

Fig. 2.1. Schematic representation of dielectric polarization.

area A of the metal electrodes with a density s as

$$Q = \int_A s\, dA. \quad (2.2)$$

We represent this true charge density s by a vector $\mathbf{D}$, the *electric flux density* (or *dielectric displacement*), such that the surface charge density shall be equal to the normal component of **D**, or

$$s\, da \equiv D\cos\alpha\, dA \equiv \mathbf{D}\cdot\mathbf{n}\, dA = D_n\, dA. \quad (2.3)$$

A positive value of the *scalar* or *dot product* of the vector **D** and the unit normal vector **n** indicates a positive charge.

Similarly, we allocate to the free charge density s/κ' a vector **E**, the *electric field strength* or *field intensity*, by defining

$$\frac{s}{\kappa'}\, dA \equiv \epsilon_0 \mathbf{E}\cdot\mathbf{n}\, dA = \epsilon_0 E_n\, dA, \quad (2.4)$$

and to the bound charge density a vector **P**, called the *polarization*, as

$$s\left(1 - \frac{1}{\kappa'}\right) dA \equiv \mathbf{P}\cdot\mathbf{n}\, dA = P_n\, dA \quad (2.5)$$

(Fig. 2.2).† Equations 2.3 and 2.4 show the relation-

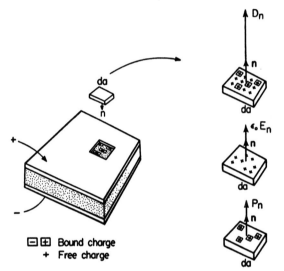

Fig. 2.2. Representation of total, free, and bound charge densities by field vectors.

ship between dielectric flux density and the field strength

$$\mathbf{D} = \epsilon' \mathbf{E}, \quad (2.6)$$

and Eqs. 2.3 to 2.5 the interrelation between the three field vectors

$$\mathbf{P} = \mathbf{D} - \epsilon_0 \mathbf{E} = (\epsilon' - \epsilon_0)\mathbf{E} \equiv \chi \epsilon_0 \mathbf{E}. \quad (2.7)$$

The factor

$$\chi = \frac{\mathbf{P}}{\epsilon_0 \mathbf{E}} = \kappa' - 1 = \frac{\text{bound charge density}}{\text{free charge density}} \quad (2.8)$$

† By postulating that the surface charge densities are equal to the normal components of the field vectors instead of 4π times their magnitude, we have optioned for a rationalized system of units.

is known as the *electric susceptibility* of the dielectric material.

Electric flux density **D** and polarization **P** have, according to their defining equations, the dimension *charge per unit area*, whereas the electric field strength **E** can have a different physical meaning because the dimensions of the dielectric constant may yet be chosen. We take advantage of this possibility and extend the concept of the electric field into space by postulating that an electric probe charge Q', placed in an electrostatic field of the intensity **E**, is subjected to a force

$$\mathbf{F} \equiv Q'\mathbf{E}. \quad (2.9)$$

Thus the electric field strength **E** becomes equivalent in magnitude and direction to the force per unit charge acting on a detector charge, and the dielectric constant obtains the dimensions

$$[\epsilon] = \left[\frac{\text{charge per unit area}}{\text{force per unit charge}}\right]. \quad (2.10)$$

Two electric charges of opposite polarity, $\pm Q$, separated by a distance d, represent a dipole of the moment

$$\boldsymbol{\mu} = Q\mathbf{d}; \quad (2.11)$$

this *electric dipole moment* is symbolized by a vector of the magnitude $|\boldsymbol{\mu}|$ pointing from the negative to the positive pole (Fig. 2.3).† The polarization vector **P**

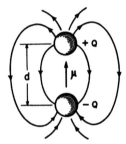

Fig. 2.3. Electric dipole of the moment $\mu = Qd$.

corresponds in magnitude to the surface charge density bound at the electrodes by the polarized dielectric, and it points in the direction of the applied field. The polarization **P** is therefore obviously identical with the *electric dipole moment per unit volume* of the dielectric material (Fig. 2.4).

The electrostatic field in space obtains physical meaning because the field strength at any point can be measured by the force acting on a detector charge (Eq. 2.9). Alternatively, it could be measured by the torque **T**

† Chemists frequently represent dipole moments by vectors pointing from the positive to the negative charge ($+ \to -$); this convention is incompatible with the definition of **P** and should be abolished.

Polarization and Magnetization

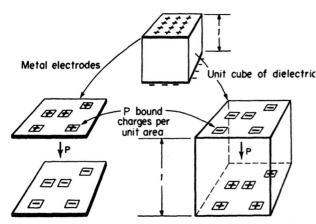

Fig. 2.4. **P**, designating both bound charge density and dipole moment per unit volume.

exercised by the electric field **E** on an electric dipole **μ** as

$$\mathbf{T} = |\mathbf{\mu}| \, |\mathbf{E}| \sin \theta \equiv \mathbf{\mu} \times \mathbf{E}, \qquad (2.12)$$

which tends to align this dipole in field direction (Fig. 2.5).

The concepts developed for the electrostatic field apply for the magnetostatic field, with the restriction

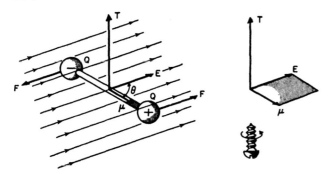

Fig. 2.5. Torque acting on electric dipole.

that individual magnetic point charges of north and south polarity are not known to exist in nature. Hence the *magnetic dipole moment*

$$\mathbf{m} = p\mathbf{d} \qquad (p = \text{pole strength}) \qquad (2.13)$$

(Fig. 2.6) is the starting point of the theory. Visualizing that under the influence of a magnetic field **H** magnetic dipole chains form in a dielectric in analogy to the electric polarization of Fig. 2.1, we can introduce a *magnetization* vector **M**, which represents the *magnetic dipole moment per unit volume* of the material. This magnetization **M**, together with the *magnetic field strength* **H** in the dielectric, determine the total *magnetic flux density* (*magnetic induction*) **B**.

Unfortunately, at this point the magnetic field theory deviates in its mathematical formulation from the electric theory, causing a great deal of confusion. Electric flux density **D** and polarization **P** have identical dimensions (*surface charge density* or *electric moment per unit volume*); and the interrelating equation

$$\mathbf{D} = \epsilon_0 \mathbf{E} + \mathbf{P} = \epsilon' \mathbf{E} \qquad (2.14)$$

allows us to define the electric field strength **E** as *force per unit charge* or *torque per unit dipole moment* by the proper choice of the dimensions of the permittivity (see Eq. 2.10). The magnetic flux density **B**, however, is defined by the equation

$$\mathbf{B} \equiv \mu_0 \mathbf{H} + \mu_0 \mathbf{M} = \mu' \mathbf{H}, \qquad (2.15)$$

where the factors μ' and μ_0 represent the *permeability* (*inductive capacity*) of the material and of vacuum, respectively. Thus the magnetic field strength **H** obtains the same dimension (*magnetic moment per unit volume*) as the magnetization **M**, and only the magnetic

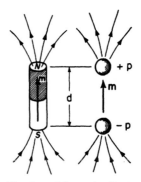

Fig. 2.6. Magnetic dipole.

induction **B** can acquire, by a proper choice of the dimensions of the permeability, the meaning of *torque per unit dipole moment*. Thus **B** appears in the force and torque equations of the magnetic field, but not **H**, and the magnetic analogue to Eq. 2.12 is

$$\mathbf{T} = |\mathbf{m}| \, |\mathbf{B}| \sin \theta \equiv \mathbf{m} \times \mathbf{B}. \qquad (2.16)$$

Rewriting Eq. 2.15 for the magnetization, we obtain

$$\mathbf{M} = \frac{1}{\mu_0} \mathbf{B} - \mathbf{H} \equiv \chi_m \mathbf{H}, \qquad (2.17)$$

and define, in analogy to the electric susceptibility of Eq. 2.8, a *magnetic susceptibility*,

$$\chi_m \equiv \frac{\mathbf{M}}{\mathbf{H}} = \frac{\mu'}{\mu_0} - 1 \equiv \kappa_m' - 1. \qquad (2.18)$$

κ_m' is the *relative permeability*.

Thus far, electricity and magnetism appear as two new phenomena independent of each other and consequently requiring the introduction of two new fundamental quantities, for example, *electric charge* and *magnetic dipole moment* for their description. Actually, they are interlinked and only one new, independent quantity may be introduced. One interrelation is given by the fact that an electric current creates a magnetic field according to *Ampère's circuital law* [2]

$$\oint \mathbf{H} \cdot dl = I \qquad (2.19)$$

(Fig. 2.7). The magnetic field encircling an electric

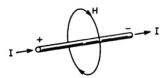

Fig. 2.7. Magnetic field encircling electric current.

current is a whirlpool field of closed lines in contrast to the magnetostatic field that originates in the free ends of dipole chains of a magnetic material. However, a ring current of the magnitude I encircling an area A produces at a distance, large in comparison to its radius, a magnetic field which is identical to that of a magnetic dipole of the moment

$$\mathbf{m} = IA\mathbf{n} \qquad (2.20)$$

(Fig. 2.8). This equivalence between the fields of

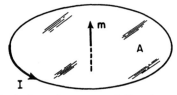

Fig. 2.8. Equivalence between curl H and current density.

magnetic dipoles and of circular currents allows us to explain the phenomenon *magnetism* and the nonexistence of magnetic monopoles on the basis that the sources of magnetism may be *molecular ring* or *Ampère currents*.

A second interrelation between electric and magnetic fields is given by *Faraday's induction law*.[3] It states that when the magnetic flux

$$\Phi = \int_A \mathbf{B} \cdot \mathbf{n} \, dA \qquad (2.21)$$

traversing a loop of wire changes, whether by a change of the magnetic induction **B** or by a change in the position or shape of the loop, an emf is created along the

[2] A. M. Ampère, *Recueil d'observation électrodynamiques*, Crochard, Paris, 1820–1833.
[3] M. Faraday, *Experimental Researches in Electricity*, Taylor, London, Vol. I, 1839, pp. 1–109.

wire causing an induced voltage $\mathcal{U}_i$ to appear between its ends,

$$\mathcal{U}_i = \int_a^b \mathbf{E} \cdot dl = -\frac{d\Phi}{dt}, \qquad (2.22)$$

proportional to the speed of this change. If the loop is closed, a current I will flow through the loop resistance R,

$$I = \frac{\mathcal{U}_i}{R} = -\frac{1}{R}\frac{d\Phi}{dt}, \qquad (2.23)$$

that causes a magnetic field opposing the change in flux.

Ampère's circuital law and Faraday's induction law were discovered and formulated for currents and voltages in wire loops. Maxwell [4] postulated that these laws are valid in space quite independent of the presence of detector loops in which currents and voltages might develop, and thus he arrived at the electromagnetic field equations.

A discussion of the current drawn by a capacitor will make the formulation of Maxwell's first field equation evident. The current of density J streaming into the electrode system is equal to the change of the true charge stored in the condenser, that is,

$$\int_A \mathbf{J} \cdot \mathbf{n} \, dA = \int_A \frac{ds}{dt} \, dA. \qquad (2.24)$$

The surface charge density s originates the electric flux density **D** (see Eq. 2.3), hence

$$\frac{ds}{dt} dA = \frac{d\mathbf{D}}{dt} \cdot \mathbf{n} \, dA; \qquad (2.25)$$

the current density may be measured as the time derivative of the electric flux density at the electrode surface,

$$\mathbf{J} = \frac{d\mathbf{D}}{dt}. \qquad (2.26)$$

By postulating that this equivalence between the temporal change of the dielectric flux density and an electric current holds also for the interior of a dielectric, that is, that this change in flux produces a magnetic field just like a conduction current, Maxwell arrived at the concept that the conduction current charging a capacitor finds its continuation in a field current traversing the dielectric,

$$\int_A \mathbf{J} \cdot \mathbf{n} \, dA = \int_{A'} \frac{d\mathbf{D}}{dt} \cdot \mathbf{n} \, dA. \qquad (2.27)$$

This field current, which extends through the cross section A' of the dielectric as far as the electric field of

[4] J. C. Maxwell, *A Treatise in Electricity and Magnetism*, Clarendon Press, Oxford, 1892, Vol. 2, pp. 247–262.

the capacitor reaches, was named by Maxwell the *displacement current*. By including this displacement current into Ampère's circuital law, *Maxwell's first field equation* results in the integral formulation

$$\oint \mathbf{H} \cdot dl = \int_A \mathbf{J} \cdot \mathbf{n} \, dA + \int_{A'} \frac{\partial \mathbf{D}}{\partial t} \cdot \mathbf{n} \, dA. \quad (2.28)$$

With the help of Stokes's theorem the line integral on the left can be transformed into a surface integral, and the differential formulation

$$\nabla \times \mathbf{H} = \mathbf{J} + \frac{\partial \mathbf{D}}{\partial t} \quad (2.29)$$

is obtained.[5]

The current of the density **J** may be a true conduction current obeying Ohm's law,

$$\mathbf{J} = \sigma \mathbf{E}; \quad (2.30)$$

however, from a more general standpoint, the conductivity σ may be interpreted as the dielectric conductivity of Eq. 1.16 representing any energy-consuming process. Thus, by introducing the complex permittivity, the first field equation may be rewritten for sinusoidal fields and isotropic, linear dielectrics † as

$$\nabla \times \mathbf{H} = \epsilon^* \frac{\partial \mathbf{E}}{\partial t}. \quad (2.31)$$

[5] For details see A. von Hippel, *Dielectrics and Waves*, John Wiley and Sons, New York, 1954.

The *second field equation* is a generalization of Faraday's induction law and asserts that the change of a magnetic flux density creates an emf in space quite independently of the presence of a loop for its detection,

$$\oint \mathbf{E} \cdot dl = -\int_A \frac{\partial \mathbf{B}}{\partial t} \cdot \mathbf{n} \, dA. \quad (2.32)$$

In differential form it becomes

$$\nabla \times \mathbf{E} = -\frac{\partial \mathbf{B}}{\partial t}, \quad (2.33)$$

or, if we recall that also magnetization may lead to energy dissipation and introduce the complex permeability, we arrive at a formulation completely symmetrical to that of the first field equation, except for the negative sign,

$$\nabla \times \mathbf{E} = -\mu^* \frac{\partial \mathbf{H}}{\partial t}. \quad (2.34)$$

Maxwell's field equations thus describe the coupling between the electric and magnetic field vectors and their interaction with matter in space and time.

† The designation "linear" dielectric signifies that the relation between **D** and **E** and between **B** and **H** is a linear one, that is, that the permittivity and permeability are independent of field strength. Anisotropic and nonlinear dielectrics are discussed in IB, Sec. 5.

3 · Description of Dielectrics by Various Sets of Parameters

An interpretation of Maxwell's field equations requires, as a first step, the separation of the field vectors **E** and **H**. This can be done by differentiating the equations with respect to time and substituting from one equation into the other. We obtain in this way the *wave equations of the electromagnetic field*, which for our purpose may be simplified by assuming that **E** and **H** are a function of x and t only:

$$\frac{\partial^2 \mathbf{E}}{\partial x^2} = \epsilon^* \mu^* \frac{\partial^2 \mathbf{E}}{\partial t^2},$$
$$\frac{\partial^2 \mathbf{H}}{\partial x^2} = \epsilon^* \mu^* \frac{\partial^2 \mathbf{H}}{\partial t^2}. \quad (3.1)$$

The solution of these differential equations concerning us here is a plane wave,

$$\mathbf{E} = \mathbf{E}_0 e^{j\omega t - \gamma x},$$
$$\mathbf{H} = \mathbf{H}_0 e^{j\omega t - \gamma x}, \quad (3.2)$$

varying periodically in time with the frequency

$$\nu = \omega/2\pi \quad (3.3)$$

and advancing in the $+x$-direction through space with a *complex propagation factor*

$$\gamma = j\omega(\epsilon^* \mu^*)^{1/2} = \alpha + j\beta; \quad (3.4)$$

α is the *attenuation factor* and β the *phase factor* of the wave. Introducing these factors, we may rewrite Eq. 3.2:

$$\mathbf{E} = \mathbf{E}_0 e^{-\alpha x_1} e^{j2\pi(\nu t - \beta x/2\pi)},$$
$$\mathbf{H} = \mathbf{H}_0 e^{-\alpha x_1} e^{j2\pi(\nu t - \beta x/2\pi)}. \quad (3.5)$$

Obviously the wave has a time period

$$T = 1/\nu \quad (3.6)$$

and a space period

$$\lambda = 2\pi/\beta. \quad (3.7)$$

Surfaces of constant phase are given by

$$vt - \frac{x}{\lambda} = \text{constant}, \quad (3.8)$$

hence propagate with the phase velocity

$$\frac{dx}{dt} = v = \nu\lambda = \frac{\omega}{\beta}. \quad (3.9)$$

For a dielectric without loss ($\epsilon^* = \epsilon'$, $\mu^* = \mu'$), we obtain from Eq. 3.4 the phase factor

$$\beta = \omega(\epsilon'\mu')^{1/2}, \quad (3.10)$$

so that the *phase velocity* in a *loss-free unbounded medium* is

$$v = 1/(\epsilon'\mu')^{1/2}. \quad (3.11)$$

To learn about the coupling between the **E** and **H** vectors, we have to return to the field equations and write out the field components. These component equations contain three statements:

(1) The x components of the field vectors, the longitudinal field components of the electromagnetic wave, are independent of space and time, hence may be assumed to be zero. The plane wave is a *transverse electromagnetic* or TEM wave.

(2) The coupled transversal components of the **E** and **H** waves are *perpendicular* to each other and form, together with the propagation direction, a right-hand co-ordinate system of the sequence $+x \to E_y \to H_z$ (Fig. 3.1).

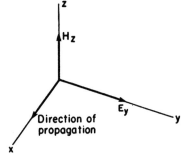

Fig. 3.1. Right-hand co-ordinate system for traveling TEM wave.

(3) The ratio of the coupled electric and magnetic field vectors follows as

$$\frac{\mathbf{E}}{\mathbf{H}} = \frac{\gamma}{j\omega\epsilon^*} \equiv Z. \quad (3.12)$$

This ratio Z, the *intrinsic impedance of the dielectric*, may be rewritten with the help of Eq. 3.4 in any one of the three versions,

$$Z = \frac{\gamma}{j\omega\epsilon^*} = \sqrt{\frac{\mu^*}{\epsilon^*}} = \frac{j\omega\mu^*}{\gamma}. \quad (3.13)$$

In Sec. 1 the response of a dielectric material to sinusoidal electric and magnetic fields was expressed by the two complex parameters ϵ^* and μ^* which determine the storage and dissipation of electric and magnetic energy in the medium. These parameters were derived from the amplitude and temporal-phase relations between voltage and current in capacitors and coils. It is now obvious that we are not restricted to using ϵ^* and μ^*, but may refer to alternate parameters that convey the same information.

The power engineer replaces the dielectric constant ϵ' and the loss factor ϵ'' by the combination of ϵ' and *power factor* $\cos\theta$; the radio engineer may choose ϵ' and the *loss tangent* $\tan\delta$, where

$$\tan\delta = \frac{\epsilon''}{\epsilon'} = \frac{\text{loss current}}{\text{charging current}}. \dagger \quad (3.14)$$

Frequently the inverse of the loss tangent, the quality factor Q of the dielectric,

$$Q = \frac{1}{\tan\delta} = \frac{\omega\epsilon' E_0^2}{\omega\epsilon'' E_0^2} = 2\pi\nu \frac{\frac{1}{2}\epsilon' E_0^2}{\frac{1}{2}\sigma E_0^2}$$

$$= 2\pi \frac{\text{av. energy stored per half cycle}}{\text{energy dissipated per half cycle}}$$

$$= \left[\frac{\text{reactive v-amp}}{\text{watts}}\right] \quad (3.15)$$

serves as the *figure of merit*, especially in wave-guide problems. An engineer interested in dielectric heating will probably refer to ϵ' and the *dielectric conductivity*

$$\sigma = \omega\epsilon'' \quad [\text{ohm}^{-1}\,\text{m}^{-1}], \quad (3.16)$$

because the power absorbed per unit volume is

$$P = \sigma\frac{E_0^2}{2} \quad [\text{watt m}^{-3}]. \quad (3.17)$$

If, instead of the time relation between current and voltage, the electromagnetic field in space is considered, new substitutes for ϵ^* and μ^* offer themselves. To derive them conveniently, we visualize the spatial electric wave train at some moment t_1

$$E_y = E_1 e^{-\gamma x} = E_1 e^{-\alpha x} e^{-j2\pi\frac{x}{\lambda}} \quad (3.18)$$

(see Eq. 3.5). The wave amplitude oscillates in space with a periodicity λ; it is enclosed between exponential envelopes determined by the attenuation constant α (Fig. 3.2a). Alternatively, in polar co-ordinates, the

---

† It should be noted that, since $\cos\theta = \sin\delta$, the power factor and loss tangent (dissipation factor) may be considered equal only for sufficiently small *loss angles* δ, when $\sin\delta \simeq \tan\delta$, because $\cos\delta \simeq 1$.

wave amplitude may be depicted as a radius vector which, rotating clockwise as the distance increases, describes a logarithmic spiral (Fig. 3.2b). The parameter x is replaced in the latter representation by the phase angle ϕ according to the relation

$$\frac{x}{\lambda} = \frac{\phi}{2\pi}, \qquad (3.19)$$

and the electric field strength is rewritten as

$$E_y = E_1 e^{-\phi\left(\frac{\alpha\lambda}{2\pi} + j\right)}. \qquad (3.20)$$

In vacuum the wavelength is λ_0 and the wave travels with the velocity of light (see Eq. 3.11),

$$c = \lambda_0 \nu = 1/(\epsilon_0 \mu_0)^{1/2}. \qquad (3.21)$$

The physicist normally uses the index of refraction as one of his parameters and pairs with it, by making use of the polar representation of the wave, the attenuation per radian called *index of absorption*,

$$k = \frac{\alpha\lambda}{2\pi} = \frac{\alpha}{\beta}. \qquad (3.25)$$

By substituting these indices of refraction and absorption for the attenuation factor α and the phase factor β of the propagation factor in Eq. 3.4, we obtain

$$\gamma = j\frac{2\pi}{\lambda_0} n(1 - jk) = j\frac{2\pi}{\lambda_0} n^*. \qquad (3.26)$$

The propagation factor γ used by the communication

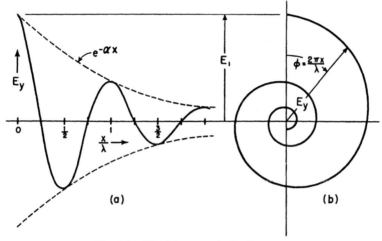

Fig. 3.2. Electric wave train in space.

In other media the wavelength normally shortens and the phase velocity slows down. The ratio of the wavelength or phase velocity in vacuum to that in the dielectric designates the *index of refraction* of the dielectric medium

$$n \equiv \frac{\lambda_0}{\lambda} = \frac{c}{v} = \frac{\lambda_0}{2\pi}\beta. \qquad (3.22)$$

For a loss-free medium this equation simplifies to

$$n = (\epsilon'\mu'/\epsilon_0\mu_0)^{1/2} \equiv (\kappa'\kappa_m')^{1/2}. \qquad (3.23)$$

If, in addition, the magnetization can be neglected ($\mu' = \mu_0$), the well-known *Maxwell relation* † results:

$$n^2 = \frac{\epsilon'}{\epsilon_0} = \kappa' \qquad (3.24)$$

† This relation has been abused frequently in predicting static dielectric constants from optical refraction data. Actually, it states only that the square of the index of refraction of a nonabsorbing, nonmagnetic material is equal to the relative permittivity at that frequency.

engineer may thus be replaced by the *complex index of refraction*

$$n^* = n(1 - jk) \qquad (3.27)$$

employed in the calculations of physical optics.

The propagation factor γ is proportional to the product $(\epsilon^*\mu^*)^{1/2}$, whereas the intrinsic impedance Z is equal to the ratio $(\mu^*/\epsilon^*)^{1/2}$. Both complex quantities have to be determined to obtain ϵ^* and μ^* individually.

From the intrinsic impedance

$$Z = \frac{E}{H} = (\mu^*/\epsilon^*)^{1/2} \qquad (3.28)$$

in polar form,

$$Z = |Z|e^{j\zeta}$$

$$= \left[\frac{(\epsilon'\mu' + \epsilon''\mu'')^2 + (\epsilon''\mu' - \epsilon'\mu'')^2}{(\epsilon'^2 + \epsilon''^2)^2}\right]^{1/4} e^{j\zeta} \qquad (3.29)$$

with

$$\tan 2\zeta = \frac{\epsilon''\mu' - \epsilon'\mu''}{\epsilon'\mu' + \epsilon''\mu''}, \qquad (3.30)$$

we can derive the phase relation between the electric and magnetic wave. It is evident that the electric field vector is advanced or retarded with respect to the magnetic vector in temporal phase, depending on the preponderance of the term pertaining to the electric or the magnetic loss. For negligible magnetic loss ($\mu'' = 0$),

$$\tan 2\zeta = \tan \delta = \frac{2k}{1 - k^2} \qquad (3.31)$$

or

$$\tan \zeta = k; \qquad (3.32)$$

the phase advance of the electric wave is equal to the arc tangent of the index of absorption. In a loss-free medium in unbounded space the electric and magnetic field vectors of an electromagnetic wave are exactly in phase (Fig. 3.3).

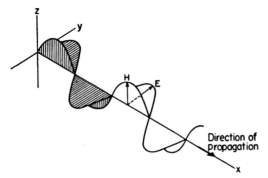

Fig. 3.3. Traveling TEM wave in loss-free dielectric.

The general characterization of a dielectric as the carrier of an electromagnetic field requires two independent complex parameters which have to be determined by four independent measurements; however, the situation fortunately simplifies in practice. Ferromagnetics exempted, the magnetic polarization is, in general, so weak that μ^* may be replaced by the permeability μ_0 of free space for all practical purposes. Thus, two measurements normally suffice to determine the dielectric response of homogeneous isotropic materials at a given frequency. Consequently, in most cases, our dielectric characteristics show only the specific permittivity

$$\kappa' = \epsilon'/\epsilon_0 \qquad (3.33)$$

and the loss tangent $\tan \delta$ (see Part V).

To allow a convenient change-over from these to other parameters, we equate the real and imaginary parts of Eq. 3.4. It follows for the attenuation factor of a transversal electromagnetic wave (TEM wave)

$$\alpha = \frac{\lambda \omega^2}{4\pi} (\epsilon'\mu'' + \epsilon''\mu'), \qquad (3.34)$$

and for the phase factor

$$\beta = \frac{2\pi}{\lambda} = \omega \left[\frac{(\epsilon'\mu' - \epsilon''\mu'')}{2} \right.$$
$$\left. \times \left\{ 1 + \sqrt{1 + \left(\frac{\epsilon'\mu'' + \epsilon''\mu'}{\epsilon'\mu' + \epsilon''\mu''}\right)^2} \right\} \right]^{1/2}. \qquad (3.35)$$

Thus we arrive at the conversion formulas:

For materials with negligible magnetic loss ($\mu'' = 0$) we obtain from Eq. 3.35 for the wavelength the simplified expression

$$\lambda = \frac{1}{\nu} \frac{1}{[\frac{1}{2}\epsilon'\mu'\{1 + \sqrt{1 + \tan^2 \delta}\}]^{1/2}}. \qquad (3.36)$$

If, in addition, the permeability is that of vacuum ($\mu' = \mu_0$), we may write for the index of refraction

$$n = \frac{\lambda_0}{\lambda} = [\tfrac{1}{2}\kappa'\{\sqrt{1 + \tan^2 \delta} + 1\}]^{1/2}. \qquad (3.37)$$

Similarly the attenuation factor becomes

$$\alpha = \frac{2\pi}{\lambda_0} [\tfrac{1}{2}\kappa'\{\sqrt{1 + \tan^2 \delta} - 1\}]^{1/2}, \qquad (3.38)$$

and the index of absorption

$$k = \frac{\alpha}{\beta} = \left[\frac{\sqrt{1 + \tan^2 \delta} - 1}{\sqrt{1 + \tan^2 \delta} + 1} \right]^{1/2}. \qquad (3.39)$$

These equations show that it is convenient to discuss the effect of the dielectric loss on other parameters by studying the three boundary cases: $\tan^2 \delta \ll 1$, $\tan^2 \delta \simeq 1$, and $\tan^2 \delta \gg 1$.

The attenuation produced by a dielectric is frequently expressed as the *attenuation distance* $1/\alpha$ through which the field strength decays to $1/e = 0.368$ of its original value,

$$\frac{1}{\alpha} = \frac{\lambda_0}{2\pi} \left[\frac{2}{\kappa'(\sqrt{1 + \tan^2 \delta} - 1)} \right]^{1/2} \text{ [m]}, \qquad (3.40)$$

or as the attenuation in *decibels per meter* produced by the material. If the field strength falls from $E(0)$ to $E(x)$, or the power from $P(0)$ to $P(x)$, over a length x of the dielectric, this decibel loss is defined as

$$20 \log \frac{E(0)}{E(x)} = 10 \log \frac{P(0)}{P(x)} = 8.686 \, \alpha x \text{ [db]}; \qquad (3.41)$$

that is, the decibel loss per meter is given as

$$8.686 \, \alpha = 8.686 \frac{2\pi}{\lambda_0}$$
$$\times [\tfrac{1}{2}\kappa'\{\sqrt{1 + \tan^2 \delta} - 1\}]^{1/2} \text{ [db/m]}. \qquad (3.42)$$

For low-loss materials ($\tan \delta \ll 1$) this loss becomes simply

$$8.686 \frac{\pi}{\lambda_0} \sqrt{\kappa'} \tan \delta = 1637 \frac{\sigma}{\sqrt{\kappa'}} \quad [db/m]. \quad (3.43)$$

To arrive at numerical values requires two steps: a measurement of the velocity of light

$$c = \frac{1}{(\epsilon_0 \mu_0)^{1/2}}, \quad (3.44)$$

which establishes the product $\epsilon_0 \mu_0$, and an agreement as to the value of ϵ_0 or μ_0. By international consent the value of μ_0 has been fixed for the rationalized mks system as

$$\mu_0 = 4\pi \times 10^{-7} = 1.257 \times 10^{-6} \quad [\text{henry m}^{-1}]. \quad (3.45)$$

The velocity of light has been measured [1] as

$$c = 2.9979 \times 10^8 \simeq 3 \times 10^8 \quad [\text{m sec}^{-1}]; \quad (3.46)$$

hence the dielectric constant of free space becomes

$$\epsilon_0 = \frac{1}{36\pi} \times 10^{-9}$$

$$= 8.854 \times 10^{-12} \quad [\text{farad m}^{-1}]. \quad (3.47)$$

The intrinsic impedance Z_0 of free space, determined by the ratio of permeability to permittivity (see Eq. 3.28), follows as

$$Z_0 = (\mu_0/\epsilon_0)^{1/2} = 120\pi \simeq 376.6 \quad [\text{ohm}]. \quad (3.48)$$

[1] The most accurate measurements of the light velocity are no longer free-space determinations over long distances such as the Michelson-Morley experiment [*Phil. Mag.* 13, 236 (1882); *ibid.* 24, 449 (1887)], but resonance measurements in wave-guide cavities [see K. D. Froome, *Nature* 169, 107 (1952)].

4 · Reflection and Refraction of Electromagnetic Waves on Boundaries; Measurement of Dielectrics by Standing Waves [1]

Electromagnetic waves are, in general, not observed in unbounded space but confined by various types of boundaries. This is no disadvantage; on the contrary, the reflection and refraction of fields by matter provide the essential means for directing and modifying such fields and for measuring the dielectric properties of materials.

The incident, reflected, and refracted waves are interconnected by the *boundary condition:* the tangential component of **E** and **H** must be continuous in traversing the interface between two media. These are really two conditions, since the amplitudes as well as the phases must be continuous. The first condition, prescribing that the sum of the amplitudes of the tangential components of the incident and reflected waves at the interface must equal the amplitude of the tangential components of the transmitted wave, leads to *Fresnel's equations.* The second condition, demanding equality of the phases at the boundary, contains *Snell's laws of reflection and refraction.* Snell's laws and Fresnel's equations together determine unambiguously for TEM waves the direction, intensity, and polarization of the reflected and refracted beams as a function of the prop-

[1] For a more extensive discussion, see A. von Hippel, *Dielectrics and Waves*, John Wiley and Sons, New York, 1954, Pt. I, Secs. 14 to 24.

erties of the incident wave and of the dielectric characteristics of the two abutting media.

In the discussion of reflection and refraction phenomena, the relative amplitudes and intensities of the partial beams are normally of importance, but not their absolute values. Defining the ratio of the reflected to the incident amplitude at the boundary as the *reflection coefficient*

$$r_E = \frac{E_1}{E_0}, \quad r_H = \frac{H_1}{H_0}, \quad (4.1)$$

and the corresponding ratio of the transmitted to incident amplitude as the *transmission coefficient*

$$t_E = \frac{E_2}{E_0}, \quad t_H = \frac{H_2}{H_0}, \quad (4.2)$$

and indicating with the subscripts n and p the field components *normal* and *parallel* to the *plane of incidence* (the plane containing the direction of incidence and the normal to the interface) (Fig. 4.1), we may formulate Fresnel's equations as follows:

$$r_{E_n} = \left(\frac{E_1}{E_0}\right)_n = \frac{Z_2 \cos \phi - Z_1 \cos \psi}{Z_2 \cos \phi + Z_1 \cos \psi} = -r_{H_p},$$

$$r_{E_p} = \left(\frac{E_1}{E_0}\right)_p = \frac{Z_2 \cos \psi - Z_1 \cos \phi}{Z_2 \cos \psi + Z_1 \cos \phi} = -r_{H_n}, \quad (4.3)$$

and

$$t_{E_n} = \left(\frac{E_2}{E_0}\right)_n = \frac{2Z_2 \cos\phi}{Z_2 \cos\phi + Z_1 \cos\psi} = \frac{Z_2}{Z_1} t_{H_p},$$

$$t_{E_p} = \left(\frac{E_2}{E_0}\right)_p = \frac{2Z_2 \cos\phi}{Z_2 \cos\psi + Z_1 \cos\phi} = \frac{Z_2}{Z_1} t_{H_n}.$$

(4.4)

Z_1 and Z_2 are the intrinsic impedances of the two dielectrics 1 and 2; ϕ is the angle of incidence and ψ the

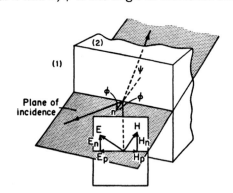

Fig. 4.1. Plane of incidence and field components.

angle of refraction, interrelated by Snell's refraction law

$$\sin\phi = \frac{\gamma_2}{\gamma_1} \sin\psi,$$ (4.5)

with γ_1 and γ_2 designating the propagation factors of the two media.

The general case of the reflection and refraction of plane waves obliquely striking the interface between two lossy materials has very complicated results. For the present purpose it suffices to discuss the special case of perpendicular incidence ($\phi = \psi = 0$), where the distinction between n and p component disappears and Fresnel's equations simplify to

$$r_E = \frac{Z_2 - Z_1}{Z_2 + Z_1} = -r_H,$$

$$t_E = \frac{2Z_2}{Z_2 + Z_1} = \frac{Z_2}{Z_1} t_H.$$

(4.6)

At normal incidence the incoming and reflected waves superpose and form, by interference, a standing-wave pattern. This standing-wave pattern in space can serve for a determination of the dielectric properties of the abutting media.

Let the incident wave $\mathbf{E}_y$, $\mathbf{H}_z$ propagate in the $+x$-direction through medium 1 and strike the boundary of medium 2 (Fig. 4.2) at $x = 0$. The reflected wave, returning in the $-x$-direction, combines with it to yield the resulting field strength

$$E_{y_1} = E_0 e^{j\omega t - \gamma_1 x} + E_1 e^{j\omega t + \gamma_1 x},$$
$$H_{z_1} = H_0 e^{j\omega t - \gamma_1 x} + H_1 e^{j\omega t + \gamma_1 x}.$$

(4.7)

By introducing the reflection coefficients at the boundary $x = 0$, r_0 and $-r_0$ for the electric and the magnetic waves, and by expressing the magnetic field in terms

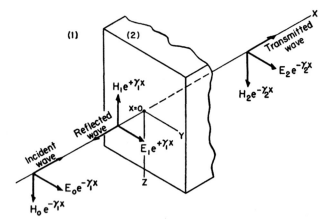

Fig. 4.2. Formation of standing waves.

of the electric field and impedance, we may rewrite this equation as

$$E_{y_1} = E_0 e^{j\omega t}(e^{-\gamma_1 x} + r_0 e^{\gamma_1 x}),$$
$$H_{z_1} = \frac{E_0}{Z_1} e^{j\omega t}(e^{-\gamma_1 x} - r_0 e^{\gamma_1 x}).$$

(4.8)

Let us assume for the present that the transmitted wave continues in medium 2 without further reflection,

$$E_{y_2} = E_2 e^{j\omega t - \gamma_2 x},$$
$$H_{z_2} = \frac{E_2}{Z_2} e^{j\omega t - \gamma_2 x}.$$

(4.9)

The reflection coefficient for normal incidence is in this case

$$r_0 = \frac{Z_2 - Z_1}{Z_2 + Z_1}.$$ (4.10)

There will be no reflection ($r_0 = 0$) when the intrinsic impedances of the two media are matched, that is, if

$$\frac{\epsilon_1^*}{\epsilon_2^*} = \frac{\mu_1^*}{\mu_2^*},$$ (4.11)

or, for loss-free media,

$$\frac{D_1}{D_2} = \frac{B_1}{B_2}.$$ (4.12)

This *invisibility condition* for a boundary at normal

incidence postulates that the electric and magnetic flux densities must change by equal ratios in traversing the interface if reflection is to be avoided.

The opposite extreme, *total reflection*, requires a complete mismatch of the two flux densities, whether by adjoining media with very low and very high permittivity or permeability, or by a mismatch in losses, as, for instance, by choosing as medium 2 a metal. In metals up to the optical frequency range, the conductivity dominates over the real part of the dielectric constant. Consequently, if magnetic loss can be neglected ($\mu'' = 0$), the *intrinsic impedance of a metal* may be written (see Eq. 3.13)

$$Z_2 = (\mu_2'/-j\epsilon_2'')^{1/2} = (j\omega\mu_2'/\sigma_2)^{1/2}. \quad (4.13)$$

Because of the high conductivity σ_2, $Z_2 \ll Z_1$; the reflection coefficient at the metal boundary thus becomes

$$r_0 = \frac{Z_2 - Z_1}{Z_2 + Z_1} \simeq -1. \quad (4.14)$$

Under this *condition of total reflection* and for a loss-free medium 1 $\left(\gamma_1 = j\frac{2\pi}{\lambda_1}\right)$, Eq. 4.8 simplifies to

$$E_{y_1} = -j2E_0 e^{j\omega t} \sin\frac{2\pi x}{\lambda_1},$$
$$H_{z_1} = \frac{2E_0}{Z_1} e^{j\omega t} \cos\frac{2\pi x}{\lambda_1}, \quad (4.15)$$

or, by changing over to the real field components,

$$\text{Re}(E_{y_1}) = 2E_0 \sin\omega t \sin\frac{2\pi x}{\lambda_1},$$
$$\text{Re}(H) = \frac{2E_0}{Z_1} \cos\omega t \cos\frac{2\pi x}{\lambda_1}. \quad (4.16)$$

This equation represents a standing wave for which the electric field strength has its minima or *nodes* at the metal surface and at distance $-x = n\frac{\lambda_1}{2}$ ($n = 1, 2, 3, \ldots$), whereas the *antinodes* (maxima) of E are found at $-x = m\frac{\lambda_1}{4}$ ($m = 1, 3, 5, \ldots$) (Fig. 4.3). The nodes and antinodes of the magnetic wave appear in the reverse order.

In comparing Figs. 3.3 and 4.3 and the corresponding equations of a progressing and a standing TEM wave in a loss-free medium, an essential difference between the two wave types becomes apparent. Whereas no phase difference exists between the electric and magnetic field in the traveling wave, in the standing wave the two fields are displaced by 90° with respect to each other in spatial and temporal phase. In the traveling wave the energy is equally divided between the electric and the magnetic fields, whereas in the standing wave it alternates between total electric and total magnetic energy storage.

To visualize the standing-wave pattern in the general case of partial reflection, we rewrite the electric wave of Eq. 4.8

$$E_y = E_0 e^{j\omega t - \gamma_1 x}(1 + r_0 e^{2\gamma_1 x}), \quad (4.17)$$

that is, we express it as the product of the incident wave and of a term in brackets which gives rise to

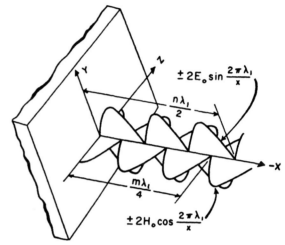

Fig. 4.3. Standing wave in loss-free dielectric in front of total reflector.

the standing-wave pattern. By writing the distance and the reflection coefficient in polar form

$$x = \frac{\lambda_1}{2\pi}\phi,$$
$$r = |r_0|e^{-j2\psi}, \quad (4.18)$$

and introducing the index of absorption k (see Eq. 3.25), we may reformulate this term as

$$1 + r_0 e^{2\gamma_1 x} = 1 + |r_0| e^{2k\phi} e^{j2(\phi-\psi)}. \quad (4.19)$$

Its second part, $r_0 e^{2\gamma_1 x}$, now represents a radius vector in the complex plane which, at the interface $\phi = 0$, has the length $|r_0|$, and points at an angle -2ψ (Fig. 4.4). As we move from the boundary into medium 1 in the $-x$-direction, the radius vector spins clockwise one full turn for each half wavelength ($\phi = \pi, 2\pi$, etc.) and shrinks simultaneously, thus describing a logarithmic spiral. The vector $1 + r_0 e^{2\gamma_1 x}$ originates at the point -1 and connects to the tip of this radius vector. Similarly, the vector $1 - r_0 e^{2\gamma_1 x}$ of the magnetic wave connects from the point -1 to the tip of the radius vector which starts its spiral 180° displaced.

Reverting from polar to linear scale (Fig. 4.5), we obtain the relative amplitude of the standing-wave pattern and may read off from both diagrams immediately the dielectric information which the standing-wave pattern contains.

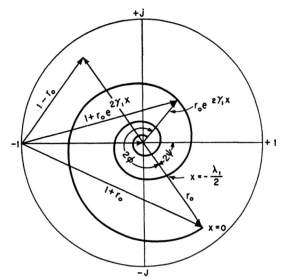

Fig. 4.4. Polar diagram of standing wave.

The ratio of the two field vectors at the boundary ($x = 0$),

$$Z(0) = \frac{E_1(0)}{H_1(0)} = Z_1 \frac{1 + r_0}{1 - r_0}, \quad (4.20)$$

known as the *terminating impedance* of medium 1, can be found by measuring the amplitude $|r_0|$ and the phase angle 2ψ of the reflection coefficient r_0. Obviously, the first minimum of the E-wave is reached when the radius vector has rotated through an angle

$$2\phi_0 = \pi - 2\psi, \quad (4.21)$$

or, in linear scale,

$$2\phi_0 = \frac{4\pi x_0}{\lambda_1}. \quad (4.22)$$

Hence, by measuring the distance x_0 of the first minimum from the boundary of the dielectric, the phase angle is obtained as

$$2\psi = 4\pi \left(\frac{1}{4} - \frac{x_0}{\lambda_1} \right). \quad (4.23)$$

The ratio of the electric to the magnetic field strength at this minimum is

$$\frac{E_{\min}}{H_{\max}} = Z_1 \frac{1 - |r_0| e^{-2\alpha_1 x_0}}{1 + |r_0| e^{-2\alpha_1 x_0}}. \quad (4.24)$$

If the attenuation in medium 1 can be neglected ($\alpha_1 \simeq 0$), $H_{\max} = \dfrac{E_{\max}}{Z_1}$ and Eq. 4.24 simplifies to

$$\frac{E_{\max}}{E_{\min}} = \frac{1 + |r_0|}{1 - |r_0|}. \quad (4.25)$$

This ratio of maximum to minimum electric field intensity is called the *voltage standing-wave ratio* (VSWR).

Thus, by measuring the ratio of maximum to minimum field strength and the distance of the first minimum from the dielectric boundary, the magnitude and phase of the reflection coefficient r_0 are determined. With it the terminating impedance $Z(0)$ is found if the intrinsic impedance Z_1 of medium 1 is known.

Since the tangential field components are continuous through the boundary, the terminating impedance for normal incidence is the same whether seen from medium 1 or 2:

$$Z(0) = \frac{E_1(0)}{H_1(0)} = \frac{E_2(0)}{H_2(0)}. \quad (4.26)$$

Consequently, if no further reflection takes place in medium 2 (see Eq. 4.9), the terminating impedance measured by the standing-wave pattern in medium 1 is directly equal to the intrinsic impedance of medium 2, $Z(0) = Z_2$.

In case the media 1 and 2 are loss free and medium 2 of infinite extent, the reflection coefficient at the

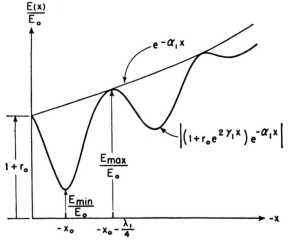

Fig. 4.5. Linear diagram of standing wave (attenuation in medium 1 overemphasized).

boundary is real (see Eq. 4.10); the phase jump 2ψ is either zero or π, and the VSWR simplifies for $\epsilon_2' > \epsilon_1'$ ($r_0 = -|r_0|$) to

$$\frac{E_{\max}}{E_{\min}} = \frac{Z_1}{Z_2} = \sqrt{\frac{\epsilon_2'}{\epsilon_1'}} = n_{21}; \quad (4.27)$$

it is equal to the *relative index of refraction* of the two media.

Medium 2 will, in general, not approximate infinite length, but will consist of a layer of thickness d, followed by a medium 3, etc. (Fig. 4.6). Additional reflections will therefore occur at the boundary 2–3 and possibly subsequent boundaries, and a part of this

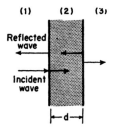

Fig. 4.6. Wave reflected on dielectric layer.

reflected energy will return into medium 1. The reflected wave in medium 1, thus composed of several partial beams, will itself be the resultant of an interference phenomenon.

This more complex situation does not affect the measurement of the standing-wave pattern in medium 1, but it changes the interpretation of the reflection coefficient r_0 and terminating impedance $Z(0)$. Before proceeding with this discussion, we shall reformulate $Z(0)$ in terms of the directly measurable parameters $E_{\min}/E_{\max}$, x_0, and λ_1, to arrive at an expression useful in actual calculations.

Setting

$$r_0 \equiv e^{-2u} \quad \text{(where } u = \rho + j\psi\text{)}, \quad (4.28)$$

we may write the terminating impedance (Eq. 4.20)

$$Z(0) = Z_1 \frac{1 + r_0}{1 - r_0} = Z_1 \coth u, \quad (4.29)$$

and the inverse standing wave ratio (Eq. 4.25)

$$\frac{E_{\min}}{E_{\max}} = \frac{1 - e^{-2\rho}}{1 + e^{-2\rho}} = \tanh \rho. \quad (4.30)$$

By expanding $\coth u$ and recalling the relation between ψ and x_0 (Eq. 4.23), we obtain the desired expression

$$Z(0) = Z_1 \frac{\tanh \rho - j \cot \psi}{1 - j \tanh \rho \cot \psi}$$

$$= Z_1 \frac{\dfrac{E_{\min}}{E_{\max}} - j \tan \dfrac{2\pi x_0}{\lambda_1}}{1 - j \dfrac{E_{\min}}{E_{\max}} \tan \dfrac{2\pi x_0}{\lambda_1}}. \quad (4.31)$$

The terminating impedance is thus determined experimentally by measurements on the standing-wave pattern in medium 1 as discussed. To obtain from it the properties of medium 2, we consider the situation at the boundary 2,3 at $x = d$.

At this boundary a reflection coefficient r_{23} produces a reflected wave which, together with the incident wave, sets up a standing wave (see Eq. 4.8) in medium 2:

$$E_{y_2} = E_2 e^{j\omega t} \{ e^{-\gamma_2(x-d)} + r_{23} e^{\gamma_2(x-d)} \},$$

$$H_{z_2} = \frac{E_2}{Z_2} e^{j\omega t} \{ e^{-\gamma_2(x-d)} - r_{23} e^{\gamma_2(x-d)} \}. \quad (4.32)$$

The terminating impedance at $x = 0$, seen from medium 2, is

$$Z(0) = \frac{E_2(0)}{H_2(0)} = Z_2 \frac{e^{\gamma_2 d} + r_{23} e^{-\gamma_2 d}}{e^{\gamma_2 d} - r_{23} e^{-\gamma_2 d}}. \quad (4.33)$$

In case the transmitted wave continues in medium 3 without further reflection, the reflection coefficient at the boundary 2,3 becomes (see Eq. 4.6)

$$r_{23} = \frac{Z_3 - Z_2}{Z_3 + Z_2}. \quad (4.34)$$

If medium 3 is a metal, approximately total reflection takes place at d,

$$r_{23} \simeq -1 \quad (4.35)$$

(cf. Eq. 4.14), and the terminating impedance for this *short-circuit measurement* of medium 2 simplifies to

$$Z(0) = Z_2 \tanh \gamma_2 d. \quad (4.36)$$

In case the shorting metal plate is placed a distance Δ behind d (Fig. 4.7), a standing wave forms in medium 3

$$E_{y_3} = E_3 e^{j\omega t} (e^{-\gamma_3(x-d-\Delta)} - e^{\gamma_3(x-d-\Delta)}),$$

$$H_{z_3} = \frac{E_3}{Z_3} e^{j\omega t} (e^{-\gamma_3(x-d-\Delta)} + e^{\gamma_3(x-d-\Delta)}). \quad (4.37)$$

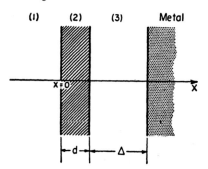

Fig. 4.7. Arrangement for measuring a dielectric 2 by standing wave in medium 1.

Since the impedance at $x = d$ is the same, whether seen from medium 2 or 3

$$(E_{y_2}/H_{z_2})_{x=d} = (E_{y_3}/H_{z_3})_{x=d}, \quad (4.38)$$

the reflection coefficient r_{23} can be expressed in its dependence on the layer thickness Δ as

$$r_{23} = \frac{Z_3(1 - e^{-2\gamma_3\Delta}) - Z_2(1 + e^{-2\gamma_3\Delta})}{Z_3(1 - e^{-2\gamma_3\Delta}) + Z_2(1 + e^{-2\gamma_3\Delta})}. \quad (4.39)$$

For $\Delta = \lambda_3/4$ and a loss-free medium 3 $\left(\gamma_3 = j\dfrac{2\pi}{\lambda_3}\right)$, we obtain

$$r_{23} = +1; \quad (4.40)$$

the combination, a quarter wavelength section terminated by a short circuit, corresponds to an *open circuit*, and the terminating impedance becomes

$$Z(0) = Z_2 \coth \gamma_2 d. \quad (4.41)$$

Thus, by performing an *open circuit* and a *short circuit* measurement of the terminating impedance, we obtain two equations for Z_2 and γ_2 and we can calculate the complex dielectric constant and the complex permeability of medium 2. For a nonmagnetic dielectric ($\mu_2^* = \mu_0$), one of the measurements suffices.[2]

[2] S. Roberts and A. von Hippel, Abstract, *Phys. Rev.* 57, 1056 (1940); *J. Appl. Phys.* 17, 610 (1946). A. von Hippel, D. G. Jelatis, and W. B. Westphal, "The Measurement of Dielectric Constant and Loss with Standing Waves in Coaxial Wave Guides," National Defense Research Committee, Contract OEM sr-191, Laboratory for Insulation Research, Massachusetts Institute of Technology, April, 1943; see also IIA, Sec. 2.

B · Molecular Properties of Dielectrics[1]

1 · Molecular Mechanisms of Polarization

A dielectric material can react to an electric field because it contains charge carriers that can be displaced. In Part A this polarization phenomenon was pictured schematically by the formation of dipole chains which line up parallel to the field and bind countercharges at the electrodes (see I, A, Fig. 2.1). The density of the neutralized surface charge is represented by the polarization vector

$$\mathbf{P} = (\epsilon' - \epsilon_0)\mathbf{E} = (\kappa' - 1)\epsilon_0\mathbf{E} \quad [\text{coul/m}^2] \quad (1.1)$$

(IA, Eq. 2.7). Alternatively, the polarization $\mathbf{P}$ proves to be equivalent to the dipole moment per unit volume of the material (IA, Fig. 2.4).

This latter interpretation of $\mathbf{P}$ provides an entrance from the macroscopic into the molecular world. The dipole moment per unit volume may be thought of as resulting from the additive action of N elementary dipole moments $\bar{\mu}$,

$$\mathbf{P} = N\bar{\mu}. \quad (1.2)$$

The average dipole moment $\bar{\mu}$ of the elementary particle, furthermore, may be assumed to be proportional to the *local electric field strength* E' that acts on the particle,

$$\bar{\mu} = \alpha \mathbf{E}'. \quad (1.3)$$

The proportionality factor α, called *polarizability*, measures the electrical pliability of the particle, that is, the average dipole moment per unit field strength. Its dimensions are

$$[\alpha] = \left[\frac{\sec^2 \text{coul}^2}{\text{kg}}\right] = [\epsilon\, \text{m}^3]$$

in the mks, or

$$[\alpha] = [\text{cm}^3] \quad (1.4)$$

in the esu system, respectively.

Equations 1.1 to 1.3 give us the two alternative expressions for the polarization:

$$\begin{aligned}\mathbf{P} &= (\kappa' - 1)\epsilon_0\mathbf{E} \\ &= N\alpha\mathbf{E}',\end{aligned} \quad (1.5)$$

linking the macroscopically measured permittivity to three molecular parameters: the number N of contributing elementary particles per unit volume; their polarizability α; and the locally acting electric field $\mathbf{E}'$. This field will normally differ from the applied field $\mathbf{E}$ due to the polarization of the surrounding dielectric medium. It is the goal of the molecular theories to evaluate these parameters and thus to arrive at an understanding of the phenomenon *polarization* and its dependence on frequency, temperature, and applied field strength.

Such theories are not the result of metaphysical speculation, but the products of experiments and their interpretation. It is beyond the scope of this book to trace in detail the interplay between experiment and theory that led to our present-day ideas about the structure of matter. We have to reverse the approach and

[1] For a more extensive discussion, see A. von Hippel, *Dielectrics and Waves*, John Wiley and Sons, New York, 1954, Pt. II.

to introduce at this point some general concepts which are the outcome of this development and which allow us to foresee various mechanisms of polarization.

Matter, electrically speaking, consists of positive atomic nuclei surrounded by negative electron clouds. Upon the application of an external electric field the electrons are displaced slightly with respect to the nuclei; *induced* dipole moments result and cause the so-called *electronic polarization* of materials. When atoms of different types form molecules, they will normally not share their electrons symmetrically, as the electron clouds will be displaced eccentrically toward the stronger binding atoms. Thus atoms acquire charges of opposite polarity, and an external field acting on these net charges will tend to change the equilibrium positions of the atoms themselves. By this displacement of charged atoms or groups of atoms with respect to each other, a second type of induced dipole moment is created; it represents the *atomic polarization* of the dielectric. The asymmetric charge distribution between the unlike partners of a molecule gives rise, in addition, to *permanent* dipole moments which exist also in the absence of an external field. Such moments experience a torque in an applied field that tends to orient them in the field direction (see IA, Eq. 2.12). Consequently, an *orientation* (or *dipole*) *polarization* can arise.

These three mechanisms of polarization, characterized by an *electronic polarizability* α_e, an *atomic polarizability* α_a, and an *orientation* or *dipole polarizability* α_d, are due to charges that are locally bound in atoms, in molecules, or in the structures of solids and liquids. In addition, charge carriers usually exist that can migrate for some distance through the dielectric. When such carriers are impeded in their motion, either because they become trapped in the material or on interfaces, or because they cannot be freely discharged or replaced at the electrodes, space charges and a macroscopic field distortion result. Such a distortion appears to an outside observer as an increase in the capacitance of the sample and may be indistinguishable from a real rise of the dielectric permittivity. Thus we have to add to our polarization mechanisms a fourth one, a *space-charge* (or *interfacial*) *polarization*, characterized by a *space-charge* (or *interfacial*) *polarizability* α_s.

Assuming at present that the four polarization mechanisms indicated schematically in Fig. 1.1 act independently of each other, we may write the total polarizability α of a dielectric material as the sum of the four terms,

$$\alpha = \alpha_e + \alpha_a + \alpha_d + \alpha_s, \quad (1.6)$$

where each term again may represent a sum of contributions. How these four effects actually shape the response characteristics of dielectrics will be discussed in subsequent sections.

In Eq. 1.5 we took for granted that the polarizability α is a real quantity. Actually, in alternating fields, a temporal phase shift may occur between the driving field and the resulting polarization, and a loss current

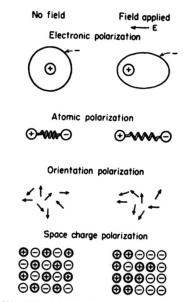

Fig. 1.1. Mechanisms of polarization.

component may appear (see IA, Sec. 1). Thus α becomes complex, and Eq. 1.5 has to be replaced by the more general formulation

$$\mathbf{P} = (\kappa^* - 1)\epsilon_0 \mathbf{E} = N\alpha \mathbf{E}'. \quad (1.7)$$

Confronted with three molecular parameters (N, α, $\mathbf{E}'$) at once, we shall tend to concentrate initially on the most instructive one and to eliminate the other two by reasonable approximations. In the present case the parameter α contains the primary information on the electric charge carriers and their polarizing action. Hence we shall try to eliminate N and $\mathbf{E}'$.

We can foresee that the locally acting field $\mathbf{E}'$ will be identical with the externally applied field $\mathbf{E}$ for gases at low pressure where the interaction between the molecules can be neglected. At high pressures, however, and especially in the condensed phases of solids and liquids, the field acting on a reference molecule A may be modified decisively by the polarization of the surroundings. To take this effect into account, we try the following model (Fig. 1.2).

Let our reference molecule A be surrounded by an imaginary sphere of such an extent that beyond it the dielectric can be treated as a continuum. If the molecules inside this sphere were removed while the polarization outside remains frozen, the field acting on A

would stem from two sources: from the free charges at the electrodes of the plate capacitor (E_1), and from the free ends of the dipole chains that line the cavity walls (E_2). Actually there are molecules inside the sphere, and they are so near to A that their individual positions

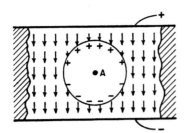

Fig. 1.2. Model for calculation of internal field.

and shapes have to be considered. This adds an additional contribution E_3 to the local field E'; hence we obtain

$$E' = E_1 + E_2 + E_3. \quad (1.8)$$

The contribution from the free charges at the electrodes is, by definition, equal to the applied field intensity

$$E_1 = E. \quad (1.9)$$

To calculate E_2 we recall that the charge density lining the cavity walls stems from bound charges and is correspondingly determined by the normal component of the polarization vector $\mathbf{P}$ (see IA, Eq. 2.5) as

$$\mathbf{P} \cdot \mathbf{n} \, dA = P \cos \theta \, dA \quad (1.10)$$

(Fig. 1.3). Each surface element dA of the sphere con-

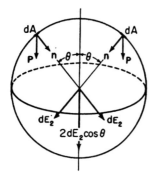

Fig. 1.3. Geometry for calculation of internal field.

tributes at A, according to Coulomb's law, a radial field intensity

$$d\mathbf{E}_2 = \frac{P \cos \theta}{\epsilon_0 4\pi r^2} dA. \quad (1.11)$$

For each surface element dA at a latitude position between θ and $\theta + d\theta$, there exists its counterpart which produces the same vertical but an equal and opposite horizontal field component. Hence only the vertical components $dE_2 \cos \theta$ count and create a field intensity

$$\mathbf{E}_2 = \oint_{\text{sphere}} \frac{P \cos^2 \theta}{\epsilon_0 4\pi r^2} dA, \quad (1.12)$$

orientated parallel to the applied field and strengthening it. By dividing the cavity walls into ring elements

$$dA = 2\pi (\sin \theta) r d\theta, \quad (1.13)$$

and integrating over θ, we obtain as the field contribution of the cavity wall charge

$$\mathbf{E}_2 = \int_0^\pi \frac{P \cos^2 \theta}{\epsilon_0 4\pi r^2} 2\pi r^2 \sin \theta \, d\theta = \frac{1}{3} \frac{\mathbf{P}}{\epsilon_0}$$

$$= \frac{\mathbf{E}}{3} (\kappa' - 1). \quad (1.14)$$

An evaluation of the field contribution E_3 which arises from the individual action of the molecules inside the sphere requires accurate information on the geometrical arrangement and polarizability of the contributing particles. Even if this information is available, the mathematical treatment may prove prohibitively difficult. In a general way we have recognized the existence of these neighboring molecules already in the calculation of E_2 by assuming that the cavity is scooped out without disturbing the state of polarization of the remaining dielectric. Hence we postulate for the present, as a simple expedient, that the additional individual field effects of the surrounding molecules on the particle at A shall mutually cancel; that is,

$$\mathbf{E}_3 = 0. \quad (1.15)$$

This assumption, first made by Mosotti[2] in 1850, is not altogether a confession of ignorance. It is a reasonable approximation when the elementary particles are neutral and without permanent dipole moment, or when they are arranged either in complete disorder or in cubic or similar highly symmetrical arrays. It allows us to substitute for the unknown molecular parameter E' the expression

$$\mathbf{E}' = \mathbf{E}_1 + \mathbf{E}_2 = \mathbf{E} + \frac{\mathbf{P}}{3\epsilon_0} = \frac{\mathbf{E}}{3} (\kappa' + 2), \quad (1.16)$$

that is, known macroscopic parameters.

By inserting this local *Mosotti field* into Eq. 1.5, we obtain the relation between the *polarizability per unit*

---

[2] O. F. Mosotti, *Mem. di math. e di fisica in Modena* [II] *24*, 49 (1850). See also P. Debye, *Polar Molecules*, Dover Publications, New York, 1945, pp. 11 ff.

volume, $N\alpha$, and the relative permittivity of the dielectric, κ',

$$\frac{N\alpha}{3\epsilon_0} = \frac{\kappa' - 1}{\kappa' + 2}. \quad (1.17)$$

For gases at low pressure, $\kappa' - 1 \ll 1$; hence, $\kappa' + 2$ may be replaced by the digit 3. This is the same as replacing the local field $\mathbf{E}'$ by the applied field $\mathbf{E}$, and Eq. 1.17 simplifies to

$$\frac{N\alpha}{\epsilon_0} = \kappa' - 1 = \chi, \quad (1.18)$$

where χ is the *electric susceptibility of the gas* (see IA, Eq. 2.8).

Frequently we shall refer to an ideal gas under standard conditions (0°C, 760 mm Hg). The number of molecules per unit volume, N, is in this case identical with the *Loschmidt number*

$$N_L = 2.687 \times 10^{25} \quad [\text{m}^{-3}]. \quad (1.19)$$

In other cases, as long as the molecules themselves are the dipole carriers, it will be convenient to eliminate the dependence of the polarization on the density of the material by referring to the *polarization per mole*. The number of molecules per mole is *Avogadro's number*

$$N_0 = \frac{NM}{\rho} = 6.023 \times 10^{23}, \quad (1.20)$$

where M designates the molecular weight in [kg] and ρ the density in [kg m^{-3}], if mks units are used. Substituting N_0 for N in Eq. 1.17, we obtain the *polarizability per mole (molar polarization)*

$$\Pi = \frac{N_0\alpha}{3\epsilon_0} = \frac{\kappa' - 1}{\kappa' + 2}\frac{M}{\rho} \quad [\text{m}^3]. \quad (1.21)$$

This is the famous *Clausius-Mosotti equation*,[3] in which we have reached our goal of eliminating N and $\mathbf{E}'$ and retain the polarizability α as the only unknown molecular parameter.

Because of the tendency to treat the electrical and optical frequency range as two separate fields of interest, the same equation was formulated independently for the optical range by Lorentz[4] in Holland and Lorenz[5] in Denmark. In this case, the relative dielectric constant κ' of Eq. 1.21 may be replaced by the square of the index of refraction, n^2, according to the Maxwell relation (IA, Eq. 3.24). Thus we arrive at the *Lorentz-Lorenz equation*

$$\Pi = \frac{N_0\alpha}{3\epsilon_0} = \frac{n^2 - 1}{n^2 + 2}\frac{M}{\rho}, \quad (1.22)$$

with Π called the *molar refraction*.

Both equations are identical and not quite general enough, because they assume that α is a real quantity. By replacing the real permittivity or index of refraction with their complex counterparts (see Eq. 1.7), we arrive at the more general formulation of the *Clausius-Mosotti-Lorentz-Lorenz equation*

$$\Pi = \frac{N_0\alpha}{3\epsilon_0} = \frac{\kappa^* - 1}{\kappa^* + 2}\frac{M}{\rho} = \frac{n^{*2} - 1}{n^{*2} + 2}\frac{M}{\rho} \quad [\text{m}^3]. \quad (1.23)$$

In using this equation, valuable information on the polarizability α can be obtained, as we shall see in the following sections. However, we should not be surprised to arrive also at quite erroneous conclusions in case the *near field* $\mathbf{E}_3$ cannot be neglected.

[3] R. Clausius, *Die Mechanische Wärmetheorie*, Vieweg, Braunschweig, 1879, Vol. 2, pp. 62 ff.
[4] H. A. Lorentz, *Ann. Physik 9*, 641 (1880).
[5] L. Lorenz, *Ann. Physik 11*, 70 (1880).

2 · Polarization and Atomic Structure

The phenomenon of electronic polarization can be observed undisturbed by other effects in *monatomic gases*. Here the elementary particles are nuclei of the positive charge $+Ze$, surrounded by a neutralizing electron atmosphere of the charge $-Ze$; the factor Z, the *order number of the atom*, designates the position of the atom in the periodic system of the elements; the *elementary charge e* has the value

$$|e| = 1.602 \times 10^{-19} \quad [\text{coul}]. \quad (2.1)$$

An external field $\mathbf{E}$ will exert a force

$$\mathbf{F} = -Ze\mathbf{E} \quad (2.2)$$

on the electron atmosphere and displace its charge center, which previously coincided with that of the nucleus, by a distance d. Here it will be balanced by the Coulomb attraction exercised by the positive nucleus $+Ze$ on the eccentric part of the electron cloud (Fig. 2.1). Let us assume tentatively that the electrons originally form a cloud of constant charge density around the nucleus, confined to a sphere of the radius r_0. After displacement, the eccentric part of the charge, Q_d, is the fraction of the cloud that fills a sphere (radius d) outside the nucleus, that is,

$$Q_d = -Ze \frac{d^3}{r_0^3}. \qquad (2.3)$$

It acts as if concentrated in the center of the small sphere, while the charge further out does not exercise

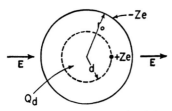

Fig. 2.1. Displacement of electron cloud by external field (overemphasized).

a force on the nucleus. Hence the Coulomb force between the nucleus of charge $+Ze$ and the negative charge Q_d,

$$\mathbf{F}_c = -(Ze)^2 \frac{d^3/r_0^3}{\epsilon_0 4\pi d^2}, \qquad (2.4)$$

has to balance $\mathbf{F}$.

The external field has induced in the atom a dipole moment

$$\boldsymbol{\mu} = (Ze)\mathbf{d} = \alpha_e \mathbf{E} \qquad (2.5)$$

(see Eq. 1.3). From the equilibrium condition

$$\mathbf{F} = \mathbf{F}_c, \qquad (2.6)$$

we obtain this dipole moment as

$$\boldsymbol{\mu} = \epsilon_0 4\pi r_0^3 \mathbf{E}; \qquad (2.7)$$

that is, the electronic polarizability of the atom is

$$\alpha_e = \epsilon_0 4\pi r_0^3. \qquad (2.8)$$

The molar polarization of our monatomic gas, defined in Eq. 1.21, becomes

$$\Pi = \frac{N_0 \alpha_0}{3\epsilon_0} = N_0 \frac{4\pi}{3} r_0^3 \quad [\text{m}^3]. \qquad (2.9)$$

This result has the graphic meaning: the molar polarization (polarizability per mole) of a monatomic gas of spherical atoms equals the volume actually filled by these spheres.

Consequently, we have only to know the radius r_0 of our atoms to arrive at numerical results. This radius can be obtained in various ways, for instance, as the van der Waals radius in gaskinetic measurements. It is of the order of one angstrom,

$$1 \text{ A} = 1 \times 10^{-10} \quad [\text{m}]. \qquad (2.10)$$

For hydrogen, consisting of one proton surrounded by one electron, we find

$$r_0 \simeq 0.53[\text{A}] \simeq 0.53 \times 10^{-10} \quad [\text{m}]. \qquad (2.11)$$

Since the dielectric constant of vacuum is $\epsilon_0 = 8.854 \times 10^{-12}$ [farad m] (see IA, Eq. 3.47) we obtain for the hydrogen atom the electronic polarizability (see Eq. 2.8),

$$\alpha_{eH} = 1.66 \times 10^{-41} \quad [\epsilon \text{ m}^3]. \qquad (2.12)$$

In the unrationalized esu system the factor $\epsilon_0 4\pi$ disappears, and the electronic polarizability is simply equal to the cube of the atomic radius,

$$\alpha_{eH} = r_0^3 = 1.5 \times 10^{-25} \quad [\text{cm}^3]. \qquad (2.13)$$

The molar polarization has the value

$$\Pi_H = 37.8 \times 10^{-8} \quad [\text{m}^3] = 37.8 \times 10^{-2} \quad [\text{cm}^3]. \qquad (2.14)$$

For atomic hydrogen under standard conditions our simple sphere model thus predicts the susceptibility (see Eqs. 1.18 and 1.22)

$$\chi = \kappa' - 1 \simeq 5.1 \times 10^{-5}. \qquad (2.15)$$

This static susceptibility, though small, can be determined with accuracy quasi-statically by a bridge or beat-frequency measurement, in which a gas-filled capacitor of about 10^{-4}-meter plate separation is used at 20 volts. The maximum field strength of 2×10^5 [volt m^{-1}] thus applied would induce a dipole moment

$$\boldsymbol{\mu} = \alpha_e \mathbf{E} = e\mathbf{d} \simeq 3.3 \times 10^{-36} \quad [\text{coul m}], \qquad (2.16)$$

that is, displace the center of the electron cloud from that of the proton by a distance

$$d \simeq 2 \times 10^{-17} \quad [\text{m}]. \qquad (2.17)$$

By measuring the dielectric constant of the gas, the electrical engineer has thus performed the amazing feat of determining this subnuclear length! However, his record for micro-distance measurements loses some of its luster when we consider that the electrical technique employed integrates over the polarization of about 10^{18} atoms. Furthermore, the actual meaning of d is not that of a precise distance.

The purpose of the preceding numerical calculation is to give a feeling for the orders of magnitude involved. A measurement of the electric susceptibility of atomic hydrogen has unfortunately not yet been carried through because the gas consists normally of diatomic

(H_2) molecules. However, we can predict [1] that the actual value will be about 4.5 times larger than that just calculated. The concept of a proton surrounded by an electron cloud of radius r_0 and constant density, used above, gives the same result as the Bohr model in which the electron revolves in a circular orbit of the radius r_0 around the proton. Quantum mechanics leads to the picture of a more extended electron cloud; and since the distant parts of the electron atmosphere are more weakly bonded to the nucleus, they contribute appreciably to the polarization in spite of their rapidly decreasing density (Fig. 2.2).

More detailed information about the structure of atoms may be obtained by polarization measurements in alternating fields, that is, by investigating the frequency response characteristics spectroscopically. Not knowing what to expect in detail, we may start with the simple model of electrons quasi-elastically bound to equilibrium positions and reacting to field changes like linear harmonic oscillators.

Such an oscillator, displaced in the z-direction by a driving force $e\mathbf{E}'$ (where $\mathbf{E}'$ represents the locally acting electric field) obeys the equation of motion

$$\frac{d^2z}{dt^2} + 2\alpha \frac{dz}{dt} + \omega_0^2 z = \frac{e}{m}\mathbf{E}'. \quad (2.18)$$

We can rewrite this expression in the form of a differential equation for the polarization $\mathbf{P}$ by assuming that the dielectric is composed of N oscillators per unit volume, each of them contributing an induced electric moment

$$\bar{\mu} = ez; \quad (2.19)$$

then (see Eq. 1.2)

$$\mathbf{P} = Nez \quad (2.20)$$

If, in addition, we assume that the locally acting field $\mathbf{E}'$ can be described by the Mosotti approximation (see Eq. 1.16) as

$$\mathbf{E}' = \mathbf{E} + \frac{\mathbf{P}}{3\epsilon_0} \quad (2.21)$$

the differential equation becomes

$$\frac{d^2\mathbf{P}}{dt^2} + 2\alpha \frac{d\mathbf{P}}{dt} + \left(\omega_0^2 - \frac{Ne^2}{3m\epsilon_0}\right)\mathbf{P} = \frac{Ne^2}{m}\mathbf{E}. \quad (2.22)$$

The effect of the polarization of the surroundings is to lower the resonance frequency of the individual oscillator from ω_0 to

$$\omega_0'' = \sqrt{\omega_0^2 - \frac{Ne^2}{3m\epsilon_0}}. \quad (2.23)$$

[1] J. H. Van Vleck, *The Theory of Electric and Magnetic Susceptibilities*, Clarendon Press, Oxford, 1932, pp. 203 ff.

The steady-state solution of Eq. 2.22 for a sinusoidal driving field

$$\mathbf{E} = \mathbf{E}_0 e^{j\omega t} \quad (2.24)$$

is

$$\mathbf{P} = \mathbf{P}_0 e^{j(\omega t + \psi)} = \frac{Ne^2/m}{\omega_0''^2 - \omega^2 + j\omega 2\alpha}\mathbf{E}. \quad (2.25)$$

Because of the friction factor 2α, a phase shift ψ occurs between the driving field and the resultant polarization; $\mathbf{P}$ becomes complex. The ratio $\mathbf{P}/\epsilon_0 \mathbf{E}$ (see Eq.

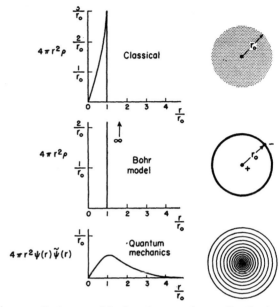

Fig. 2.2. Various models for electron charge distribution in the hydrogen atom: (a) electron cloud of uniform density, (b) Bohr model, (c) quantum mechanics (normalized to unit electronic charge).

1.1) determines the complex relative permittivity of the medium in molecular terms as

$$\kappa^* = 1 + \frac{\mathbf{P}}{\epsilon_0 \mathbf{E}} = 1 + \frac{Ne^2/\epsilon_0 m}{\omega_0^2 - \omega^2 + j\omega 2\alpha}. \quad (2.26)$$

So far we have assumed that the dielectric contains only one oscillator type. In the more general case of s oscillator types which contribute to κ^* without mutual coupling, the equation may be generalized as

$$\kappa^* = 1 + \sum_s \frac{N_s e^2/\epsilon_0 m_s}{\omega_s^2 - \omega^2 + j\omega 2\alpha_s}. \quad (2.27)$$

For nonmagnetic media,

$$\kappa^* = n^{*2}, \quad (2.28)$$

where n^* is the complex index of refraction of the medium (see IA, Eq. 3.27). Equation 2.27 represents the *dispersion formula of classical physics*; N_s designates the *number of dispersion electrons* per unit volume of the oscillator type s.

Far below its resonance frequency ($\omega \ll \omega_s$), each oscillator type adds a constant contribution

$$\frac{N_s e^2/\epsilon_0 m_s}{\omega_s^2} \qquad (2.29)$$

to the static dielectric constant of the medium, whereas far above the resonance frequency its contribution vanishes. To follow the behavior of the lowest oscillator type r through its resonance region, we lump the effect of vacuum and of the remaining resonator types in a constant contribution

$$A = 1 + \sum_s \frac{N_s e^2/m_s}{\omega_s^2} \qquad \text{for } s \neq r. \qquad (2.30)$$

Furthermore, we introduce in place of ω the deviation from resonance

$$\Delta\omega \equiv \omega_r - \omega \qquad (2.31)$$

as the variable, approximate

$$\omega_r + \omega \simeq 2\omega_r, \qquad (2.32)$$
$$\frac{\omega}{\omega_r} \simeq 1,$$

and thus rewrite Eq. 2.27 as

$$\kappa^* = n^{*2} = A + \frac{B}{\Delta\omega + j\alpha}. \qquad (2.33)$$

The factor B stands for

$$B \equiv \frac{N_r e^2/\epsilon_0 m_r}{2\omega_r}. \qquad (2.34)$$

The frequency dependence of the real part of the relative permittivity,

$$\kappa' = n^2(1 - k^2) = A + \frac{B \Delta\omega}{(\Delta\omega)^2 + \alpha^2}, \qquad (2.35)$$

describes the *dispersion characteristic* of the dielectric medium (Fig. 2.3). It rises hyperbolically from the low-frequency value $A + \dfrac{2B}{\omega_r}$ to a maximum

$$\kappa'_{\text{max}} = A + \frac{B}{2\alpha} \qquad (2.36)$$

at $\Delta\omega = +\alpha$, falls with a linear slope $-\dfrac{B}{\alpha^2}$ through the value A at the resonance frequency ($\Delta\omega = 0$), reaches a minimum

$$\kappa'_{\text{min}} = A - \frac{B}{2\alpha} \qquad (2.37)$$

at $\Delta\omega = -\alpha$, and then rises again asymptotically to the constant value A for very high frequencies ($\omega \gg \omega_r$).

The absorption characteristic of the dielectric, identified by the relative loss factor

$$\kappa'' = 2n^2 k = \frac{B\alpha}{(\Delta\omega)^2 + \alpha^2}, \qquad (2.38)$$

starts from zero at low frequencies, traverses its maximum B/α at resonance, and falls again symmetrically to zero at high frequencies. The half-value points $B/2\alpha$ of this bell-shaped absorption characteristic are reached at the deviation from resonance $\Delta\omega = \pm\alpha$.

Since the real dielectric constant and index of refraction rise with increasing frequency over the major part

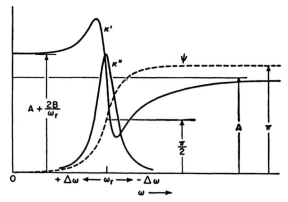

Fig. 2.3. Anomalous dispersion and resonance absorption.

of the dispersion characteristic, this behavior has been called *normal* dispersion in contrast to the *anomalous* dispersion in the half-width region of the spectral line, where the characteristic falls toward shorter wavelengths. This combination of normal and anomalous dispersion is typical for resonance phenomena.

For the phase relation between applied field and induced dipole moment, we obtain the relation

$$\tan \psi = \frac{\kappa''}{\kappa' - A} = \frac{\alpha}{\Delta\omega}. \qquad (2.39)$$

At low frequencies ($\omega \ll \omega_r$) the moment follows in phase, at resonance it lags by $\pi/2$, and at very high frequencies by π. In between, it passes the values $\pi/4$ and $3\pi/4$ at the half-width points $\Delta\omega = \pm\alpha$ (Fig. 2.3). The loss tangent itself,

$$\tan \delta = \frac{\kappa''}{\kappa'} = \frac{\alpha B}{A(\Delta\omega^2 + \alpha^2) + B\Delta\omega}, \qquad (2.40)$$

rises from zero through the value $B/\alpha A$ at resonance to a maximum at $\Delta\omega = -B/2A$ and then falls again to zero.

While the dispersion formula of classical physics predicts correctly the general shape of a spectral line, neither the intensity nor the resonance frequency can be derived from classical considerations. An electron trapped in the field of a positive nucleus is equivalent to a rotating dipole composed of two linear dipoles that oscillate out of phase by 90° in space and time (Fig. 2.4). Classically, the trapped electron represents a light source that transforms electrostatic energy through kinetic energy into radiation, and, in so doing, must spiral into the nucleus. For the hydrogen atom,

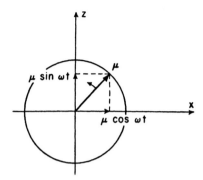

Fig. 2.4. Rotating electric dipole.

the total kinetic energy initially available in the orbit r_0 would be dissipated by radiation in ca. 10^5 rotations or 10^{-11} sec.

In contrast to this prediction, spectroscopy proves that atoms are stable, and that their radiation consists of series of sharply defined spectral lines governed by the empirical *Rydberg-Ritz combination principle*.[2] It shows furthermore that the temperature radiation of black bodies violates the equipartition principle, and that the energy of photoelectrons depends on the frequency and not on the intensity of the light. These last two facts compelled physics to break with its classical continuum ideas and to conclude that radiation is parcelled out in *light quanta* $h\nu$, where h is a universal constant of action, the *Planck constant*[3]

$$h = 6.623 \times 10^{-34} \quad \text{[joule sec]}. \quad (2.41)$$

Maxwell's theory thus loses its absolute validity and becomes a theory of photon statistics.

The remaining problem of linking the quantum theory of radiation to the structure of atoms was solved in a first approximation by Bohr's quantum theory.[4] It postulated that electrons can move in certain stable orbits around the nuclei without radiation (Rutherford atom) and that electron transitions between these stable energy states produce the spectral lines by the emission or absorption of light quanta. Since the total energy of the system (atom-light quantum) has to be

[2] J. R. Rydberg, *Phil. Mag.* **29**, 331 (1890); W. Ritz, *Physik. Z.* **9**, 521 (1908); *idem*, *Astrophys. J.* **28**, 237 (1908).
[3] M. Planck, *Ann. Physik* **4**, 553 (1901).
[4] N. Bohr, *Phil. Mag.* **26**, 1, 476, 857 (1913).

conserved, the frequency of the radiation is given by the energy difference between the initial ($\mathcal{E}_1$) and the final state ($\mathcal{E}_2$) as

$$h\nu = \mathcal{E}_1 - \mathcal{E}_2. \quad (2.42)$$

Bohr's frequency condition, Eq. 2.42, immediately explains the empirical statement of the Rydberg-Ritz combination principle that the frequency of spectral lines can be represented as the difference of two terms.

Bohr added as a third postulate the *correspondence principle*, which states in essence that, for very small energy differences, the laws of quantum physics approach asymptotically those of classical physics. Thus he derived the *quantization condition of the rotator*, which prescribes that an angular momentum has to be quantized as an integral multiple of

$$\hbar \equiv h/2\pi, \quad (2.43)$$

or

$$p' = mvr = n\hbar \quad \text{with } n = 1, 2, 3, \ldots. \quad (2.44)$$

An electron circling a nucleus corresponds to a ring current and consequently produces an orbital magnetic moment (see IA, Eq. 2.20), which, for a rotation frequency ν in an orbit of radius r, equals

$$\mathbf{m} = IF\mathbf{n} = -e\nu r^2 \pi \mathbf{n}. \quad (2.45)$$

The electron has simultaneously an angular momentum

$$p' = mvr = m\nu 2r\pi r, \quad (2.46)$$

which can be represented by a vector $\mathbf{p}'$ orientated antiparallel to $\mathbf{m}$ (Fig. 2.5). Hence, the magnetic and

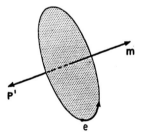

Fig. 2.5. Angular momentum and magnetic moment of orbital electron.

mechanical moments of circling electrons are interrelated and have the classical ratio

$$\gamma \equiv \frac{|\mathbf{m}|}{|\mathbf{p}'|} = \frac{e}{2m}. \quad (2.47)$$

The quantity γ (gamma) is called the *gyromagnetic ratio*. The angular momentum is quantized according to Eq. 2.44; in consequence, the orbital magnetic moment is quantized as a multiple of an elementary magnetic moment, the *Bohr magneton*,

$$m_B = \frac{e}{2m}\hbar = 9.27 \times 10^{-24} \quad [\text{joule/weber m}^{-2}]$$
$$= 9.27 \times 10^{-21} \quad [\text{erg/gauss}]. \quad (2.48)$$

By measuring magnetic moments in this unit, the ratio of magnetic to angular moment, called the *g factor*, is, for orbital electrons, $g = 1$.

In addition to causing orbital moments, each electron itself carries a magnetic spin moment of 1 Bohr magneton, coupled to a mechanical spin of only ½ quantum of action, as Goudsmit and Uhlenbeck[5] concluded from spectroscopic data. *Free electrons* have therefore a gyromagnetic ratio $g = 2$.

Bohr's theory, while eminently successful for one-electron systems, led to serious difficulties in the analysis of multi-electron systems and in the fine structure of spectral lines. Obviously, the planetary atom with its mysterious stabilization of electrons in precisely defined orbits was an intermediary concept that would have to yield to some more deeply founded description.

The quantum theory of radiation had introduced the *dualism* that the behavior of light might be discussed either from the particle standpoint by the energy $\mathcal{E}$ and the momentum p of *photons*, or from the wave standpoint by the frequency ν and wavelength λ of the corresponding electromagnetic wave, by the interrelating equations

$$\mathcal{E} = h\nu,$$
$$p = mv = \frac{h}{\lambda}. \quad (2.49)$$

De Broglie[6] now postulated that this dualism for photons traveling with the velocity of light also holds for particles of a much smaller velocity v and possessing a true rest mass m_0. According to the *de Broglie equations* (Eqs. 2.49), the motion of any particle is correlated to a wave phenomenon which statistically prescribes the motion of the particle by its wave pattern. This revolutionary speculation was soon confirmed by the electron diffraction experiment of Davisson and Germer.[7]

Newtonian mechanics, on which the Rutherford atom was based, assumed that once the initial conditions are given, the position and momentum of each particle are precisely defined at all values of time. It

[5] G. E. Uhlenbeck and S. Goudsmit, *Naturwiss.* **13**, 953 (1925); *Nature* **117**, 264 (1926).
[6] L. de Broglie, *Nature* **112**, 540 (1923); *Thesis*, Paris, 1924; *Ann. Physik* **3**, 22 (1925).
[7] C. Davisson and L. H. Germer, *Phys. Rev.* **30**, 705 (1927); *Proc. Natl. Acad. Sci.* **14**, 317 (1928).

became evident then that this concept is limited to the macroscopic world. Just as geometrical optics yields to wave optics, Newtonian mechanics is superseded by the new wave mechanics based on de Broglie's equations whenever the wavelength becomes comparable to some significant dimension of the problem. In the experience of daily life, the meter, kilogram, and second are appropriate yardsticks for the phenomena encountered; for example, a mass of 1 kg may move with the velocity of 1 m/sec. In this case the wavelength

$$\lambda = \frac{h}{m} = 6.623 \times 10^{-34} \quad [\text{m}] \quad (2.50)$$

proves negligible. In the realm of the molecular world, on the other hand, particles of very small mass, like protons (m_+) and electrons (m),

$$\left.\begin{array}{l}m_+ = 1.6725 \times 10^{-27} \quad [\text{kg}]\\ m = 9.107 \times 10^{-31} \quad [\text{kg}]\end{array}\right\} \frac{m_+}{m} = 1836.5, \quad (2.51)$$

move about with much larger velocities, and the interference phenomena caused by their wavelengths become decisive.

Electromagnetic waves contain information about the statistical behavior and the energy of *many* photons. A de Broglie wave, in contrast, is associated with the location of an *individual* particle. The intensity of the de Broglie wave at any point in space represents the probability of finding the particle at that point.

This wave aspect gives the stationary states of the Rutherford atom a reasonable interpretation. If the

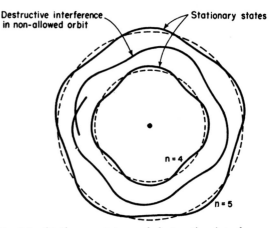

Fig. 2.6. Stationary states and destructive interference.

position of a circling electron is described by a probability wave, it can obviously persist only in orbits in which the de Broglie wave can set up a standing-wave pattern instead of destroying itself by interference (Fig. 2.6). For the circular orbits of the hydrogen atom we thus arrive at the stability condition

$$2r\pi = n\lambda \quad \text{with } n = 1, 2, 3, \ldots, \quad (2.52)$$

which reverts to Bohr's quantization condition of the rotator (Eq. 2.44), if the momentum is introduced in place of the wavelength.

Although this confirmation is gratifying, it should not be misunderstood. The stringent assignment of an electron of a precise momentum to a precise orbit brought Bohr's theory to grief and contradicts the spirit of wave mechanics. There is an uncertainty in the simultaneous definition of such correlated parameters as position and wavelength; this is a typical feature of any wave phenomenon. In the case of quantum mechanics, the uncertainty in the simultaneous definition of the momentum p of a particle at a location q, or of its total energy $\mathcal{E}$ at a time t, is given by the *uncertainty relations*, first formulated by Heisenberg,[8]

$$\Delta q \, \Delta p \geq \hbar,$$
$$\Delta t \, \Delta \mathcal{E} \geq \hbar. \qquad (2.53)$$

De Broglie's theory had been largely of a qualitative nature. To make use of the wave concept for a quantitative discussion of molecular phenomena requires the formulation of a basic wave equation, a task solved by Schrödinger.[9] In Maxwell's theory there appears the electric plane wave

$$\mathbf{E} = \mathbf{E}_0 e^{j\left(\omega t - \frac{2\pi}{\lambda} x\right)} \qquad (2.54)$$

as a solution of the wave equation

$$\frac{\partial^2 \mathbf{E}}{\partial x^2} + \left(\frac{2\pi}{\lambda}\right)^2 \mathbf{E} = 0. \qquad (2.55)$$

If a de Broglie probability wave traveling in the $+x$-direction is described in analogy by a wave function

$$\psi = A e^{j\left(\omega t - \frac{2\pi}{\lambda} x\right)}, \qquad (2.56)$$

it obviously is a solution of the differential equation

$$\frac{\partial^2 \psi}{\partial x^2} + \left(\frac{2\pi}{\lambda}\right)^2 \psi = 0. \qquad (2.57)$$

The wavelength of the particle can be replaced, according to Eq. 2.49, by its kinetic energy expressed as the difference of its total energy $\mathcal{E}$ and potential energy U,

$$E_{\text{kin}} = \mathcal{E} - U = \frac{h^2}{2m\lambda^2}. \qquad (2.58)$$

If, in addition, we remove the restriction that the particle is confined to the $+x$-direction, Eq. 2.57 is transformed into *Schrödinger's equation*

$$\nabla^2 \psi + \frac{2m}{\hbar^2}(\mathcal{E} - U)\psi = 0. \qquad (2.59)$$

[8] W. Heisenberg, *Z. Physik* **33**, 879 (1925).
[9] E. Schrödinger, *Ann. Physik* **79**, 361, 489 (1926).

Schrödinger's equation, unlike the electromagnetic field equations, can be solved accurately in very few cases only. The inherent difficulty lies in the fact that the photons of the electromagnetic field do not interact, whereas the particles of wave mechanics are in general charged and interact with strong forces. Thus we face, in a new disguise, the old difficulty of classical mechanics: the two-body problem can be solved rigorously, the multi-body problem only by approximations. Simultaneously, if a charged particle moves in the electrostatic field of others, as electrons do in atoms, the kinetic energy and thus the particle wavelength vary from point to point. Hence, while the electromagnetic theory deals normally with homogeneous media, the media of wave mechanics are extremely dispersive. Quantum mechanics is therefore compelled to strive for solutions by approximations; we have to be on guard continuously lest the approximations chosen neglect essential aspects of the physical situation.

The two-body problem of the hydrogen atom can be solved accurately, and wave mechanics arrives thus at a precise designation of the various stationary states or orbitals in which the electron can move by the use of *three quantum numbers*. The *principal* or *total quantum number* n determines the distance of the electron from the nucleus and prescribes a sequence of successive shells ($n = 1$ or K shell; $n = 2$ or L shell; $n = 3$ or M shell, etc.) in the electronic structure. The *azimuthal quantum number* l describes the eccentricity of the electron clouds; it may assume the values

$$l = 0, 1, 2, \ldots \quad (n-1), \qquad (2.60)$$

where $l = 0$ designates spherical or s orbitals; $l = 1$, p orbitals; $l = 2$, d orbitals; etc. (Fig. 2.7).[10] Eccentric

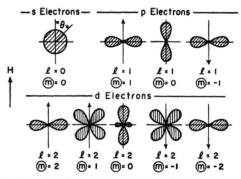

Fig. 2.7. Angular dependence of the electron density $\psi\tilde{\psi}$ of some s, p, and d orbitals. (After White.[10])

electronic states have a net angular momentum and thus a magnetic dipole moment. The torque of a

[10] H. E. White, *Phys. Rev.* **37**, 1416 (1931). See also G. Herzberg, *Atomic Spectra and Atomic Structure*, Dover Publications, New York, 1944, p. 44.

TABLE 2.1. PERIODIC SYSTEM OF THE

n	Shell	I	II
1	K	1 H Hydrogen 1.0081 1 $^2S_{1/2}$	
1 2	K L	3 Li Lithium 6.940 2 1 $^2S_{1/2}$	4 Be Beryllium 9.02 2 2 $^1S_0$
1 2 3	K L M	11 Na Sodium 22.997 2 2 6 1 $^2S_{1/2}$	12 Mg Magnesium 24.32 2 2 6 2 $^1S_0$

		I_a	II_a	III_a	IV_a	V_a	VI_a	VII_a	VIII		
1 2 3 4	K L M N	19 K Potassium 39.096 2 2 6 2 6 1 $^2S_{1/2}$	20 Ca Calcium 40.08 2 2 6 2 6 2 $^1S_0$	21 Sc Scandium 45.10 2 2 6 2 6 1 2 $^2D_{3/2}$	22 Ti Titanium 47.90 2 2 6 2 6 2 2 $^3F_2$	23 V Vanadium 50.95 2 2 6 2 6 3 2 $^4F_{3/2}$	24 Cr Chromium 52.01 2 2 6 2 6 4 2 $^7S_3$	25 Mn Manganese 54.93 2 2 6 2 6 5 2 $^6S_{5/2}$	26 Fe Iron 55.84 2 2 6 2 6 6 2 $^5D_4$	27 Co Cobalt 58.94 2 2 6 2 6 7 2 $^4F_{9/2}$	
1 2 3 4 5	K L M N O	37 Rb Rubidium 85.48 2 2 6 2 6 10 2 6 1 $^2S_{1/2}$	38 Sr Strontium 87.63 2 2 6 2 6 10 2 6 2 $^1S_0$	39 Y Yttrium 88.92 2 2 6 2 6 10 2 6 1 2 $^2D_{3/2}$	40 Zr Zirconium 91.22 2 2 6 2 6 10 2 6 2 2 $^3F_2$	41 Nb Niobium 92.91 2 2 6 2 6 10 2 6 4 1 $^6D_{1/2}$	42 Mo Molybdenum 95.95 2 2 6 2 6 10 2 6 5 1 $^7S_3$	43 Tc Technetium 99 2 2 6 2 6 10 2 6 6 1	44 Ru Ruthenium 101.7 2 2 6 2 6 10 2 6 7 1 $^5F_5$	45 Rh Rhodium 102.91 2 2 6 2 6 10 2 6 8 1 $^4F_{9/2}$	
1 2 3 4 5 6	K L M N O P	55 Cs Cesium 132.91 2 2 6 2 6 10 2 6 10 2 6 1 $^2S_{1/2}$	56 Ba Barium 137.36 2 2 6 2 6 10 2 6 10 2 6 2 $^1S_0$	57 La Lanthanum 138.92 2 2 6 2 6 10 2 6 10 2 6 1 2 $^2D_{3/2}$	58 Ce Cerium 140.13 2 2 6 2 6 10 2 6 10 1 2 6 1 2	59 Pr Praseodymium 140.92 2 2 6 2 6 10 2 6 10 2 2 6 1 2	60 Nd Neodymium 144.27 2 2 6 2 6 10 2 6 10 3 2 6 1 2	61 Pm Prometheum 147 2 2 6 2 6 10 2 6 10 4 2 6 1 2 $^5I_4$	62 Sm Samarium 150.43 2 2 6 2 6 10 2 6 10 5 2 6 1 2 $^7F_0$	63 Eu Europium 152.0 2 2 6 2 6 10 2 6 10 6 2 6 1 2 $^8S_{7/2}$	
1 2 3 4 5 6	K L M N O P				72 Hf Hafnium 178.6 2 2 6 2 6 10 2 6 10 14 2 6 2 2 $^3F_2$	73 Ta Tantalum 180.88 2 2 6 2 6 10 2 6 10 14 2 6 3 2 $^4F_{3/2}$	74 W Tungsten 183.92 2 2 6 2 6 10 2 6 10 14 2 6 4 2 $^5D_0$	75 Re Rhenium 186.31 2 2 6 2 6 10 2 6 10 14 2 6 5 2 $^6S_{5/2}$	76 Os Osmium 190.2 2 2 6 2 6 10 2 6 10 14 2 6 6 2 $^5D_4$	77 Ir Iridium 193.1 2 2 6 2 6 10 2 6 10 14 2 6 9 2 $^4F_{9/2}$	
1 2 3 4 5 6 7	K L M N O P Q	87 Fr Francium 223 2 2 6 2 6 10 2 6 10 14 2 6 10 2 6 1 $^2S_{1/2}$	88 Ra Radium 226.05 2 2 6 2 6 10 2 6 10 14 2 6 10 2 6 2 $^1S_0$	89 Ac Actinium 227 2 2 6 2 6 10 2 6 10 14 2 6 10 2 6 1 2 $^2D_{3/2}$	90 Th Thorium 232.12 2 2 6 2 6 10 2 6 10 14 2 6 10 2 6 2 2 $^3F_2$	91 Pa Protoactinium 231 2 2 6 2 6 10 2 6 10 14 2 6 10 2 6 3 2	92 U Uranium 238.07 2 2 6 2 6 10 2 6 10 14 2 6 10 2 6 4 2 $^5L_6$	93 Np Neptunium 237 2 2 6 2 6 10 2 6 10 14 2 6 10 1 2 6 4 2	94 Pu Plutonium 239 2 2 6 2 6 10 2 6 10 14 2 6 10 2 2 6 4 2	95 Am Americium 241 2 2 6 2 6 10 2 6 10 14 2 6 10 3 2 6 4 2	
		s p d f	s p d f	s p d f	s p d f	s p d f	s p d f	s p d f	s p d f	s p d f	

Polarization and Atomic Structure

ELEMENTS				III	IV	V	VI	VII	O	Shell	n
									2 He Helium 4.003 2 $^1S_0$	K	1
				5 B Boron 10.82 2 2 1 $^2P_{1/2}$	6 C Carbon 12.010 2 2 2 $^3P_0$	7 N Nitrogen 14.008 2 2 3 $^4S_{3/2}$	8 O Oxygen 16.0000 2 2 4 $^3P_2$	9 F Fluorine 19.00 2 2 5 $^2P_{3/2}$	10 Ne Neon 20.138 2 2 6 $^1S_0$	K L	1 2
				13 Al Aluminum 26.97 2 2 6 2 1 $^2P_{1/2}$	14 Si Silicon 28.06 2 2 6 2 2 $^3P_0$	15 P Phosphorus 30.98 2 2 6 2 3 $^4S_{3/2}$	16 S Sulfur 32.06 2 2 6 2 4 $^3P_2$	17 Cl Chlorine 35.457 2 2 6 2 5 $^2P_{3/2}$	18 A Argon 39.944 2 2 6 2 6 $^1S_0$	K L M	1 2 3
	I_b	II_b	III_b	IV_b	V_b	VI_b	VII_b				
28 Ni Nickel 58.69 2 2 6 2 6 8 2 $^3F_4$	29 Cu Copper 63.57 2 2 6 2 6 10 1 $^2S_{1/2}$	30 Zn Zinc 65.38 2 2 6 2 6 10 2 $^1S_0$	31 Ga Gallium 69.72 2 2 6 2 6 10 2 1 $^2P_{1/2}$	32 Ge Germanium 72.60 2 2 6 2 6 10 2 2 $^3P_0$	33 As Arsenic 74.91 2 2 6 2 6 10 2 3 $^4S_{3/2}$	34 Se Selenium 78.96 2 2 6 2 6 10 2 4 $^3P_2$	35 Br Bromine 79.916 2 2 6 2 6 10 2 5 $^2P_{3/2}$	36 Kr Krypton 83.7 2 2 6 2 6 10 2 6 $^1S_0$		K L M N	1 2 3 4
46 Pd Palladium 106.7 2 2 6 2 6 10 2 6 10 $^1S_0$	47 Ag Silver 107.88 2 2 6 2 6 10 2 6 10 1 $^2S_{1/2}$	48 Cd Cadmium 112.41 2 2 6 2 6 10 2 6 10 2 $^1S_0$	49 In Indium 114.76 2 2 6 2 6 10 2 6 10 2 1 $^2P_{1/2}$	50 Sn Tin 118.70 2 2 6 2 6 10 2 6 10 2 2 $^3P_0$	51 Sb Antimony 121.76 2 2 6 2 6 10 2 6 10 2 3 $^4S_{3/2}$	52 Te Tellurium 127.61 2 2 6 2 6 10 2 6 10 2 4 $^3P_2$	53 I Iodine 126.92 2 2 6 2 6 10 2 6 10 2 5 $^2P_{3/2}$	54 Xe Xenon 131.3 2 2 6 2 6 10 2 6 10 2 6 $^1S_0$		K L M N O	1 2 3 4 5
64 Gd Gadolinium 156.9 2 2 6 2 6 10 2 6 10 7 2 6 1 2 $^9D_2$	65 Tb Terbium 159.2 2 2 6 2 6 10 2 6 10 8 2 6 1 2	66 Dy Dysprosium 162.46 2 2 6 2 6 10 2 6 10 9 2 6 1 2	67 Ho Holmium 163.5 2 2 6 2 6 10 2 6 10 10 2 6 1 2	68 Er Erbium 167.2 2 2 6 2 6 10 2 6 10 11 2 6 1 2	69 Tm Thulium 169.4 2 2 6 2 6 10 2 6 10 12 2 6 1 2 $^2F_{7/2}$	70 Yb Ytterbium 173.04 2 2 6 2 6 10 2 6 10 13 2 6 1 2 $^1S_0$	71 Lu Lutecium 175.0 2 2 6 2 6 10 2 6 10 14 2 6 1 2 $^2D_{3/2}$			K L M N O P	1 2 3 4 5 6
78 Pt Platinum 195.23 2 2 6 2 6 10 2 6 10 14 2 6 10 1 $^3D_3$	79 Au Gold 197.2 2 2 6 2 6 10 2 6 10 14 2 6 10 1 $^2S_{1/2}$	80 Hg Mercury 200.61 2 2 6 2 6 10 2 6 10 14 2 6 10 2 $^1S_0$	81 Tl Thallium 204.39 2 2 6 2 6 10 2 6 10 14 2 6 10 2 1 $^2P_{1/2}$	82 Pb Lead 207.21 2 2 6 2 6 10 2 6 10 14 2 6 10 2 2 $^3P_0$	83 Bi Bismuth 209.00 2 2 6 2 6 10 2 6 10 14 2 6 10 2 3 $^4S_{3/2}$	84 Po Polonium 210 2 2 6 2 6 10 2 6 10 14 2 6 10 2 4 $^3P_2$	85 At Astatine 211 2 2 6 2 6 10 2 6 10 14 2 6 10 2 5 $^2P_{3/2}$	86 Rn Radon 222 2 2 6 2 6 10 2 6 10 14 2 6 10 2 6 $^1S_0$		K L M N O P	1 2 3 4 5 6
96 Cm Curium 242 2 2 6 2 6 10 2 6 10 14 2 6 10 7 2 6 1 2	97 Bk Berkelium 2 2 6 2 6 10 2 6 10 14 2 6 10 8 2 6 1 2	98 Cf Californium 2 2 6 2 6 10 2 6 10 14 2 6 10 9 2 6 1 2								K L M N O P Q	1 2 3 4 5 6 7
s p d f	s p d f	s p d f	s p d f	s p d f	s p d f	s p d f	s p d f	s p d f			

magnetic field causes a precession of the momentum vector around the field axis, that is, an additional angular momentum of precession that has to be quantized. The *magnetic quantum number* $\textcircled{m}$ designates the quantized orientations of the moment in respect to the field axis; it varies from parallel to antiparallel as

$$\textcircled{m} = l, l-1, \ldots, -(l-1), -l. \qquad (2.61)$$

A p state splits thus into three orbitals (1, 0, −1), a d state into five orbitals (2, 1, 0, −1, −2), etc., when the hydrogen atom is brought into a magnetic field. The corresponding splitting of the spectral lines is the *Zeeman effect*.[11]

The description of the stationary states of the hydrogen atom can be used to derive a first understanding of atomic structures in general and of the *periodic system of the elements*. The building principle is to increase the nuclear charge in successive steps of one elementary charge and simultaneously to add one compensating electron at each step to the electronic system in hydrogen-like orbitals. The question how many electrons an orbital can accommodate, is decided by the *Pauli exclusion principle*[12] which states that each electron in an atom must be described by its individual set of quantum numbers and that no two sets can be alike. Since the electrons themselves have a quantized angular momentum

$$p' = s\hbar \qquad \text{with } s = \pm \frac{1}{2}, \qquad (2.62)$$

each *orbital* designated by its set of quantum numbers $(n, l, \textcircled{m})$ can take two electrons with antiparallel spins.

With this proviso, by filling in succession the orbitals of lowest energy, the *Bohr-Stoner arrangement of the periodic table*[13] results (Table 2.1). For the rare gas helium the K shell is filled with two electrons in antiparallel spin position. Since this is the $n = 1$ state and an s orbital ($l = 0$), the electron structure of the helium atom can be identified by the symbol $1s^2$, where the exponent signifies the number of electrons occupying the orbital type. The situation is repeated for neon, where filling of the L shell produces the constellation $1s^2 2s^2 2p^6$, and for argon where the occupation of the $3s$ and $3p$ orbitals ($1s^2 2s^2 2p^6 3s^2 3p^6$) is completed.

The electron clouds of the hydrogen atom illustrate the fact that the s electrons approach the nucleus closest on the average, whereas the p, d, and f electrons reach farther and farther into space. Hence, we expect that first the s orbital and then the p orbitals will be filled as indicated up to the element argon. For farther outlying levels it becomes a complicated problem to decide which orbital represents the lower energy state, and the sequence may shift as the nuclear charge increases. For the element 19(K) the $4s$ orbital lies below the $3d$ level and fills first, but at chromium and at copper one $4s$ electron is lost to the $3d$ orbitals. The magnetic behavior of the iron group arises from the unpaired spin moments of the filling $3d$ shell. A similar situation occurs for the rare earth elements, where the $5s$ and $4d$ orbitals compete, and finally in the transuranium elements in a complicated competition between $6d$ and $5f$.

The chemical reactivity of atoms arises from the *outer* or *valence electrons* in uncompleted shells or subshells and reflects therefore the periodicity of the electronic cloud structure.

[11] P. Zeeman, *Phil. Mag.* **43**, 226 (1897).
[12] W. Pauli, Jr., *Z. Physik* **31**, 765 (1925).
[13] E. C. Stoner, *Phil. Mag.* **48**, 719 (1924). For special arrangement of transuranic elements see C. D. Coryell, *J. Chem. Educ.* **29**, 62 (1952); *Record Chem. Progress* **12**, 55 (1951).

3 · Structure and Dielectric Response of Molecules

In the preceding section we have been concerned with the structure of the electronic atmosphere that surrounds *one* nucleus. We found that the electrons assume stationary states prescribed by quantum mechanics as the standing wave patterns of probability waves. The mutual interaction of the electrons is, in the main, of an electrostatic nature; however, the electron clouds may have, in addition, orbital magnetic moments and the electrons themselves have magnetic spin moments.

In extending the discussion[1] to molecules, we may expect additional difficulties. The electronic systems surrounding *several* nuclei are of lower symmetry, and the possible wave modes will depend critically on the distance of the nuclei and their mutual orientation in space.

As two atoms approach each other, they come first

[1] For a more detailed quantum-mechanical discussion, see J. C. Slater, *Quantum Theory of Matter*, McGraw-Hill Book Co., New York, 1951, Chapters VIII and IX.

under the influence of the *van der Waals attraction*. This force, as London showed, arises because the outer electrons of the approaching atoms tend to repel each other electrostatically. Thus the motion of these electrons becomes correlated and a dipole-dipole coupling force by fluctuating induced moments results. The van der Waals force between neutral atoms is accordingly rather short-reaching; it is inversely proportional to the seventh power of the distance.

As the atoms draw closer together, their electron clouds begin to overlap and a *true Coulomb attraction* sets in; the nuclei start binding the electrons of the other partner. What happens next is a question of quantum mechanics, since bonding may occur only when the interpenetrating electron clouds can form joint standing-wave patterns. This is simple when the interacting orbitals are partly filled; such orbitals may link together by contributing electrons from each partner and producing *electron pair bonds* with antiparallel spins. Such pair bonds, proposed by Lewis on chemical grounds ten years before the advent of quantum mechanics, are the *normal covalent (homopolar) bonds* of chemistry. If such bonding is not possible, the Pauli principle can be satisfied only if one of the partners goes over into an excited state. Usually the collision energy will not suffice to bring this about, and a strong repulsion takes place. A third possibility is that the partners give and take electrons, thus forming ions of opposite polarity which attract each other by Coulomb forces (*ionic bonds*).

Actually, the van der Waals, the covalent, and the ionic bond are extreme cases and all kinds of intermediate types of binding may be realized. But it is convenient to establish these limiting cases and to visualize the size of atoms in the different stages of binding by *characteristic radii* of a *van der Waals, covalent* or *ionic size*.

When unlike atoms form molecules, the attraction of the two partners for electrons, their *electronegativity*, is different, and *permanent electric dipole moments* result. Since Debye[2] was the first to realize the significance of such permanent moments for molecular physics and chemistry, a convenient unit for measuring such moments has been named in his honor: 1 debye = 1×10^{-18} [esu] = 3.33×10^{-30} [coul m]. The dipole moments of molecules, like the bond strength, the intermolecular distance, and other parameters, cannot be calculated with accuracy at present but must be measured.

The dipole moments[3] of gas molecules have been determined, in general, by a quasi-static measurement of the polarization of the dielectric as a function of temperature. This method is based on a statistical theory of orientation, first developed for the permanent magnetic moments of paramagnetic substances by Langevin[4] and applied to permanent electric moments by Debye.[5]

Molecules carrying a permanent dipole moment suffer a torque in an electric field that tends to align the dipole axis in the field direction (see IA, Eq. 2.12). Thermal agitation, on the other hand, tends to maintain a random distribution. The outcome of these counteracting influences is a statistical equilibrium that can be calculated without reference to the actual rotation of the molecules and their interaction as long as the electric field changes so slowly that the equilibrium is reached with certainty.

The assumptions for a simplified calculation are: the permanent dipole moment μ of the molecules is unaffected by the field; the density of the gas is so low that the dipolar interaction energy is small in comparison to the thermal equilibrium energy kT; finally, the dipoles can assume any direction with respect to the field axis.

The potential energy U of a dipole pointing at an angle θ to the field is $-|\mu||E|\cos\theta$. Consequently, the number pointing at a space angle

$$d\Omega = 2\pi \sin\theta \, d\theta \quad (3.1)$$

(Fig. 3.1) is, according to Boltzmann's statistics,

$$N = A \exp\left(-\frac{U}{kT}\right) d\Omega$$
$$= A \exp\left(\frac{|\mu||E|\cos\theta}{kT}\right) d\Omega. \quad (3.2)$$

Each of these dipoles contributes to the orientation polarization the component $|\mu|\cos\theta$ in the field direction; hence it appears to the outside observer as if each molecule carries the average moment

$$\bar{\mu}_d = \frac{\int_0^\pi A \exp\left(\frac{|\mu||E|\cos\theta}{kT}\right)(\mu\cos\theta)2\pi\sin\theta \, d\theta}{\int_0^\pi A \exp\left(\frac{|\mu||E|\cos\theta}{kT}\right) 2\pi\sin\theta \, d\theta}. \quad (3.3)$$

---

[2] P. Debye, *Physik. Z.* **13**, 97 (1912).

[3] An extensive compilation has been made by L. G. Wesson, *Tables of Electric Dipole Moments*, Technology Press, Cambridge, Mass., 1948.

[4] M. P. Langevin, *J. physique* **4**, 678 (1905); *Ann. chim. phys.* **5**, 70 (1905).

[5] P. Debye, *Polar Molecules*, Dover Publications, New York, 1945, pp. 27–30.

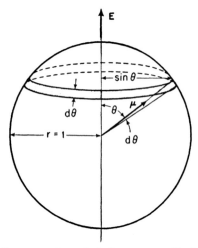

Fig. 3.1. Geometry for calculating average dipole moment.

The integration over the space angle θ can be carried out by introducing the parameters

$$x \equiv \frac{|\mu||\mathbf{E}|}{kT},$$
$$\zeta \equiv \cos\theta, \qquad (3.4)$$

and rewriting Eq. 3.3 as

$$\frac{\bar{\mu}_d}{\bar{\mu}} = \frac{\int_{-1}^{+1} e^{x\zeta}\zeta\, d\zeta}{\int_{-1}^{+1} e^{x\zeta}\, d\zeta}. \qquad (3.5)$$

The denominator of this equation is

$$\int_{-1}^{+1} e^{x\zeta}\, d\zeta = \frac{e^x - e^{-x}}{x}, \qquad (3.6)$$

whereas the numerator equals the differential quotient of this denominator after the parameter x, or

$$\int_{-1}^{+1} e^{x\zeta}\zeta\, d\zeta = \frac{x(e^x + e^{-x}) - (e^x - e^{-x})}{x^2}. \qquad (3.7)$$

Hence

$$\bar{\mu}_d/\bar{\mu} = \coth x - \frac{1}{x} \equiv L(x). \qquad (3.8)$$

The function $L(x)$ which describes the ratio of the average to actual moment of a gas molecule in its dependence on temperature and field strength was first derived by Langevin in the theory of paramagnetism and is therefore known as the *Langevin function*. Figure 3.2 shows the characteristic $L(x)$ as applied by Debye for the electric case; the field scale refers to the orientation of polar molecules of unit moment, $\mu = 1$ debye, against the thermal agitation at room temperature.

Obviously very high field strengths are required to produce a deviation from linearity and an approach towards saturation. For the normal case of relatively small fields ($x \ll 1$) the Langevin function can be approximated by its tangent $x/3$; that is, the average moment due to orientation becomes simply

$$\bar{\mu}_d \simeq \frac{|\mu|^2}{3kT}\mathbf{E}. \qquad (3.9)$$

The applied field will, in addition, induce a moment in the gas molecule by deforming its electron cloud (α_e) and by changing the spacing of the nuclei (α_a). Thus deformation and orientation polarization together produce the total moment per dipole molecule

$$\mu_t = \left(\alpha_e + \alpha_a + \frac{\mu^2}{3kT}\right)\mathbf{E}. \qquad (3.10)$$

Our derivation assumed a gas of low pressure. The polarizability per unit volume and with it the relative permittivity for static fields, κ_s', follows in this case from Eq. 1.18 as

$$\frac{N\mu_t}{\epsilon_0 \mathbf{E}} = \kappa_s' - 1 = \frac{N}{\epsilon_0}\left(\alpha_e + \alpha_a + \frac{|\mu|^2}{3kT}\right). \qquad (3.11)$$

We will designate by κ_∞' the relative dielectric constant due to the induced moments only,

$$\kappa_\infty' \equiv 1 + \frac{N}{\epsilon_0}(\alpha_e + \alpha_a); \qquad (3.12)$$

it can be measured at frequencies so high that the orientation polarization has no time to develop. The con-

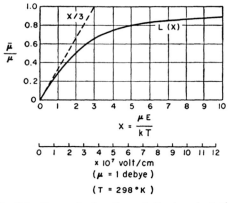

Fig. 3.2. Langevin function of dipole orientation.

tribution of the permanent dipoles to the static permittivity then becomes

$$\kappa_s' - \kappa_\infty' = \frac{N|\mu|^2}{\epsilon_0 3kT}. \qquad (3.13)$$

The fact that the orientation of the permanent dipole moments is strongly temperature-dependent in contrast to the practically temperature-independent contributions of the induced moments gives a convenient method of determining $\kappa_s' - \kappa_\infty'$ and with it the permanent dipole moment μ of the molecules. To eliminate the effect of varying gas density, we refer to the polarizability per mole Π (see Eq. 2.9) in the formulation for low gas pressure

$$\Pi = \frac{N_0 \alpha}{3\epsilon_0} = \frac{\kappa_s' - 1}{3} \frac{M}{\rho}. \quad (3.14)$$

If Π is plotted as a function of $1/T$, we can distinguish easily between polar and nonpolar molecules (Fig. 3.3). For the former a straight line is obtained inter-

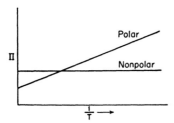

Fig. 3.3. Molar polarization of polar and nonpolar molecules as $f(1/T)$.

secting the $1/T$ axis at a finite angle, whereas the polarization of a nonpolar gas is temperature-independent, hence produces a horizontal line.

If we write the molar polarization of the gas in the form

$$\Pi = A + \frac{B}{T} \quad [m^3], \quad (3.15)$$

where

$$A = \frac{N_0}{3\epsilon_0}(\alpha_e + \alpha_a) = 2.27 \times 10^{34}(\alpha_e + \alpha_a) \quad (3.16)$$

contains the effect of the deformation polarization, and

$$B = \frac{N_0}{3\epsilon_0}\frac{|\mu|^2}{3k} = 5.48 \times 10^{56}|\mu|^2 \quad (3.17)$$

that of the orientation polarization, we find the dipole moment as

$$\begin{aligned}\mu &= 4.27 \times 10^{-29}\sqrt{B} \quad [\text{coul m}] \\ &= 12.7\sqrt{B} \quad [\text{debye}].\end{aligned} \quad (3.18)$$

Figure 3.4 shows the results of measurements on some polar and nonpolar gases. Extrapolating the straight lines back to the point $1/T = 0$, we obtain the A values, whereas their slope determines the B values as

$$B = \frac{\Pi_1 - \Pi_2}{1/T_1 - 1/T_2}. \quad (3.19)$$

Since the electronic polarization of these compounds can be determined from measurements of the refractive index in the visible region according to the Lorentz-Lorenz equation (Eq. 1.22), the molar polarization Π can be separated into its electronic, atomic, and dipole contributions. As a rule of thumb, the atomic polarization Π_a can be estimated as amounting to about 10 percent of the electronic polarization Π_e for dipole moments greater than 1 debye.[6]

The frequency response of molecular gases, like that of atoms, is characterized by resonance spectra. In addition to the electronic excitation limited in general to the visible and ultraviolet spectral range, the vibration of nuclei relative to each other (*atomic polarization*) produces resonance states in the near infrared; and the rotation of molecules as a whole or of molecular groups around internuclear axes (*orientation* or *dipole polarization*) gives rise to resonances in the far infrared down into the microwave region. The classical model for a discussion of these additional modes of motion for diatomic molecules is the dumbbell molecule, assumed to oscillate as a harmonic oscillator or to rotate as a rigid rotator.[7]

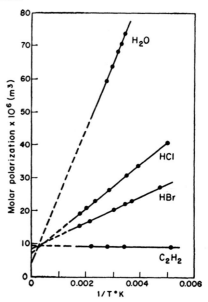

Fig. 3.4. Polarization of polar and nonpolar gases. (Adapted in part from Smyth.[6])

A vibrating molecule will be optically active, that is, emit or absorb radiation, if its dipole moment changes during the oscillation $\left(\frac{\partial \mu}{\partial r} \neq 0\right)$. The variation of the

---

[6] See C. P. Smyth, *Dielectric Constant and Molecular Structure*, Chemical Catalog Co., New York, 1931. See also R. J. W. LeFèvre, *Dipole Moments*, Methuen and Co., Ltd., London, 1938.

[7] For a detailed discussion of molecular spectroscopy, see G. Herzberg, *Molecular Spectra and Molecular Structure*, Vol. I, *Diatomic Molecules*, Van Nostrand, New York, 1950; Vol. II, *Infrared and Raman Spectra*, Prentice-Hall, New York, 1945.

dipole moment is classically equivalent to an alternating current traversing a dipole antenna. The harmonic oscillator has only one resonance frequency (ν_{osc}) and its total energy is quantized in integral multiples of $h\nu_{osc}$ or, more accurately,

$$\mathcal{E}_{osc} = h\nu_{osc}(v + \tfrac{1}{2}), \qquad (3.20)$$

where v is the *quantum number of vibration;* at absolute zero, $v = 0$, a *zero-point energy* of one-half vibrational quantum still remains, as required by the uncertainty relation.

The dumbbell molecule, as a rigid rotator, will be optically active if it carries a permanent dipole moment, since an electric field can exercise a torque on such a moment. Classically it can rotate with any frequency, whereas quantum-mechanically the rotator goes through a sequence of resonance frequencies because its angular momentum is quantized (see Eq. 2.44) as

$$p' = J\hbar; \qquad (3.21)$$

J is the *quantum number of rotation.*

The energy of interaction between *two* atoms, when plotted as a function of their nuclear separation, is represented by a potential curve as shown in Fig. 3.5. Its minimum corresponds to the mean distance of separation at absolute zero; its depth measures the

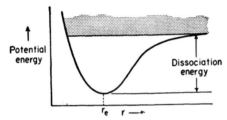

Fig. 3.5. Potential curve of diatomic molecule.

energy of dissociation. When covalent-bonded, the molecule separates into neutral atoms; when ionic-bonded, into a positive and negative ion.

As atoms combine, well-defined molecules may form by the saturation of valence bonds; at the other extreme, atoms may flock together without limit and build up the structures of liquids and solids. The Coulomb field of the Na^+ and Cl^- ion, for example, by attracting additional ionic partners builds up the ionic sodium chloride crystal and the metallic bond between mercury atoms leads to the formation of mercury metal. Intermediate stages of condensation are realized by the polymerization of molecules.

The possible shapes and sizes of true valence molecules depend on the type of orbitals that are used in the formation of electron-pair bonds. There are available not only the s, p, d, etc., orbitals of the atoms on which the periodic system of the elements was based (see Table 2.1), but new *hybrid* orbitals can be realized by linear combinations of these wave functions. The tetrahedral s, p- and the octahedral d, s, p-hybrid bonds (Fig. 3.6) explain many structures of stereochemistry, as Pauling[8] has shown in detail.

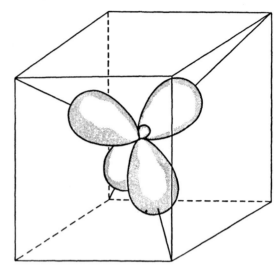

Fig. 3.6. Tetrahedral s, p-hybrid orbitals.

The interrelation between bond type, molecule structure, and dipole moment, and the ambiguities of the quantum-mechanical interpretation become clearly apparent in the sequence: $HCl \rightarrow H_2O \rightarrow NH_3 \rightarrow CH_4$. Reasoning theoretically we might represent the diatomic HCl molecule either as completely ionic or completely covalent. In the former case an electron is transferred, a closed octet shell surrounds the chlorine ion, and the proton is bound to it by Coulomb forces. For the homopolar molecule, on the other hand, the octet is completed by an electron-pair bond joining the partners as atoms without polarity (Fig. 3.7). In

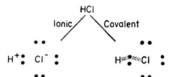

Fig. 3.7. Possible representations of HCl.

how far either standpoint is justified a dipole moment measurement can obviously decide. Since the nuclear separation distance $H \rightarrow Cl$ is known from spectroscopic data as $d = 1.28$ A, a completely ionic H^+Cl^- molecule should have a dipole moment of about 6.1 debyes; the measurement gives 1.03 debyes, thus assigning to the molecule about 17 percent ionic character.

[8] L. Pauling, *The Nature of the Chemical Bond*, Cornell University Press, Ithaca, N. Y., 1940, Chapter III.

The oxygen atom of the water molecule with its six outer electrons ($2s^2 2p^4$) has two p orbitals still half filled; it might therefore use two p bonds to combine with two hydrogen atoms in a covalent H₂O molecule. Because p bonds are orientated perpendicularly to each other, a valence angle of 90° is expected; the actually measured bond angle is ca. 104° (Fig. 3.8).

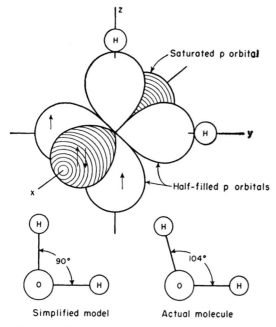

Fig. 3.8. Formation of H_2O molecule by p bonds.

This might be explained on the basis that the hydrogen atoms are partly positively charged and repel one another. Similarly, the nitrogen atom with three p bonds might form a pyramidal ammonia molecule with these bonds at right angles. Actually, the pyramid is much flatter and the angles are enlarged to 108°. Again electrostatic repulsion of the hydrogen atoms might be blamed for this fact, and the strong dipole moment of NH₃ (1.46 debyes) testifies as to its polarity. Alternatively, however, s,p-hybrid bonds could have been used placing the hydrogen atoms at three of the four corners of a regular tetrahedron with a bond angle H—N—H of about 109°. In the case of the carbon atom, this hybridization certainly has taken place; four equivalent tetrahedral bonds are used in forming the methane molecule, as the organic chemists inferred long ago.

The existence of four equal tetrahedral orbitals suggests representing the aliphatic carbon atom itself by the model of a tetrahedron. Two such carbon atoms may then combine by sharing one, two, or three corners. By attaching hydrogen atoms to the remaining free corners, we arrive at models of the ethane, ethylene, and acetylene molecule characterized by a single, double, and triple carbon bond [9] (Fig. 3.9). These models make it graphically clear that only a *single bond*

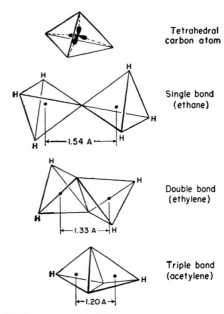

Fig. 3.9. Various ways of joining two tetrahedral carbon atoms.

allows *free rotation* of the partners; a double bond, by defining a preferential plane, and still more a triple bond, by specifying a preferred direction, prove much more restraining.

Frequently, a good first approximation of the energy content and dipole moment of polyatomic molecules

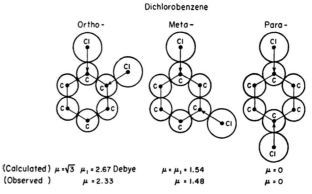

Fig. 3.10. Vector addition of dipole moments.

can be obtained by visualizing such molecules as composed of diatomic groups, for which the bond strengths and dipole moments are known. The total energy is

---

[9] Actually, ethylene and acetylene are derived by a different kind of hybridization between the s and p electrons, the so-called trigonal and digonal bond types. See A. von Hippel, *Dielectrics and Waves*, John Wiley and Sons, New York, 1954, p. 153.

found by adding arithmetically the individual bond energies, whereas the overall dipole moment results by a *vector addition* of the individual group moments.

Dichlorobenzene in its ortho, meta, and para forms is the classical example by which Debye first illustrated this procedure (see Fig. 3.10).

4 · Relaxation Polarization in Liquids and Solids

The general type of response of gases to alternating fields is a sequence of resonance states, each line described as to shape correctly by the dispersion formula of classical physics (see Eq. 2.27). Quantum mechanics interprets the spectral lines as transitions between various energy levels, thus determining their location in the frequency spectrum, and accounts for the intensity of the lines by the *statistical weights* of the terms and the *transition probabilities*.[1]

Progressing to liquids and solids we still find resonance spectra of electronic excitation and vibration. To be sure, many of them do not belong any more to individual particles but characterize electronic and vibrational states of the condensed phase. The characteristic colors of materials, for example, arise from their electronic absorption spectra in the visible spectral range, whereas the infrared absorption of ionic crystals is caused by vibrations of the positive against the negative ions of the crystal lattice.

In comparison to the spectral lines of gases, these absorptions are greatly broadened, and, in consequence, much detailed information is lost that would be of help in their interpretation. This is not surprising when we consider that the density of the condensed phases corresponds to that of normal gases under several thousand atmospheres of pressure. Thus any quantum state will suffer many interrupting collisions during its natural lifetime, and its energy, according to the uncertainty principle (see Eq. 2.53), will be ill defined.

It may be useful to illustrate this situation once more from a different standpoint. We saw that atoms and molecules carry permanent magnetic moments that assume quantized orientations in an external magnetic field; the spectral lines split into multiplets by the Zeeman effect (see p. 30). An electric field induces electric dipole moments and thus creates similarly a splitting of the energy states; the spectral lines assume a multiplet structure by the *Stark effect*.[2] If the condensed phases showed an ideal periodic order, the spectral lines might still be sharp as in the gaseous state. However, matter is disordered by thermal agitation and imperfections of all kinds. The electric fields of the surroundings produce, therefore, Stark effects varying from point to point and from instant to instant, and broaden the spectral states into continua, as is observed.

Whereas the electronic excitation and vibration states thus keep a modified resonance character, the rotation states are altered completely. To turn molecules or molecular groups requires space, and space is at a premium in liquids and solids. Simultaneously, the permanent dipole moments of molecular groups are characteristic construction elements in the formation of the condensed phases. They are being built into their surroundings; free rotation as in gases becomes impossible, and the quantized rotation spectra disappear.

The dispersion formula of classical physics pictures, electrically speaking, resonating atoms or molecules in the gaseous state as LRC circuits shunted by a capacitance. The classical approach to the treatment of dipoles in the condensed phases of liquids and solids

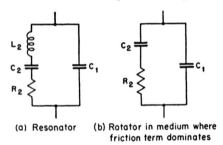

(a) Resonator (b) Rotator in medium where friction term dominates

Fig. 4.1. Equivalent circuits of free resonator molecule and of rotator molecule in medium of dominating friction.

is to consider the polar molecules as rotating in a medium of dominating friction (Debye). This corresponds to a neglect of the acceleration term, that is, to the reduction of the LRC equivalent circuit to the RC circuit of Fig. 4.1.

The admittance of such RC circuit is

with
$$Y = \left(j\omega C_1 + \frac{1}{Z_2}\right)$$
$$Z_2 = R_2 + \frac{1}{j\omega C_2}.$$
(4.1)

---

[1] A. Einstein, *Physik. Z.* **18**, 121 (1917); J. H. Van Vleck, *Phys. Rev.* **24**, 330, 347 (1924).

[2] J. Stark, *Berliner Sitzungsber.*, November, 1913; *Ann. Physik* **43**, 965, 983, 991 (1914).

If its capacitor C_2 is charged to a voltage V_0 and then the circuit shorted across the $R_2 - C_2$ combination, a transient is observed

$$V_2 = V_0 e^{-\frac{t}{R_2 C_2}}. \quad (4.2)$$

The voltage across the capacitor drops exponentially with a *relaxation time*

$$\tau \equiv R_2 C_2. \quad (4.3)$$

By introducing this time constant into Eq. 4.1 and recalling that the admittance of a capacitor of the geometrical capacitance C_0 can be expressed by the complex permittivity as (see IA, Eq. 1.10)

$$Y = j\omega \kappa^* C_0, \quad (4.4)$$

we obtain for the specific complex permittivity of our RC circuit

$$\kappa^* = \frac{C_1}{C_0} + \frac{C_2}{C_0} \frac{1}{1 + j\omega\tau}. \quad (4.5)$$

Finally, by introducing the specific permittivities for zero frequency

$$\kappa_s' \equiv \frac{C_1}{C_0} + \frac{C_2}{C_0}, \quad (4.6)$$

and for infinite frequency $\left(\omega \gg \frac{2\pi}{\tau}\right)$

$$\kappa_\infty' = \frac{C_1}{C_0}, \quad (4.7)$$

the equation for the complex dielectric constant assumes the general formulation

$$\kappa^* = \kappa_\infty' + \frac{\kappa_s' - \kappa_\infty'}{1 + j\omega\tau} \quad (4.8)$$

without reference to a specific equivalent picture.

The resonance spectrum of rotation has thus been replaced by a relaxation spectrum. It is the task of molecular physics to reinterpret the static and optical permittivities κ_s' and κ_∞' and the relaxation time τ by molecular quantities.

The optical dielectric constant κ_∞' represented in the electric circuit analogue by the by-pass capacitor corresponds obviously to the electronic and atomic resonance polarization of the dielectric. In the low-frequency range of the relaxation polarization, this deformation polarization contributes constant induced moments (see Eq. 1.3)

$$\boldsymbol{\mu}_i = (\alpha_e + \alpha_a)\mathbf{E}'. \quad (4.9)$$

The static dielectric constant κ_s' contains, in addition, the full contribution resulting from the orientation of the permanent moments. These dipoles do not line up in the field direction but contribute an average moment (see Eq. 3.9)

$$\bar{\boldsymbol{\mu}}_d = \frac{\mu^2}{3kT} \mathbf{E}' \quad (4.10)$$

as a statistical compromise between the torque action of the acting electric field and the randomizing thermal agitation.

If the local field E' represents the applied field E, as in gases at low pressure, the static and optical permittivities would be given by the moments as discussed and the complex permittivity could be rewritten (see Eq. 3.11)

$$\kappa^* = 1 + \frac{N}{\epsilon_0}\left\{(\alpha_e + \alpha_a) + \frac{\mu^2}{3kT}\frac{1}{1 + j\omega\tau}\right\}. \quad (4.11)$$

Actually, the local field will differ from the applied field and its evaluation will require a detailed structure analysis of the material in question. To escape this necessity, while making some reference to the influence of the polarized surroundings, we introduce at this point usually the Mosotti field of Eq. 1.16

$$\mathbf{E}' = \mathbf{E} + \frac{\mathbf{P}}{3\epsilon_0} = \frac{\mathbf{E}}{3}(\kappa^* + 2). \quad (4.12)$$

In this case we obtain for the polarizability per unit volume not the κ^* of Eq. 4.11, but the more involved equation

$$\frac{N\alpha}{3\epsilon_0} = \frac{\kappa^* - 1}{\kappa^* + 2}$$
$$= \frac{N}{3\epsilon_0}\left\{(\alpha_e + \alpha_a) + \frac{\mu^2}{3kT}\frac{1}{1 + j\omega\tau}\right\}. \quad (4.13)$$

This modified polarizability has the static and the optical value

$$\frac{\kappa_s' - 1}{\kappa_s' + 2} = \frac{N}{3\epsilon_0}\left\{(\alpha_e + \alpha_a) + \frac{\mu^2}{3kT}\right\},$$
$$\frac{\kappa_\infty' - 1}{\kappa_\infty' + 2} = \frac{N}{3\epsilon_0}(\alpha_e + \alpha_a). \quad (4.14)$$

By introducing the expressions on the left into Eq. 4.13, we return to the formulation for the complex permittivity given in Eq. 4.8,

$$\kappa^* = \kappa_\infty' + \frac{\kappa_s' - \kappa_\infty'}{1 + j\omega\tau_e}, \quad (4.15)$$

with the new time constant

$$\tau_e = \tau \frac{\kappa_s' + 2}{\kappa_\infty' + 2}. \quad (4.16)$$

Hence, by replacing the applied field **E** with the Mosotti field **E'**, the interpretation but not the shape of the relaxation spectrum has changed. The relaxation time has lengthened from τ to τ_e, and the molecular meaning of the static and optical dielectric constants has been altered by the introduction of the denominators $\kappa_s' + 2$ and $\kappa_\infty' + 2$ in Eq. 4.14.

The remaining problem is the molecular interpretation of the relaxation time τ. According to the assumption of dominating friction, we have to picture the polar molecules as rotating under the torque **T** of the electric field with an angular velocity $\dfrac{d\theta}{dt}$ proportional to this torque, or

$$\mathbf{T} = \zeta \frac{d\theta}{dt}. \tag{4.17}$$

The friction factor ζ will depend on the shape of the molecule and on the type of interaction it encounters. If we visualize the molecule as a sphere of the radius a rotating in a liquid of the viscosity η according to Stokes's law,[3] classical hydrodynamics leads to the value

$$\zeta = 8\pi\eta a^3. \tag{4.18}$$

In a static field, the spherical dipole carriers will have a slight preferential orientation parallel to this field and thus contribute the average moment of Eq. 4.10. A sudden removal of the external field will cause an exponential decay of this ordered state due to the randomizing agitation of the Brownian movement. The relaxation time τ (or τ_e) measures the time required to reduce the order to $1/e$ of its original value. Debye[4] was able to calculate this time statistically by deriving the space orientation under the counteracting influences of the Brownian motion and of a time-dependent electric field and found

$$\tau = \zeta/2kT. \tag{4.19}$$

Combining Eqs. 4.18 and 4.19 Debye obtained for the spherical molecule, if it behaves like a ball rotating in oil, the relaxation time

$$\tau = \frac{4\pi a^3 \eta}{kT} = V \frac{3\eta}{kT}. \tag{4.20}$$

The time constant is proportional to the volume of the sphere and to the macroscopic viscosity of the solution. Water at room temperature has a viscosity $\eta = 0.01$ poise; with a radius of ca. 2A for the water molecule, a time constant of $\tau \simeq 0.25 \times 10^{-10}$ sec results.

[3] G. Stokes, *Trans. Cambridge Phil. Soc. 9*, 8 (1851).
[4] P. Debye, *Polar Molecules*, Dover Publications, New York, 1945, pp. 77–89.

Figure 4.2 shows that indeed the relaxation time of water is located near the wavelength of 1 cm. The agreement, however, is somewhat marred by the realization that experimentally we should have determined $\tau_e \simeq 20\tau$ instead of τ itself. Obviously, the sphere model should not be taken too seriously; the essence of Debye's approach is to postulate that the orientation of polar molecules in liquids and solids leads spectroscopically to a simple relaxation spectrum.

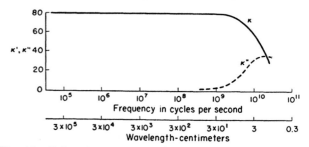

Fig. 4.2. Relaxation spectrum of water at room temperature.

The *Debye equation* (Eq. 4.15) may be written in various forms which have their special merits for the evaluation of experimental characteristics. Separating it into its real and imaginary parts, we obtain the standard version

$$\kappa' - \kappa_\infty' = \frac{\kappa_s' - \kappa_\infty'}{1 + \omega^2 \tau_e^2},$$

$$\kappa'' = \frac{(\kappa_s' - \kappa_\infty')\omega\tau_e}{1 + \omega^2 \tau_e^2}, \tag{4.21}$$

$$\tan\delta = \left|\frac{\kappa''}{\kappa'}\right| = \frac{(\kappa_s' - \kappa_\infty')\omega\tau_e}{\kappa_s' + \kappa_\infty'\omega^2\tau_e^2}.$$

If we introduce as a new variable

$$z = \ln\omega\tau_e, \tag{4.22}$$

Eqs. 4.21 may be rewritten in a normalized form [5]

$$\frac{\kappa' - \kappa_\infty'}{\kappa_s' - \kappa_\infty'} = \frac{1}{1 + e^{2z}} = \frac{e^{-z}}{e^z + e^{-z}},$$

$$\frac{\kappa''}{\kappa_s' - \kappa_\infty'} = \frac{1}{e^z + e^{-z}}, \tag{4.23}$$

$$\frac{\tan\delta}{\kappa_s' - \kappa_\infty'} = \frac{1}{\kappa_\infty' e^z + \kappa_s' e^{-z}}.$$

Figure 4.3 shows this logarithmic plot of the dispersion and absorption characteristic, and, added to them, as a third curve, the relative dielectric conductivity (see IA, Eq. 1.16)

$$\sigma = \omega\kappa'' \tag{4.24}$$

[5] H. Fröhlich, *Theory of Dielectrics*, Clarendon Press, Oxford, 1949, p. 73.

in the normalized form

$$\frac{\sigma \tau_e}{\kappa_s' - \kappa_\infty'} = \frac{e^z}{e^z + e^{-z}}. \quad (4.25)$$

The conductivity curve is the mirror image of the κ' characteristic; that is, the orientation polarization leads to a constant maximum conductivity contribution beyond the range of the dispersion region.

It may seem surprising that, after the polarizing action of the permanent dipoles has disappeared, their existence is still noted with full force as a conduction effect, but the explanation is simple: as the frequencies range so high that the molecules have no time to turn, we do not notice that the two opposite dipole charges

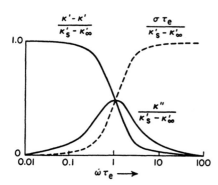

Fig. 4.3. Dielectric constant, loss factor, and conductivity of simple relaxation spectrum in normalized form.

are coupled together; their effect on the conduction is therefore the full contribution of two ions of opposite polarity moving in the electric field according to Ohm's law.

Figure 4.3 shows clearly the frequency spread of the dispersion phenomenon. According to the decade scale, it is practically limited to one decade for κ' and to two decades for κ'' above and below the center frequency. One further graphical representation of the Debye equation proves of value in analyzing and extrapolating experimental data. If we plot κ'' against κ' in the complex plane, points obeying the Debye equation fall on a semicircle with its center at $\dfrac{\kappa_s' + \kappa_\infty'}{2}$ (Fig. 4.4), as Cole and Cole[6] first pointed out. This becomes evident when we rewrite Eq. 4.15 in the form

$$(\kappa^* - \kappa_\infty') + j(\kappa^* - \kappa_\infty')\omega\tau_e = \kappa_s' - \kappa_\infty'. \quad (4.26)$$

The first member on the left side corresponds to a vector **u**; the second term, as the factor j indicates, adds perpendicular to it and represents a vector **v**; the sum is the diagonal of the circle.

[6] K. S. Cole and R. H. Cole, *J. Chem. Phys.* **9**, 341 (1941).

The loss factor κ'' reaches its maximum at the critical frequency

$$\omega_m = 1/\tau_e, \quad (4.27)$$

that is, at the critical wavelength

$$\lambda_m = 2\pi c \tau_e \quad (4.28)$$

at which the dipole polarization has fallen to its half value. Furthermore,

$$\kappa''_{\max} = \frac{\kappa_s' - \kappa_\infty'}{2} \equiv \frac{S}{2}. \quad (4.29)$$

The relaxation time and the contribution S of the orientation polarization to the permittivity can be determined by these relations from the absorption characteristic of a dielectric as long as the Debye equation is valid.

The assumption of a simple relaxation spectrum fits satisfactorily the frequency response of a number of

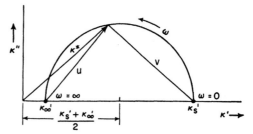

Fig. 4.4. Cole-Cole circle diagram of κ^* in complex plane.

dielectrics, especially of dilute solutions of polar materials in nonpolar solvents. This fact, however, should not be construed as a confirmation of the special Debye equation (Eq. 4.15), based on the Mosotti field

$$\mathbf{E}' = \mathbf{E} + \frac{\mathbf{P}}{3\epsilon_0}. \quad (4.30)$$

On the contrary, by introducing this local field in the equation for the polarization

$$\mathbf{P} = (\kappa' - 1)\epsilon_0 \mathbf{E} = N\alpha \mathbf{E}', \quad (4.31)$$

we invite catastrophic consequences. The polarization and susceptibility become

$$\mathbf{P} = \frac{N\alpha \mathbf{E}}{1 - \dfrac{N\alpha}{3\epsilon_0}},$$

$$\chi = \frac{N\alpha/\epsilon_0}{1 - \dfrac{N\alpha}{3\epsilon_0}}, \quad (4.32)$$

that is, they must approach infinity when the polarizability term of the denominator $N\alpha/3\epsilon_0$ approaches 1.

Obviously this is bound to happen at a critical or *Curie temperature* T_c, when permanent moments contribute an orientation polarizability

$$\alpha_d = \mu^2/3kT. \qquad (4.33)$$

Forgetting about the deformation polarization, we may replace α by α_d, and obtain

$$\chi = \frac{3T_c}{T - T_c}, \qquad (4.34)$$

where

$$T_c = \frac{N\mu^2}{9\epsilon_0 k}. \qquad (4.35)$$

Equation 4.34 is the famous *Curie-Weiss law* of ferromagnetism,[7] here derived for permanent electric instead of magnetic dipole moments. It predicts the spontaneous polarization (or magnetization) of dielectrics containing such moments.

Actually, a Mosotti-type catastrophe happens under very special conditions only, and not at all as foreseen in the preceding derivation. The reason is that permanent electric dipole moments are anchored in molecular groups that tend to lose their freedom of

[7] P. Weiss, *J. physique* 6, 661 (1907).

orientation in condensed phases through association and steric hindrance. Even if these groups could rotate like spheres in a medium of high friction, the Clausius-Mosotti formula would not apply, since the reference molecule A in the cavity B of Fig. 1.2 is not a mathematical point but itself a dipole carrier. As soon as the cavity is visualized as a molecular sphere in which a mathematical dipole is centered, calculations carried through by Onsager[8] show that spontaneous polarization does not occur.

In Onsager's treatment the surroundings of the dipole are still considered as a continuum. An improved model of Kirkwood[9] visualizes the dipole molecule with its first layer of neighbors as a structural unit, known in its statistical arrangement from X-ray patterns. The behavior of such a molecular island, floating in a dielectric continuum, is examined for static fields; the permittivities thus found are in fair agreement with experiment. To obtain still better results the molecular island is extended to include the second nearest neighbors, but the mathematical problem becomes increasingly formidable.

[8] L. Onsager, *J. Am. Chem. Soc.* 58, 1486 (1936).
[9] J. G. Kirkwood, *J. Chem. Phys.* 7, 911 (1939).

5 · Piezoelectricity and Ferroelectricity

Permanent electric dipole moments arise when unlike atoms form molecules. These moments tend to orientate in an external electric field, but in gases the effect is a minor one, because the torque of the field imposes only a weak restraining action on the thermal agitation (Fig. 3.2). A local field of the Mosotti type (Eq. 4.30), where the inner polarization supports the external field, should cause a spontaneous orientation of the dipoles at a critical temperature as the Curie-Weiss law indicates (Eq. 4.34). However, the tacit assumption in predicting such a "catastrophe" is that the dipoles are free to orientate, a situation generally not encountered in condensed phases.

Debye's model introduces a viscous friction characterized by *one* time constant (Eq. 4.20); this hindrance of orientation leaves the position of the dipoles still unrestricted. The actual relaxation spectra of liquids and polymers are frequently characterized by a *distribution* of relaxation times (Fig. 5.1) that indicates a coupling of the dipole moments to their surroundings, a fixation in certain prescribed positions. Instead of smoothly rotating spheres we have to visualize the dipoles as statistically jumping over potential walls, whenever the activation energy becomes available.[1] In crystals, finally, permanent electric dipole moments are, in general, completely immobilized as far as any

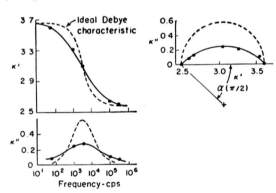

Fig. 5.1. Relaxation spectrum and Cole-Cole diagram of vulcanized rubber. (After Kauzmann.[1])

individual rotation is concerned. They are structural elements, co-determining the crystal lattice and therefore firmly anchored in their surroundings. Such dipoles cannot respond to alternating fields by relaxation

[1] See. W. Kauzmann, *Revs. Mod. Phys.* 14, 12 (1942).

spectra; but, if properly arranged, they may cause mechanical resonance vibrations of the crystal as a whole (Fig. 5.2).

An electric field polarizes any material by *inducing* dipole moments. This displacement of charges from their equilibrium positions, discussed in preceding sections as electronic and atomic polarization, alters the

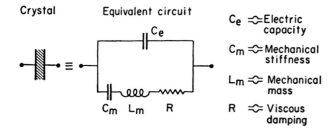
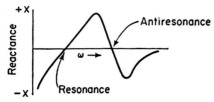

Fig. 5.2. Equivalent circuit and frequency response near resonance of a piezoelectric crystal.

mechanical dimensions of a solid; it causes *electrostriction*. Mechanical stress, on the other hand, since it acts on the mass points of a material without discerning their charges, can obviously not induce dipole moments; that is, *electrostriction has no inverse*. If a mechanical distortion creates a voltage, the effect must be caused by permanent dipole moments anchored in the structure without a center of symmetry. This *piezoelectric effect* was discovered in 1880 by the brothers Curie[2] on certain asymmetrical crystals like quartz, tourmaline, and rochelle salt. Compressed in specific directions, the materials develop a potential difference, and, vice versa, the application of an electric voltage creates a mechanical distortion. *Piezoelectricity is characterized by a one-to-one correspondence of direct and inverse effect*; it causes the electromechanical resonance spectrum shown in Fig. 5.2.

Thus far our discussion has been restricted to isotropic materials, where an electric field produces a polarization of the same orientation, as the simple vector relation

$$\mathbf{P} = \chi \epsilon_0 \mathbf{E} \quad (5.1)$$

indicates. A piezoelectric effect exists only in anisotropic crystals, where the polarization will, in general,

[2] J. and P. Curie, *Compt. rend.* **91**, 294 (1880); **93**, 1137 (1881).

not point parallel to **E**. Hence, Eq. 5.1 has to be replaced by the more general expressions

$$\begin{aligned}
\mathbf{P}_1 &= \chi_{11}\epsilon_0\mathbf{E}_1 + \chi_{12}\epsilon_0\mathbf{E}_2 + \chi_{13}\epsilon_0\mathbf{E}_3, \\
\mathbf{P}_2 &= \chi_{21}\epsilon_0\mathbf{E}_1 + \chi_{22}\epsilon_0\mathbf{E}_2 + \chi_{23}\epsilon_0\mathbf{E}_3, \quad (5.2) \\
\mathbf{P}_3 &= \chi_{31}\epsilon_0\mathbf{E}_1 + \chi_{32}\epsilon_0\mathbf{E}_2 + \chi_{33}\epsilon_0\mathbf{E}_3.
\end{aligned}$$

The subscripts 1, 2, 3, refer to orthogonal directions x, y, z. The relation between the vectors is still assumed to be a linear one, but each component of **P** (and also of **D**) has become a linear function of the three components of **E**.

More complex still are the relations between the mechanical stress $\overset{\leftrightarrow}{T}$ applied to such crystals and the resulting strain $\overset{\leftrightarrow}{S}$, since a material may respond to stress not only by expansion and contraction but also by shear (Fig. 5.3). Here we have to deal with equations such as

$$\overset{\leftrightarrow}{T}_i = \sum_j c_{ij} \overset{\leftrightarrow}{S}_j, \quad (5.3)$$

where the subscripts i and j vary from 1 to 6; that is, stress and strain are correlated by 36 coefficients. The existence of a piezoelectric effect implies, in addition, a coupling between the mechanical and electrical parameters and leads to equations as

$$\mathbf{P}_i = \sum_j d_{ij} \overset{\leftrightarrow}{T}_j \quad (i = 1, 2, 3; j = 1, \ldots, 6), \quad (5.4)$$

with 18 electromechanical coupling coefficients. How many of this formidable array of coefficients really

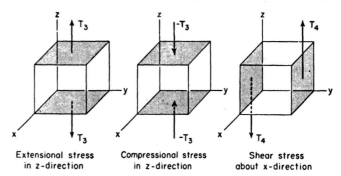

Fig. 5.3. Polar and axial stresses.

exist, and are independent of each other, is prescribed by the macroscopic crystal symmetry (see IIIC, Sec. 2).[3]

Whereas this macroscopic treatment of piezoelectricity is complicated and requires specialization in group

[3] A detailed discussion of the macroscopic aspects of piezoelectricity and of the applications of piezoelectric materials may be found in W. G. Cady, *Piezoelectricity*, McGraw-Hill Book Co., New York, 1946, and W. P. Mason, *Piezoelectric Crystals and Their Applications to Ultrasonics*, D. Van Nostrand Co., New York, 1950.

theory, a graphical molecular insight into piezoelectric phenomena can be obtained if we picture polar crystals as networks of permanent moments and investigate the possible arrangements and reactions of such moments.[4] In this way, the 20 piezoelectric crystal classes may be characterized by dipole configurations and these net constellations reinterpreted by the actual molecular groups of the crystal in question.

Piezoelectricity does not require that the crystal have a polar axis. Such an axis will exist only when the individual momentum vectors in the unit cell, instead of canceling mutually, add up to a permanent resultant moment. Since the magnitude of this resultant moment depends on the separation distance of the particles, temperature changes will produce a voltage across such a crystal, opposite in sign for heating and cooling. This effect, characterizing piezoelectric crystals with a polar axis, is known as *pyroelectricity*.

The piezoelectricity discussed thus far stems from permanent electric dipole moments firmly built into the crystal structure. *Ferroelectricity*, the spontaneous alignment of electric dipoles by mutual interaction, was not observed until recently, and few materials are, as yet, known to be true ferroelectrics. Obviously, very special conditions must prevail if polar groups can act as "free" dipoles in solids. Characteristic representatives are rochelle salt; the tetrahydrate of potassium sodium tartrate, recognized as a ferroelectric by Valasek[5] in 1921; potassium dihydrogen phosphate and arsenate by Busch and Scherrer[6] in 1935; and barium titanate, noticed for its unusual dielectric properties by Wainer and Salomon[7] in 1942–1943 and established as a new ferroelectric material[8] at M.I.T. in 1943–1944. Additional ferroelectrics, related to the titanates, have been found by Matthias.[9] They are $NaCbO_3$, $KCbO_3$, $NaTaO_3$, and $KTaO_3$ of the perovskite structure; WO_3, in which the WO_6 octohedra correspond in their arrangement to the TiO_6 octohedra; and $LiTaO_3$ and $LiCbO_3$ of the ilmenite structure.

The ferroelectric range of rochelle salt is very narrow and that of the phosphates and arsenates is limited to low temperatures. Both crystal types, furthermore, have a relatively complicated crystal structure; are piezoelectric above the Curie point, and develop ferroelectricity in one axis direction only. Barium titanate, in contrast, crystallizes in the simple perovskite structure (Fig. 5.4), is cubic with a center of

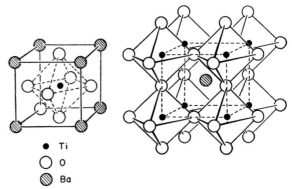

Fig. 5.4. Ideal perovskite structure.

symmetry and hence not piezoelectric above its Curie point near 120°C. It may be employed as a single crystal or as a rugged ceramic material which can be formed into any shape desired. Thus this substance lends itself better to fundamental investigations and applications than the earlier-discovered groups. Since also most of our own studies on ferroelectrics were concerned with $BaTiO_3$ we shall illustrate the pertinent phenomena on this material.

When a multicrystalline sample of $BaTiO_3$ cools down through the temperature region near 120°C, a number of properties undergo rapid changes. The dielectric constant and loss traverse a sharp maximum and minimum, respectively, the slope of the thermal expansion characteristic alters, and ferroelectric hysteresis loops appear (Fig. 5.5). The X-ray diagram of

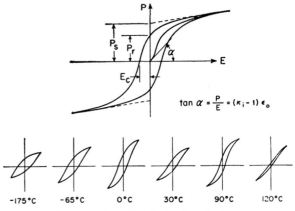

Fig. 5.5. Ferroelectric hysteresis loops.

the cubic structure transforms progressively into that of a tetragonal crystal.[10] When the initial permittiv-

[4] A. von Hippel, *Z. Physik* **133**, 158 (1952).
[5] J. Valasek, *Phys. Rev.* **17**, 475 (1921).
[6] G. Busch and P. Scherrer, *Naturwiss.* **23**, 737 (1935).
[7] E. Wainer and A. N. Salomon, *Titanium Alloy Mfg. Co. Elec. Rep.* **8** (1942); **9** and **10** (1943).
[8] A. von Hippel and co-workers, *N.D.R.C. Rep.* 300 (August, 1944) and 540 (October, 1945); A. von Hippel, R. G. Breckenridge, F. G. Chesley, and L. Tisza, *Ind. Eng. Chem.* **38**, 1097 (1946).
[9] B. T. Matthias, *Phys. Rev.* **75**, 1771 (1949); **76**, 175, 430, 1886 (1949); J. K. Hulm, B. T. Matthias, and E. A. Long, *Phys. Rev.* **79**, 885 (1950).
[10] H. D. Megaw, *Proc. Roy. Soc.* (London), **A189**, 261 (1947).

ity is plotted as $f(T)$, two lower transition points become visible near 0° and −75°C (Fig. 5.6), but the material remains ferroelectric from the Curie tempera-

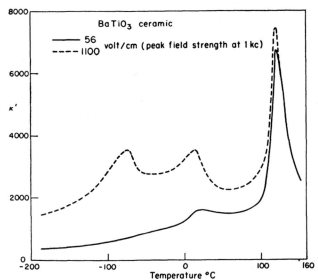

Fig. 5.6. Dielectric constant of barium titanate ceramic as function of temperature. (Measurements of W. B. Westphal, Laboratory for Insulation Research.)

ture near 120°C downwards, as the hysteresis loops indicate.

The field-strength dependence of the dielectric constant (Fig. 5.7) makes the ceramics useful in nonlinear devices; and a biasing field transforms them into piezoelectric transducers and resonators. The

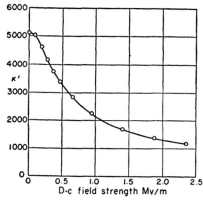

Fig. 5.7. Field-strength dependence of dielectric constant for barium-strontium titanate ceramic (above Curie point). (After Roberts.[11])

piezoelectric response persists afterwards without an external field because of the remanence of the polarization.[11]

A single crystal of $BaTiO_3$ is transparent above its Curie temperature, but with a yellowish tinge. As it

[11] S. Roberts, Phys. Rev. 71, 890 (1947).

cools, a variety of shaded areas appears (Fig. 5.8), a domain pattern which can be altered by temperature, pressure, or the application of an electric field.[12] A careful analysis reveals that the polar axis of the crystal develops at the Curie point in any one of the cube-edge directions, whereas at the two lower transitions it changes into the face-diagonal and space-diagonal direction, respectively [13] (Fig. 5.9).

To understand how the spontaneous polarization arises, we return to the structure of Fig. 5.4. The titanium ions of $BaTiO_3$ are surrounded by six oxygen ions in an octahedral configuration. This co-ordination is to be expected from the radius ratio of the partners when visualized as ionic spheres; alternatively, the TiO_6 groups may be explained as resulting from covalent binding (by octahedrals s, p, d-hybrid bonds). Actually, a compromise between these two points of view is in order. The TiO_6 constellation leads to a high dielectric constant in all crystals containing it; the temperature coefficient of the permittivity is strongly negative outside the ferroelectric range and reaches its high value through the infrared vibrations. We find ourselves, therefore, in a transition region between polar and nonpolar binding where slight changes in internuclear separation produce large changes in the electric dipole moments.

The regular TiO_6 octahedron has a center of symmetry; its permanent moments cancel in antiparallel pairs. A mechanical distortion per se does not impair this mutual cancellation and therefore leads neither to piezo- nor to ferroelectricity. However, in the perovskite structure, all octahedra are placed in identical orientation, joined only at their corners and fastened in position by barium ions. Thus the stage is set for an effective coupling between them as soon as a disturbance arises.

Above the Curie point $BaTiO_3$ is isotropic, and the equilibrium position of the titanium ions is in the center of their octahedra; however, strong fluctuations of the net moments occur by thermal agitation. Any displacement of one titanium ion towards a specific oxygen ion creates, by a kind of feedback coupling through the oxygen lattice, a tendency for the other titanium ions to move in the same direction. Figure 5.10 explains this situation.

The dipole moment of a completely ionic compound, plotted as function of the separation distance of the partners, is represented by an ascending straight line; a nonpolar compound without dipole moment follows the abscissa. Compounds of mixed bonding, but separating into atoms, follow a characteristic traversing a

[12] B. Matthias and A. von Hippel, Phys. Rev. 73, 1378 (1948).
[13] P. W. Forsbergh, Jr., Phys. Rev. 76, 1187 (1949).

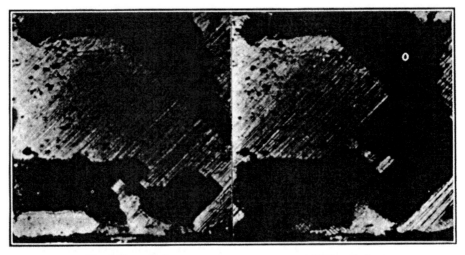

No field applied   +2000 volts/cm

Field removed   −2000 volts/cm

Fig. 5.8. Domain areas in BaTiO$_3$ crystal and effect of electric field.

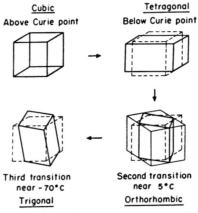

Fig. 5.9. Phase transitions of barium titanate. (After Forsbergh.[13])

maximum and then falling asymptotically to the zero line. BaTiO$_3$ seems to correspond to this latter type and has its equilibrium distance, according to the strongly negative temperature coefficient of the dielectric constant, beyond the ferroelectric range on the far side of the maximum. If the titanium ion A moves towards the oxygen ion O$_I$, the dipole moment O$_I \to$ A becomes stronger and the moment O$_{II} \to$ A weaker. Hence O$_I$ moves towards A and O$_{II}$ away from it with the result that the titanium ions B and C follow the motion of the titanium ion A. The side

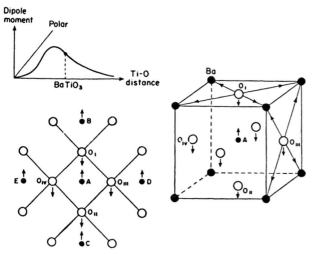

Fig. 5.10. Feedback coupling and displacement of ions in barium titanate.

titanium ions D, E, and their counterparts in front and back of the figure-plane tend to follow suit because the coupling oxygen ions O$_{III}$, O$_{IV}$, etc., move downwards, repelled by O$_I$. The whole action is akin to that of a Mosotti field: an applied field or thermal motion creates a net dipole moment in an octahedron by displacing the titanium ion against its oxygen surroundings; this displacement in turn produces through the oxygen coupling an increase of the displacement, that is, of the locally acting field, until at a critical temperature the thermal agitation can be overcome and the Mosotti catastrophe occurs. The postulated displacement directions of the titanium and oxygen ions has been checked by X-ray analysis, as far as thermal motion of the oxygen ions permits.[14]

Until now, the barium ions have been treated as inert spacers that permit the proper joining of the TiO$_6$ octahedra. Actually, the role of these divalent ions is much more important, as their influence on the

[14] H. T. Evans, Jr., *Tech. Rep.* 58, ONR Contracts N5ori-07801 and N5ori-07858, Laboratory for Insulation Research, Massachusetts Institute of Technology, January, 1953.

Curie point indicates. By substituting strontium for barium, the Curie point can be lowered systematically from 120°C to a theoretical temperature beyond the absolute zero point. A replacement of Ba^{2+} by Pb^{2+}, on the other hand, raises the Curie point, until, for pure PbTiO$_3$, it reaches about 490°C.[15] This dependence is not explainable on the basis of ionic radii since Pb^{2+} at 1.21A lies between Ba^{2+} (1.35) and Sr^{2+} (1.13); it becomes understandable when one draws the dipole moments from the oxygen ions to their four divalent metal-ion neighbors (see Fig. 5.3). Obviously, this set of moments tends to hold the oxygens in place against the action of the titanium ions. Hence the divalent ions reduce the feedback effect that leads to the Mosotti catastrophe, the more so the higher their bond strength to oxygen. The bond distance Ba—O in barium titanate is practically that of barium oxide; the melting points of the oxides are therefore a measure of the bond strength of the divalent ions to oxygen. The melting points increase from Pb—O (888°C) to Ba—O (1933°C) and Sr—O (2430°C); the feedback action and Curie points decrease in this order.

This interpretation of the onset of spontaneous polarization and its dependence on structure parameters clarifies some of the prerequisites for the formation of a ferroelectric state. The old idea that the rotation of permanent moments leads to a Mosotti catastrophe has to be discarded; such moments are built in and not available for free rotation. Ferroelectricity arises, not from rotation, but from vibration states; the displacement of certain ions from their equilibrium positions strongly unbalances the equilibrium of the permanent moments. By a proper structural arrangement this upset induces a motion of the neighboring ions in a supporting sense which increases the original displacement by feedback. The tendency to bring the vibrations of neighboring ions into ordered phase relations prevails at the Curie point against the random agitation, and the equilibrium position of the critical ions shifts to one side since the whole effect was made possible only by the displacement of these ions. The old balance of the permanent moments is destroyed and a polar axis created by the transformation of induced moments into additional permanent moments.

The preceding discussion of the formation of the ferroelectric state in BaTiO$_3$ has clarified by implication the relation between ferro- and piezoelectricity. A piezoelectric crystal, when pyroelectric, has a polar axis like a ferroelectric crystal, but the arrow direction of its axis is prescribed by the arrangement of the ions

[15] G. Shirane, S. Hoshino and K. Suzuki, *Phys. Rev.* 80, 1105 (1950).

and cannot be reversed. For the ferroelectric crystal, in contrast, the possibility of reversal is inherent because the axis evolves at the Curie point from a state of higher symmetry. Only in ferroelectric crystals, therefore, can domain structures appear and the moment be inverted by a sufficiently strong opposing field. However, the response to pressure and voltage of a ferroelectric crystal is a truly piezoelectric one, involving the change of permanent moments.[16]

[16] A more extensive discussion of piezo- and ferroelectricity and a survey on para- and ferromagnetism, space-charge polarization, conduction, and electric breakdown may be found in the companion volume to this book, A. von Hippel, *Dielectrics and Waves*, John Wiley and Sons, New York, 1954, Pt. II, Secs. 29 to 32.

II · DIELECTRIC MEASURING TECHNIQUES

A · Permittivity

1 · Lumped Circuits

By ROBERT F. FIELD

METHODS OF MEASUREMENT

Quantities measured

Lumped circuits are used in the determination of the complex permittivity of dielectrics over the frequency range from zero to ca. 2×10^8 cps. Neither the complex permittivity nor its rectangular components, the dielectric constant and loss factor (see IA, Eq. 1.11), are determined directly; which parameters are measured depends on the frequency. In the medium-frequency range the capacitance and the dissipation factor of a dielectric specimen are obtained by a null method, involving some type of capacitance bridge; dielectric constant and loss factor are calculated therefrom (see IA, Eqs. 1.12, 1.13). At higher frequencies, where the errors of a capacitance bridge become excessive, a resonant circuit is used. The capacitance of the specimen is still measured by a substitution method, but the loss component is found from the width of the resonance curve. At frequencies below that at which a capacitance bridge can be balanced, the shape of a current-time curve is observed while the dielectric specimen is being charged or discharged.

Low-frequency range (0 to 10 cps)

A voltmeter-ammeter method is used to measure the charge or preferably the discharge current of a dielectric specimen as a function of time. The necessary current range of 1 $\mu\mu$a to 1 amp requires a sensitive d-c amplifier. Sometimes the current must be observed over times as long as two weeks corresponding to a lower frequency limit of 10^{-7} cps. A short time limit is set by the response time of the indicator, seconds for a recording meter or milliseconds for a cathode-ray tube, corresponding to frequencies of 10^{-1} to 10^2 cps. This short response time cannot always be utilized because of the longer time constant of the circuit given by the product of the high-frequency capacitance of the specimen and the standard resistance R_s of the d-c amplifier.

Medium-frequency range (1 to 10^7 cps)

Resistive-arm and inductive-arm capacitance bridges and parallel-T networks are used to measure the capacitance and dissipation factor of the dielectric by null methods. The equal-arm Schering bridge, with or without a guard circuit, is most frequently chosen since it can cover the entire frequency range without serious impairment of accuracy. A lower frequency limit is given by one or several of the following factors: excessive harmonics in the generator, insufficient selectivity or sensitivity in the detector, and excessive time required for balancing the bridge. Improvements in instrument design now make, in general, the last factor the limiting one. A certain number of cycles must pass to establish equilibrium after any adjustment of a bridge circuit. Therefore, a lower frequency limit between 10^{-2} and 1 cps is set, below which it becomes too laborious to proceed. The upper frequency limit is determined by the residual impedances in the leads to the specimen, in the standards and the arms of the bridge. Although the use of micrometer electrodes eliminates the effect of the first of these impedances and partly the second, the remaining residuals set a limit around 10^7 cps.

High-frequency range (10^5 to 2×10^8 cps)

Resonant circuits are used with resonance indicated by a vacuum-tube voltmeter. This is therefore a deflection method. Micrometer electrodes are required to avoid the errors arising from residual impedances. The dissipation factor of the circuit is usually obtained from the width of the resonance curve, determined by either susceptance or frequency variation. The lower frequency limit (ca. 10^4 cps) overlaps that of the null methods and is given by the input resistance of the vacuum-tube voltmeter. An upper limit (ca. 2×10^8 cps) is set by the smallest inductor that can be built with sufficiently low dissipation factor, since the tuning capacitance must be somewhat larger than that of the dielectric specimen. By replacing the conventional inductor with a re-entrant cavity surrounding the micrometer electrodes, the frequency limit may be raised to 10^9 cps. Additional limitations are imposed by the diameters of the specimen and its electrodes which must be very small compared to the wavelength, and by the radial resistance of the foil electrodes applied to the specimen.

Specimens and Electrodes for Solid Dielectrics

The specimen must be of such shape that its vacuum capacitance can be calculated (see Eqs. 1.4 and 1.5). The preferable shape is a flat plate with plane and parallel surfaces, either circular or square. When micrometer electrodes are employed, the specimen should be a disk of the same diameter as the electrodes or smaller, and $\frac{1}{16}$ to $\frac{1}{4}$ in. thick.

Electrode arrangements

When electrodes are applied to the two sides of the sample, the capacitance between them is greater than that defined by the electrodes proper owing to the edge capacitance beyond the electrodes and the capacitance to ground of the high electrode. The magnitude of the edge capacitance depends on the size of the electrodes with respect to the size of the specimen and is proportional to the electrode perimeter (the smaller one if they are unequal). The magnitude of the ground capacitance depends on the distance of the high electrode from the grounded surfaces of the equipment. The effect of both capacitances can be eliminated by the use of micrometer electrodes or a suitably shielded three-electrode system. The former method, however, does not remove the effects caused by d-c surface conductivity and by surface polarization resulting from moisture films.[1]

In the three-electrode system an additional guard electrode surrounds one of the measuring electrodes (Fig. 1.1) and is connected to a suitable terminal of the measuring circuit. In order to be fully effective, the width of the guard electrode must be at least twice the thickness of the specimen, and the unguarded electrode must extend to the outer edge of the guard electrode. The gap between the guarded and the guard

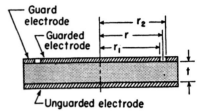

Fig. 1.1. Three-electrode system.

electrodes should be as small as possible. The effective diameter of the guarded electrode is greater than its actual diameter by approximately the width of this gap.

The exact value of this increase depends on the ratio of gap width $g = r_2 - r_1$ to thickness t. Amey[2] has shown that the effective radius r of the guarded electrode is

$$r = r_1 + \frac{g}{2} - \delta, \quad (1.1)$$

where

$$\frac{\delta}{t} = \frac{2}{\pi} \ln \cosh \frac{\pi}{4} \frac{g}{t} = 1.466 \log \cosh 0.7858 \frac{g}{t}. \quad (1.2)$$

A guard electrode has little effect in decreasing the ground capacitance of the high electrode, except when the guard terminal is grounded. When the unguarded electrode is grounded, the entire three-electrode system must be surrounded by a shield connected to the guard terminal in order to eliminate ground capacitance.

Electrode materials

For the accurate measurement of dielectric constant it is necessary to apply some type of thin metallic electrode to a specimen before it is placed between the rigid electrodes of a measuring cell. Any airgap between specimen and electrode provides a series air capacitance that decreases both the measured capacitance and the dissipation factor. This error varies inversely with thickness and may become excessive for thin films. Lead or tin foil (0.0003 to 0.003 in. thick), pressed on by a roller and made adherent by a minimum quantity of refined petrolatum, silicone grease, or some other suitable low-loss adhesive, is the standard electrode material. If the electrode is to extend to the edge, it should be made larger than the specimen and cut back to the edge with a finely ground

[1] R. F. Field. *J. Appl. Phys.* 17, 318 (1946).

[2] W. G. Amey and F. Hamburger, Jr., *Proc. Am. Soc. Testing Materials* 49, 1079 (1949).

blade. A guarded and guard electrode can be prepared simultaneously by cutting out a narrow strip (≤ 0.02 in.) by means of a compass provided with a narrow cutting edge.

Conducting silver paints, either air-drying or low-temperature-baking varieties, are commercially available for use as electrode material. They are sufficiently porous to permit diffusion of moisture and conditioning of the specimen after application of the electrodes. These paints have disadvantages in that all solvent must be removed by air drying or baking, that masks must be used to obtain sharply defined electrode edges, and that their comparatively low conductivity may increase the measured dissipation factor at the highest frequencies. Fired-on silver electrodes are suitable for glass and ceramics which can withstand the firing temperature (ca. 350°C). Their high conductivity allows the use of such electrodes on low-loss materials, such as fused quartz, even at the highest frequencies, and their tight bonding makes them satisfactory for high-dielectric-constant materials, such as the titanates.

Low-melting metal applied with a spray-gun provides a spongy electrode of roughly the same electrical conductivity and moisture porosity as conducting paints. The metal conforms readily to a rough surface, for example, cloth, and does not penetrate very small holes in a thin film to produce a short circuit. Its adhesion to some surfaces is poor, especially after exposure to high humidity or to water immersion. Advantages over conducting paint are freedom from effects of solvents and readiness for use immediately after application.

Evaporated metal as an electrode material has such low conduction because of its thinness that it usually must be backed with electroplated copper or sheet metal.

Mercury electrodes, applied by floating the specimen on a pool of mercury and using confining rings with sharp edges for holding the metal for the guarded and guard electrodes, are a health hazard in some ways, but convenient in special cases.

Measuring cells for solids

In the simple two-terminal system the measuring cell consists of a grounded support for the specimen and a stiff, small-diameter wire connecting the high terminal of the measuring circuit to the high electrode (Fig. 1.2). This lead has an appreciable capacitance to ground, which changes when it is disconnected at the high electrode. When disconnected, it also has some capacitance to the high electrode, which decreases rapidly as the end of the lead is moved away from the high electrode. At some critical distance these two effects will cancel and reduce the connection error to zero. The critical distance is usually about $\frac{1}{16}$ in., but depends somewhat on wire size. The resulting

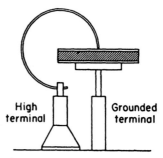

Fig. 1.2. Stiff wire connector for two-terminal measurements.

error in capacitance[3] should not be greater than 0.02 $\mu\mu$f.

The connecting leads have both inductance and resistance, which at high frequencies increase the measured capacitance and dissipation factor. These increases are

$$\Delta C = \omega^2 L C^2 \quad \text{and} \quad \Delta D = R\omega C, \quad (1.3)$$

where L is the series inductance, R the series resistance of the leads, and C the true capacitance of the capacitor. Although it is desirable to have these leads as short as possible, it is difficult to reduce their inductance and resistance below 0.1 μh and 0.05 ohm at 1 Mc; the resistance increases with the square root of the frequency. Hence lead corrections become important above 10 Mc.

The *micrometer-electrode system* (Fig. 1.3) was developed by Hartshorn and Ward,[4] to eliminate the errors caused by series inductance and resistance of the connecting leads and of the measuring capacitor at high frequencies. It accomplishes this by maintaining the inductance and resistance constant regardless of whether the specimen is in circuit or not. The specimen may be the same size as the electrodes, or smaller. Upon removal, the electrode system is adjusted to the previous capacitance by moving the electrodes closer together. The vernier capacitor measures the width of the resonance curves. Corrections for edge capacitance, ground capacitance, and connection capacitance are practically eliminated.

One disadvantage of a micrometer-electrode system is that its capacitance calibration is not as accurate

[3] R. F. Field, *Gen. Radio Exp. 21* (May, 1947), p. 1.
[4] L. Hartshorn and W. H. Ward, *J. Inst. Elec. Engrs.* (London) **79**, 597 (October, 1936).

as that of a standard precision capacitor and also not direct-reading. This disadvantage can be eliminated at low frequencies by connecting a standard capacitor in parallel with the micrometer-electrode system and measuring the change in capacitance with the specimen in and out in terms of this capacitor. A

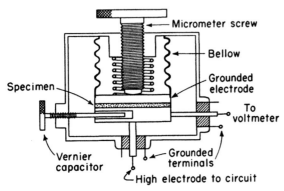

Fig. 1.3. Micrometer electrode system.

second defect is that the edge capacitance of the electrode is slightly increased when the specimen is inserted. This increase varies inversely with thickness and can be practically eliminated by making the diameter of the specimen less than that of the electrodes by two times its thickness.

In the *three-terminal system* the measuring cell is usually a shallow metal box provided with two insulated terminals (Fig. 1.4). The box itself is connected to the guard terminal and is grounded if the guard terminal is grounded. The lead for the guarded electrode must be shielded back to the measuring circuit. When the measuring circuit requires the unguarded

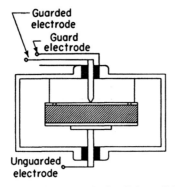

Fig. 1.4. Three-terminal cell for solids.

terminal to be grounded, the entire box must be insulated from ground, placed in a grounded shield to eliminate induced voltages, and the lead from the guarded electrode must be doubly shielded.

Measuring Cells for Liquids and Gases

A *three-terminal system* consisting of an outer cylindrical case and an inner cylinder, with a cylindrical guard electrode, is shown in Fig. 1.5. The opposing surfaces of the measuring electrodes should be carefully finished to give a polished surface and a uniform

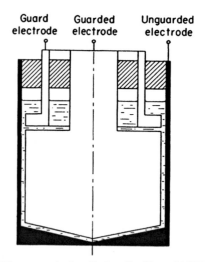

Fig. 1.5. Three-terminal cell for liquids. (ASTM designation: D150.)

spacing between 0.05 and 0.1 in. The lower surfaces are usually conical in order to eliminate air bubbles during filling. The insulation should maintain the alignment of the electrodes at the highest temperatures used and still allow easy disassembling and thorough cleaning. Neither the insulation nor the electrode surfaces should be affected by the test liquid or cleaning solvents.

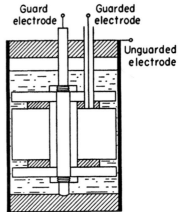

Fig. 1.6. Alternative three-terminal cell for liquids.

A type of cell which is easier to construct is shown in Fig. 1.6, having cylindrical electrodes supported at both ends. A second guard electrode at the lower end is required, and the insulation at this end is immersed in the test liquid. These cells may also be constructed with an outer envelope of low-loss glass, with both the measuring electrodes made of metal tubing and supported on a glass press.

For the measurement of gases the cell must obviously be gas-tight and fitted with suitable connections for the introduction of the gas and its evacuation.

The three-terminal design of Fig. 1.5 provides a satisfactory *two-terminal cell* when the guard electrode is omitted, provided that the upper portion of the inner electrode is sufficiently small in diameter for small changes in the height of the test liquid to have a negligible effect on the measured capacitance.

Micrometer electrodes can be used to measure liquids by clamping a ring of low-loss material such as Teflon, between the electrodes (Fig. 1.7). The

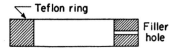

Fig. 1.7. Teflon spacer ring (for use between micrometer electrodes).

liquid is introduced through a small radial hole. The ring can also be faced on both sides with metal to make it a self-contained cell. An extra set of measurements is required on the empty cell.

Calculations on Measuring Cells

The *vacuum capacitance* C_0 for flat parallel plates with electrodes having an area A (area of the smaller, if unequal) and spacing t is

$$C_0 = 0.08854 \frac{A}{t} \mu\mu\text{f (cm)} = 0.2249 \frac{A}{t} \mu\mu\text{f (in.)}. \quad (1.4)$$

When the electrodes are circular and have a diameter d (diameter of the smaller, if unequal) this becomes

$$C_0 = 0.06954 \frac{d^2}{t} \mu\mu\text{f (cm)} = 0.1766 \frac{d^2}{t} \mu\mu\text{f (in.)}. \quad (1.5)$$

When one of the electrodes is guarded, the effective radius of the guarded electrode must be used (see Eq. 1.1).

The *edge capacitance* C_e for infinitely thin electrodes,[5] expressed as the capacitance per inch of perimeter P of the electrodes (perimeter of the smaller, if unequal), is given for the three cases: (a) electrodes to the edge of the specimen,

[5] A.S.T.M. Designation D150-54T for dielectric constant and loss characteristics.

(b) equal electrodes smaller than the specimen, and (c) unequal electrodes:

(a) $\quad \dfrac{C_e}{P} = -0.094 \quad\quad\quad\quad + 0.114 \log \dfrac{P}{t} \mu\mu\text{f/in.,}$

(b) $\quad \dfrac{C_e}{P} = -0.225 + 0.050\kappa' + 0.153 \log \dfrac{P}{t} \mu\mu\text{f/in.,} \quad (1.6)$

(c) $\quad \dfrac{C_e}{P} = -0.229 + 0.105\kappa' + 0.214 \log \dfrac{P}{t} \mu\mu\text{f/in.}$

The *ground capacitance* C_g of the high electrode[5] differs from one half its capacitance to free space because of the shielding effect of the edge capacitance C_e and the proximity of grounded shields. It depends on the thickness t of the specimen and on its height h above surrounding grounds.

$$C_g = 0.134 \frac{A}{(h)^{1/2}} + 0.65t \ \mu\mu\text{f, (in.)}, \quad (1.7)$$

where A is the area of the high electrode. For a circular electrode of diameter d this becomes

$$C_g = 0.105 \frac{d^2}{h} + 0.65t \ \mu\mu\text{f (in.)}. \quad (1.8)$$

Calculation of Dielectric Constant and Loss Factor

Corrections for solids

When extra capacitances have been included, such as edge capacitance C_e and ground capacitance C_g in two-terminal measurements, the observed parallel capacitance C_x will be increased and the observed dissipation factor D_x decreased. The correct values will be

$$\begin{aligned} C_p &= C_x - C_e - C_g, \\ D &= \frac{C_x D_x}{C_p} = \frac{C_x D_x}{C_x - C_e - C_g} = \frac{D_x}{1 - \dfrac{C_e + C_g}{C_x}}. \end{aligned} \quad (1.9)$$

The expression for dissipation factor assumes that the extra capacitances are free from loss. This is essentially true for the ground capacitance and also for the edge capacitance if the electrodes extend to the edge of the specimen and the stray field is in air. The dielectric constant and loss factor become

$$\begin{aligned} \kappa' &= \frac{C_p}{C_0} = \frac{C_x - C_e - C_g}{C_0}, \\ \kappa'' &\equiv D\kappa' = \frac{C_x D_x}{C_0}. \end{aligned} \quad (1.10)$$

When one or both of the electrodes are smaller than the specimen, a loss in the capacitance is associated with the flux lines that pass through the uncovered dielectric. Designating this capacitance by $\kappa' C_d$ and the loss-free capacitance by C_f (so that $C_e = \kappa' C_d + C_f$), the dissipation factor becomes

$$D = \frac{C_x D_x}{C_p + \kappa' C_d} = \frac{C_x D_x}{C_x - C_f - C_g} = \frac{D_x}{1 - \dfrac{C_f + C_g}{C_x}}, \quad (1.11)$$

and, in consequence,

$$\kappa' = \frac{C_p}{C_0} = \frac{C_x - C_f - C_g}{C_0 + C_d},$$

$$\kappa'' = D\kappa' = \frac{C_x D_x}{C_0 + C_d}. \qquad (1.12)$$

Corrections for liquids and gases

When liquids and gases are measured in a three-terminal measuring cell, there are no corrections to the observed values of capacitance and dissipation factor. The vacuum capacitance C_0 is measured by evacuating the cell, or, in close approximation, by measuring the cell filled with air, since the dielectric constant of dry air at 25°C is only 0.054 percent greater than unity.

When liquids are measured in a two-terminal measuring cell, the extra capacitances between electrodes and to ground cannot be directly measured or calculated. The value of their combined capacitances, $C_e + C_g$, can be found by measuring the cell when filled to the same level (C_s) with a liquid of known dielectric constant κ_s', and when empty (C_0), as

$$C_e + C_g = \frac{\kappa_s' C_0 - C_s}{\kappa_s' - 1}. \qquad (1.13)$$

Since these extra capacitances are usually free from loss, the capacitance, dissipation factor, dielectric constant and loss factor can be found from Eqs. 1.9 to 1.12.

Accuracy

The accuracy attainable in the determination of dielectric constant and loss factor depends upon the accuracy of the observations for capacitance and dissipation factor, upon the magnitude and accuracy of the corrections to these quantities caused by the electrode arrangements used, and finally upon the accuracy of the calculation of the vacuum capacitance. Under favorable conditions capacitance can be measured with an accuracy of $\pm(0.2 \text{ percent} + 0.04 \mu\mu f)$ and dissipation factor to $\pm(2 \text{ percent} + 0.00005)$.†

Specimens provided with a guard electrode are subject to an error only in the calculation of the vacuum capacitance. The error caused by too wide a gap between the guarded and guard electrodes will generally amount only to a few tenths of a percent and the correction can be determined to a few percent. The error in measuring the thickness of the specimen can amount to a few tenths of a percent for an average thickness of 1/16 in., on the assumption that the thickness can be measured to 0.0002 in.

Specimens with electrodes carried to the edge and measured with micrometer electrodes have no corrections other than that for thickness provided that they are smaller in diameter than the electrodes by two times the thickness. When two-terminal specimens are measured in any other manner, the calculation of edge and ground capacitances will involve considerable error, since each may be 2 to 10 percent of the specimen capacitance. With the present knowledge of these capacitances there may occur a 10 percent error in calculating the edge capacitance and a 20 to 30 percent error for the ground capacitance. Hence the total error may vary from several tenths to a few percent.

† The fixed value quoted is the minimum value given by stray parameters.

TYPICAL MEASURING CIRCUITS

Low-Frequency Range

Current-time curves

When a steady voltage is applied to a dielectric exhibiting either dipole or interfacial polarization, the current decreases with time from its initial value to a steady leakage current determined by the leakage resistance. The value of the current remaining after subtracting this leakage current, together with its rate of decrease, defines the polarization parameters from which the complex permittivity contributed by this polarization can be determined.

An expression for the charging current as a function of time is obtained by integrating the complex admittance of the circuit that represents the polarization, such as the circuit for the circular arc with depressed center (IB, Fig. 5.1). Expressions for the complex admittance can also be found for other distribution functions, even when an equivalent circuit cannot be simply formed. Cole and Cole[6] have carried through this integration for the circular arc and have calculated the current for unit voltage and unit capacitance change for a wide range of time ratios. The family of current-time curves, as amplified by Field,[7] is shown in Fig. 1.8 for all values of storage coefficient α in steps

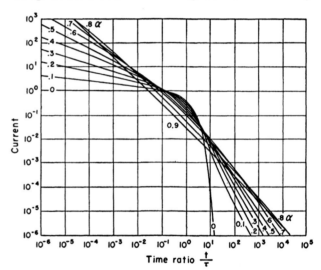

Fig. 1.8. Charge (or discharge) current for unit voltage and unit capacitance change.

of tenths over a time range of a millionth to ten-thousand times the relaxation time. An equal log-log plot is used to give equal weight to the different time-ratios. A sort of symmetry about the point (1,1) is also ob-

[6] K. S. Cole and R. H. Cole, *J. Chem. Phys.* 10, 98 (1942).

[7] R. F. Field, Boston District Meeting, American Institute of Electrical Engineers, April, 1944.

tained. The curve for $\alpha = 0$ would be a straight line on an arith-log plot, and refers to the current in a simple RC circuit. Its value at a time ratio of unity is of course $1/e = 0.368$. All the curves have slanting asymptotes, whose slopes are functions of the storage coefficient, $-\alpha$ at zero time and $-(2-\alpha)$ at infinite time.

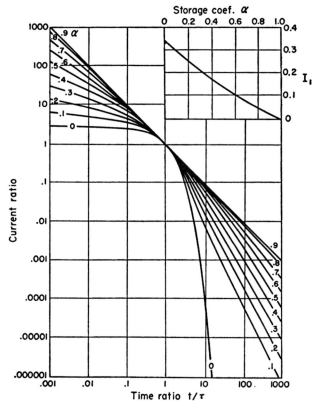

Fig. 1.9. Charge (or discharge) current ratio.

The current-time curve of a dielectric can be analyzed by matching it to one of the family of curves of Fig. 1.8. This task is facilitated by raising these curves so that all pass through the point (1,1) as shown in Fig. 1.9. The curve in the upper right corner shows the factor I_1, by which all current values must be multiplied in order to return them to their original positions. When the unknown curve has been matched, a value of α is defined, together with values of current I and relaxation time τ corresponding to the point (1,1). The change in capacitance is

$$\Delta C = C_s - C_\infty = \frac{I\tau}{EI_1}, \qquad (1.14)$$

where E is the applied voltage. The relaxation frequency ν_m is determined by the relaxation time as (cf. IB, Eq. 4.27)

$$\nu_m = \frac{1}{2\pi\tau}. \qquad (1.15)$$

The change in dielectric constant is

$$\Delta\kappa' = \frac{\Delta C}{C_0}. \qquad (1.16)$$

The charging-current curve is generally useless for matching to the theoretical curves as long as the leakage current has not been subtracted. The leakage current can be obtained as the discharge current after a time that is long compared to the relaxation time of the polarization. Also the difference between charge and discharge current at equal times is the d-c leakage current, and this is probably the most accurate method of determining it.

Measuring circuits

Any measuring circuit designed for the measurement of d-c resistance is suitable provided it has sufficient sensitivity (see the circuits described in A.S.T.M. designation D 257). The most sensitive ones are the d-c amplifier arrangements of Figs. 1.10a and b and the Wheatstone bridge of Fig. 1.11. A current of 1 $\mu\mu$a can be easily detected. The standard resistance R_s is in series with the dielectric specimen and reduces the current unless it is negligible compared to the instantaneous impedance of the specimen. Since this is frequently not the case, it is advisable, whenever possible, to degenerate it by suitable connection to the

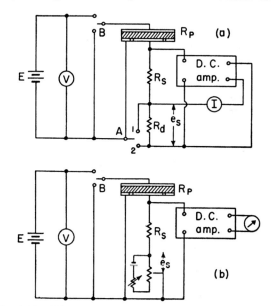

Fig. 1.10. Direct-current amplifier circuits for resistivity measurements.

negative feedback resistance used to stabilize the d-c amplifier. Two possible connections for the feedback resistance are shown in Fig. 1.10a, where the amplifier serves as an indicating instrument. In position 1 the entire resistance R_s is in circuit, and any a-c voltage appearing across the specimen is amplified only as much as the d-c voltage produced across resistance R_s by the charging current. In position 2 the effective resistance placed in the measuring circuit is less than R_s in the ratio of the degenerated gain of the amplifier to its intrinsic gain. Since this ratio usually approaches 0.0001, even a standard resistance of 1 kilomegohm places only 100 kilo-ohm in the circuit. Unfortunately any a-c voltage across the specimen is amplified by the intrinsic gain of the amplifier (10,000 times in this example) so that this connection cannot be used unless the specimen is completely shielded and batteries provide the applied voltage.

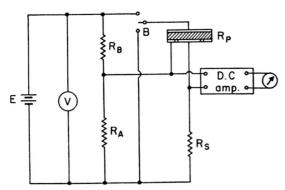

Fig. 1.11. Wheatstone bridge for resistivity measurements.

In the circuits of Figs. 1.10b and 1.11 the d-c amplifier is used merely as a null indicator. For the Wheatstone bridge there is no way in which the standard resistance R_s can be decreased by degeneration. For the circuit of Fig. 1.10b the voltage drop across the standard resistance R_s is opposed by an equal voltage obtained from a separate potentiometer. At balance, indicated by a null deflection of the d-c amplifier, the resistance R_s is effectively reduced to zero. Off balance, the fraction of R_s left in the circuit is the ratio of the change in potentiometer voltage from balance to the voltage at balance. Any a-c voltage appearing across the specimen is amplified only by the net gain of the amplifier.

The a-c voltage usually encountered is large enough to require the use of a one- or two-stage resistance-capacitance filter across the amplifier. Its input resistance shunts the standard resistance R_s and must therefore be high compared with it. This is an added reason why R_s should be degenerated whenever possible.

In the circuits of Fig. 1.10 the standard resistance R_s is shunted by the resistance between the guarded and guard electrodes. In order to make the resulting error negligible, it may be necessary to make the gap between these electrodes wider than is desirable from the standpoint of accurate calculation of the vacuum capacitance. In the Wheatstone bridge of Fig. 1.11 the resistance shunted is usually not greater than 1 megohm.

Since a complete current-time curve requires observations for hours and even days, the indicating meter of the d-c amplifier is usually replaced by a recording meter of either the deflection or the null-balance type. The zero of the meter must be checked periodically for drift if the amplifier is of the conductively coupled type, whereas the zero drift of a degenerative a-c amplifier with input and output converters is negligible.

After the charging current has been observed for a sufficient time, the discharge current is measured in the circuits of Fig. 1.10 by shorting out the voltage source (switch B) and reversing the indicating or recording meter. For the Wheatstone bridge of Fig. 1.11 the specimen is connected directly across the standard resistance R_s (switch B) and the indicating or recording meter is reversed. The voltage source remains connected to provide a comparison voltage for the null balance.

The specimen is shown with a guard electrode to assure freedom from leakage over the edge of the specimen under conditions of high humidity. Two-terminal specimens can be measured by merely omitting the guard connection. The circuits are grounded at the guard terminal or at the junction of voltage source and detector.

A reasonable estimate of the error in both current I and relaxation time τ in Eq. 1.14 is 10 percent; hence if a 2 percent voltmeter is employed the error for the change in capacitance ΔC is ca. 25 percent.

Alternative methods

The very wide range of current or resistance that must be measured in obtaining a current-time curve, when a steady d-c voltage is applied to the specimen, demands a set of at least eighteen standard resistors to cover the range of nine decades. Davidson, Auty, and Cole [8] have used a modification in which the applied voltage is increased linearly. The resulting current increases instead of decreasing. Although the sensitivity of the current measuring device is the same as for a steady voltage, the range required is greatly reduced.

The standard method of measuring the static dielectric constant of a dielectric consists in charging

---

[8] D. W. Davidson, R. P. Auty, and R. H. Cole, *Rev. Sci. Instr.* **22**, 678 (1951).

the capacitor and then discharging it through a ballistic galvanometer of a period long compared with the time of discharge. Since the period of a ballistic galvanometer is rarely over 20 seconds, this method is limited to short-time discharge phenomena.

Medium-Frequency Range (Bridges)
Balance

The capacitance bridges used for dielectric measurements usually have equal ratio arms R_A and R_B, a standard variable air capacitor C_N, and a balancing capacitor C_T (Fig. 1.12). A substitution method of

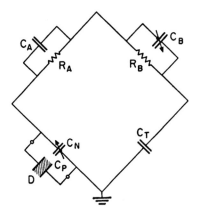

Fig. 1.12. Schering bridge.

measurement is used because of its greater accuracy. The dielectric specimen is connected in parallel with the standard capacitor and the bridge is balanced, then disconnected and the bridge rebalanced. The parallel capacitance C_P of the specimen is the difference of the two readings of the standard capacitor (readings with the specimen disconnected being primed),

$$C_P = \Delta C = C_N' - C_N. \quad (1.17)$$

The resistive balance of the bridge can be made in three ways: by a variable resistor of low value in series with the standard capacitor C_N (series-resistance bridge), by a variable resistor of high value in parallel with the adjacent capacitor C_T (parallel-resistance bridge), and by a variable capacitor in parallel with the ratio arm R_B (Schering bridge). The resistance bridges may be subject to considerable error, particularly at high frequencies, because of the relatively large and varying residual impedances (series resistance and parallel capacitance) of variable resistors. The Schering bridge is generally used, both because it can be carried to high frequency with negligible error and because the resistive control can be calibrated directly in dissipation factor.

For the Schering bridge the dissipation factor D of the specimen is

$$D = \frac{C_N'}{\Delta C} \Delta D, \quad (1.18)$$

where $\Delta D = D - D' = R_B \omega (C_B - C_B')$. The total capacitance C_N' in the capacitance arms should be as small as possible so that the change in dissipation factor ΔD may be as large as possible, especially for low-loss dielectrics.

The balance equations of the simple-Schering bridge are concerned with the series capacitances rather than the parallel capacitances of the two capacitance arms. This fact, combined with the existence of fixed stray capacitance C_A across ratio arm R_A, which produces a dissipation factor reading D_A, causes cross terms to appear in the balance equations when high losses are measured. The complete equations are

$$C_p = \Delta C \frac{1 + aD}{1 + D^2},$$
$$D = \frac{C'}{\Delta C} \frac{\Delta D}{1 + aD}, \quad (1.19)$$

where $a = D' - \dfrac{C}{\Delta C} \Delta D = D \dfrac{C'}{\Delta C} \Delta D$, and the dissipation factor readings D and D' have been increased by D_A.

When a calibrated micrometer-electrode system is used, it may be connected either in parallel with the standard capacitor or in the adjacent capacitance arm T, since the standard capacitor is used only as a balancing capacitor (see Fig. 1.12). Placing it in arm T minimizes the total capacitance C_N', but the dissipation factor capacitor C_B' must then be placed across the ratio arm A. The bridge is first balanced with the dielectric specimen clamped in the micrometer electrodes (C_s measurement). The specimen is then removed and the bridge rebalanced by increasing the capacitance of the micrometer electrodes (C' measurement). The capacitance of the specimen is

$$C_p = C' - C_s + C_a, \quad (1.20)$$

where C' is the calibration capacitance (in the fringing field) of the micrometer electrodes for the "sample out" balance, C_s is the calibration capacitance of the micrometer electrodes for the sample spacing,† and C_a is the geometrical air capacitance without fringing field for the specimen thickness. Since an air capacitance is only 0.054 percent greater than a vacuum

† Foil or other secondary electrodes may require a correction to this setting.

capacitance, Eqs. 1.4 and 1.5 can be used for this geometrical calculation. The true thickness t of the specimen must be known in calculating the dielectric constant.

At frequencies below 1 Mc, where the effect of series inductance and resistance in the leads is negligible, the capacitance calibration of the micrometer-electrode system can be replaced by that of the standard capacitor. When the specimen is removed, the micrometer electrodes are returned to the spacing, which they had with the specimen in place and the bridge rebalanced by adjustment of the standard capacitor to a value C_{Ns}'. The capacitance of the specimen is

$$C_p = C_{Ns}' - C_N + C_a, \qquad (1.21)$$

where C_N is the reading of the standard capacitor with the specimen in place.

Shielding

A two-terminal bridge is usually grounded at the junction of its capacitance arms (Fig. 1.12). The capacitances to

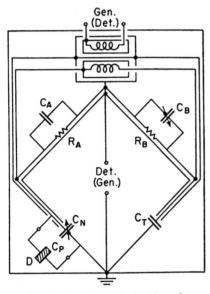

Fig. 1.13. Schering bridge with shields and transformer.

ground of the ratio arms and of the transformer which is connected to them are placed across the capacitance arms. This introduces no error in the determination of capacitance in substitution measurements because only capacitance difference enters. However, in the dissipation factor the total capacitance C_N' enters, and it is difficult to determine this added capacitance. It is therefore customary to enclose the ratio arms and transformer in a shield which is connected to the junction of the ratio arms. The capacitances thus placed across the ratio arms must be adjusted so that the resulting dissipation factors are equal. These capacitances are kept as small as possible in order to reduce the cross terms in the exact bridge equations. The entire bridge is placed in a grounded shield, both to eliminate the capacitance effects of surrounding objects, including the operator, and to prevent external a-c, static, and magnetic fields from inducing voltages in the bridge arms and transformer. Such induced voltages are particularly serious when measurements are made at power frequency. A Schering bridge with these shields is shown in Fig. 1.13.

Transformer

A shielded transformer is connected across the pair of bridge arms whose ends are not grounded, because most generators and detectors have one terminal grounded and because external connections can change the capacitances appearing across the bridge arms and thus alter the adjustment of the bridge. The shielding of such a transformer is shown in Fig. 1.13. Both windings of the transformer should be completely shielded and their shields connected to one end of the winding. The primary shield is grounded and the secondary shield connected to the ratio arm which does not have the dissipation-factor capacitor, in order to allow the capacitance of this shield to balance the zero capacitance of this capacitor. All insulation outside the winding shields should be of low-loss material, such as polystyrene, since it is the conductive component of any remaining leakage admittance which can cause an error in the dissipation-factor reading. A third shield between the two winding shields connected to the junction of the ratio arms removes the effect of the grounded primary-winding shield.

The generator or detector may be connected to the shielded transformer; the choice depends on voltage and sensitivity. Since the power dissipation of the ratio arms is usually small, only low voltages can be connected across the ratio arms and therefore across the transformer. This is also advantageous because bridge sensitivity is a maximum when the generator is connected across arms of equal impedance. High voltages must be connected across the capacitance and resistance arms in series, where the major part of the voltage will appear across the capacitors. The loss in sensitivity caused by the inequality in impedance of these arms is partially compensated by the increased voltage.

Residual impedances

At high frequencies residual impedances (series inductance and resistance and shunt capacitance of the circuit elements and their connecting leads) produce errors in the calibrations of these circuit elements. These errors are generally so large for decade resistors that only the Schering bridge, which uses variable air capacitors as calibrated elements, can serve above 100 kc. The effect of shunt capacitance across the ratio arms of the Schering bridge is negligible in substitution measurements because it is customary to decrease the value of the ratio arms in proportion to the increase in frequency, though usually in decade steps. Series inductance will increase the capacitance of the standard and the dissipation-factor capacitors in proportion to the square of the frequency (see Eq. 1.3). In a Schering bridge designed for low frequencies, the residual impedances are of the order of 0.1 μh and 100 $\mu\mu$f; these values cause an error of 0.1 percent at 1.5 Mc. Even with a value for L of 0.006 μh, the lowest so far attainable, the limiting frequency for 0.1 percent error is only 6 Mc. For higher frequencies a correction must be made, or the total capacitance reduced. Series resistance R will increase the losses in the standard capacitor and decrease the dissipation-factor reading in proportion to the 3/2 power of the frequency (see Eq. 1.3), since this resistance increases with the 1/2 power of frequency. A reasonable value of R is 0.05 ohm at 1 Mc, producing an error of 0.0001 at 2 Mc.

Guard circuit

A guard terminal G for a three-terminal capacitor can be provided by the addition of a guard circuit across one pair of bridge arms. The circuit elements of the guard circuit

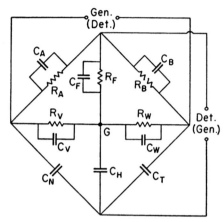

Fig. 1.14. Guard circuits (type V-W across arms A and B; type F-H across arms A and N).

should match those of the bridge arms across which they are placed. For convenience both types of guard circuit are shown in Fig. 1.14. Although generally only one of them is provided, the capacitive portion of the other exists as a coupling circuit, because of the terminal capacitances of the three-terminal capacitor. The bridge balance depends on that of the guard circuits according to the two sets of balance equations:[9]

$$\frac{Z_A}{Z_N} = \frac{Z_B}{Z_T} = \frac{Z_F}{Z_H}; \quad \frac{Z_A}{Z_B} = \frac{Z_N}{Z_T} = \frac{Z_V}{Z_W}. \qquad (1.22)$$

These balances are interdependent, and the bridge and guard circuit must be balanced successively. The accuracy with which the guard circuit must be balanced in order that no appreciable error occurs in the bridge balance depends on both the magnitude of the coupling capacitances and the degree of the coupling-circuit balance. It is frequently useful to provide a partial balance for the coupling circuit in addition to a balance for the guard circuit in order to decrease the accuracy with which the guard circuit must be balanced.

The type of guard circuit chosen depends on several factors: the method of connecting the detector and three-terminal capacitor, and the frequency range of the bridge. The guard-circuit balance can be tested by transferring one of the detector terminals to the guard circuit. This, of necessity, demands that the guard circuit be connected across the generator. In an alternative method without this limitation the detector is left connected to the bridge circuit and the guard terminal is connected to a junction of the bridge arms that places the guard circuit in parallel with similar circuit elements of the

[9] J. C. Balsbaugh and A. Herzenberg, J. Franklin Inst. 218, 49 (1934).

bridge. The guard circuit can then be across either generator or detector, and hence of either type. Although the direct capacitances of a three-terminal capacitor are independent of the voltages placed across them, except when the material forming these capacitances is voltage-sensitive, this possibility, together with the necessity of keeping the voltage between the guarded and guard electrodes low because of their close spacing (see Fig. 1.1), makes it desirable to use that type of guard circuit and detector connection which brings the guarded and guard electrodes to the same potential at balance.

When the guard circuit of the type F-H (Fig. 1.14) is used, a large dielectric loss or conductance in the terminal capacitance which becomes part of the capacitance C_N may make it impossible to balance the guard circuit because the capacitances C_A and C_B in the ratio arms are too small; this is frequently the case. Extra capacitance must then be added to the ratio arms, or an inductance placed in series with the resistive arm R_F of the guard circuit. Neither method is really satisfactory. The added capacitances increase the value of the cross terms in the complete balance equations of the Schering bridge (see Eq. 1.19), and the inductance cannot be easily varied, particularly over a wide frequency range. The guard circuit of the type V-W in Fig. 1.14, the so-called *Wagner ground*, is free from this limitation because any terminal capacitance placed across one of its arms can be directly balanced by added capacitance across its other arm. The combination of the guard circuit V-W and the method of detector connection whereby the guard circuit is placed in parallel with two bridge arms, provides the best design, especially over a wide frequency range.

Grounding

Although a guard-circuit bridge can be grounded at any junction, the choice is really between the guard terminal, the

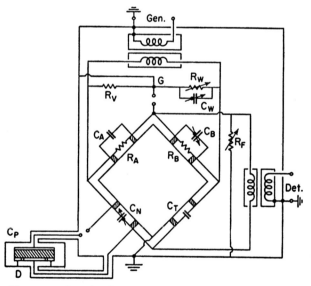

Fig. 1.15. Schering bridge with grounded guard terminal.

junction of the capacitance arms, and the junction of the ratio arms. When neither measuring electrode is grounded, ground should be placed at the guard terminal (Fig. 1.15). Only a grounded shield around the bridge and guard circuit is required. Two transformers are needed because no bridge junction is grounded. All circuit elements of the bridge must be three-terminal impedances with their insulated parts mounted on the grounded panel. The three-terminal measuring cell requires merely a grounded shield. It is shown connected to the bridge in such a way that the generator should be connected across the ratio arms and guard circuit for the guarded and guard electrodes to be at the same potential. Switches are provided for connecting the specimen to the bridge, and the guard and coupling circuits to the proper bridge junction. If generator and detector are interchanged, the connections to the measuring electrodes should be interchanged, in order to keep guarded and guard electrodes at the same potential.

When either the junction of the capacitance arms or the ratio arms is grounded, the shield, on which the three-terminal circuit elements of the bridge are mounted, is connected to the guard terminal, but cannot be grounded. Instead, it must be surrounded by a grounded shield, as in Fig. 1.16, where the junction of the capacitance arms is grounded. Only one transformer is required because one bridge junction is grounded. The three-terminal measuring cell must have a grounded shield, and the high lead to it must be doubly shielded. It is shown connected to the bridge in such a way that the generator

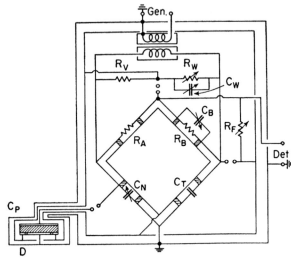

Fig. 1.16. Schering bridge with grounded junction of capacitance arms.

should be connected across the ratio arms and guard circuit, for the guarded and guard electrodes to be at the same potential.

Accuracy

Over the frequency range of 20 c to 1 Mc the capacitance of dielectric specimens can be measured with the Schering bridge circuit to an accuracy of $\pm(0.2$ percent $+ 0.04$ $\mu\mu f)$ and dissipation factor to $\pm(2$ percent $+ 0.00005)$, when a standard precision capacitor of suitable capacitance range is used. When the direct calibration of a micrometer-electrode system is used, the error in capacitance may increase to $\pm(0.5$ percent $+ 0.1$ $\mu\mu f)$, but this increase in error is more than offset at frequencies above 1 Mc by the freedom of the micrometer-electrode system from errors caused by residual impedances.

Inductive-ratio-arm capacitance bridges

One method of obtaining a guard point without the necessity of balancing a guard circuit is to replace the resistive ratio arms of a capacitance bridge with two inductive ratio arms having very nearly unity coupling (Fig. 1.17). When the

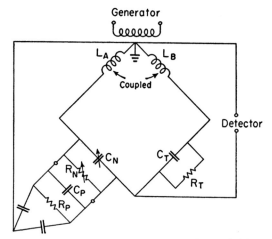

Fig. 1.17. Inductive-ratio-arm capacitance bridge.

junction of the ratio arms is used as the guard point G, any impedance placed across the A arm by a three-terminal dielectric specimen appears also across the B arm because of the nearly unity coupling between the two inductive arms. Hence the balance of the bridge is not affected.

The resistive balance of the bridge must be made by a variable resistor placed across one of the capacitance arms (Fig. 1.17) in the manner usually adopted in parallel-resistance capacitance bridges. This use of a variable resistor prevents accurate measurements of dissipation factor above 1 Mc unless specially designed resistors are used. Cole and Gross,[10] Clark and Vanderlyn,[11] and Wilhelm[12] have described bridges of this nature.

High-Frequency Range

Resonant circuits

The various resonant circuits used for the measurement of a dielectric specimen are composed of an inductor and a standard capacitor, which together form a closed circuit. This circuit is loosely coupled to a generator by either capacitive, inductive, or resistive coupling. Resonance is indicated by a vacuum-tube voltmeter connected in parallel with the tuning capacitor. Since resonant circuits are used only when the frequency is so high that a Schering bridge cannot be used because of the errors caused by residual impedances, a micrometer-electrode system is always used for holding the dielectric specimen.

[10] R. H. Cole and P. M. Gross, Jr., *Rev. Sci. Instr.* **20**, 252 (1949).

[11] H. A. M. Clark and P. B. Vanderlyn, *Proc. Inst. Elec. Engrs.* (London) **96**, Pt. III, 189 (May, 1949).

[12] H. T. Wilhelm, *Bell System Tech. J.* **31**, 999 (1952).

The circuits are first tuned to resonance with the dielectric specimen clamped in the micrometer electrodes. The specimen is then removed and the circuit retuned by increasing the capacitance of the micrometer electrodes. The parallel capacitance of the specimen is, as in Eq. 1.20,

$$C_p = C' - C_s + C_a. \qquad (1.23)$$

The various resonant circuits differ in the method of measuring the resistive component. Five different methods are available: the *resistance-variation* and *conductance-variation* methods, in which the reading of the vacuum-tube voltmeter with specimen disconnected is reduced to its reading with the specimen connected by introducing a suitable resistance in series or in parallel with the standard capacitor; the *susceptance-variation* and *frequency-variation* methods, in which the a-c loss of the entire circuit is measured by a process of detuning that measures the width of the resonance curve; and the *resonance-rise* method, in which the voltage injected into the circuit through suitable series coupling is compared with the resonant voltage.

For the greatest accuracy in determining the resistive component, the tuning capacitance should be no larger than necessary to take care of the specimen capacitance. Hence the inductance should decrease as the frequency increases. At least two values for each decade of frequency should be provided. Adjacent inductances will then differ by a factor of 10. Throughout the frequency range over which an inductor is used, its dissipation factor should not be greater than 0.005 ($Q = 200$), so that the increase in dissipation factor of the circuit, when the specimen is connected, is as large as possible. A dissipation factor of 0.002 ($Q = 500$) is attainable at the higher frequencies and should be approached when a dielectric with a dissipation factor as low as 0.0002 must be measured.

As the frequency approaches 100 Mc, an error can occur in the measurement of dissipation factor if the micrometer electrodes do not make complete contact with the metallic electrodes applied to the specimen. There will then be a radial resistance in these electrodes, due to the concentration of current at the contact surface, which will increase with the square root of the frequency. Under these conditions an increase in dissipation factor of 0.0005 at 100 Mc is not unusual. Hartshorn and Ward[4] have suggested that if the measurements are repeated with no metallic electrodes applied to the specimen, the correct value of dissipation factor can be found by increasing the value obtained from the bare specimen in the ratio of the two capacitances obtained for the coated and the bare specimens.

In the *resistance-variation method* (Fig. 1.18) the added series resistor R prevents both the inductor L and the standard capacitor C from being grounded. The capacitance placed across the added resistance when the standard capacitor is grounded introduces errors which increase with frequency and make this method undesirable.

In the *conductance-variation method* (Fig. 1.19) all circuit elements (inductor L, added parallel resistor R, vacuum-tube voltmeter V, and micrometer-electrode system C) are in parallel, and one terminal of each can be grounded. The parallel resistor R cannot be continuously variable or even variable in decade steps because of the change in resistance and capacitance of such resistors with frequency. It should consist of a

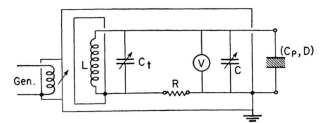

Fig. 1.18. Resistance-variation method.

set of resistors ranging in resistance from 100 ohms to 10 megohms, all of the same physical size and made of deposited carbon. The parallel resistance of such resistors decreases with increasing frequency owing to the so-called Boella effect, first discussed by Howe.[13] This effect is produced when the resistance and capacitance are distributed and when the capacitance has a dielectric loss arising from the dielectric material on which the carbon is deposited. Even in the best resistors there will be a decrease of 1 percent in parallel resistance when the product of frequency and d-c resistance is 1 Mc-megohm.

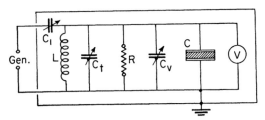

Fig. 1.19. Conductance-variation method.

That portion of the parallel capacitance which is distributed also decreases with increasing frequency. For these reasons this method is not much used.

In spite of these difficulties Weaver and McCool[14] have developed a conductance-variation method that at least equals any of the other methods in accuracy and sensitivity. The greatest error in dissipation factor comes from the process of matching the voltmeter deflection with the specimen removed to that with the specimen between the electrodes, since a small difference between two large quantities has to be determined. Weaver and McCool use a null method in which for both resonance conditions the voltage developed across the tuned circuit is compared with the generator voltage. Since this method demands a continuously variable resistance, they use a diode connected in series with a decade resistor. For frequencies

[13] G. W. O. Howe, *Wireless Engr.* **12**, 291 (1935).
[14] E. O. Weaver and W. A. McCool, Naval Research Laboratory Report R-3133 (June, 1947).

up to 10 Mc a voltage resolution of 1 mv in 50 volts is obtained, which allows a dissipation factor of 0.000001 to be detected. With this equipment it was possible to measure the dissipation factor of dry air to ±20 percent. McCool [15] has extended the frequency range to 100 Mc, with some loss in accuracy.

In the *susceptance-variation method* (Fig. 1.20) all circuit elements are in parallel, and one terminal of each can be grounded. The susceptance-variation method makes use of the fact that the shape, in particular the width, of the resonance curve of a resonant circuit is a measure of the dissipation factor of the circuit. The circuit is tuned to resonance, and the

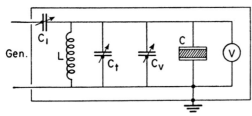

Fig. 1.20. Susceptance- and frequency-variation method.

capacitance C and maximum voltage V_r noted. The circuit is then detuned on each side of resonance to a lower voltage V for capacitance settings C_1 and C_2. The dissipation factor of the circuit is

$$D = \frac{P \, \Delta C}{2(C + C_e)(1 - P^2)^{1/2}}, \quad (1.24)$$

where $P = V/V_r$ and $\Delta C = C_2 - C_1$ with $C_2 > C_1$ and C_e the extra capacitance in circuit, so that $C + C_e$ is the total tuning capacitance including capacitance C_t and the distributed capacitance of inductor L. When the voltage ratio P is made equal to $(1/2)^{1/2} = 0.707$, the simpler expression is obtained:

$$D = \frac{\Delta C}{2(C + C_e)}. \quad (1.25)$$

The change in capacitance ΔC needed to detune the circuit is generally so small, less than 5 $\mu\mu$f, that a special capacitor C_v is required. In the micrometer-electrode system it is usually a cylindrical capacitor built into the fixed electrode (see Fig. 1.3). Although this capacitor can also be used to tune the circuit when the specimen is clamped between the electrodes, it is generally preferable to provide a separate tuning capacitor C_t for this purpose. The zero capacitances of these extra capacitors should be kept as small as possible in order not to lower materially the upper frequency limit of the circuit.

[15] W. A. McCool, Naval Research Laboratory Report R-3310 (July, 1948).

Measuring Procedure. With the dielectric specimen clamped between the micrometer electrodes, the circuit is tuned to resonance and the coupling to the generator adjusted to give somewhat less than full-scale deflection of the voltmeter. The circuit is retuned if necessary, and the exact reading V of the voltmeter noted. The circuit is then detuned on both sides of resonance by adjusting the vernier capacitor C_v to reduce the voltmeter reading to 0.707 V, and the two readings of the vernier capacitor C_1 and C_2 are noted. The specimen is then removed, the circuit retuned, and the new voltmeter reading V' noted. The circuit is again detuned on both sides of resonance, and two new readings of the vernier capacitor C_1' and C_2' are noted. For each of these tunings such capacitance settings are taken as to allow the calculation of the capacitance of the specimen by using either Eq. 1.23 or 1.20. The capacitance setting at resonance is not as exact as the settings off resonance because the resonance curve at its maximum is relatively broad. Hence the capacitance setting at resonance can be improved by using a value halfway between the two detuned capacitance settings. The dissipation factor of the specimen is

$$D = \frac{\Delta C - \Delta C'}{2C_p}. \quad (1.26)$$

When use is made of the fact that the resonant voltage is inversely proportional to the dissipation factor of the circuit, only one detuning is necessary. The dissipation factor of the specimen is

$$D = \frac{\Delta C'}{2C_p} \frac{V' - V}{V} = \frac{\Delta C}{2C_p} \frac{V' - V}{V'}. \quad (1.27)$$

This method of using voltage differences is less accurate than that using capacitance differences alone, but, since it involves fewer adjustments, it is usually to be preferred. The choice as to when the circuit should be detuned depends on other factors. Detuning the circuit with the specimen connected is usually preferable (second half of Eq. 1.27), because both the change of capacitance ΔC and resonant voltage V' are larger.

The *frequency-variation method* uses the same circuit as the susceptance-variation method (Fig. 1.20) except that the vernier capacitor C_V is not required. Similarly this method makes use of the fact that the shape of the resonance curve of a resonant circuit is a measure of the dissipation factor of the circuit. The circuit is tuned to resonance, and the frequency ν_0 and maximum voltage V_r are noted. The circuit is then detuned on each side of resonance to a lower voltage V for frequency values of ν_1 and ν_2. The dissipation factor of the circuit is

$$D = \frac{P \, \Delta \nu}{\nu (1 - P^2)^{1/2}}, \quad (1.28)$$

where $P = V/V_r$ and $\Delta \nu = \nu_2 - \nu_1$, with $\nu_2 > \nu_1$. When the voltage ratio P is made equal to $(1/2)^{1/2} = 0.707$, the equation simplifies to

$$D = \frac{\Delta \nu}{\nu}. \quad (1.29)$$

Since the required fractional change of frequency is

the dissipation factor D of the circuit (0.002 or less), the frequency calibration of the generator must be such that frequency differences can be read to a hundredth of this value or better, that is, 0.002 percent or less. Usually special means for reading frequency differences to this accuracy must be provided, and the generator must be stabilized for frequency against changes in oscillator amplitude. When it is not necessary to measure at any given frequency, the tuning capacitor C_t may be omitted, and the upper frequency limit of the circuit correspondingly raised.

Measuring Procedures. With the dielectric specimen clamped between the micrometer electrodes, the circuit is tuned to resonance and the coupling to the generator adjusted to give somewhat less than full-scale deflection of the voltmeter. The circuit is retuned, if necessary, and the exact reading V of the voltmeter noted. The circuit is then detuned on both sides of resonance by changing the frequency of the generator to reduce the voltmeter reading to $0.707V$, and the two frequencies ν_1 and ν_2 are noted. The specimen is then removed, the frequency of the generator returned to its original value ν, the circuit retuned, and the voltmeter reading V' noted. The circuit is again detuned on both sides of resonance, and the two new frequencies ν_1' and ν_2' are noted. For each of these tunings such capacitance settings are taken as to allow the calculation of the capacitance of the specimen by either Eq. 1.23 or 1.20. The capacitance or frequency settings at resonance are not as exact as the frequency settings off resonance because the resonance curve at its maximum is relatively broad. Hence the capacitance setting at resonance can be improved by using a value for capacitance such that the corresponding frequency is halfway between the two detuned frequencies. The dissipation factor of the specimen is

$$D = \frac{C' + C_e}{C_p} \frac{\Delta\nu - \Delta\nu}{\nu}, \quad (1.30)$$

where $C' + C_e$ is the total capacitance of the circuit at resonance, including tuning capacitance C_t and the distributed capacitance of inductor L. The value of the stray capacitance C_e of the circuit can be measured only by an indirect method. Its appearance in the expression for the dissipation factor of the specimen (Eq. 1.30) instead of in that of the circuit (Eq. 1.29), as for the susceptance-variation method (Eqs. 1.25 and 1.26), must be balanced against the advantages of the frequency-variation method.

When use is made of the fact that the resonant voltage is inversely proportional to the dissipation factor of the circuit, only one detuning is necessary, and the dissipation factor of the specimen becomes

$$D = \frac{C' + C_e}{C_p} \frac{\Delta\nu'}{\nu} \frac{V' - V}{V} = \frac{C' + C_e}{C_p} \frac{\Delta\nu}{\nu} \frac{V' - V}{V'}. \quad (1.31)$$

This method of using voltage differences is less accurate than that using frequency differences alone but, since it involves fewer adjustments, it is usually to be preferred. The choice as to when the circuit should be detuned depends on other factors. Detuning the circuit with the specimen connected is usually preferable (second half of Eq. 1.32), because both the change in frequency $\Delta\nu$ and the resonant voltage V' are larger.

As the measuring frequency is increased, the inductance must decrease as the square of the frequency, since the capacitance in the circuit is approximately constant. Below a limiting value of perhaps 10 mμh the dissipation factor of the total inductance, which must necessarily include the connecting leads, increases rapidly to a value that renders the circuit useless for measurement purposes. This situation can be avoided up to 500 Mc and perhaps 1000 Mc by using a re-entrant cavity, as shown schematically in Fig. 1.21.

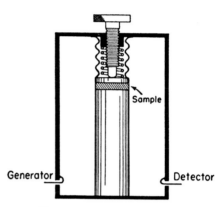

Fig. 1.21. Re-entrant cavity.

The micrometer-electrode system is placed along the axis of the cylindrical cavity, thus producing a coaxial line shorted at both ends. The resonant frequency depends on the size of the line, mainly its length, and on the loading capacitance of the micrometer-electrode system. Its dissipation factor is generally smaller than 0.001, increases with the square root of the frequency, and is limited by the conductivity of the metal of which the cavity is made, together with the smoothness of finish. The generator and detector are coupled to the cavity either by loops or probes, placed conveniently on opposite sides so that direct coupling between them is avoided by the shielding effect of the electrode system. Works, Dakin, and Boggs[16] describe a satisfactory measuring equipment, using a re-entrant cavity at 500 Mc. The use of a re-entrant cavity is the first step in the transition to a measuring system having distributed impedances. The inductance and part of the tuning capacitance are distributed, whereas the measuring capacitance is lumped. The upper frequency limit appears when the diameter of the electrodes is no longer negligible compared to the wavelength.

In the *resonance-rise method* (Fig. 1.22) the coupling resistor R prevents inductor L from being

---

[16] C. N. Works, T. W. Dakin, and F. W. Boggs, *Trans. Am. Inst. Elec. Engrs.* 63, 1092 and 1452 (1944).

grounded. The latter is shielded so that only the ground capacitance of its shield is placed across the coupling resistor. The generator current I through the coupling resistor is measured by a thermocouple ammeter A. The coupling resistor must have a very small inductance, in order that its impedance shall not differ appreciably from its resistance. A commonly used value of resistance is 0.04 ohm, for which the inductance can be reduced to 0.07 mμh. The ratio of impedance to resistance is then 1.006 at 10 Mc, which becomes the upper frequency limit. This frequency

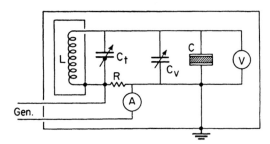

Fig. 1.22. Resonance-rise method.

limitation may be raised to 100 Mc or as high as the minimum inductance of the circuit will allow by replacing resistor R with a very small inductor. Extra inductance is placed in the generator circuit, and the voltage across both inductors is measured with a vacuum-tube voltmeter. The ratio of the dissipation factors of these two inductors must remain constant with frequency.

The resonance-rise method makes use of the fact that, when a voltage is introduced into a resonant circuit, the ratio of this voltage to the voltage appearing across the capacitance is approximately the dissipation factor of the circuit. The circuit is tuned to resonance, and the generator current I and maximum voltage V_r are noted. The dissipation factor of the circuit is

$$D = IR/V_r. \quad (1.32)$$

By choosing a standard generator current I, the voltmeter can be calibrated directly in dissipation factor D. Since this is a reciprocal scale as compared to the voltage calibration, it is usual to calibrate the meter in terms of Q, the quality factor, the reciprocal of the dissipation factor D.

Measuring Procedures. With the dielectric specimen clamped between the micrometer electrodes, the circuit is tuned to resonance and the generator current is adjusted to its standard value. The circuit is retuned if necessary, and the exact reading, V or D or Q, of the voltmeter is noted. The specimen is then removed, the circuit retuned, and the reading, V' or D' or Q', noted. For each of these tunings such capacitance readings are taken as to allow the calculation of the capacitance of the specimen by using either Eq. 1.23 or Eq. 1.20. The dissipation factor of the specimen is

$$D = \frac{C' + C_e}{C_p} IR \left(\frac{1}{V} - \frac{1}{V'}\right) = \frac{C' + C_e}{C_p} \Delta D$$
$$= \frac{C' + C_e}{C_p}\left(\frac{1}{Q} - \frac{1}{Q'}\right), \quad (1.33)$$

where $\Delta D = D - D'$ and C_e is the extra capacitance in the circuit in excess of that given by the calibration of the micrometer-electrode system, but not including capacitance C_t when connected as shown in Fig. 1.22. Its value should be kept as small as possible.

Accuracy. Over the frequency range of 1 to 100 Mc the capacitance of dielectric specimens can be measured with the susceptance- and frequency-variation methods and the resonance-rise methods, all using a micrometer-electrode system, to an accuracy of $\pm (0.5$ percent $+ 0.1$ $\mu\mu$f), the same as at lower frequencies. By using an inductor having a dissipation factor of 0.003, the dissipation factor of the specimen can be measured to $\pm (2$ percent $+ 0.0002)$. The lower limit of 0.0002 is set more by the final results obtained in a series of observations than by consideration of the accuracy of the component measurements. The resonance-rise method is subject to greater errors when the frequency is increased than the variation methods, because of the method of coupling the generator to the resonant circuit.

The null method used by Weaver and McCool [14] greatly decreases this error. Brodhun [17] has developed a null method for the frequency-variation method, for which a lower limit of 0.00003 is claimed; it could also be applied to the susceptance-variation method.

[17] C. Brodhun, National Research Council Conference, October, 1951.

2 · Distributed Circuits

By WILLIAM B. WESTPHAL

Many of the measuring schemes for high frequencies are extensions of lumped-circuit methods used at low frequencies. As the frequency in a lumped-circuit reactance variation equipment (Fig. 2.1a) is increased, skin effect and radiation of the coil lower the Q of the circuit. A first remedy is the substitution of a reentrant cavity for the coil (Fig. 2.1b; see also IIA, Sec. 1, Fig. 1.21). As the frequency is increased

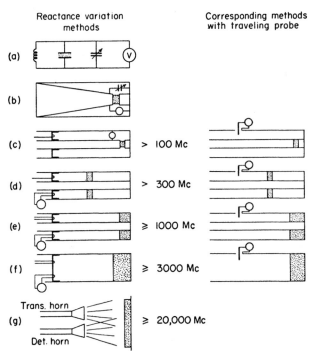

Fig. 2.1. Distributed circuits.

further we substitute for the fixed cavity a section of coaxial line with movable short circuit (Fig. 2.1c). The inductance of the electrodes for a disk sample involves serious corrections above 300 Mc. It is preferable in this range to use a coaxial sample located near the middle of a half-wavelength line (Fig. 2.1d). At still higher frequencies the sample extends over an appreciable part of a wavelength, and may be measured with distributed parameters in the short circuit method (Fig. 2.1e). When the distance between the coaxial conductors is approximately one wavelength, a hollow wave guide is substituted (Fig. 2.1f). Finally, in the millimeter wave region we may measure the disk specimen in space, using beams and optical methods (Fig. 2.1g).

Transmission-Line Methods

Various approaches

High-frequency methods need not be based on lumped-circuit resonance measurements. The principles of transmission-line methods [1-6] follow directly from the consideration of ϵ^* and μ^* in transmission-line or wave equations (see IA, Eq. 3.1). In the term *transmission line measurements* we include all measurements in which parallel beams of electromagnetic waves impinge perpendicularly on dielectric boundaries (see IA, Sec. 4). Thus, in this category, fall methods utilizing a sample in the form of a lumped parallel-plate capacitor loading a coaxial line, as well as methods using a distributed dielectric medium in bounded or unbounded space. In general, both the propagation

[1] S. Roberts and A. von Hippel, "A New Method for Measuring Dielectric Constant and Loss in the Range of Centimeter Waves," Electrical Engineering Dept., Massachusetts Institute of Technology, March, 1941; *J. Appl. Phys.* 17, 610 (1946).

[2] A. von Hippel, D. G. Jelatis, and W. B. Westphal, "The Measurement of Dielectric Constant and Loss with Standing Waves in Coaxial Wave Guides," N.D.R.C. Report, Laboratory for Insulation Research, Massachusetts Institute of Technology, April, 1943.

[3] W. B. Westphal, "Techniques and Calculations Used in Dielectric Measurements on Shorted Lines," N.D.R.C. Report 490, Laboratory for Insulation Research, Massachusetts Institute of Technology, August, 1945.

[4] M. G. Haugen and W. B. Westphal, "The Design of Equipment for Measurement of Dielectric Constant and Loss with Standing Waves in Waveguides," N.D.R.C. Report 541, Laboratory for Insulation Research, Massachusetts Institute of Technology, October, 1945.

[5] W. B. Westphal, Tech. Report XVII, ONR Contract N5ori-07801, Laboratory for Insulation Research, Massachusetts Institute of Technology, February, 1949.

[6] W. B. Westphal, Tech. Report XXXVI, ONR Contract N5ori-07801, Laboratory for Insulation Research, Massachusetts Institute of Technology, July, 1950.

constant and the intrinsic impedance must be measured to determine ϵ^* and μ^*, that is, two independent measurements of complex quantities are required, but for the present purposes we assume $\mu^* = \mu_0$.

If the material is nonmagnetic, as stated, the basic measurement methods are those of Fig. 2.2. The first four of these are mainly of academic interest and will not be considered further; the other five have practical

electric constant and loss only for a narrow range of materials, namely, those having low dielectric constants and very high losses. For low or moderate losses the magnitude of the impedance gives the dielectric constant, but the polar angle is small.

The transmission method with long samples (Fig. 2.2g) is a powerful method for measuring high-loss or medium-loss samples when available in relatively large

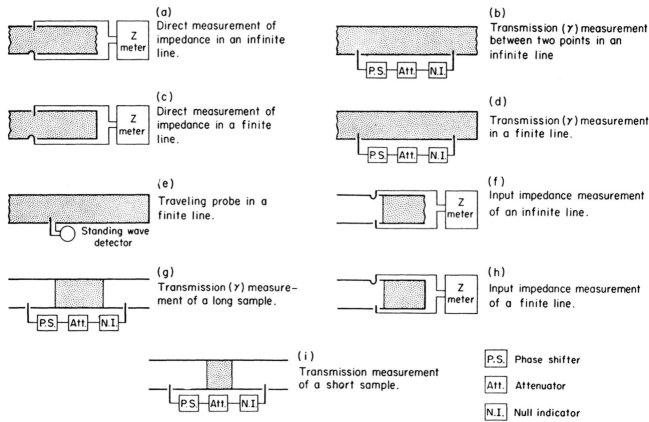

Fig. 2.2. General methods for dielectric measurements in distributed circuits.

uses. They will be briefly discussed, and their limitations given.

A traveling probe within the sample (Fig. 2.2e) is used mainly for liquids. For gases, sealing the probe is difficult whereas in a slotted solid sample the probe signal is decreased because of reflections at the air-to-dielectric boundary; small variations in the slot cause irregularities in the standing-wave pattern. These effects are particularly noticeable with high dielectric constant materials. With liquids the major problems are leakage or evaporation from the slot, but, in principle, accurate measurements on low-loss liquids can be made. For high-loss liquids, the standing-wave pattern is not symmetrical and the method is poor.

Input impedance measurements with long samples (Fig. 2.2f) yield accurate measurements of both di-

volume. It is superior for these materials because the accuracy is limited in principle only by the total attenuation range. This can be 80 db or greater, and accurate to a fraction of a decibel. Samples with this range of loss show several hundred degrees phase shift; thus β can be determined with very high accuracy. This method has been used for water and soils at 3 cm by Straiton and Tolbert.[7] A similar arrangement has been used by Collie, Hasted, and Ritson[8] for water and aqueous salt solutions at 3 and 1.25 cm wavelength. By using tapered samples the power loss of even very low-loss materials have been meas-

[7] A. W. Straiton and C. W. Tolbert, *J. Franklin Inst.* 246, 13 (1948).

[8] C. H. Collie, J. B. Hasted, and D. M. Ritson, *Proc. Phys. Soc. (London)* 60, 145 (1948).

ured and recorded as a function of temperature by Cumming.[9]

The input impedance or transmission measurements with short samples (Figs. 2.2h and 2.2i) are capable of high accuracy with moderate amounts of medium- or low-loss substances of any dielectric constant (see Appendix I, p. 74). For high-loss materials these methods are very useful, but usually not precise because of difficulties in preparing samples of optimum thickness (often less than 10 mils) and in making good contact with them. The input impedance measurement has been more widely used, partly because of greater simplicity in the calculations for the general case. The transmission measurement, with finite length of sample and frequency variation, was used by Powles[10] to measure high-constant, low- or medium-loss ceramics at 10 and 3 cm.

In some cases *combination methods* offer advantages in accuracy and simplicity. For example, a combination of input impedance measurement (Fig. 2.2g) to determine ϵ' and attenuation measurement (Fig. 2.2g) to determine ϵ'' has been used by Powles[10] to evaluate $BaTiO_3$ and $(Ba-Sr)TiO_3$ high-constant, high-loss ceramics at 3 cm. The Laboratory for Insulation Research has used a combination of method f, Fig. 2.2 (on a very long sample to determine ϵ') with method h (on a very short sample to determine ϵ'') for such ceramics at various frequencies.

Calculations for standing waves

The basic equations for a standing wave represent it as the sum of two components traveling in opposite directions parallel to the direction of propagation (see IA, Eqs. 4.7):

$$E = E_0 e^{-\gamma_1 x} + E_1 e^{\gamma_1 x}, \quad (2.1)$$

$$H = H_0 e^{-\gamma_1 x} + H_1 e^{\gamma_1 x}. \quad (2.2)$$

Here E and H are the transverse field components (with time dependence omitted), in medium 1 at a variable distance x from the boundary; E_0 and H_0 are incident fields at $x = 0$; E_1 and H_1 are the reflected fields at $x = 0$; and γ_1 is the propagation constant for medium 1. In terms of the reflection coefficients r_0 and $-r_0$ for the E and H waves, respectively, at the boundary $x = 0$ and Z_1 the intrinsic impedance of medium 1,

$$E = E_0(e^{-\gamma_1 x} + r_0 e^{\gamma_1 x}), \quad (2.3)$$

$$H = \frac{E_0}{Z_1}(e^{-\gamma_1 x} - r_0 e^{\gamma_1 x}). \quad (2.4)$$

[9] W. A. Cumming, *J. Appl. Phys.* **23**, 768 (1952).
[10] J. G. Powles and W. Jackson, *Proc. Inst. Elec. Engrs.* (London) **96**, Pt. III, 383 (1949).

When the boundary is a perfectly reflecting plane, the amplitude of the incident and reflected waves is the same, and

$$r_0 = -1. \quad (2.5)$$

Then

$$E = E_0(e^{-\gamma_1 x} - e^{\gamma_1 x}). \quad (2.6)$$

A plot of the magnitude of E^2 versus x is given in Figs. 2.3 and 2.4 for different values of loss. For low

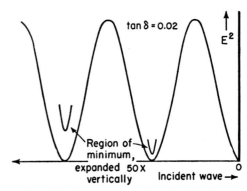

Fig. 2.3. Standing wave in a low-loss line.

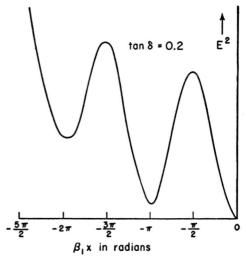

Fig. 2.4. Standing wave in a high-loss line.

or medium loss the distance l between successive minima is a half wavelength,

$$l = \frac{\gamma_1}{2} = \frac{\pi}{\beta_1}, \quad (2.7)$$

and

$$\frac{\epsilon'}{\epsilon_0} = \left(\frac{\lambda_0}{2l}\right)^2 \quad (2.8)$$

for coaxial line or free space. At the first minimum the amplitudes of the incident and reflected waves are $|E_0|(1 + \alpha_1 l)$ and $|E_0|(1 - \alpha_1 l)$, respectively. The field strength at the minimum is the difference

$|E_0|2\alpha_1 l$. The field strength at the first or second maximum is the sum of the two waves and is equal to $2|E_0|$ with good approximation. The inverse standing-wave ratio $E_{\min}/E_{\max}$ is thus a measure of α_1:

$$\frac{E_{\min}}{E_{\max}} = \alpha_1 l \qquad (2.9)$$

and

$$\tan \delta = \frac{2\alpha_1}{\beta_1} = \frac{2}{\pi}\frac{E_{\min}}{E_{\max}}. \qquad (2.10)$$

The curve for high loss is asymmetrical, and the position of the extrema is shifted. Hence accurate measurements cannot be made with an amplitude detector alone (see method 5); a combination of phase and amplitude measurements would be required.

The solutions to Eqs. 2.1 and 2.2 for three media in sequence (Fig. 2.5) have been derived previously in

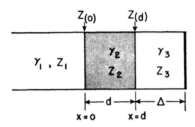

Fig. 2.5. Three-layer medium.

terms of reflection coefficients at the boundaries (see IA, Sec. 4). An alternative method for deriving Z and γ from hyperbolic functions is well established in transmission-line theory.[11] The ratio of input impedance at $x = 0$ to intrinsic impedance is given as (see IA, Eq. 4.31)

$$\frac{Z(0)}{Z_1} = \frac{\dfrac{E_{\min}}{E_{\max}} - j \tan \dfrac{2\pi x_0}{\lambda_1}}{1 - j \dfrac{E_{\min}}{E_{\max}} \tan \dfrac{\pi 2 x_0}{\lambda_1}}. \qquad (2.11)$$

When medium 2 is terminated in a shorting plate ($\Delta = 0$) (see IA, Eq. 4.36)

$$Z(0) = Z_2 \tanh \gamma_2 d. \qquad (2.12)$$

Dividing by Z_1 and transposing gives

$$\frac{Z_2}{Z_1} \tanh \gamma_2 d = \frac{Z(0)}{Z_1}. \qquad (2.13)$$

For all TE modes or TEM modes, the relation between intrinsic impedance and propagation factor is (see IA, Eq. 3.13)

$$Z = \frac{j\omega\mu_0}{\gamma} \quad (\text{for } \mu^* = \mu_0). \qquad (2.14)$$

Hence

$$\frac{Z_2}{Z_1} = \frac{\gamma_1}{\gamma_2}. \qquad (2.15)$$

Substituting this relation in Eq. 2.13 and dividing by d, we obtain

$$\frac{\tanh \gamma_2 d}{\gamma_2 d} = \frac{1}{\gamma_1 d} \frac{Z(0)}{Z_1}. \qquad (2.16)$$

When the loss in medium 1 can be neglected ($\gamma_1 = j2\pi/\lambda_1$), this expression simplifies to

$$\frac{\tanh \gamma_2 d}{\gamma_2 d} = \frac{-j\lambda_1}{2\pi d} \frac{Z(0)}{Z_1}. \qquad (2.17)$$

When Δ is a quarter wavelength of medium 3 and $\alpha_3 = 0$, we have the open-circuit condition (see IA, Eq. 4.41)

$$Z(0) = Z_2 \coth \gamma_2 d, \qquad (2.18)$$

and for a loss free medium 1

$$\frac{\coth \gamma_2 d}{\gamma_2 d} = \frac{-j\gamma_1}{2\pi d} \cdot \frac{Z(0)}{Z_1}. \qquad (2.19)$$

Thus Eqs. 2.17 and 2.19 are expressions relating the unknown γ to the measured quantities d, λ_1, $E_{\min}/E_{\max}$, and x_0 for the sample mounted in front of a shorting plate or a quarter wavelength away. These relations are valid as long as the bounding conductor walls are uniform throughout the three media.

Charts [6] for determining $\gamma_2 d$ for either position of the sample are given in Appendices IV and V, pp. 86 and 102 in the forms $(\tanh T\underline{/\tau})/T\underline{/\tau} = C\underline{/-\zeta}$ and $(\coth T\underline{/\tau})/T\underline{/\tau} = C\underline{/-\zeta}$. Thus $T\underline{/\tau}$ is the polar form of $\gamma_2 d$, and C and $-\zeta$ are the magnitudes and angle, respectively, of $(-j\lambda_1/2\pi d)(Z(0)/Z_1)$. The functions of T, that is, $\gamma_2 d$, are multivalued. Ambiguity in the choice of T is removed when the approximate magnitude of ϵ'/ϵ_0 or $\gamma_2 d$ is known, or when measurements are made on two different lengths of sample.

The relationship between $\epsilon_2{}^*$ and γ_2 depends on the cut-off wavelength of the particular wave guide.[12] The cut-off wavelength γ_c is the longest free-space wavelength for which propagation in a particular mode is possible. It may be expressed in terms of the cross-section dimensions of the wave guide: for a rectangular guide (and a TE_{01} wave) $\gamma_c = 2.0000$ times the

---

[11] See A. E. Kennelly, *The Application of Hyperbolic Functions to Electrical Engineering Problems*, McGraw-Hill Book Co., New York, 1925; see H. E. Hartig, *Physics 1*, 380 (1931).

[12] For example, see R. I. Sarbacher and W. A. Edson, *Hyper and Ultrahigh Frequency Engineering*, John Wiley and Sons, New York, 1943.

width; for a circular hollow guide (and a TE_{11} wave) $\gamma_c = 3.4129$ times the radius; and for coaxial line, or free space (TEM wave), $\gamma_c = \infty$. For TE modes in hollow wave guides

$$\frac{\epsilon_2'}{\epsilon_1'} - j\frac{\epsilon_2''}{\epsilon_1'} = \frac{(1/\lambda_c)^2 - (\gamma_2 d/2\pi d)^2}{(1/\lambda_1)^2 + (1/\lambda_c)^2}$$
$$\equiv \frac{u - (\gamma_2 d/2\pi d)^2}{w}, \quad (2.20)$$

where $u = (\lambda_1/\lambda_c)^2$, $w = 1 + u$.

In free space or coaxial line Eq. 2.20 reduces to

$$\frac{\epsilon_2'}{\epsilon_1'} - j\frac{\epsilon_2''}{\epsilon_1'} = -\left(\frac{\lambda_1 \gamma_2 d}{2\pi d}\right)^2. \quad (2.21)$$

Alternative methods for obtaining ϵ^* from the measurement of input impedance. The ambiguity in finding $\gamma_2 d$ can also be avoided by making both open and short-circuit measurements on the same sample; then the product on the input impedances yields the impedance of the material (a single-valued function) instead of $\gamma_2 d$. This calculation procedure is considered in Appendix IX, p. 121.

The calculation of x_0 is an intermediate step between the measured quantity (node shift with introduction of the sample) and the determination of ϵ'. Charts have been prepared which give simple functions of ϵ' and ϵ'' in terms of minimum width and node shift.[13]

Although the equations discussed thus far are valid for any type of equipment, the method using a traveling detector to which Eq. 2.11 refers is outstanding because it covers an especially wide range of impedances.

Measurement of standing-wave ratio with square law detector. In dielectric measurements high standing-wave ratios are common. To avoid probe interference effects the pick-ups must work in a very low-power region and are usually square law detectors. The equation for E^2 at any point in a lossless line in terms of the angular displacement of a traveling probe from the minimum is

$$E_x^2 = (E_{max})^2 \sin^2 \theta + (E_{min})^2 \cos^2 \theta, \quad (2.22)$$

for

$$\theta = \frac{\pi \Delta x}{\lambda_1}; \quad (2.23)$$

$\frac{\Delta x}{2}$ is the distance from the minimum to some point x (Fig. 2.6). The inverse standing-wave ratio is

$$\frac{E_{min}}{E_{max}} = \frac{\sin \theta}{\left[\left(\frac{E_x}{E_{min}}\right)^2 - \cos^2 \theta\right]^{1/2}}. \quad (2.24)$$

When Δx corresponds to the distance between the points of twice minimum power,

$$\frac{E_{min}}{E_{max}} = \frac{\sin \theta}{(2 - \cos^2 \theta)^{1/2}}. \quad (2.25)$$

If, simultaneously, $\frac{\pi \Delta x}{\lambda_1}$ is ≤ 0.1,

$$\frac{E_{min}}{E_{max}} = \frac{\pi \Delta x}{\lambda_1}. \quad (2.26)$$

[13] R. M. Redheffer, R. C. Wildman, and V. O'Gorman, J. Appl. Phys. **23**, 505 (1952).

For values of $\frac{\pi \Delta x}{\lambda_1}$ between 0.075 and 0.625, the value of E_{min}/E_{max} can be calculated conveniently by subtracting the factor indicated in Appendix III, p. 86. For higher values of E_{min}/E_{max} the ratio is usually measured directly.

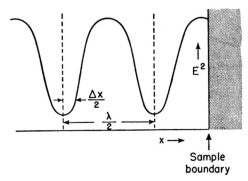

Fig. 2.6. Wave pattern in lossless standing-wave section.

In principle the measurement of E_x at any two points in the standing wave suffices to determine E_{min}/E_{max}. In practice the most accurate measurements at low inverse standing-wave ratios are made in the region of the minimum.

For low-loss samples, the loss in the wave-guide walls between the minimum and the sample boundary may be appreciable. The loss in the empty line is determined by a measurement of Δx_e taken at a distance d_e from the end of the line (Fig. 2.7).

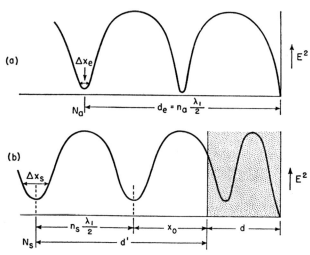

Fig. 2.7. Notation for standing-wave measurements: (a) wave pattern in empty line; (b) wave pattern with sample.

With sample-in a value $\Delta x'$ is subtracted which is Δx_e multiplied by the ratio of the distance d' between sample and probe to d_e.

$$(\Delta x_s)_{corrected} = (\Delta x_s)_{measured} - \frac{d'}{d_e}\Delta x_e. \quad (2.27)$$

The loss in the shorting plate can usually be neglected.

Measurement of x_0. When the traveling probe is direct-reading in distance from the shorting plane, x_0 is simply the node distance minus the length of the sample d (or $d + \lambda_1/4$ for the open-circuit method). Usually the probe is calibrated in terms of distance from an arbitrary point; then x_0 is determined by the change in node readings with sample in and out. From Fig. 2.7 the expression for this distance of the first minimum from the dielectric boundary follows:

$$x_0 = N_a - N_s + n_a \frac{\lambda_1}{2} + n_s \frac{\lambda_1}{2} - d. \quad (2.28)$$

Here N_a is the node reading in air-filled line, N_s the node reading with sample, n_a the number of half wavelengths to probe with empty line, and n_s the number of half wavelengths to probe from first node outside sample.

Simplification of the theory.

(a) *Medium-loss samples.* The general expression for $\gamma_2 d$, using the short-circuit method, is

$$\frac{\tanh \gamma_2 d}{\gamma_2 d} = -j \frac{\lambda_1}{2\pi d} \frac{\frac{E_{\min}}{E_{\max}} - j \tan \frac{2\pi x_0}{\lambda_1}}{1 - j \frac{E_{\min}}{E_{\max}} \tan \frac{2\pi x_0}{\lambda_1}}. \quad (2.29)$$

It simplifies for a loss-free medium 2 to

$$\frac{\tan \beta_2 d}{\beta_2 d} = -\frac{\lambda_1}{2\pi d_2} \tan \frac{2\pi x_0}{\lambda_1}. \quad (2.30)$$

In this case ϵ' is given by x_0 independently of $E_{\min}/E_{\max}$. The same expression results when we equate the real parts of both sides of Eq. 2.29 for low-loss. Equating the imaginary parts leads to a relation connecting the sample loss with the width Δx of the minimum (see Eq. 2.26)

$$\tan \delta = \frac{\Delta x}{d} \cdot \frac{\beta_2 d \left(1 + \tan^2 \left(\frac{2\pi x_0}{\lambda_1}\right)\right)}{\beta_2 d (1 + \tan^2 \beta_2 d) - \tan \beta_2 d} \cdot \frac{w - u \frac{1}{\epsilon'/\epsilon_0}}{w}. \quad (2.31)$$

For coaxial lines the right-hand factor reduces to unity (see Eq. 2.21). This expression for $\tan \delta$ was first used by Dakin [14] and has been extensively employed.[3,6]

The corresponding relations for the open-circuit method are

$$\frac{\cot \beta_2 d}{\beta_2 d} = \frac{\lambda_1}{2\pi d} \tan \frac{2\pi x_0}{\lambda_1} \quad (2.32)$$

$$\tan \delta = \frac{\Delta x}{d} \cdot \frac{\beta_2 d \left(1 + \tan^2 \frac{2\pi x_0}{\lambda_1}\right)}{\beta_2 d (1 + \cot^2 \beta_2 d) + \cot \beta_2 d} \cdot \frac{w - u \frac{1}{\epsilon'/\epsilon_0}}{w}. \quad (2.33)$$

Tables of the functions $\frac{\tan x}{x}$ and $\frac{\cot x}{x}$ are included in Appendices VI and VII, pp. 104 and 106.

For one percent error in ϵ_2' these simplified methods are restricted by: (1) $\tan \delta < 0.1$, (2) $n \frac{\epsilon''}{\epsilon_0} \leq 0.2$, where n is the number of wavelengths in the sample. Usually, the unabridged calculation is required whenever the measured $E_{\min}/E_{\max}$ is greater than 0.15.

Additional simplifications result if the sample length is a multiple of $\lambda_2/4$ (see Appendix VIII, p. 120). These relations prove especially useful in production control.

(b) *Very low-loss samples.* The wall losses are not negligible when very low-loss samples are measured. In addition to the wall loss between sample and probe (Eq. 2.27), in this case also the loss $\tan \delta_{ws}$ in the sample-filled section has to be considered

[14] T. W. Dakin and C. N. Works, *J. Appl. Phys.* **18**, 789 (1947).

(see Appendix II, p. 86). Also in writing Eq. 2.17 from 2.16, the loss associated with λ_1 is neglected, but this correction is usually minor (see Ref. 6, p. 93).

(c) *Short samples.* Here the open-circuit method is used to place the sample in the region of high field strength. Equation 2.18 may be written

$$\frac{Z(0)}{Z_1} = \frac{Z_2}{Z_1} \coth \gamma_2 d = \frac{\gamma_1}{\gamma_2} \coth \gamma_2 d. \quad (2.34)$$

When $\gamma_2 d \leq 0.3$, for 3 percent error the $\tanh \gamma_2 d$ reduces to $\gamma_2 d$, and we obtain

$$\frac{Z(0)}{Z_1} = \frac{\gamma_1}{\gamma_2^2 d}. \quad (2.35)$$

Neglecting the loss in medium 1 and noting that $E_{\min}/E_{\max}$ will be small, we obtain

$$\frac{E_{\min}}{E_{\max}} \left(1 + \tan^2 \frac{2\pi x_0}{\lambda_1}\right) - j \tan \frac{2\pi x_0}{\lambda_1} = \frac{\lambda_0}{j 2\pi d \epsilon_2^*/\epsilon_0}. \quad (2.36)$$

From the imaginary part,

$$\frac{\epsilon_2'}{\epsilon_0} = \frac{\lambda_0}{2\pi d} \cot \frac{2\pi x_0}{\lambda_1}, \quad (2.37)$$

or, in terms of the node shift ΔN,

$$\frac{\epsilon_2'}{\epsilon_0} = 1 + \frac{\Delta N}{d}. \quad (2.38)$$

From the real parts and with $\frac{\pi \Delta x}{\lambda}$ substituted for $E_{\min}/E_{\max}$,

$$\frac{\epsilon_2''}{\epsilon_0} = \frac{\Delta x}{2d}. \quad (2.39)$$

For hollow wave guides (TE modes) the corresponding relations are

$$\frac{\epsilon_2'}{\epsilon_0} = \frac{\frac{\Delta N + d}{d} + u}{w} \quad (2.40)$$

$$\frac{\epsilon_2''}{\epsilon_0} = \frac{\Delta x}{2dw}. \quad (2.41)$$

(d) *Small samples.* Reducing the cross section of the sample leads to further simplification in the sense that results may be expressed in terms of fractional volume of the sample and its surrounding space. For example, we may consider a rodlike sample extending between the two conductors of a coaxial line that have radii large in comparison to the sample dimensions (Fig. 2.8).

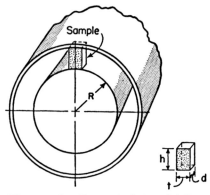

Fig. 2.8. Small sample in coaxial line.

The node shift, compared to a ring sample, will be multiplied by the ratio $t/2\pi R$ so that

$$\frac{\epsilon_2'}{\epsilon_0} = 1 + \frac{\Delta N}{d} \cdot \frac{2\pi R}{t}. \quad (2.42)$$

The volume of the line between probe and short circuits is its length $n\lambda_1/2$ (n is the number of half wavelengths) times the cross-section area $2\pi Rh$. The right-hand term of Eq. 2.42 may be multiplied by $\dfrac{h}{h} \cdot \dfrac{n\lambda_1/2}{n\lambda_1/2}$, and then Eq. 2.42 becomes

$$\frac{\epsilon_2'}{\epsilon_0} = 1 + \frac{\Delta N}{d} \cdot \frac{2\pi R}{t} \cdot \frac{h}{h} \cdot \frac{n\lambda_1/2}{n\lambda_1/2} = 1 + \frac{\Delta N}{n\lambda_1/2} \cdot \frac{V}{v}. \quad (2.43)$$

Likewise

$$\frac{\epsilon_2''}{\epsilon_0} = \frac{\Delta x}{2d} \cdot \frac{2\pi R}{t} \cdot \frac{h}{h} \cdot \frac{n\lambda_1/2}{n\lambda_1/2} = \frac{\Delta x}{n\lambda_1} \cdot \frac{V}{v}. \quad (2.44)$$

These equations intrinsically assume a uniform field strength throughout the cross section t, that is, we have a transmission line with small "lumped capacitor" loading.

(e) *Disk samples in coaxial line.* For the sample arrangement shown in Fig. 2.9, we can, by neglecting the fringing field, estab-

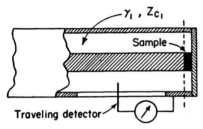

Fig. 2.9. Coaxial line terminated by disk sample.

lish a boundary as before and apply the transmission line equations

$$\frac{Z(0)}{Z_1} = \frac{1}{Z_1} \cdot \frac{1}{j\omega C^*}. \quad (2.45)$$

The value of the complex capacitance $C^* = \epsilon^* C_0$ is

$$C^* = \frac{1}{j\omega Z_1(Z(0)/Z_1)} \frac{1}{j867\nu(Z(0)/Z_1) \log_{10} \dfrac{D_2}{D_1}} \quad [\text{fd}], \quad (2.46)$$

if the loss in the line is neglected. D_1 and D_2 are the conductor diameters, and ν is the frequency in cycles per second. For low-loss samples, x_0 and Δx are independently related to the dielectric constant and loss as before:

$$\epsilon_2' C_0 = \frac{1}{867 f \log_{10} \dfrac{D_2}{D_1} \tan \dfrac{2\pi x_0}{\lambda_1}} \quad (2.47)$$

$$\tan \delta = \frac{\pi \Delta x}{\lambda_1} \left[\tan \frac{2\pi x_0}{\lambda_1} + \cot \frac{2\pi x_0}{\lambda_1} \right]. \quad (2.48)$$

Transmission measurements

In the simplest transmission measurements we determine the attenuation $\Delta \alpha$ and phase shift $\Delta \phi$ produced by an additional section of sample of the length Δl (Fig. 2.10). The attenuation and phase constants of the sample-filled line are, if $\Delta \alpha$ is given in decibels and $\Delta \phi$ in degrees,

$$\alpha = \frac{\Delta \alpha}{8.68 \Delta l} \quad [\text{nepers/unit length}], \quad (2.49)$$

$$\beta = \frac{\Delta \phi}{57.296 \Delta l} + \frac{2\pi}{\lambda_1 \Delta l} \quad [\text{radians/unit length}]. \quad (2.50)$$

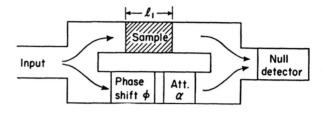

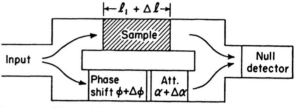

Fig. 2.10. Transmission measurements with two lengths of sample.

From α and β, λ_1 (wavelength in air-filled guide), and λ_c (cut-off wavelength), the dielectric constant and loss tangent are calculated:

$$\frac{\epsilon_2'}{\epsilon_1'} = \frac{\dfrac{1}{\lambda_c^2} + \dfrac{\beta^2 - \alpha^2}{4\pi^2}}{\dfrac{1}{\lambda_c^2} + \dfrac{1}{\lambda_1^2}} \quad (2.51)$$

and

$$\tan \delta = \frac{2\alpha\beta}{\beta^2 - \alpha^2 + \dfrac{4\pi^2}{\lambda_c^2}} \quad \text{for TE modes.} \quad (2.52)$$

For TEM modes (coaxial line or free space),

$$\frac{\epsilon_2'}{\epsilon_1'} = \frac{\lambda_1^2}{4\pi^2} (\beta^2 - \alpha^2) \quad (2.53)$$

$$\tan \delta = \frac{2\alpha\beta}{\beta^2 - \alpha^2}. \quad (2.54)$$

Equations 2.49 and 2.50 assume that the standing-wave pattern in the dielectric can be neglected. This can be accomplished in one of the following ways (Fig. 2.11):

(a) When a matched load and tapered dielectric boundaries are used (Fig. 2.11a), the initial length of sample may consist of the tapered matching sections only. (b) With unmatched load (Fig. 2.11b) the initial sample length must be chosen to reduce the reflected signals reaching the front of the sample to about one percent of the incident signal. (c) With perpendicular sample boundaries and matched load (Fig. 2.11c), the initial transmission loss of the shortest sample should be 20 db in field strength to make the rear reflection negligible. The distance x must remain constant as the sample length is increased. (d) With unmatched load and sample (Fig. 2.11d), the distances x, y, and z must remain fixed; these conditions are fulfilled only by adding a sample-filled section for the second measurement, and the term $\frac{2\pi}{\lambda_1 \Delta l}$ is dropped in Eq. 2.50.

A plot of α and β versus sample length will be linear if the prerequisites for satisfactory sample lengths are fulfilled.

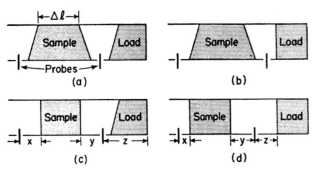

Fig. 2.11. Various sample and load arrangements for transmission measurements.

Resonance Methods

When measuring with a *traveling* detector, we may find it desirable for reasons of power transfer to tune the wave-guide section to resonance, but it is not essential. Alternatively, we can use *fixed* detectors and resonance methods by either varying the cavity dimensions (line-length variation) or the frequency (frequency variation). Figure 2.12 shows the physical arrangement of the *line-length variation method;* it is closely related to the standing-wave method with traveling detector. Only TE and TEM waves will be considered. As in all resonance methods, the generator is loosely coupled to the line, in this case through a

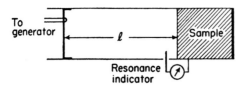

Fig. 2.12. Line-length variation method.

small loop which is attached to the moving plunger. The current in the loop creates in the line in front of the plunger a voltage e which is independent of line tuning. The resonance indicator, a voltmeter E located, for example, at the sample boundary, gives the resonance curve of Fig. 2.13 as the line length l is changed.

It can be shown that the length of line for resonance, l_0, is equal to x_0 in the standing-wave method, and the width of the resonance curve at the half-power points, Δl, is equal to Δx of Eq. 2.26. Therefore, all previous relations for calculating ϵ_2' and ϵ_2'' in the traveling detector methods apply equally well to line-length variation methods, provided that the systems are low-loss. For high-loss, the resonance methods are not useful because the resonance curves are not symmetrical.

In the *frequency variation method* l_0/λ_1 is replaced by $l_0\nu_0/c$, where l_0 is the length of the air section of

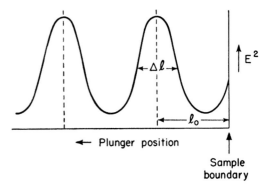

Fig. 2.13. Resonance curves (line-length variation method).

the coaxial line and ν_0 the resonance frequency. Correspondingly, $\Delta l/\lambda_1$ is replaced by $l_0\Delta\nu/c$, where $\Delta\nu$ is the width of the resonance curve at half-power points in the frequency domain.

Thus the three methods employing a standing-wave detector, line-length variation, or frequency variation are related by the conversion formulas

$$\frac{2\pi x_0}{\lambda} = \frac{2\pi l_0}{\lambda} = \frac{2\pi l_0 \nu_0}{c} \qquad (2.55)$$

and

$$\frac{\pi \Delta x}{\lambda} = \frac{\pi \Delta l}{\lambda} = \frac{\pi l \Delta\nu}{c}. \qquad (2.56)$$

The overall Q, if d is the geometrical sample length, may be written

$$Q = \frac{(x_0 + d) \equiv D}{\Delta x} = \frac{(l_0 + d) \equiv L}{\Delta l} = \frac{\nu_0}{\Delta\nu}. \qquad (2.57)$$

Equations 2.55–2.57 hold for coaxial lines (TEM mode). The correlation between the traveling detector and line-length variation methods applies for all TE modes.

Equation 2.57 holds for small samples regardless of the mode, so that Eqs. 2.43 and 2.44 may be expressed in terms Δx, Δl, or Δf (measured between half-power points) and ΔN, ΔL, $\Delta\nu_0$, the shift in node, resonant length, and resonant frequency, respectively. Thus we have the alternative expressions

$$\frac{\epsilon_2'}{\epsilon_1'} - 1 = \frac{\Delta N}{D}\frac{V}{v} = \frac{\Delta L}{L}\frac{V}{v} = \frac{\Delta\nu}{2\nu_0}\frac{V}{v}, \qquad (2.58)$$

$$\frac{\epsilon_2''}{\epsilon_1'} = \frac{\Delta x V}{2Dv} = \frac{\Delta l}{2L}\frac{V}{v} = \frac{\Delta\nu}{4\nu_0}\frac{V}{v}. \qquad (2.59)$$

This technique of using small samples has been applied mainly in resonance systems, for example, by Birnbaum[15] at the National Bureau of Standards. This first-order perturbation theory implies the same conditions as lumped capacitance loading: the sample must be small enough not to disturb the magnitude of the electric field, and the field must be uniform throughout the cross section of the sample.

In addition to cavities using TE or TEM modes, the E_{010} mode resonator has received considerable attention. The electric field is axial, as shown in Fig. 2.14, and therefore well suited for rod-shaped samples placed coaxially. Frequency

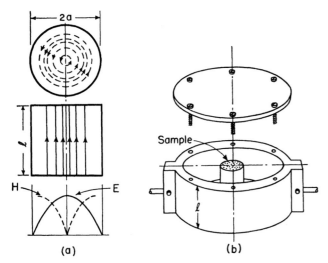

Fig. 2.14. E_{010} mode: (a) field distribution; (b) sample in cavity.

is the only practical variable. The derivation of the equations for calculating dielectric constant and loss is given in Appendix X, p. 121.

The standing-wave method with coaxial line and disk samples (Fig. 2.9) can be replaced by a resonance method using re-entrant cavities, as previously mentioned in IIA, Sec. 1. In either case calibration with known materials is required to evaluate the fringing field.

Impedance bridges

During the last two years bridges for *high* frequencies† have been developed, but they are not designed especially for dielectric measurements (nominal accuracy, 5 percent). When used as impedance comparators with a variable, parallel-plate sample holder, accuracies of the order of 2 percent in dielectric constant can be obtained. In this service the bridge is used simply to indicate when the holder with sample-out is tuned to have the same impedance as with sample-in.

Beljers and van de Lindt[16] have described a bridge for dielectric measurements at 3.2 cm wavelength. Their equipment uses two precision magic tees connected as shown in Fig. 2.15. The tee (B) serves as a bridge, making use of the relation that with input at arm I, the output at arm IV is a minimum when the impedances in arms II and III are matched. The

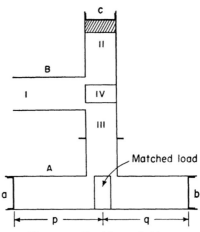

Fig. 2.15. Double-tee bridge.

sample is mounted on a plunger in arm II, and x_0 is determined by the change in this plunger position when the bridge is balanced to a minimum with sample in and out. Arm III connects to the second tee, which has a matched output load and movable plungers at a and b. The inverse standing-wave ratio in front of the sample is

$$\frac{E_{\min}}{E_{\max}} = \tan^2\left(\frac{\pi \Delta p}{\lambda_{1g}}\right), \quad (2.60)$$

in which Δp is the difference between the two plunger readings p and q at balance. With sample out, unbalanced losses in wave-guide arms and plungers cause a detectable minimum. This is corrected for in an approximate manner by using

$$\frac{E_{\min}}{E_{\max}} = \left(\frac{\pi}{\lambda_{1g}}\right)^2 [\Delta p^2 - \Delta p_0^2], \quad (2.61)$$

where Δp_0 is the difference between the two plunger readings p and q at balance with sample out. The advantage of the method over a traveling detector is that Δp is a larger number than Δx. It is not clear from their description whether Δp can be measured with the same percent error as Δx.

---

[15] G. Birnbaum and J. Franeau, *J. Appl. Phys.* **20**, 817 (1949).

† Type 1602-A U.H.F. Admittance Meter, 66 to 1000 Mc (*General Radio Experimenter*, May, 1950), and Type 803A V.H.F. Bridge, 50 to 500 Mc (Hewlett-Packard Co.).

[16] H. G. Beljers and W. J. van de Lindt, *Philips Research Reports*, **6**, 96 (1951).

Typical equipments

Two standing-wave equipments are shown in Figs. 2.16 and 2.17. The first, using a modulated oscillator and a-c detection system, is more common. The second has a d-c pick-up system, and the line is resonated by movement of sample and tuning plunger. Either system is suited to resonance methods as well.

A block diagram of transmission measuring equipment is given in Fig. 2.18.

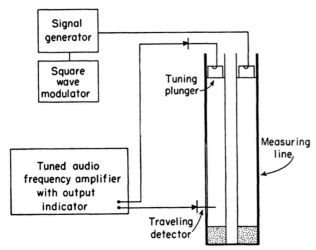

Fig. 2.16. "M.I.T. Coax Instrument" type (300 to 3000 Mc).

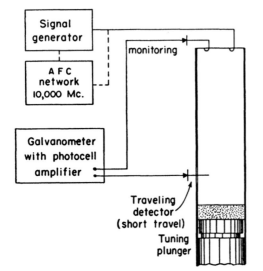

Fig. 2.17. Hollow wave guide instrument (3000 to 10,000 Mc, M.I.T.).

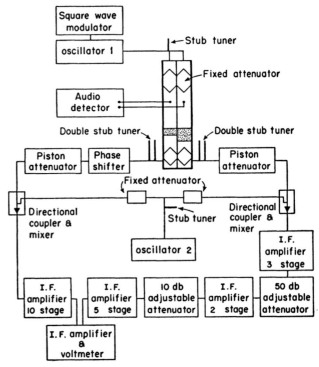

Fig. 2.18. Block diagram of 10 cm transmission equipment (M.I.T.).

Figure 2.19 shows an arrangement for cavity measurements by a frequency-variation method.

For small samples, a comparison scheme such as used by Birnbaum and others is essential. A block diagram of one type of equipment for this purpose is shown in Fig. 2.20.

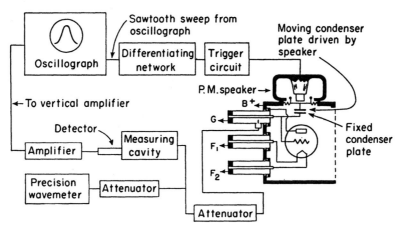

Fig. 2.19. Measurement of dielectrics by frequency modulation in the decimeter range (Works, Dakin, and Boggs [17]).

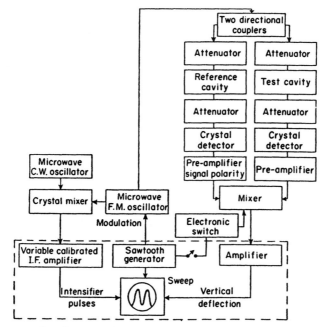

Fig. 2.20. Measurement of cavity Q and resonance frequency in the centimeter range (Birnbaum [15]).

APPENDICES FOR "DISTRIBUTED CIRCUITS"

I. Accuracy of Input Impedance Measurements
II. Wall Losses in Wave Guides
III. Correction Term for E_{min}/E_{max}
IV. Charts of the Complex Function $\dfrac{\tanh x}{x}$
V. Charts of the Complex Function $\dfrac{\coth x}{x}$
VI. Tables of $\dfrac{\tan x}{x}$
VII. Tables of $\dfrac{\cot x}{x}$
VIII. Simplified Calculations for Input Impedance Method
IX. The Two-Position Method
X. Derivation of the Dielectric Field of the E_{010} Cavity

I. Accuracy of Input Impedance Measurements

The accuracy in measuring ϵ'/ϵ_0 for low-loss samples depends on the rate of change of x_0 with dielectric constant and on the absolute accuracy of measuring x_0. In Fig. 2.21 are plotted values of the ratio x_0/λ versus the fractional wavelength in the sample. The slope of the curves can be considered independent of dielectric constant for small changes in ϵ'/ϵ_0, and therefore they can be used to determine the error in measuring the electrical length of the sample and ϵ'/ϵ_0. For example, if x_0/λ can be measured to 0.001, a quarter-wavelength sample of polystyrene ($\epsilon'/\epsilon_0 \simeq 2.6$) could be measured to 0.4 percent as the following considerations show: The slope of the x_0/λ curve is approximately 2. Thus, with $x_0/\lambda = 0.25 \pm 0.001$, the wavelength in the sample is 0.25 ± 0.0005. That is, $\dfrac{d(\epsilon'/\epsilon_0)^{1/2}}{\lambda} = 0.25 \pm 0.2$ percent, or the error in ϵ'/ϵ_0 is 0.4 percent if errors in measuring d and λ are neglected. The error in measuring a three-quarter wavelength sample is one-third this value for the same error in x_0/λ. Obviously, the higher the value of ϵ'/ϵ_0 being measured, the greater is the necessity for using odd quarter-wavelength samples.

The determination of percent error in ϵ'/ϵ_0 for samples in other positions follows the same principles. Figures 2.22 to 2.27 show the changes in x_0/λ curves with positioning of various samples, and thus aid the selection of best measuring conditions.

In practice, the accuracy of dielectric constant measurements on distributed low-loss samples (tan δ < 0.01) is limited by faulty geometry or nonhomogeneity of the sample rather than accuracy in determining node position. In coaxial lines the fit between conductors and sample is important, and even when corrections are made using the idealized correction charts of Figs. 2.28 and 2.29 errors of 1 percent are common. For circular hollow guides, errors less than ½ percent are not difficult to achieve on well-made samples.

The accuracy of measuring tan δ values in the range 0.005 to 0.1 is limited mainly by errors in the measurement of standing-wave ratio usually in the order of 2 percent. In coaxial lines the error due to imperfect sample fit may be corrected theoretically using Fig. 2.30.

The sensitivity for measuring low-loss tangents depends on the error in measuring the loss in an air-filled guide. If the fractional error e made in measuring Δx_a is known or assumed, the accuracy with which the loss tangent of a dielectric material can be measured may be calculated. It depends not only on the performance of the instrument proper, expressed by e, but also on the dielectric constant and loss of the sample, its length and position in the wave guide, and on the amount of empty guide in front of the sample. All these influences together may be lumped into the term *sensitivity* S_δ, symbolizing the smallest loss tangent which can be detected. Tables 2.1 and 2.2 give S_δ values for a number of typical arrangements and samples. They contain also a tabulation of the e values allowable for sensitivities of tan δ = 0.00001 and the product of sensitivity and sample volume, $S_\delta V$, which is a figure of merit for both instrument and method.

The optimum sample lengths are indicated by peaks in the E_{min}/E_{max} curves (Figs. 2.21 to 2.27).

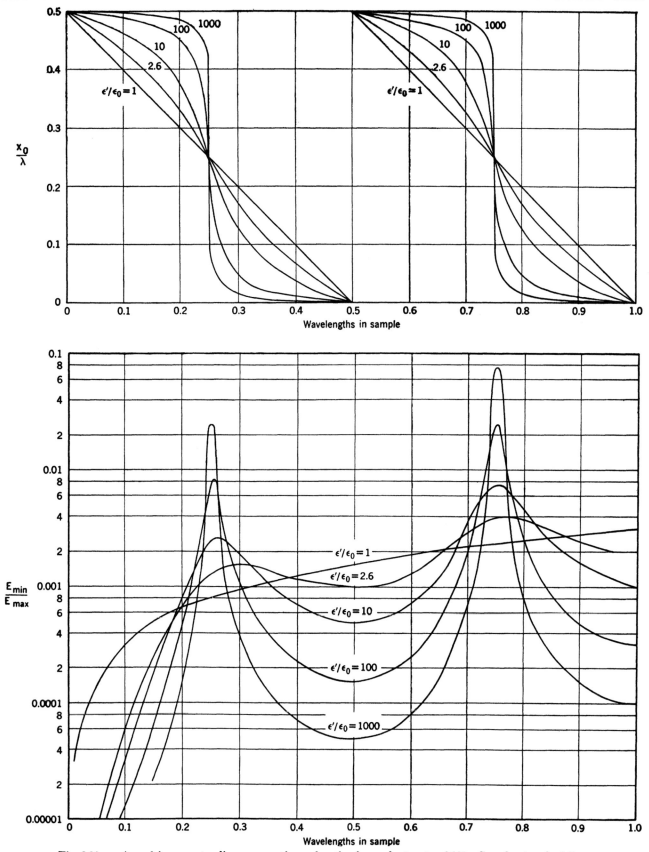

Fig. 2.21. x_0/λ and inverse standing-wave ratio vs. length of sample, $\tan \delta = 0.001$. Sample at end of line.

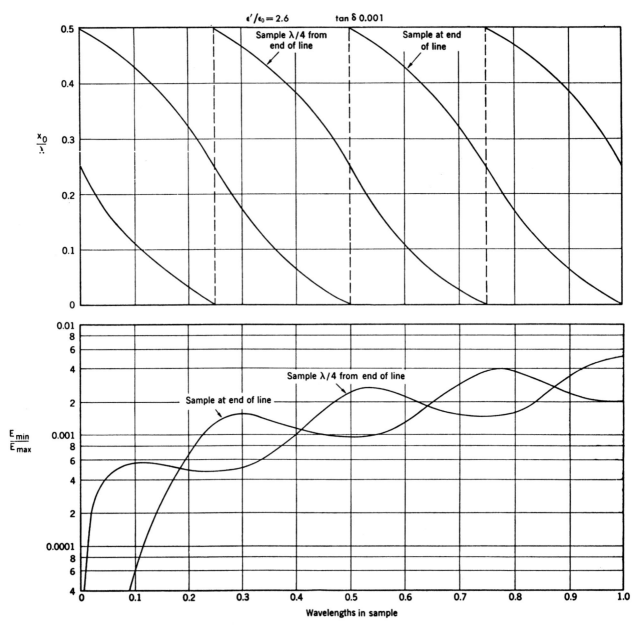

Fig. 2.22. Comparison of two sample positions showing x_0/λ and inverse standing-wave ratio vs. length of sample.

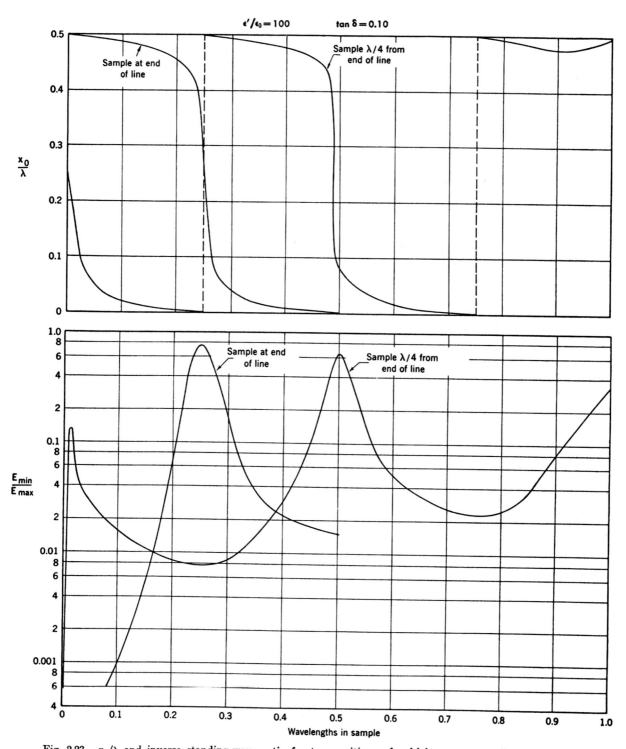

Fig. 2.23. x_0/λ and inverse standing-wave ratio for two positions of a high-constant, medium-loss material.

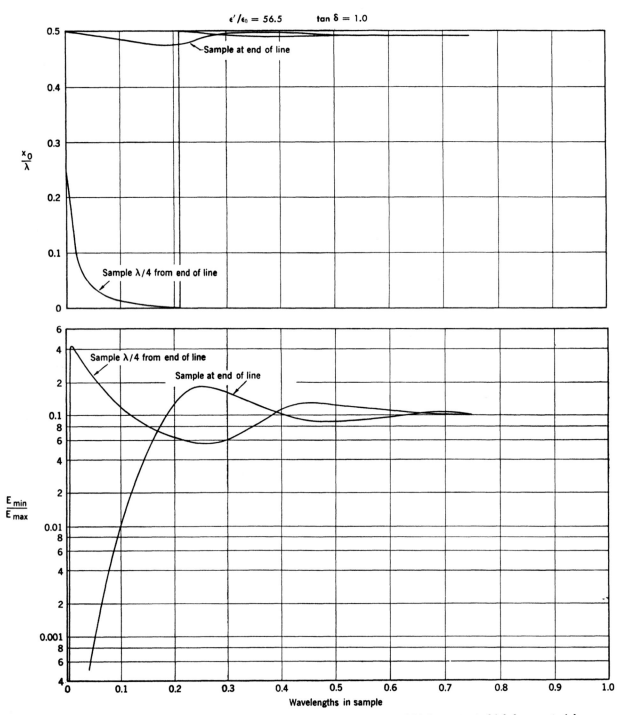

Fig. 2.24. x_0/λ and inverse standing-wave ratio for two positions of high-constant, high-loss material.

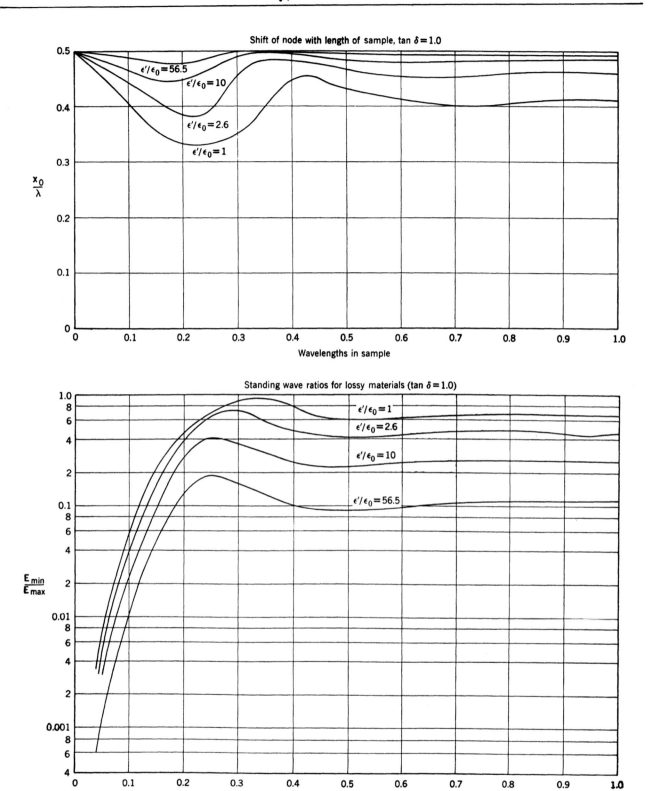

Fig. 2.25. x_0/λ and inverse standing-wave ratio for various materials with tan $\delta = 1.0$. Sample at end.

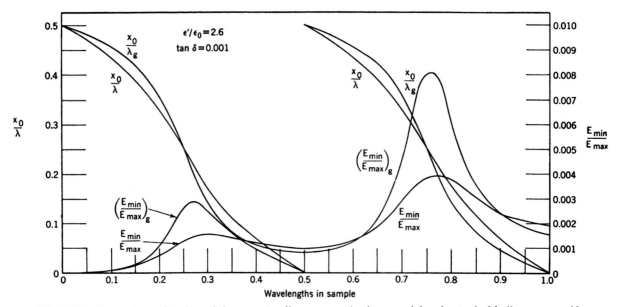

Fig. 2.26. Comparison of x_0/λ and inverse standing-wave ratios in a coaxial and a typical hollow wave guide.

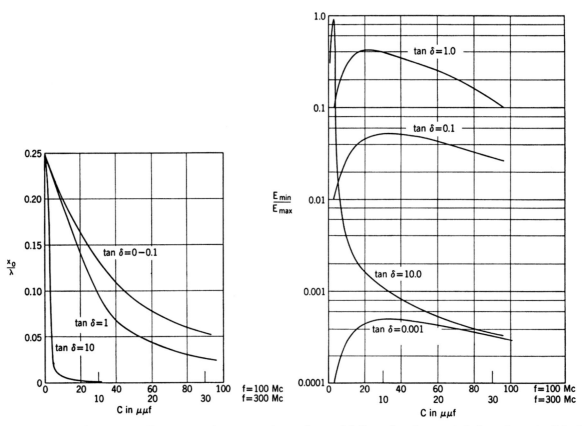

Fig. 2.27. x_0/λ and inverse standing-wave ratio vs. capacitance in coaxial line; the characteristic impedance is 49.5 ohms.

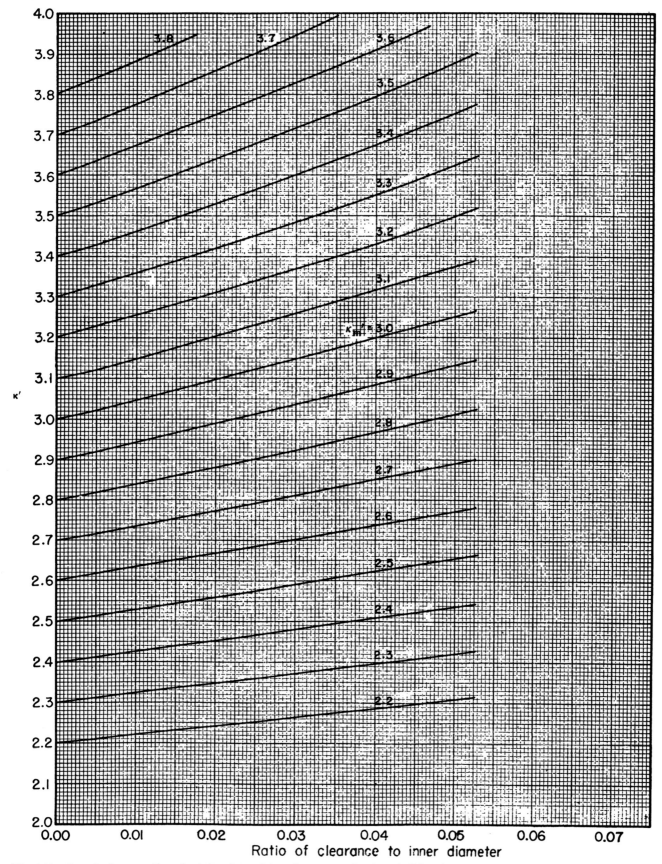

Fig. 2.28. Sample fit correction chart for determining the corrected value of dielectric constant κ' from the measured value κ_m' in coaxial line; conductor ratio = 3.64 ($2.0 < \kappa' < 4.0$).

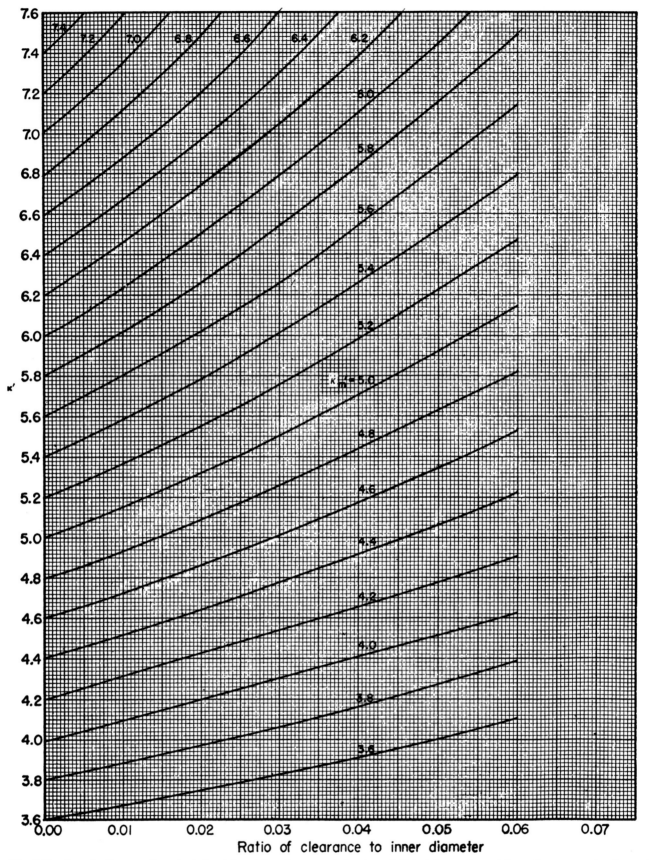

Fig. 2.29. Sample fit correction chart for determining the corrected value of dielectric constant κ' from the measured value κ_m' in coaxial line; conductor ratio = 3.64 ($3.6 < \kappa' < 7.6$).

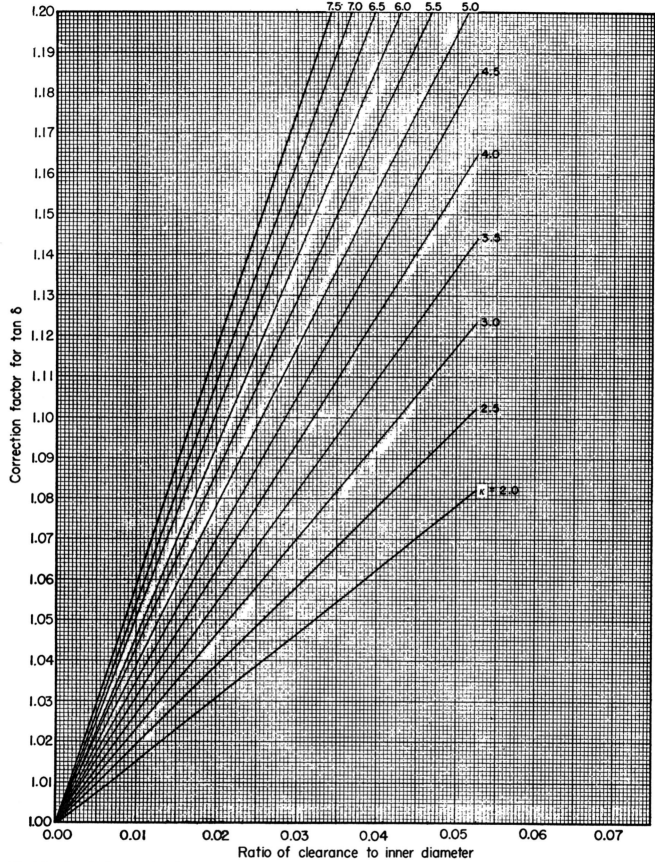

Fig. 2.30. Sample fit correction chart for determining factor by which the measured value must be multiplied to obtain corrected value of loss tangent; conductor ratio = 3.64 (κ', 2 to 7.5).

Table 2.1. Wall loss and sensitivity factors for samples at shorted end of copper line †

Distributed load at shorted end. $\epsilon'/\epsilon_o = 2.6$
n = number of half wave-lengths between probe and end of empty line.
m = number of quarter wave-lengths between probe and sample.

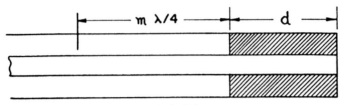

Coaxial Line

Frequency Mc.	D_1 inch	D_2 inch	$\tan \delta_w$	S_δ (n = 1, m = 1)	S_δ (n = 2, m = 1)	% e for $S_\delta = 10^{-5}$	Volume c.c.	$S_\delta V$
300	1.	3.6	.000173	.000495e		2.02	940	.465e
3000	0.316	1.14	.000173	.000495e		2.02	94	.0465e
3000	0.316	1.14	.000173		.00035e	2.86	282	.0985e
300	0.172	0.625	.00095	.00272e		0.37	27.3	.0741e
3000	0.172	0.625	.00030	.000857e		1.17	2.73	.00741e
3000	0.172	0.625	.00030		.00061e	1.64	8.45	.00515e

Circular Hollow Guides

Frequency Mc.	Diameter inches	$\tan \delta_w$	S_δ (n = 2, m = 3)	S_δ (n = 2, m = 1)	% e for $S_\delta = 10^{-5}$	Volume c.c.	$S_\delta V$
3000	2.870	.0000357	.00025e		4.00	74.5	.0186e
3000	2.870			.000076e	13.2	224.	.0170e
9500	0.870	.0000691	.000515e		1.94	2.21	.00114e
9500	0.870			.00016e	6.25	6.61	.00106e
24000	0.360	.0000993	.00069e		1.45	0.147	.000101e
24000	0.360			.00021e	4.84	0.441	.0000928e

Rectangular Guides

Frequency Mc.	Z inch	Y inch	$\tan \delta_{wa}$	S_δ (n = 2, m = 3)	S_δ (n = 2, m = 1)	% e for $S_\delta = 10^{-5}$	Volume c.c.	$S_\delta V$
3000	2.34	1.35	.000062	.00024e		4.16	3.71	.0088e
3000	2.34	1.35			.000073e	13.7	111.4	.0082e
9500	0.900	0.400	.000091	.00028e		3.57	1.26	.00035e
9500	0.900	0.400			.000085e	11.8	3.78	.00032e
24000	0.42	0.17	.0000127	.00035e		2.86	0.0958	.000134e
24000	0.42	0.17			.00011e	9.10	0.288	.000032e

† For the definition of symbols, see p. 74.

Permittivity, Distributed Circuits

Table 2.2. Wall loss and sensitivity factors for lumped samples and distributed samples located $\lambda/4$ from end of copper line †

Capacitance loaded line, $\epsilon'/\epsilon_o = 2.6$, thickness = 1/16"

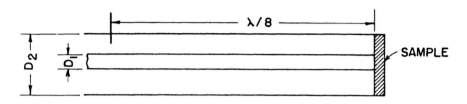

Coaxial Line

Frequency Mc.	D_1 inch	D_2 inch	$\tan \delta_w$	S_δ	% e for $S_\delta = 10^{-5}$	Volume c.c.	$S_\delta V$
300	1	3.6	.000173	.00027e	3.7	9.75	.00263e
3000	0.316	1.14	.000173	.00027e	3.7	0.975	.000263e
300	1	2	.00026	.00041e	2.4	3.00	.00123e

Short distributed load $\lambda/4$ from shorted end, $\epsilon'/\epsilon_o = 2.6$

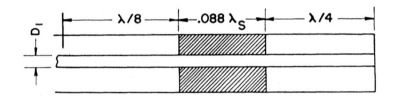

Coaxial Line

Frequency Mc.	D_1 inch	D_2 inch	$\tan \delta_w$	S_δ	% e for $S_\delta = 10^{-5}$	Volume c.c.	$S_\delta V$
300	1	3.6	.000173	.00124e	0.81	332	.41e
3000	0.316	1.14	.000173	.00124e	0.81	3.32	.0041e
300	1	2	.00026	.00187e	0.54	83	.155e
300	0.172	0.625	.00095	.00682e	0.147	10	.0682e
3000	0.172	0.625	.00030	.00216e	0.46	1	.00216e

† For the definition of symbols, see p. 74.

II. Wall Losses in Wave Guides

The propagation factor for a dielectric-filled section of wave guide is

$$\gamma = \alpha + \alpha' + j\beta, \quad (2.62)$$

where α' represents the attenuation due to metallic losses in the walls. Theoretical values of α' are given in several textbooks.[17] For dielectric measurement work experimental values are more accurate since the effect of surface finish on microwave loss is taken into account. The wall loss is conveniently expressed as the loss tangent $\tan \delta_w$. Values of theoretical loss tangents for empty guides are given in Tables 2.1 and 2.2. In general, the wall loss is a function of the dielectric constant and permeability of the dielectric. The theoretical ratio of wall loss with sample in to the wall loss of the empty guide is given below for common wave guides.

For the $H_1(\mathrm{TE}_{11})$ wave in a circular guide:

$$\frac{\tan \delta_{w_s}}{\tan \delta_w} = \frac{\mu_0}{\mu'} \frac{0.42 + \left(\frac{\lambda_0}{\lambda_c}\right)^2 \frac{\mu_0 \epsilon_0}{\mu' \epsilon'}}{0.42 + \left(\frac{\lambda_0}{\lambda_c}\right)^2}. \quad (2.63)$$

For the $H_{01}(\mathrm{TE}_{01})$ wave in a rectangular guide with dimensions Y, Z, with Z the larger:

$$\frac{\tan \delta_{w_s}}{\tan \delta_w} = \frac{\mu_0}{\mu'} \frac{\frac{Z}{2Y} + \left(\frac{\lambda_0}{\lambda_c}\right)^2 \frac{\mu_0 \epsilon_0}{\mu' \epsilon'}}{\frac{Z}{2Y} + \left(\frac{\lambda_0}{\lambda_c}\right)^2}. \quad (2.64)$$

For the TEM mode of the coaxial line,

$$\frac{\tan \delta_{w_s}}{\tan \delta_w} = \frac{\mu_0}{\mu'}. \quad (2.65)$$

Experimentally, $\tan \delta_w$ is determined from the value of Δx taken at distance d from the short-circuited end of an empty guide:

$$\tan \delta_w = \frac{\Delta x}{d\left[1 + \left(\frac{\lambda_1}{\lambda_c}\right)^2\right]} = \frac{\Delta x}{d} \frac{\lambda_0^2}{\lambda_c^2 - \lambda_0^2}. \quad (2.66)$$

For a coaxial line, this reduces to

$$\tan \delta_w = \frac{\Delta x}{d}. \quad (2.67)$$

[17] For example, R. I. Sarbacher and W. A. Edson, *Hyper and Ultrahigh Frequency Engineering*, John Wiley and Sons, New York, 1943.

III. Correction Term for $E_{\min}/E_{\max}$

In preference to Eq. 2.25, the correct value of $E_{\min}/E_{\max}$ for values above 0.1 and less than 0.5 may be found by subtracting a correction term indicated in Table 2.3 from the value of $\pi \Delta x/\lambda$.

Table 2.3. Subtraction term to obtain correct $E_{\min}/E_{\max}$

$\pi \Delta x / \lambda$	Subtraction Factor	$\pi \Delta x / \lambda$	Subtraction Factor
0.075	0.0000	0.375	0.0308
0.100	0.0002	0.400	0.0367
0.125	0.0010	0.425	0.0436
0.150	0.0019	0.450	0.0506
0.175	0.0032	0.475	0.0585
0.200	0.0050	0.500	0.0670
0.225	0.0072	0.525	0.0762
0.250	0.0098	0.550	0.0862
0.275	0.0130	0.575	0.0969
0.300	0.0164	0.600	0.1082
0.325	0.0207	0.625	0.1200
0.350	0.0255		

IV. Charts of the Complex Function $\dfrac{\tanh x}{x}$

These charts are based on the work of S. Roberts. His original charts are of two forms: a survey chart [18] of

$$\frac{\tanh T \underline{/\tau}}{T \underline{/\tau}} = C \underline{/-\zeta}$$

in which the values of T range from 0 to 7.0 as the abscissa and τ from 40° to 90° as the ordinate; a group of five charts [19] in which C or $1/C$ is the abscissa scale and ζ the ordinate covering values of T from 0 to 7.0 and τ from 70° to 90°. This group of charts is useful but lacks accuracy because of their small size. These have been enlarged and extended as follows:

Chart I: T, 0 to 2.0; τ, 70° to 90°.

Part   A: $1/C$, 0 to 0.75; ζ, 0° to 100°.
          B: $1/C$, 0 to 0.75; ζ, 80° to 180°.
          C: $1/C$, 0.6 to 1.35; ζ, 80° to 180°.
          D: $1/C$, 0.6 to 1.35; ζ, 0° to 100°.

Chart II: T, 2.0 to 2.4; τ, 70° to 90°.
             C, 0.35 to 1.1; ζ, 80° to 180°.

Chart III: T, 2.3 to 4.0; τ, $\simeq$70° to 90°.

Part   A: C, 0 to 0.3; ζ, 0° to 100°.
          B: C, 0 to 0.3; ζ, 80° to 180°.
          C: C, 0.25 to 0.51; ζ, 80° to 180°.

[18] S. Roberts and A. von Hippel, *J. Appl. Phys.* **17**, 610 (1946).
[19] A. von Hippel, D. G. Jelatis, and W. B. Westphal, "The Measurement of Dielectric Constant and Loss with Standing Waves in Coaxial Wave Guides," N.D.R.C. Contract OEMsr-191, Laboratory for Insulation Research, Massachusetts Institute of Technology, April, 1943, pp. 50–54.

Chart IV: T, 4.0 to 5.4; τ, $\simeq 70°$ to $90°$.

Part A: $1/C$, 0 to 3.0; ζ, $0°$ to $100°$.
          B: $1/C$, 0 to 3.0; ζ, $80°$ to $180°$.
          C: $1/C$, 2.5 to 5.3; ζ, $80°$ to $180°$.
          D: $1/C$, 2.5 to 5.1; ζ, $0°$ to $100°$.

Chart V: T, 5.4 to 7.0; τ, $70°$ to $90°$.
          C, 0 to 0.23; ζ, $0°$ to $180°$.

Chart VI: T, 7.0 to 8.8; τ, $80°$ to $90°$.
          $1/C$, 0 to 13; ζ, $0°$ to $180°$.

Other charts of this function are given in the literature. The charts of Lowan et al.[20] give x in rectangular form, are designed for use in acoustic measurements, and therefore cannot be read accurately for the loss range most useful in dielectric measurements. The corresponding range of T is 0 to 10.

Benoit[21] has shown that u and w (where $x = u + j2\pi w$) can be plotted as circles. The practical difficulty of making charts by his method is that the circles for medium- and low-loss dielectrics are too large for book form.

Nomograms[22] and tables[23] of the function $\tanh(a + jb)$ have been given by Rybner. They are useful in preparing charts of the function $\dfrac{\tanh x}{x}$.

Kennelly has given a chart covering T from 0 to 3.0.[24]

[20] A. N. Lowan, P. M. Morse, H. Feshbach, and E. Haurwitz, "Tables for Solutions of the Wave Equation for Rectangular and Circular Boundaries Having Finite Impedance," AMP Note No. 18, N.D.R.C. Section 6.1-sr1046–2043, June, 1945.

[21] J. Benoit, *Ann. télécommunications* 4, 27 (1949).

[22] J. Rybner, *Nomograms of Complex Hyperbolic Functions*, Jul. Gjellerups, Copenhagen, 1947.

[23] J. Rybner, *Tabeller til Brug red Løsning af Ligningen tgh* $(b + ja) = r/\underline{\Phi}$, Jul. Gjellerups, Copenhagen, 1943.

[24] A. E. Kennelly, *Chart Atlas of Complex Hyperbolic and Circular Functions*, Harvard University Press, Cambridge, 1914, pp. 28–33.

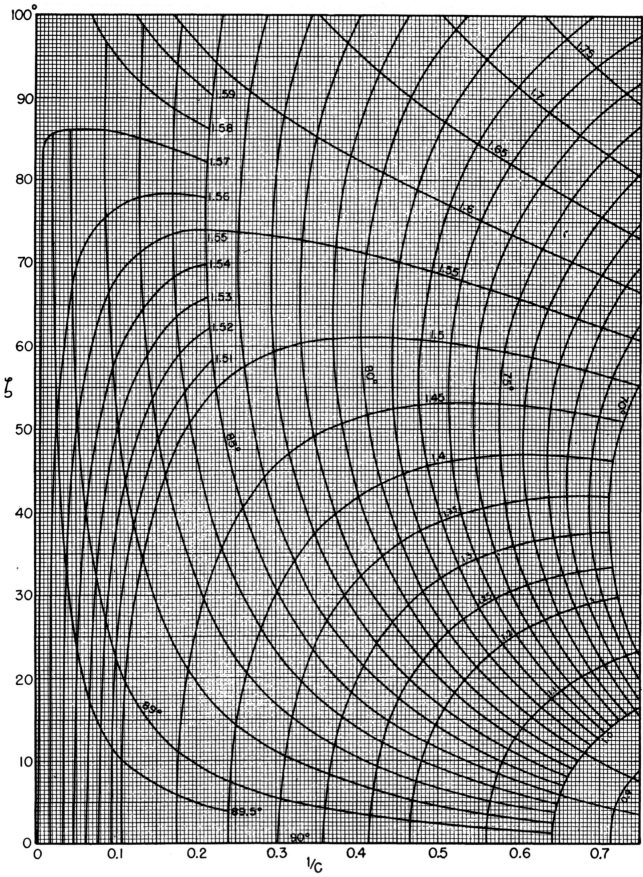

Chart IA. $\dfrac{\tanh T/\tau}{T/\tau} = \dfrac{1}{C}\underline{/-\zeta}$; T, 0 to 2.0; τ, 70° to 90°; Part A, $1/C$, 0 to 0.75; ζ, 0° to 100°.

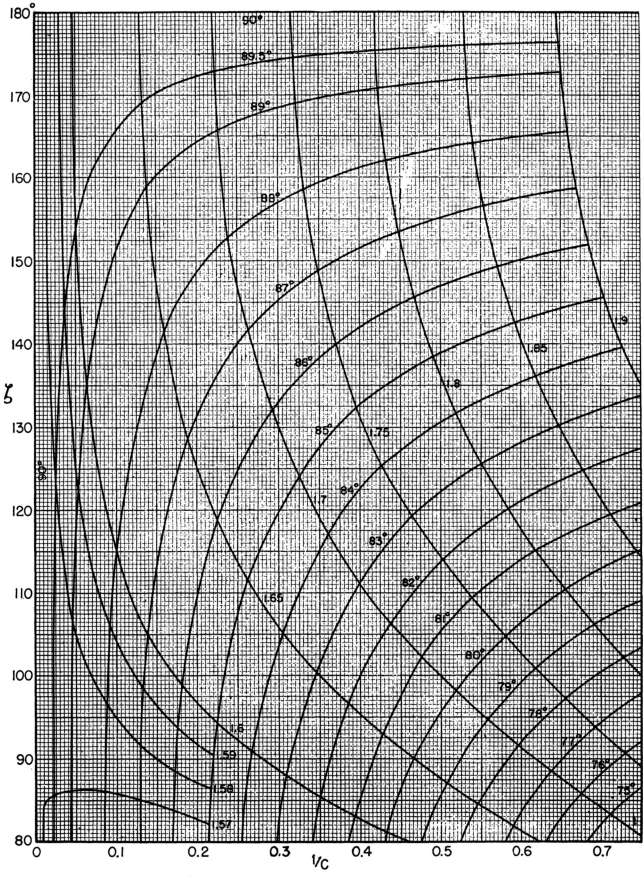

Chart 1B. $\dfrac{\tanh T/\tau}{T/\tau} = \dfrac{1}{C}\underline{/-\zeta}$; T, 0 to 2.0; τ, 70° to 90°; Part B, $1/C$, 0 to 0.75; ζ, 80° to 180°

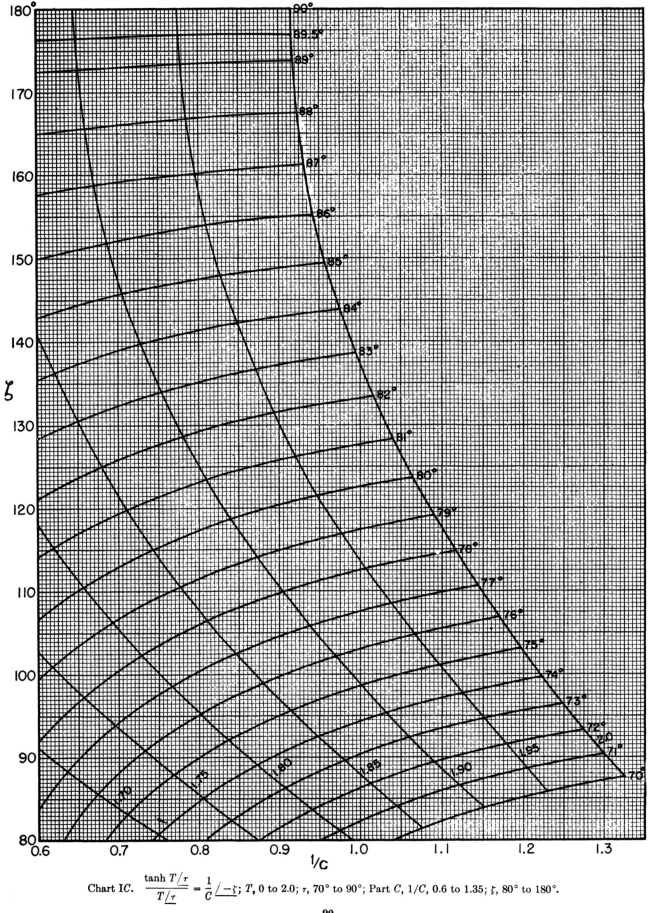

Chart IC. $\dfrac{\tanh T/\tau}{T/\tau} = \dfrac{1}{C}\underline{/-\zeta}$; T, 0 to 2.0; τ, 70° to 90°; Part C, $1/C$, 0.6 to 1.35; ζ, 80° to 180°.

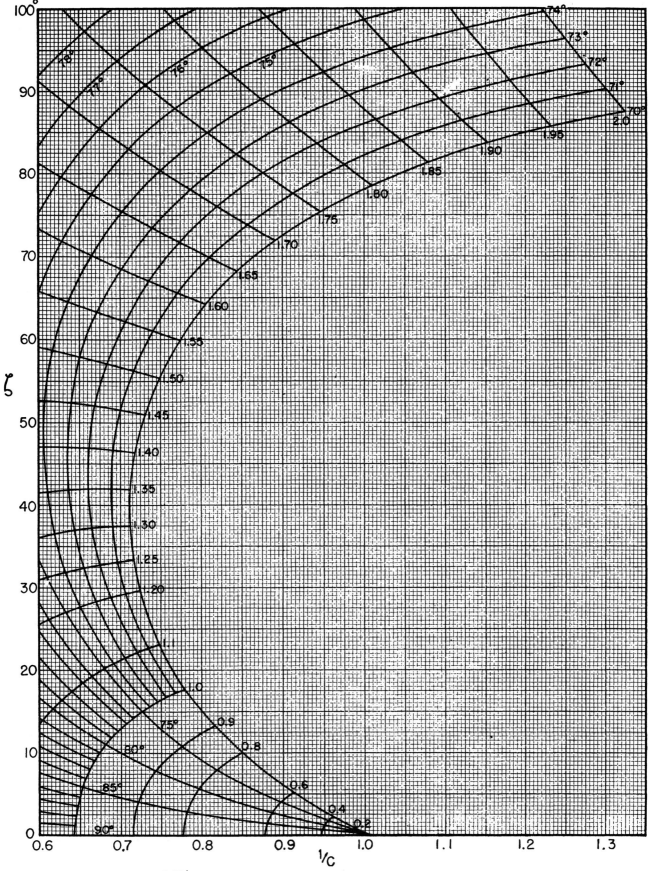

Chart ID. $\dfrac{\tanh T/\tau}{T/\tau} = \dfrac{1}{C} \underline{/-\zeta}$; T, 0 to 2.0; τ, 70° to 90°; Part D, $1/C$, 0.6 to 1.35; ζ, 0° to 100°.

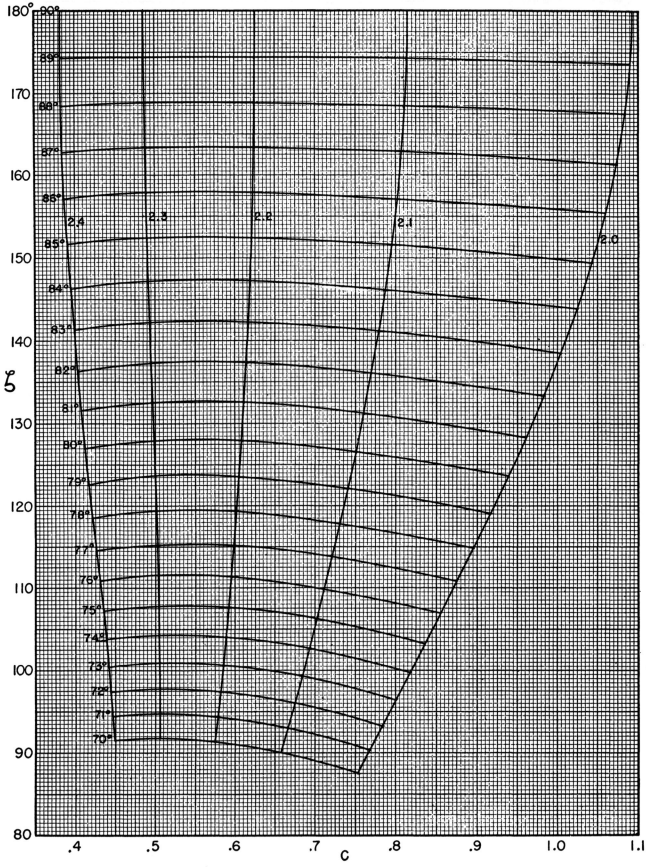

Chart II. $\dfrac{\tanh T/\tau}{T/\tau} = C \underline{/-\zeta}$; T, 2.0 to 2.4; τ, 70° to 90°; C, 0.35 to 1.1; ζ, 80° to 180°.

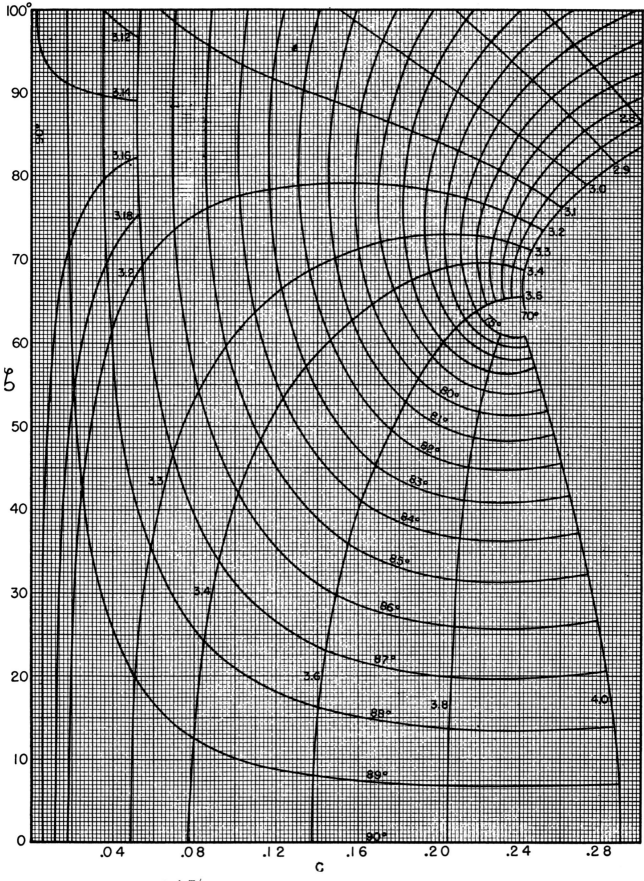

Chart IIIA. $\dfrac{\tanh T/\tau}{T/\tau} = C\underline{/}-\zeta$; T, 2.3 to 4.0; τ, 70° to 90°; Part A, C, 0 to 0.3; ζ, 0° to 100°.

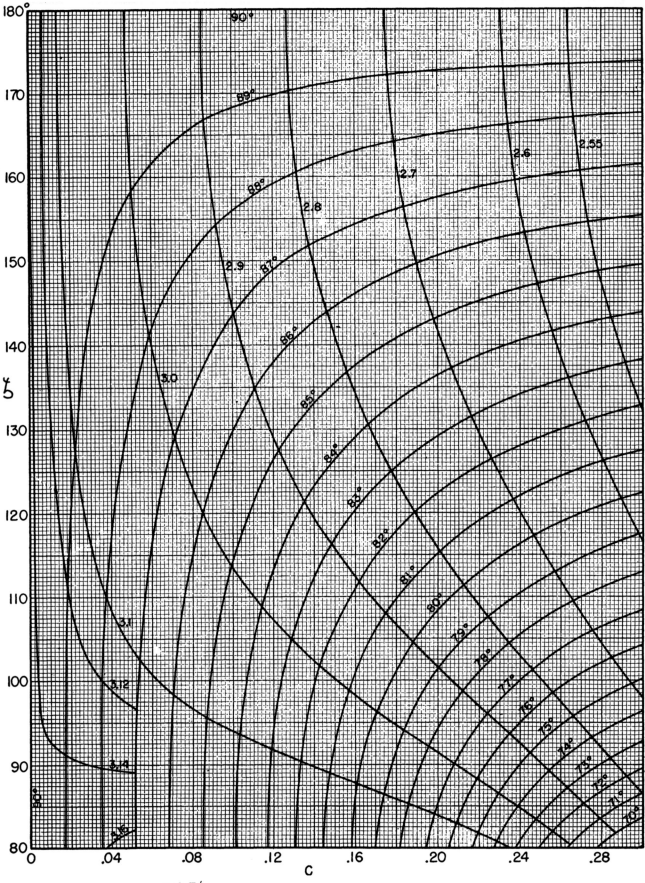

Chart IIIB. $\dfrac{\tanh T/\underline{\tau}}{T/\underline{\tau}} = C\underline{/-\zeta}$; T, 2.3 to 4.0; τ, 70° to 90°; Part B, C, 0 to 0.3; ζ, 80° to 180°.

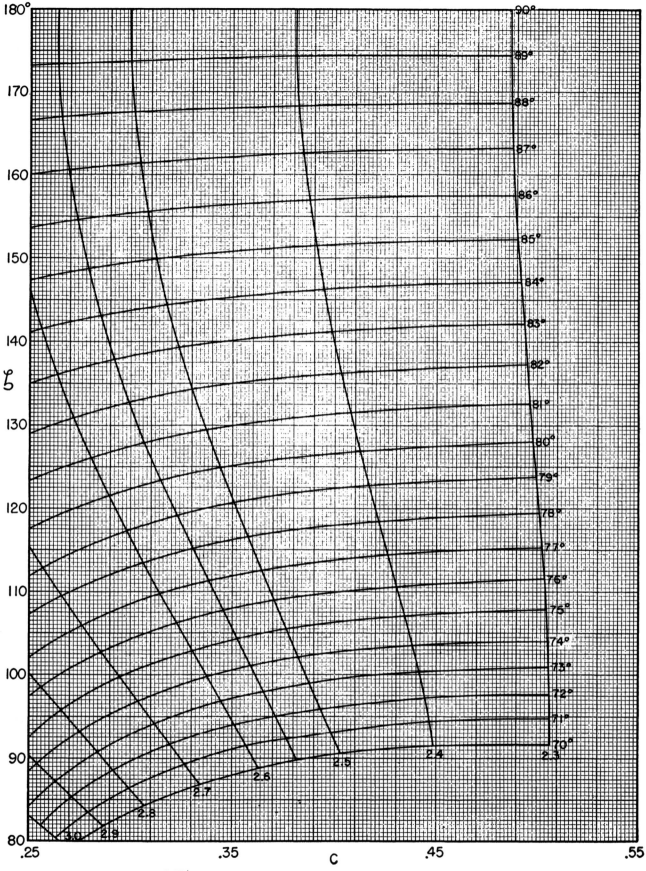

Chart IIIC. $\dfrac{\tanh T/\tau}{T/\tau} = C\underline{/-\zeta}$; T, 2.3 to 4.0; τ, 70° to 90°; Part C, C, 0.25 to 0.51; ζ, 80° to 180°.

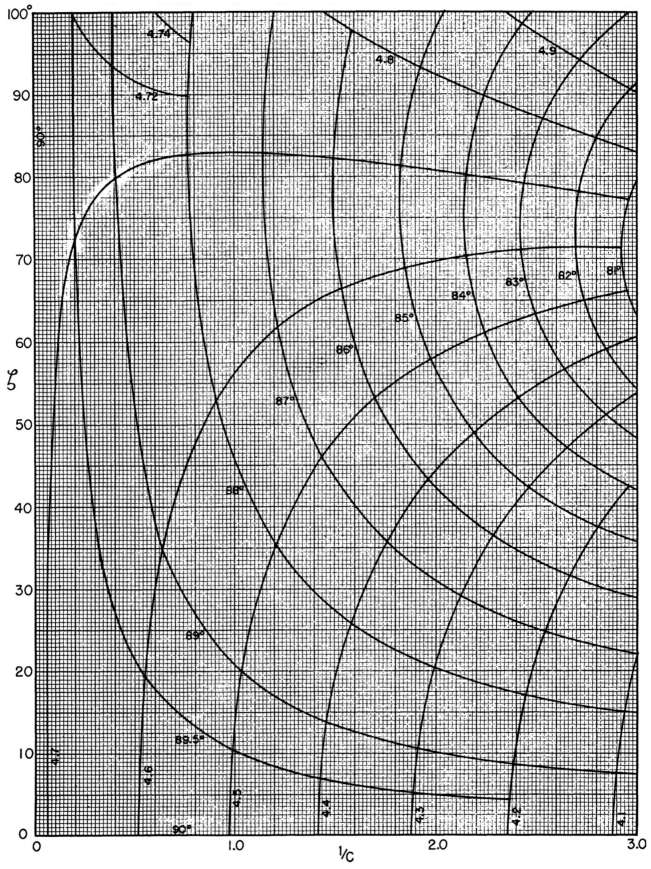

Chart IVA. $\dfrac{\tanh T/\tau}{T/\tau} = \dfrac{1}{C}\underline{/-\zeta}$; T, 4.0 to 5.4; τ, 70° to 90°; Part A, $1/C$, 0 to 3.0; ζ, 0° to 100°.

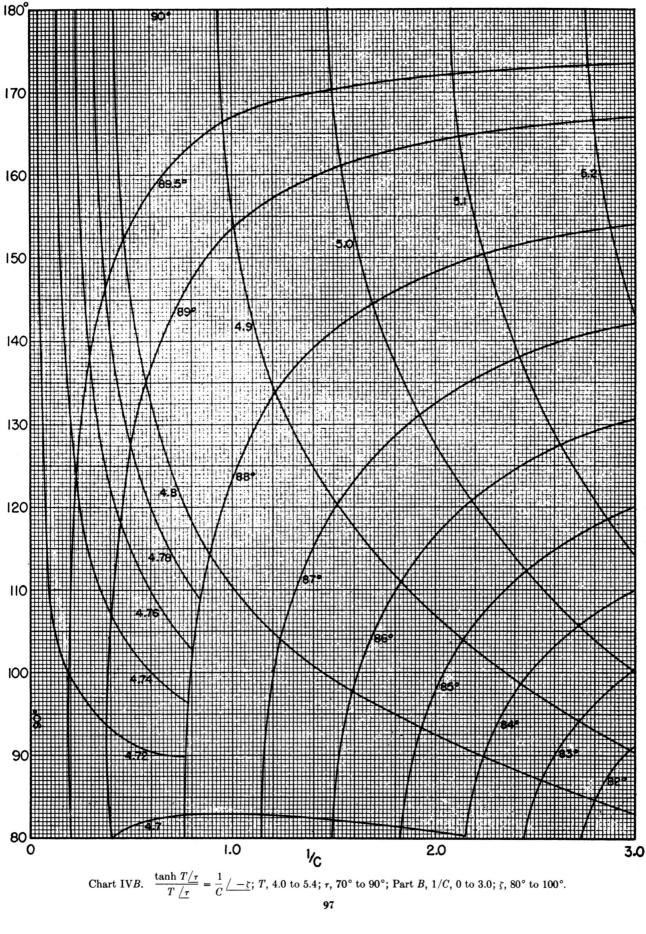

Chart IVB. $\dfrac{\tanh T\underline{/\tau}}{T\underline{/\tau}} = \dfrac{1}{C}\underline{/-\zeta}$; T, 4.0 to 5.4; τ, 70° to 90°; Part B, $1/C$, 0 to 3.0; ζ, 80° to 100°.

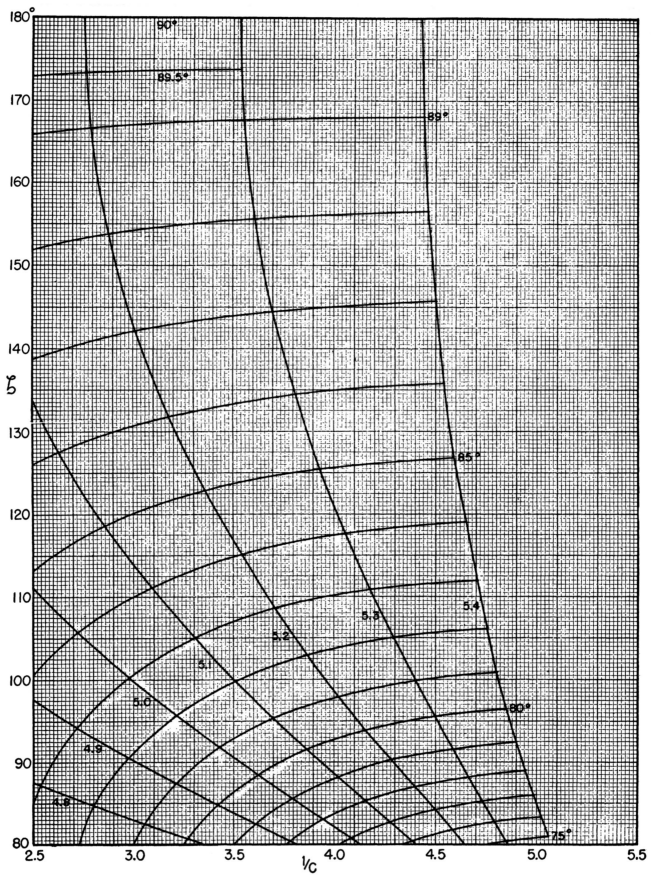

Chart IVC. $\dfrac{\tanh T/\tau}{T/\tau} = \dfrac{1}{C}\underline{/-\zeta}$; T, 4.0 to 5.4; τ, 70° to 90°; Part C, $1/C$, 2.5 to 5.3; ζ, 80° to 180°.

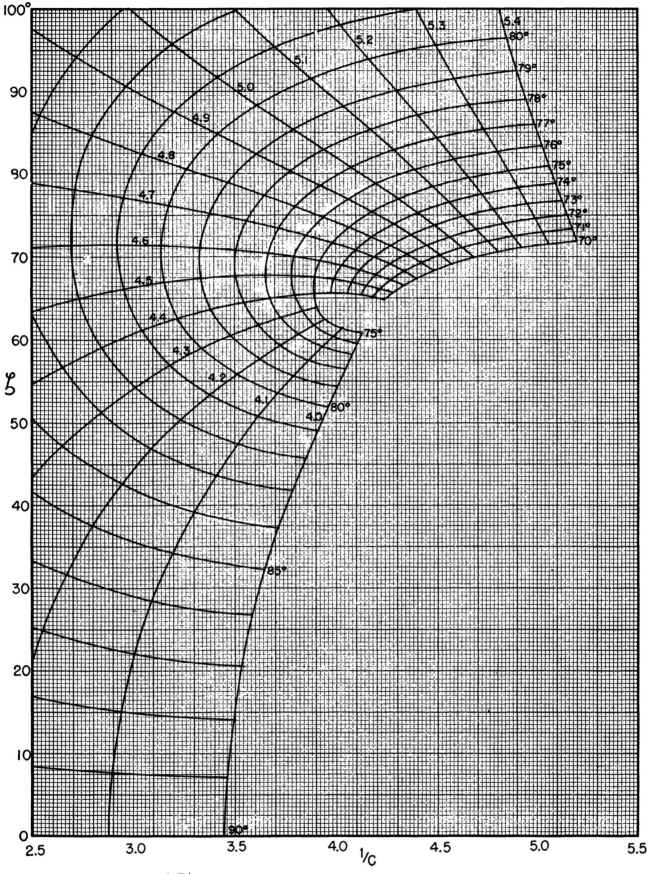

Chart IVD. $\dfrac{\tanh T/\tau}{T/\tau} = \dfrac{1}{C}\,\underline{/-\zeta}$; T, 4.0 to 5.4; τ, 70° to 90°; Part D, $1/C$, 2.5 to 5.1; ζ, 0° to 100°.

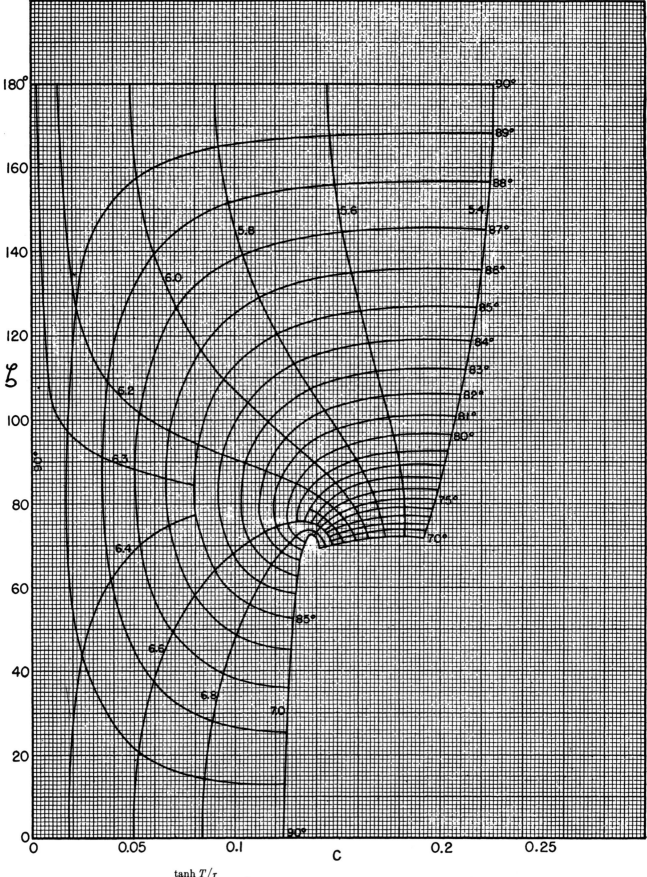

Chart V. $\dfrac{\tanh T/\tau}{T/\tau} = C\underline{/}-\zeta$; T, 5.4 to 7.0; τ, 70° to 90°; C, 0 to 0.23; ζ, 0° to 180°.

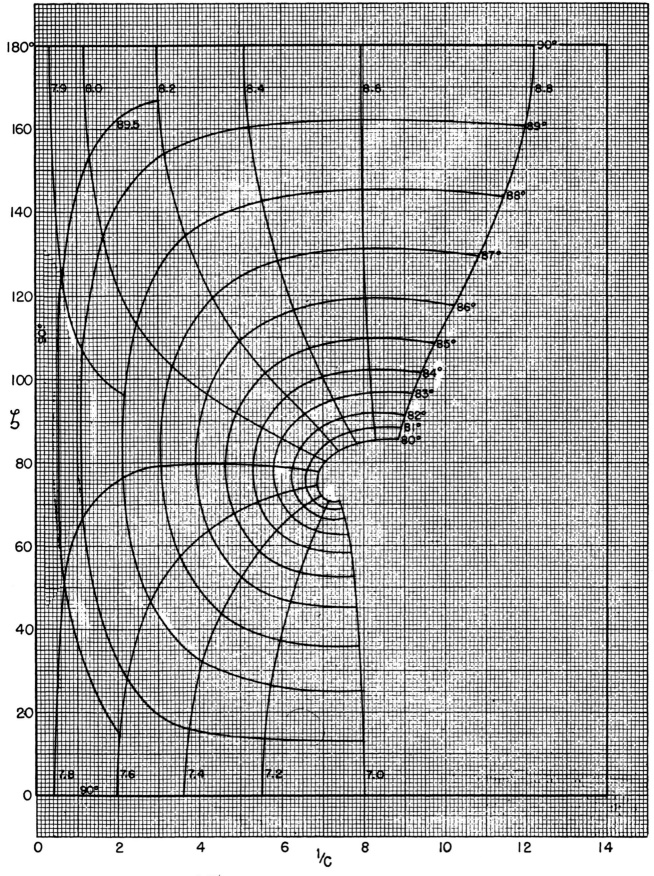

Chart VI. $\dfrac{\tanh T/\tau}{T/\tau} = \dfrac{1}{C}\,\underline{/-\zeta}$; T, 7.0 to 8.8; τ, 80° to 90°; $1/C$, 0 to 13; ζ, 0° to 180°.

V. Charts of the Complex Function
$$\frac{\coth x}{x}$$

These charts have been replotted since they were reported [25] and show values of the function

$$\frac{\coth T\underline{/\tau}}{T\underline{/\tau}} = C\left[\frac{1}{C}\right]\underline{/-\varsigma}.$$

[25] W. B. Westphal, Report IX, N.D.R.C. Contract OEMsr-191, Laboratory for Insulation Research, Massachusetts Institute of Technology, August, 1945.

The ranges are as follows:

Chart VII T, 0 to 0.8; τ, 50° to 90°.
Chart VIII T, 0.8 to 1.5; τ, 50° to 90°.

The charts of Lowan et al.[20] give x in rectangular form. The corresponding range of T is 0 to 10.

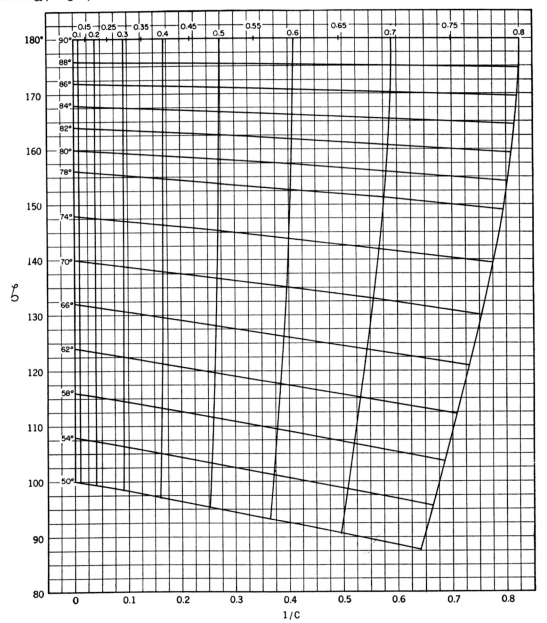

Chart VII. $\dfrac{\coth T\underline{/\tau}}{T\underline{/\tau}} = \dfrac{1}{C}\underline{/-\varsigma}$; T, 0 to 0.8; τ, 50° to 90°; $1/C$, 0 to 0.8; ς, 80° to 180°.

Permittivity, Distributed Circuits

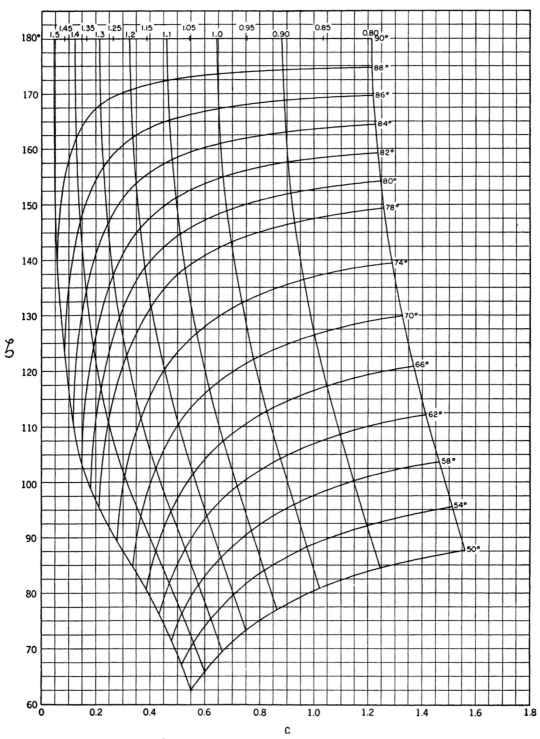

Chart VIII. $\dfrac{\coth T/\tau}{T/\tau} = C\underline{/-\zeta}$; T, 0.8 to 1.5; τ, 50° to 90°; C, 0 to 1.6; ζ, 60° to 180°.

VI. Tables of $\frac{\tan x}{x}$

The range of x is 0 to 26.9 radians in increments of 0.005 from 0 to 11.000 and in increments of 0.1 from 11.0 to 26.9.[26]

x	$\frac{\tan x}{x}$	x	$\frac{\tan x}{x}$	x	$\frac{\tan x}{x}$	x	$\frac{\tan x}{x}$	x	$\frac{\tan x}{x}$
0.000	1.0000	0.200	1.0136	0.400	1.0570	0.600	1.1402	0.800	1.2870
.005	1.0000	.205	1.0142	.405	1.0585	.605	1.1430	.805	1.2919
.010	1.0000	.210	1.0150	.410	1.0601	.610	1.1458	.810	1.2969
.015	1.0001	.215	1.0157	.415	1.0617	.615	1.1486	.815	1.3019
.020	1.0001	.220	1.0165	.420	1.0633	.620	1.1515	.820	1.3070
0.025	1.0002	0.225	1.0172	0.425	1.0649	0.625	1.1544	0.825	1.3121
.030	1.0003	.230	1.0180	.430	1.0666	.630	1.1573	.830	1.3174
.035	1.0004	.235	1.0188	.435	1.0682	.635	1.1603	.835	1.3227
.040	1.0005	.240	1.0197	.440	1.0700	.640	1.1634	.840	1.3281
.045	1.0007	.245	1.0205	.445	1.0717	.645	1.1664	.845	1.3336
0.050	1.0008	0.250	1.0214	0.450	1.0735	0.650	1.1695	0.850	1.3392
.055	1.0010	.255	1.0223	.455	1.0752	.655	1.1727	.855	1.3449
.060	1.0012	.260	1.0232	.460	1.0771	.660	1.1759	.860	1.3506
.065	1.0014	.265	1.0241	.465	1.0789	.665	1.1792	.865	1.3565
.070	1.0016	.270	1.0250	.470	1.0808	.670	1.1825	.870	1.3624
0.075	1.0019	0.275	1.0260	0.475	1.0827	0.675	1.1858	0.875	1.3685
.080	1.0021	.280	1.0270	.480	1.0846	.680	1.1892	.880	1.3746
.085	1.0024	.285	1.0280	.485	1.0866	.685	1.1926	.885	1.3809
.090	1.0027	.290	1.0290	.490	1.0885	.690	1.1961	.890	1.3872
.095	1.0030	.295	1.0301	.495	1.0906	.695	1.1997	.895	1.3936
0.100	1.0033	0.300	1.0311	0.500	1.0926	0.700	1.2033	0.900	1.4002
.105	1.0037	.305	1.0322	.505	1.0947	.705	1.2069	.905	1.4068
.110	1.0041	.310	1.0333	.510	1.0968	.710	1.2106	.910	1.4136
.115	1.0044	.315	1.0344	.515	1.0989	.715	1.2144	.915	1.4205
.120	1.0048	.320	1.0356	.520	1.1011	.720	1.2182	.920	1.4275
0.125	1.0052	0.325	1.0368	0.525	1.1033	0.725	1.2220	0.925	1.4346
.130	1.0057	.330	1.0380	.530	1.1055	.730	1.2259	.930	1.4418
.135	1.0061	.335	1.0392	.535	1.1078	.735	1.2299	.935	1.4492
.140	1.0066	.340	1.0404	.540	1.1101	.740	1.2339	.940	1.4566
.145	1.0071	.345	1.0417	.545	1.1124	.745	1.2380	.945	1.4642
0.150	1.0076	0.350	1.0429	0.550	1.1147	0.750	1.2421	0.950	1.4720
.155	1.0081	.355	1.0442	.555	1.1171	.755	1.2463	.955	1.4799
.160	1.0086	.360	1.0456	.560	1.1196	.760	1.2506	.960	1.4879
.165	1.0092	.365	1.0469	.565	1.1220	.765	1.2549	.965	1.4960
.170	1.0097	.370	1.0483	.570	1.1245	.770	1.2593	.970	1.5043
0.175	1.0103	0.375	1.0497	0.575	1.1270	0.775	1.2638	0.975	1.5128
.180	1.0109	.380	1.0511	.580	1.1296	.780	1.2683	.980	1.5214
.185	1.0116	.385	1.0525	.585	1.1322	.785	1.2729	.985	1.5301
.190	1.0122	.390	1.0540	.590	1.1348	.790	1.2775	.990	1.5391
.195	1.0129	.395	1.0555	.595	1.1375	.795	1.2823	.995	1.5482

[26] Similar tables ("Tables of $\frac{\tan x}{x}$ for Radian Measure," T. W. Dakin and M. Rutter, Res. Rep. R-9940-7-A, Westinghouse Research Laboratories, East Pittsburgh, Pa.) were published in 1945. The intervals in x in these tables are: 0.001 from 0 to 0.1 radian, 0.001 from 0.1 to 3.15 radians, 0.01 from 3.15 to 6.3 radians, and 0.1 from 6.3 to 10 radians.

$$\frac{\tan x}{x} \text{ (continued)}$$

x	$\frac{\tan x}{x}$	x	$\frac{\tan x}{x}$	x	$\frac{\tan x}{x}$	x	$\frac{\tan x}{x}$	x	$\frac{\tan x}{x}$
1.000	1.5574	1.200	2.1435	1.400	4.1413	1.600	-21.3953	1.800	-2.3813
.005	1.5668	.205	2.1666	.405	4.2535	.605	-18.2089	.805	-2.3221
.010	1.5764	.210	2.1904	.410	4.3726	.610	-15.8353	.810	-2.2655
.015	1.5862	.215	2.2148	.415	4.4994	.615	-13.9987	.815	-2.2111
.020	1.5962	.220	2.2400	.420	4.6346	.620	-12.5354	.820	-2.1590
1.025	1.6064	1.225	2.2659	1.425	4.7791	1.625	-11.3421	1.825	-2.1089
.030	1.6167	.230	2.2925	.430	4.9339	.630	-10.3504	.830	-2.0608
.035	1.6273	.235	2.3200	.435	5.1001	.635	- 9.5132	.835	-2.0144
.040	1.6381	.240	2.3483	.440	5.2790	.640	- 8.7970	.840	-1.9698
.045	1.6491	.245	2.3775	.445	5.4722	.645	- 8.1773	.845	-1.9269
1.050	1.6603	1.250	2.4077	1.450	5.6814	1.650	- 7.6359	1.850	-1.8854
.055	1.6717	.255	2.4387	.455	5.9087	.655	- 7.1588	.855	-1.8455
.060	1.6834	.260	2.4708	.460	6.1566	.660	- 6.7353	.860	-1.8069
.065	1.6953	.265	2.5040	.465	6.4279	.665	- 6.3567	.865	-1.7696
.070	1.7075	.270	2.5383	.470	6.7261	.670	- 6.0163	.870	-1.7336
1.075	1.7199	1.275	2.5737	1.475	7.0555	1.675	- 5.7086	1.875	-1.6988
.080	1.7326	.280	2.6104	.480	7.4212	.680	- 5.4290	.880	-1.6651
.085	1.7456	.285	2.6484	.485	7.8296	.685	- 5.2174	.885	-1.6325
.090	1.7588	.290	2.6878	.490	8.2885	.690	- 4.9404	.890	-1.6009
.095	1.7723	.295	2.7285	.495	8.8080	.695	- 4.7256	.895	-1.5703
1.100	1.7861	1.300	2.7708	1.500	9.4009	1.700	- 4.5274	1.900	-1.5406
.105	1.8003	.305	2.8148	.505	10.0840	.705	- 4.3440	.905	-1.5118
.110	1.8147	.310	2.8604	.510	10.8795	.710	- 4.1738	.910	-1.4838
.115	1.8295	.315	2.9078	.515	11.8176	.715	- 4.0155	.915	-1.4567
.120	1.8446	.320	2.9571	.520	12.9405	.720	- 3.8677	.920	-1.4304
1.125	1.8601	1.325	3.0084	1.525	14.3086	1.725	- 3.7295	1.925	-1.4048
.130	1.8759	.330	3.0619	.530	16.0120	.730	- 3.6001	.930	-1.3799
.135	1.8921	.335	3.1176	.535	18.1915	.735	- 3.4785	.935	-1.3557
.140	1.9087	.340	3.1758	.540	21.0787	.740	- 3.3641	.940	-1.3321
.145	1.9256	.345	3.2366	.545	25.0852	.745	- 3.2563	.945	-1.3092
1.150	1.9430	1.350	3.3002	1.550	31.0184	1.750	- 3.1545	1.950	-1.2869
.155	1.9609	.355	3.3667	.555	40.7078	.755	- 3.0583	.955	-1.2652
.160	1.9791	.360	3.4364	.560	59.3721	.760	- 2.9671	.960	-1.2440
.165	1.9979	.365	3.5094	.565	110.2371	.765	- 2.8806	.965	-1.2234
.170	2.0171	.370	3.5862	.570	799.8507	.770	- 2.7985	.970	-1.2033
1.175	2.0368	1.375	3.6668	1.575	-151.0386	1.775	- 2.7205	1.975	-1.1837
.180	2.0570	.380	3.7518	.580	- 68.7653	.780	- 2.6461	.980	-1.1646
.185	2.0778	.385	3.8413	.585	- 44.4161	.785	- 2.5753	.985	-1.1459
.190	2.0991	.390	3.9357	.590	- 32.7465	.790	- 2.5076	.990	-1.1277
.195	2.1210	.395	4.0356	.595	- 25.8984	.795	- 2.4430	.995	-1.1099

$$\frac{\tan x}{x} \text{ (continued)}$$

x	$\frac{\tan x}{x}$	x	$\frac{\tan x}{x}$	x	$\frac{\tan x}{x}$	x	$\frac{\tan x}{x}$	x	$\frac{\tan x}{x}$
2.000	-1.0925	2.200	- .6245	2.400	- .3817	2.600	- .2314	2.800	- .1270
.005	-1.076	.205	- .6165	.405	- .3771	.605	- .2283	.805	- .1247
.010	-1.059	.210	- .6087	.410	- .3725	.610	- .2253	.810	- .1225
.015	-1.043	.215	- .6011	.415	- .3680	.615	- .2223	.815	- .1203
.020	-1.027	.220	- .5935	.420	- .3636	.620	- .2193	.820	- .1181
2.025	-1.011	2.225	- .5861	2.425	- .3592	2.625	- .2164	2.825	- .1160
.030	- .9963	.230	- .5787	.430	- .3549	.630	- .2135	.830	- .1138
.035	- .9814	.235	- .5715	.435	- .3506	.635	- .2106	.835	- .1117
.040	- .9669	.240	- .5644	.440	- .3463	.640	- .2077	.840	- .1098
.045	- .9527	.245	- .5574	.445	- .3421	.645	- .2049	.845	- .1074
2.050	- .9388	2.250	- .5505	2.450	- .3380	2.650	- .2021	2.850	- .1053
.055	- .9252	.255	- .5437	.455	- .3339	.655	- .1993	.855	- .1032
.060	- .9118	.260	- .5370	.460	- .3298	.660	- .1965	.860	- .1011
.065	- .8988	.265	- .5304	.465	- .3258	.665	- .1937	.865	- .09908
.070	- .8860	.270	- .5239	.470	- .3218	.670	- .1910	.870	- .09703
2.075	- .8734	2.275	- .5174	2.475	- .3179	2.675	- .1883	2.875	- .09499
.080	- .8611	.280	- .5111	.480	- .3140	.680	- .1856	.880	- .09296
.085	- .8490	.285	- .5048	.485	- .3101	.685	- .1829	.885	- .09094
.090	- .8372	.290	- .4987	.490	- .3063	.690	- .1803	.890	- .08894
.095	- .8256	.295	- .4926	.495	- .3025	.695	- .1777	.895	- .08695
2.100	- .8142	2.300	- .4866	2.500	- .2988	2.700	- .1751	2.900	- .08497
.105	- .8030	.305	- .4807	.505	- .2951	.705	- .1725	.905	- .08300
.110	- .7921	.310	- .4749	.510	- .2915	.710	- .1699	.910	- .08104
.115	- .7813	.315	- .4691	.515	- .2878	.715	- .1674	.915	- .07909
.120	- .7707	.320	- .4634	.520	- .2843	.720	- .1649	.920	- .07715
2.125	- .7604	2.325	- .4578	2.525	- .2807	2.725	- .1624	2.925	- .07523
.130	- .7502	.330	- .4523	.530	- .2772	.730	- .1599	.930	- .07331
.135	- .7401	.335	- .4468	.535	- .2737	.735	- .1574	.935	- .07141
.140	- .7303	.340	- .4414	.540	- .2703	.740	- .1550	.940	- .06951
.145	- .7206	.345	- .4361	.545	- .2669	.745	- .1526	.945	- .06763
2.150	- .7111	2.350	- .4308	2.550	- .2635	2.750	- .1502	2.950	- .06575
.155	- .7018	.355	- .4256	.555	- .2601	.755	- .1478	.955	- .06389
.160	- .6926	.360	- .4205	.560	- .2568	.760	- .1454	.960	- .06203
.165	- .6836	.365	- .4154	.565	- .2535	.765	- .1430	.965	- .06019
.170	- .6747	.370	- .4104	.570	- .2503	.770	- .1407	.970	- .05835
2.175	- .6660	2.375	- .4055	2.575	- .2471	2.775	- .1384	2.975	- .05652
.180	- .6574	.380	- .4006	.580	- .2439	.780	- .1361	.980	- .05470
.185	- .6490	.385	- .3958	.585	- .2407	.785	- .1338	.985	- .05289
.190	- .6407	.390	- .3910	.590	- .2376	.790	- .1315	.990	- .05109
.195	- .6325	.395	- .3863	.595	- .2345	.795	- .1292	.995	- .04930

$$\frac{\tan x}{x} \text{ (continued)}$$

x	$\frac{\tan x}{x}$	x	$\frac{\tan x}{x}$	x	$\frac{\tan x}{x}$	x	$\frac{\tan x}{x}$	x	$\frac{\tan x}{x}$
3.000	− .04751	3.200	.01827	3.400	.07774	3.600	.1371	3.800	.2036
.005	− .04574	.205	.01933	.405	.07920	.605	.1386	.805	.2054
.010	− .04397	.210	.02134	.410	.08066	.610	.1402	.810	.2073
.015	− .04221	.215	.02288	.415	.08212	.615	.1417	.815	.2091
.020	− .04046	.220	.02440	.420	.08358	.620	.1433	.820	.2110
3.025	− .03872	3.225	.02592	3.425	.08504	3.625	.1448	3.825	.2129
.030	− .03698	.230	.02744	.430	.08650	.630	.1464	.830	.2148
.035	− .03525	.235	.02896	.435	.08796	.635	.1479	.835	.2167
.040	− .03353	.240	.03047	.440	.08942	.640	.1495	.840	.2186
.045	− .03182	.245	.03198	.445	.09088	.645	.1511	.845	.2206
3.050	− .03011	3.250	.03349	3.450	.09234	3.650	.1527	3.850	.2225
.055	− .02841	.255	.03499	.455	.09380	.655	.1543	.855	.2245
.060	− .02672	.260	.03649	.460	.09527	.660	.1559	.860	.2265
.065	− .02504	.265	.03799	.465	.09673	.665	.1575	.865	.2285
.070	− .02336	.270	.03949	.470	.09820	.670	.1591	.870	.2305
3.075	− .02169	3.275	.04098	3.475	.09967	3.675	.1607	3.875	.2325
.080	− .02002	.280	.04247	.480	.1011	.680	.1623	.880	.2346
.085	− .01836	.285	.04396	.485	.1026	.685	.1639	.885	.2366
.090	− .01671	.290	.04544	.490	.1041	.690	.1656	.890	.2387
.095	− .01506	.295	.04693	.495	.1056	.695	.1672	.895	.2408
3.100	− .01342	3.300	.04841	3.500	.1070	3.700	.1688	3.900	.2429
.105	− .01179	.305	.04989	.505	.1085	.705	.1705	.905	.2451
.110	− .01016	.310	.05137	.510	.1100	.710	.1722	.910	.2472
.115	− .008538	.315	.05284	.515	.1115	.715	.1738	.915	.2494
.120	− .006921	.320	.05432	.520	.1129	.720	.1755	.920	.2516
3.125	− .005309	3.325	.05579	3.525	.1144	3.725	.1772	3.925	.2538
.130	− .003703	.330	.05726	.530	.1159	.730	.1789	.930	.2560
.135	− .002102	.335	.05873	.535	.1174	.735	.1806	.935	.2582
.140	− .000506	.340	.06020	.540	.1189	.740	.1823	.940	.2605
.145	.001084	.345	.06166	.545	.1204	.745	.1840	.945	.2628
3.150	.002670	3.350	.06313	3.550	.1219	3.750	.1857	3.950	.2651
.155	.004251	.355	.06459	.555	.1234	.755	.1875	.955	.2674
.160	.005827	.360	.06606	.560	.1249	.760	.1892	.960	.2698
.165	.007398	.365	.06752	.565	.1264	.765	.1910	.965	.2721
.170	.008965	.370	.06898	.570	.1279	.770	.1928	.970	.2745
3.175	.01052	3.375	.07044	3.575	.1294	3.775	.1945	3.975	.2770
.180	.01208	.380	.07190	.580	.1310	.780	.1963	.980	.2794
.185	.01364	.385	.07336	.585	.1325	.785	.1981	.985	.2819
.190	.01519	.390	.07482	.590	.1340	.790	.1999	.990	.2844
.195	.01673	.395	.07628	.595	.1355	.795	.2017	.995	.2869

$$\frac{\tan x}{x} \text{ (continued)}$$

x	$\frac{\tan x}{x}$	x	$\frac{\tan x}{x}$	x	$\frac{\tan x}{x}$	x	$\frac{\tan x}{x}$	x	$\frac{\tan x}{x}$
4.000	.2895	4.200	.4233	4.400	.7037	4.600	1.926	4.800	-2.372
.005	.2920	.205	.4278	.405	.7151	.605	2.014	.805	-2.241
.010	.2946	.210	.4323	.410	.7269	.610	2.111	.810	-2.123
.015	.2973	.215	.4370	.415	.7391	.615	2.218	.815	-2.017
.020	.2999	.220	.4417	.420	.7516	.620	2.336	.820	-1.920
4.025	.3026	4.225	.4465	4.425	.7646	4.625	2.468	4.825	-1.833
.030	.3054	.230	.4515	.430	.7780	.630	2.616	.830	-1.752
.035	.3081	.235	.4565	.435	.7919	.635	2.782	.835	-1.678
.040	.3109	.240	.4616	.440	.8063	.640	2.972	.840	-1.610
.045	.3137	.245	.4668	.445	.8212	.645	3.190	.845	-1.547
4.050	.3166	4.250	.4721	4.450	.8367	4.650	3.443	4.850	-1.489
.055	.3195	.255	.4775	.455	.8528	.655	3.739	.855	-1.434
.060	.3224	.260	.4830	.460	.8694	.660	4.093	.860	-1.384
.065	.3254	.265	.4886	.465	.8868	.665	4.520	.865	-1.336
.070	.3284	.270	.4944	.470	.9048	.670	5.049	.870	-1.292
4.075	.3314	4.275	.5003	4.475	.9236	4.675	5.719	4.875	-1.250
.080	.3345	.280	.5063	.480	.9432	.680	6.595	.880	-1.211
.085	.3376	.285	.5124	.485	.9636	.685	7.791	.885	-1.174
.090	.3408	.290	.5186	.490	.9849	.690	9.522	.890	-1.139
.095	.3440	.295	.5251	.495	1.007	.695	12.25	.895	-1.106
4.100	.3472	4.300	.5316	4.500	1.031	4.700	17.17	4.900	-1.075
.105	.3505	.305	.5383	.505	1.055	.705	28.76	.905	-1.045
.110	.3538	.310	.5451	.510	1.081	.710	88.87	.910	-1.017
.115	.3572	.315	.5522	.515	1.110	.715	-81.23	.915	- .9904
.120	.3606	.320	.5593	.520	1.136	.720	-27.84	.920	- .9649
4.125	.3641	4.325	.5667	4.525	1.166	4.725	-16.78	4.925	- .9406
.130	.3677	.330	.5742	.530	1.197	.730	-12.00	.930	- .9173
.135	.3712	.335	.5820	.535	1.230	.735	- 9.337	.935	- .8952
.140	.3749	.340	.5899	.540	1.265	.740	- 7.637	.940	- .8739
.145	.3787	.345	.5980	.545	1.302	.745	- 6.459	.945	- .8536
4.150	.3823	4.350	.6063	4.550	1.342	4.750	- 5.594	4.950	- .8341
.155	.3861	.355	.6149	.555	1.383	.755	- 4.932	.955	- .8154
.160	.3900	.360	.6237	.560	1.428	.760	- 4.409	.960	- .7975
.165	.3939	.365	.6327	.565	1.476	.765	- 3.985	.965	- .7803
.170	.3979	.370	.6420	.570	1.526	.770	- 3.635	.970	- .7637
4.175	.4020	4.375	.6516	4.575	1.581	4.775	- 3.340	4.975	- .7477
.180	.4061	.380	.6614	.580	1.640	.780	- 3.089	.980	- .7323
.185	.4103	.385	.6715	.585	1.703	.785	- 2.873	.985	- .7175
.190	.4145	.390	.6819	.590	1.771	.790	- 2.684	.990	- .7032
.195	.4189	.395	.6927	.595	1.845	.795	- 2.519	.995	- .6894

$$\frac{\tan x}{x} \text{ (continued)}$$

x	$\dfrac{\tan x}{x}$	x	$\dfrac{\tan x}{x}$	x	$\dfrac{\tan x}{x}$	x	$\dfrac{\tan x}{x}$	x	$\dfrac{\tan x}{x}$
5.000	- .6761	5.200	- .3626	5.400	- .2255	5.600	- .1453	5.800	- .09046
.005	- .6632	.205	- .3579	.405	- .2230	.605	- .1437	.805	- .08929
.010	- .6507	.210	- .3533	.410	- .2205	.610	- .1421	.810	- .08812
.015	- .6387	.215	- .3488	.415	- .2181	.615	- .1406	.815	- .08696
.020	- .6270	.220	- .3444	.420	- .2157	.620	- .1390	.820	- .08581
5.025	- .6157	5.225	- .3401	5.425	- .2133	5.625	- .1375	5.825	- .08467
.030	- .6047	.230	- .3358	.430	- .2110	.630	- .1359	.830	- .08353
.035	- .5941	.235	- .3316	.435	- .2087	.635	- .1344	.835	- .08240
.040	- .5838	.240	- .3275	.440	- .2064	.640	- .1329	.840	- .08128
.045	- .5738	.245	- .3235	.445	- .2041	.645	- .1314	.845	- .08016
5.050	- .5641	5.250	- .3195	5.450	- .2019	5.650	- .1299	5.850	- .07906
.055	- .5546	.255	- .3156	.455	- .2001	.655	- .1284	.855	- .07795
.060	- .5454	.260	- .3117	.460	- .1975	.660	- .1270	.860	- .07686
.065	- .5365	.265	- .3080	.465	- .1954	.665	- .1255	.865	- .07577
.070	- .5278	.270	- .3043	.470	- .1933	.670	- .1241	.870	- .07469
5.075	- .5194	5.275	- .3006	5.475	- .1912	5.675	- .1227	5.875	- .07361
.080	- .5111	.280	- .2970	.480	- .1891	.680	- .1213	.880	- .07254
.085	- .5031	.285	- .2935	.485	- .1870	.685	- .1199	.885	- .07148
.090	- .4953	.290	- .2900	.490	- .1850	.690	- .1185	.890	- .07042
.095	- .4877	.295	- .2866	.495	- .1830	.695	- .1171	.895	- .06937
5.100	- .4803	5.300	- .2833	5.500	- .1810	5.700	- .1157	5.900	- .06832
.105	- .4730	.305	- .2799	.505	- .1790	.705	- .1144	.905	- .06728
.110	- .4660	.310	- .2767	.510	- .1771	.710	- .1130	.910	- .06625
.115	- .4591	.315	- .2735	.515	- .1752	.715	- .1117	.915	- .06522
.120	- .4523	.320	- .2703	.520	- .1733	.720	- .1104	.920	- .06420
5.125	- .4457	5.325	- .2672	5.525	- .1714	5.725	- .1091	5.925	- .06318
.130	- .4393	.330	- .2641	.530	- .1695	.730	- .1078	.930	- .06216
.135	- .4330	.335	- .2611	.535	- .1677	.735	- .1065	.935	- .06116
.140	- .4269	.340	- .2581	.540	- .1659	.740	- .1052	.940	- .06015
.145	- .4209	.345	- .2552	.545	- .1641	.745	- .1039	.945	- .05916
5.150	- .4150	5.350	- .2523	5.550	- .1623	5.750	- .1026	5.950	- .05817
.155	- .4093	.355	- .2494	.555	- .1605	.755	- .1014	.955	- .05718
.160	- .4036	.360	- .2466	.560	- .1588	.760	- .1001	.960	- .05620
.165	- .3981	.365	- .2439	.565	- .1570	.765	- .09890	.965	- .05522
.170	- .3928	.370	- .2411	.570	- .1553	.770	- .09767	.970	- .05424
5.175	- .3875	5.375	- .2384	5.575	- .1536	5.775	- .09645	5.975	- .05328
.180	- .3823	.380	- .2358	.580	- .1519	.780	- .09523	.980	- .05231
.185	- .3772	.385	- .2331	.585	- .1503	.785	- .09403	.985	- .05135
.190	- .3723	.390	- .2305	.590	- .1486	.790	- .09283	.990	- .05040
.195	- .3674	.395	- .2280	.595	- .1470	.795	- .09164	.995	- .04945

$$\frac{\tan x}{x} \text{ (continued)}$$

x	$\frac{\tan x}{x}$	x	$\frac{\tan x}{x}$	x	$\frac{\tan x}{x}$	x	$\frac{\tan x}{x}$	x	$\frac{\tan x}{x}$
6.000	- .04850	6.200	- .01345	6.400	.01834	6.600	.04968	6.800	.08358
.005	- .04756	.205	- .01263	.405	.01911	.605	.05048	.805	.08449
.010	- .04662	.210	- .01181	.410	.01989	.610	.05128	.810	.08541
.015	- .04569	.215	- .01099	.415	.02067	.615	.05209	.815	.08633
.020	- .04476	.220	- .01017	.420	.02145	.620	.05289	.820	.08726
6.025	- .04383	6.225	- .009357	6.425	.02222	6.625	.05370	6.825	.08819
.030	- .04291	.230	- .008544	.430	.02299	.630	.05451	.830	.08913
.035	- .04199	.235	- .007733	.435	.02378	.635	.05533	.835	.09007
.040	- .04107	.240	- .006924	.440	.02455	.640	.05614	.840	.09101
.045	- .04016	.245	- .006117	.445	.02533	.645	.05696	.845	.09196
6.050	- .03926	6.250	- .005311	6.450	.02611	6.650	.05778	6.850	.09292
.055	- .03835	.255	- .004506	.455	.02688	.655	.05860	.855	.09388
.060	- .03745	.260	- .003704	.460	.02766	.660	.05942	.860	.09484
.065	- .03656	.265	- .002902	.465	.02844	.665	.06024	.865	.09582
.070	- .03566	.270	- .002102	.470	.02922	.670	.06107	.870	.09679
6.075	- .03477	6.275	- .001304	6.475	.02999	6.675	.06190	6.875	.09777
.080	- .03389	.280	- .0005064	.480	.03077	.680	.06273	.880	.09876
.085	- .03300	.285	.0002896	.485	.03155	.685	.06357	.885	.09975
.090	- .03212	.290	.001084	.490	.03233	.690	.06440	.890	.1008
.095	- .03124	.295	.001878	.495	.03311	.695	.06524	.895	.1018
6.100	- .03037	6.300	.002670	6.500	.03389	6.700	.06608	6.900	.1028
.105	- .02950	.305	.003461	.505	.03467	.705	.06693	.905	.1038
.110	- .02863	.310	.004251	.510	.03545	.710	.06778	.910	.1048
.115	- .02777	.315	.005040	.515	.03623	.715	.06863	.915	.1058
.120	- .02690	.320	.005829	.520	.03702	.720	.06948	.920	.1069
6.125	- .02604	6.325	.006616	6.525	.03780	6.725	.07034	6.925	.1079
.130	- .02518	.330	.007402	.530	.03858	.730	.07119	.930	.1090
.135	- .02433	.335	.008187	.535	.03937	.735	.07206	.935	.1100
.140	- .02348	.340	.008972	.540	.04016	.740	.07292	.940	.1111
.145	- .02263	.345	.009756	.545	.04094	.745	.07379	.945	.1122
6.150	- .02178	6.350	.01054	6.550	.04173	6.750	.07466	6.950	.1133
.155	- .02094	.355	.01132	.555	.04252	.755	.07554	.955	.1143
.160	- .02010	.360	.01210	.560	.04331	.760	.07642	.960	.1154
.165	- .01926	.365	.01288	.565	.04410	.765	.07730	.965	.1165
.170	- .01842	.370	.01366	.570	.04489	.770	.07818	.970	.1176
6.175	- .01759	6.375	.01444	6.575	.04569	6.775	.07907	6.975	.1188
.180	- .01676	.380	.01522	.580	.04648	.780	.07997	.980	.1199
.185	- .01593	.385	.01600	.585	.04728	.785	.08086	.985	.1210
.190	- .01510	.390	.01678	.590	.04808	.790	.08177	.990	.1222
.195	- .01434	.395	.01756	.595	.04888	.795	.08267	.995	.1233

$$\frac{\tan x}{x} \text{ (continued)}$$

x	$\frac{\tan x}{x}$	x	$\frac{\tan x}{x}$	x	$\frac{\tan x}{x}$	x	$\frac{\tan x}{x}$	x	$\frac{\tan x}{x}$
7.000	.1245	7.200	.1812	7.400	.2769	7.600	.5069	7.800	2.373
.005	.1257	.205	.1830	.405	.2803	.605	.5172	.805	2.614
.010	.1268	.210	.1847	.410	.2837	.610	.5279	.810	2.910
.015	.1280	.215	.1866	.415	.2872	.615	.5390	.815	3.281
.020	.1292	.220	.1884	.420	.2908	.620	.5506	.820	3.762
7.025	.1305	7.225	.1902	7.425	.2945	7.625	.5627	7.825	4.410
.030	.1317	.230	.1921	.430	.2982	.630	.5753	.830	5.324
.035	.1329	.235	.1916	.435	.3020	.635	.5885	.835	6.724
.040	.1342	.240	.1960	.440	.3059	.640	.6023	.840	9.123
.045	.1354	.245	.1979	.445	.3099	.645	.6168	.845	14.19
7.050	.1367	7.250	.1999	7.450	.3140	7.650	.6319	7.850	31.99
.055	.1379	.255	.2019	.455	.3182	.655	.6478	.855	-125.0
.060	.1392	.260	.2040	.460	.3225	.660	.6645	.860	- 21.14
.065	.1405	.265	.2060	.465	.3268	.665	.6821	.865	- 11.54
.070	.1418	.270	.2081	.470	.3313	.670	.7007	.870	- 7.932
7.075	.1432	7.275	.2103	7.475	.3359	7.675	.7202	7.875	- 6.041
.080	.1445	.280	.2124	.480	.3407	.680	.7409	.880	- 4.876
.085	.1459	.285	.2146	.485	.3455	.685	.7627	.885	- 4.087
.090	.1472	.290	.2169	.490	.3505	.690	.7859	.890	- 3.517
.095	.1486	.295	.2191	.495	.3556	.695	.8105	.895	- 3.086
7.100	.1500	7.300	.2214	7.500	.3608	7.700	.8368	7.900	- 2.749
.105	.1514	.305	.2238	.505	.3662	.705	.8647	.905	- 2.478
.110	.1528	.310	.2262	.510	.3717	.710	.8946	.910	- 2.255
.115	.1542	.315	.2286	.515	.3774	.715	.9266	.915	- 2.068
.120	.1557	.320	.2310	.520	.3833	.720	.9610	.920	- 1.910
7.125	.1572	7.325	.2335	7.525	.3893	7.725	.9981	7.925	- 1.774
.130	.1586	.330	.2361	.530	.3955	.730	1.038	.930	- 1.656
.135	.1601	.335	.2387	.535	.4019	.735	1.082	.935	- 1.552
.140	.1616	.340	.2413	.540	.4084	.740	1.129	.940	- 1.461
.145	.1632	.345	.2440	.545	.4152	.745	1.180	.945	- 1.379
7.150	.1647	7.350	.2467	7.550	.4222	7.750	1.237	7.950	- 1.306
.155	.1663	.355	.2495	.555	.4294	.755	1.300	.955	- 1.240
.160	.1679	.360	.2523	.560	.4369	.760	1.367	.960	- 1.181
.165	.1695	.365	.2552	.565	.4446	.765	1.444	.965	- 1.126
.170	.1711	.370	.2581	.570	.4526	.770	1.529	.970	- 1.077
7.175	.1727	7.375	.2611	7.575	.4609	7.775	1.625	7.975	- 1.031
.180	.1744	.380	.2641	.580	.4694	.780	1.734	.980	- .9892
.185	.1760	.385	.2672	.585	.4783	.785	1.859	.985	- .9504
.190	.1777	.390	.2704	.590	.4875	.790	2.004	.990	- .9145
.195	.1795	.395	.2736	.595	.4970	.795	2.173	.995	- .8811

$$\frac{\tan x}{x} \text{ (continued)}$$

x	$\frac{\tan x}{x}$	x	$\frac{\tan x}{x}$	x	$\frac{\tan x}{x}$	x	$\frac{\tan x}{x}$	x	$\frac{\tan x}{x}$
8.000	- .8500	8.200	- .3383	8.400	- .1959	8.600	- .1258	8.800	- .08195
.005	- .8209	.205	- .3328	.405	- .1936	.605	- .1245	.805	- .08104
.010	- .7937	.210	- .3275	.410	- .1913	.610	- .1232	.810	- .08014
.015	- .7682	.215	- .3224	.415	- .1891	.615	- .1219	.815	- .07925
.020	- .7441	.220	- .3174	.420	- .1869	.620	- .1206	.820	- .07836
8.025	- .7215	8.225	- .3125	8.425	- .1848	8.625	- .1193	8.825	- .07749
.030	- .7002	.230	- .3078	.430	- .1826	.630	- .1181	.830	- .07661
.035	- .6801	.235	- .3031	.435	- .1806	.635	- .1168	.835	- .07575
.040	- .6609	.240	- .2986	.440	- .1785	.640	- .1156	.840	- .07489
.045	- .6428	.245	- .2942	.445	- .1765	.645	- .1144	.845	- .07404
8.050	- .6256	8.250	- .2899	8.450	- .1745	8.650	- .1132	8.850	- .07319
.055	- .6093	.255	- .2857	.455	- .1725	.655	- .1120	.855	- .07235
.060	- .5937	.260	- .2816	.460	- .1706	.660	- .1108	.860	- .07152
.065	- .5788	.265	- .2776	.465	- .1687	.665	- .1096	.865	- .07069
.070	- .5647	.270	- .2737	.470	- .1668	.670	- .1085	.870	- .06986
8.075	- .5512	8.275	- .2699	8.475	- .1649	8.675	- .1073	8.875	- .06904
.080	- .5382	.280	- .2661	.480	- .1631	.680	- .1062	.880	- .06824
.085	- .5258	.285	- .2625	.485	- .1613	.685	- .1051	.885	- .06743
.090	- .5140	.290	- .2589	.490	- .1595	.690	- .1040	.890	- .06663
.095	- .5026	.295	- .2554	.495	- .1578	.695	- .1029	.895	- .06584
8.100	- .4917	8.300	- .2520	8.500	- .1560	8.700	- .1018	8.900	- .06505
.105	- .4812	.305	- .2486	.505	- .1543	.705	- .1007	.905	- .06426
.110	- .4711	.310	- .2453	.510	- .1527	.710	- .0996	.910	- .06348
.115	- .4613	.315	- .2421	.515	- .1510	.715	- .09858	.915	- .06271
.120	- .4520	.320	- .2390	.520	- .1494	.720	- .09753	.920	- .06194
8.125	- .4430	8.325	- .2359	8.525	- .1478	8.725	- .09649	8.925	- .06118
.130	- .4343	.330	- .2329	.530	- .1462	.730	- .09546	.930	- .06042
.135	- .4259	.335	- .2299	.535	- .1446	.735	- .09444	.935	- .05966
.140	- .4177	.340	- .2270	.540	- .1430	.740	- .09343	.940	- .05892
.145	- .4099	.345	- .2241	.545	- .1415	.745	- .09243	.945	- .05817
8.150	- .4023	8.350	- .2213	8.550	- .1400	8.750	- .09144	8.950	- .05743
.155	- .3950	.355	- .2186	.555	- .1385	.755	- .09045	.955	- .05669
.160	- .3879	.360	- .2159	.560	- .1370	.760	- .08947	.960	- .05596
.165	- .3810	.365	- .2132	.565	- .1355	.765	- .08851	.965	- .05523
.170	- .3743	.370	- .2106	.570	- .1341	.770	- .08755	.970	- .05451
8.175	- .3679	8.375	- .2081	8.575	- .1327	8.775	- .08659	8.975	- .05379
.180	- .3616	.380	- .2055	.580	- .1313	.780	- .08565	.980	- .05308
.185	- .3555	.385	- .2031	.585	- .1299	.785	- .08471	.985	- .05237
.190	- .3496	.390	- .2006	.590	- .1285	.790	- .08378	.990	- .05166
.195	- .3438	.395	- .1982	.595	- .1272	.795	- .08286	.995	- .05096

$$\frac{\tan x}{x} \text{ (continued)}$$

x	$\frac{\tan x}{x}$	x	$\frac{\tan x}{x}$	x	$\frac{\tan x}{x}$	x	$\frac{\tan x}{x}$	x	$\frac{\tan x}{x}$
9.000	- .05026	9.200	- .02485	9.400	- .002637	9.600	.01844	9.800	.04019
.005	- .04956	.205	- .02427	.405	- .002103	.605	.01897	.805	.04076
.010	- .04887	.210	- .02369	.410	- .001571	.610	.01950	.810	.04133
.015	- .04818	.215	- .02310	.415	- .001039	.615	.02003	.815	.04191
.020	- .04750	.220	- .02253	.420	- .0005074	.620	.02055	.820	.04248
9.025	- .04682	9.225	- .02195	9.425	.00002334	9.625	.02108	9.825	.04306
.030	- .04614	.230	- .02137	.430	.0005536	.630	.02161	.830	.04364
.035	- .04547	.235	- .02080	.435	.001083	.635	.02215	.835	.04422
.040	- .04480	.240	- .02023	.440	.001612	.640	.02268	.840	.04480
.045	- .04413	.245	- .01966	.445	.002141	.645	.02321	.845	.04539
9.050	- .04347	9.250	- .01909	9.450	.002669	9.650	.02374	9.850	.04597
.055	- .04281	.255	- .01852	.455	.003197	.655	.02428	.855	.04656
.060	- .04215	.260	- .01796	.460	.003725	.660	.02481	.860	.04716
.065	- .04149	.265	- .01739	.465	.004252	.665	.02534	.865	.04775
.070	- .04084	.270	- .01683	.470	.004778	.670	.02588	.870	.04835
9.075	- .04020	9.275	- .01627	9.475	.005305	9.675	.02642	9.875	.04894
.080	- .03955	.280	- .01571	.480	.005831	.680	.02695	.880	.04955
.085	- .03891	.285	- .01515	.485	.006357	.685	.02749	.885	.05015
.090	- .03827	.290	- .01460	.490	.006882	.690	.02803	.890	.05076
.095	- .03763	.295	- .01404	.495	.007408	.695	.02857	.895	.05136
9.100	- .03700	9.300	- .01349	9.500	.007933	9.700	.02911	9.900	.05197
.105	- .03637	.305	- .01293	.505	.008458	.705	.02965	.905	.05259
.110	- .03574	.310	- .01238	.510	.008983	.710	.03020	.910	.05321
.115	- .03512	.315	- .01183	.515	.009508	.715	.03074	.915	.05382
.120	- .03449	.320	- .01128	.520	.01003	.720	.03129	.920	.05445
9.125	- .03387	9.325	- .01074	9.525	.01056	9.725	.03183	9.925	.05507
.130	- .03326	.330	- .01019	.530	.01108	.730	.03238	.930	.05570
.135	- .03264	.335	- .009643	.535	.01161	.735	.03293	.935	.05633
.140	- .03203	.340	- .009099	.540	.01213	.740	.03348	.940	.05696
.145	- .03142	.345	- .008555	.545	.01266	.745	.03403	.945	.05760
9.150	- .03081	9.350	- .008013	9.550	.01318	9.750	.03458	9.950	.05824
.155	- .03020	.355	- .007471	.555	.01371	.755	.03514	.955	.05889
.160	- .02960	.360	- .006931	.560	.01423	.760	.03569	.960	.05953
.165	- .02900	.365	- .006391	.565	.01476	.765	.03625	.965	.06018
.170	- .02840	.370	- .005852	.570	.01528	.770	.03681	.970	.06084
9.175	- .02780	9.375	- .005314	9.575	.01581	9.775	.03737	9.975	.06149
.180	- .02721	.380	- .004777	.580	.01633	.780	.03793	.980	.06216
.185	- .02662	.385	- .004241	.585	.01686	.785	.03849	.985	.06282
.190	- .02603	.390	- .003705	.590	.01739	.790	.03906	.990	.06349
.195	- .02544	.395	- .003171	.595	.01791	.795	.03962	.995	.06416

$$\frac{\tan x}{x} \text{ (continued)}$$

x	$\frac{\tan x}{x}$	x	$\frac{\tan x}{x}$	x	$\frac{\tan x}{x}$	x	$\frac{\tan x}{x}$	x	$\frac{\tan x}{x}$
10.000	.06484	10.200	.09606	10.400	.1419	10.600	.2259	10.800	.4674
.005	.06552	.205	.09698	.405	.1434	.605	.2290	.805	.4797
.010	.06620	.210	.09791	.410	.1449	.610	.2322	.810	.4927
.015	.06689	.215	.09884	.415	.1464	.615	.2355	.815	.5065
.020	.06758	.220	.09979	.420	.1479	.620	.2388	.820	.5210
10.025	.06828	10.225	.1007	10.425	.1495	10.625	.2422	10.825	.5363
.030	.06898	.230	.1017	.430	.1510	.630	.2458	.830	.5526
.035	.06968	.235	.1027	.435	.1527	.635	.2494	.835	.5698
.040	.07039	.240	.1037	.440	.1543	.640	.2531	.840	.5882
.045	.07110	.245	.1047	.445	.1560	.645	.2569	.845	.6077
10.050	.07182	10.250	.1057	10.450	.1576	10.650	.2608	10.850	.6286
.055	.07255	.255	.1067	.455	.1594	.655	.2648	.855	.6510
.060	.07327	.260	.1077	.460	.1611	.660	.2690	.860	.6750
.065	.07401	.265	.1087	.465	.1629	.665	.2732	.865	.7009
.070	.07475	.270	.1098	.470	.1647	.670	.2776	.870	.7287
10.075	.07549	10.275	.1108	10.475	.1665	10.675	.2821	10.875	.7589
.080	.07624	.280	.1119	.480	.1684	.680	.2868	.880	.7917
.085	.07699	.285	.1130	.485	.1703	.685	.2916	.885	.8274
.090	.07775	.290	.1141	.490	.1722	.690	.2965	.890	.8665
.095	.07852	.295	.1152	.495	.1742	.695	.3016	.895	.9095
10.100	.07929	10.300	.1163	10.500	.1762	10.700	.3069	10.900	.9570
.105	.08006	.305	.1174	.505	.1782	.705	.3124	.905	1.010
.110	.08084	.310	.1186	.510	.1803	.710	.3180	.910	1.069
.115	.08163	.315	.1197	.515	.1824	.715	.3239	.915	1.135
.120	.08243	.320	.1209	.520	.1846	.720	.3299	.920	1.209
10.125	.08323	10.325	.1221	10.525	.1868	10.725	.3361	10.925	1.295
.130	.08403	.330	.1233	.530	.1890	.730	.3426	.930	1.393
.135	.08485	.335	.1245	.535	.1913	.735	.3494	.935	1.508
.140	.08567	.340	.1258	.540	.1936	.740	.3563	.940	1.643
.145	.08649	.345	.1270	.545	.1960	.745	.3636	.945	1.805
10.150	.08732	10.350	.1283	10.550	.1985	10.750	.3712	10.950	2.007
.155	.08816	.355	.1295	.555	.2009	.755	.3790	.955	2.248
.160	.08901	.360	.1308	.560	.2035	.760	.3872	.960	2.564
.165	.08987	.365	.1322	.565	.2061	.765	.3957	.965	2.982
.170	.09073	.370	.1335	.570	.2087	.770	.4046	.970	3.564
10.175	.09160	10.375	.1348	10.575	.2114	10.775	.4139	10.975	4.428
.180	.09247	.380	.1362	.580	.2142	.780	.4236	.980	5.848
.185	.09336	.385	.1376	.585	.2170	.785	.4338	.985	8.609
.190	.09425	.390	.1390	.590	.2199	.790	.4445	.990	16.32
.195	.09515	.395	.1405	.595	.2229	.795	.4556	.995	158.4

$$\frac{\tan x}{x} \text{ (continued)}$$

x	$\frac{\tan x}{x}$	x	$\frac{\tan x}{x}$	x	$\frac{\tan x}{x}$	x	$\frac{\tan x}{x}$
11.0	-20.66	15.0	- .05707	19.0	.00797	23.0	.06775
.1	- .8598	.1	- .04609	.1	.01339	.1	.08696
.2	- .4307	.2	- .03663	.2	.01904	.2	.1139
.3	- .2817	.3	- .02825	.3	.02506	.3	.1601
.4	- .2050	.4	- .02066	.4	.03163	.4	.2617
11.5	- .1575	15.5	- .01361	19.5	.03902	23.5	.6864
.6	- .1248	.6	- .00700	.6	.04757	.6	- 1.142
.7	- .1006	.7	- .00051	.7	.05783	.7	- .3035
.8	- .08158	.8	.005842	.8	.07069	.8	- .1732
.9	- .06608	.9	.01223	.9	.08768	.9	- .1190
12.0	- .05298	16.0	.01879	20.0	.1118	24.0	- .08894
.1	- .04160	.1	.02568	.1	.1500	.1	- .06952
.2	- .03145	.2	.03309	.2	.2209	.2	- .05572
.3	- .02218	.3	.04126	.3	.4072	.3	- .04524
.4	- .01355	.4	.05053	.4	2.403	.4	- .03688
12.5	- .00532	16.5	.06141	20.5	- .6112	24.5	- .02993
.6	.00267	.6	.07468	.6	- .2672	.6	- .02397
.7	.01058	.7	.09163	.7	- .1682	.7	- .01870
.8	.01859	.8	.1147	.8	- .1205	.8	- .01394
.9	.02687	.9	.1487	.9	- .09198	.9	- .00952
13.0	.03562	17.0	.2055	21.0	- .07273	25.0	- .00534
.1	.04510	.1	.3236	.1	- .05865	.1	- .00130
.2	.05566	.2	.7360	.2	- .04771	.2	.00267
.3	.06777	.3	-2.726	.3	- .03884	.3	.00668
.4	.08219	.4	- .4719	.4	- .03136	.4	.01078
13.5	.1001	17.5	- .2541	21.5	- .02487	25.5	.01509
.6	.1234	.6	- .1708	.6	- .01909	.6	.01971
.7	.1562	.7	- .1261	.7	- .01381	.7	.02479
.8	.2067	.8	- .09785	.8	- .008876	.8	.03053
.9	.2976	.9	- .07806	.9	- .00417	.9	.03724
14.0	.5174	18.0	- .06319	22.0	.00004	26.0	.04534
.1	1.906	.1	- .05143	.1	.00495	.1	.05559
.2	-1.120	.2	- .04174	.2	.00955	.2	.06929
.3	- .4258	.3	- .03347	.3	.01431	.3	.08906
.4	- .2581	.4	- .02623	.4	.01934	.4	.1209
14.5	- .1817	18.5	- .01971	22.5	.02480	26.5	.1829
.6	- .1373	.6	- .01370	.6	.03085	.6	.3620
.7	- .1078	.7	- .00806	.7	.03778	.7	10.70
.8	- .08655	.8	- .00264	.8	.04597	.8	- .3855
.9	- .07021	.9	.00267	.9	.05604	.9	- .1868

VII. Tables of $\frac{\cot x}{x}$

The range of x is 0 to 3.99 radians in increments of 0.005 from 0 to 0.800, and increments of 0.01 from 0.80 to 3.99 radians.

x	$\frac{\cot x}{x}$	x	$\frac{\cot x}{x}$	x	$\frac{\cot x}{x}$	x	$\frac{\cot x}{x}$	x	$\frac{\cot x}{x}$
0.000		0.250	15.66	0.500	3.661	0.750	1.431	1.200	.3240
.005	40000	.255	15.04	.505	3.582	.755	1.408	.210	.3118
.010	10000	.260	14.46	.510	3.505	.760	1.384	.220	.2999
.015	4444	.265	13.91	.515	3.431	.765	1.362	.230	.2883
.020	2500	.270	13.38	.520	3.359	.770	1.339	.240	.2769
0.025	1600	0.275	12.89	0.525	3.289	0.775	1.318	1.250	.2658
.030	1111	.280	12.42	.530	3.220	.780	1.296	.260	.2549
.035	816.0	.285	11.98	.535	3.154	.785	1.275	.270	.2443
.040	624.8	.290	11.56	.540	3.089	.790	1.254	.280	.2338
.045	493.6	.295	11.16	.545	3.027	.795	1.234	.290	.2236
0.050	399.6	0.300	10.78	0.550	2.965	0.800	1.214	1.300	.2135
.055	330.2	.305	10.41	.555	2.906	.810	1.175	.310	.2037
.060	277.5	.310	10.07	.560	2.848	.820	1.138	.320	.1941
.065	236.3	.315	9.743	.565	2.792	.830	1.102	.330	.1847
.070	203.7	.320	9.430	.570	2.737	.840	1.067	.340	.1754
0.075	177.5	0.325	9.132	0.575	2.684	0.850	1.034	1.350	.1663
.080	155.9	.330	8.847	.580	2.631	.860	1.001	.360	.1574
.085	138.1	.335	8.574	.585	2.581	.870	.9698	.370	.1485
.090	123.1	.340	8.315	.590	2.531	.880	.9394	.380	.1399
.095	110.4	.345	8.066	.595	2.483	.890	.9101	.390	.1315
0.100	99.67	0.350	7.827	0.600	2.436	0.900	.8818	1.400	.1232
.105	90.37	.355	7.599	.605	2.390	.910	.8543	.410	.1150
.110	82.31	.360	7.380	.610	2.346	.920	.8277	.420	.1070
.115	75.28	.365	7.170	.615	2.302	.930	.8019	.430	.09909
.120	69.11	.370	6.968	.620	2.259	.940	.7764	.440	.09132
0.125	63.66	0.375	6.775	0.625	2.218	0.950	.7527	1.450	.08372
.130	58.84	.380	6.589	.630	2.177	.960	.7293	.460	.07623
.135	54.53	.385	6.410	.635	2.137	.970	.7065	.470	.06878
.140	50.69	.390	6.238	.640	2.099	.980	.6844	.480	.06149
.145	47.23	.395	6.072	.645	2.061	.990	.6630	.490	.05436
0.150	44.11	0.400	5.913	0.650	2.024	1.000	.6421	1.500	.04726
.155	41.29	.405	5.735	.655	1.988	.010	.6158	.510	.04033
.160	38.73	.410	5.612	.660	1.952	.020	.6022	.520	.03342
.165	36.40	.415	5.469	.665	1.918	.030	.5830	.530	.02667
.170	34.27	.420	5.332	.670	1.884	.040	.5644	.540	.02000
0.175	32.32	0.425	5.199	0.675	1.851	1.050	.5463	1.550	.01342
.180	30.53	.430	5.071	.680	1.819	.060	.5287	.560	.00692
.185	28.89	.435	4.947	.685	1.787	.070	.5115	.570	.00051
.190	27.37	.440	4.828	.690	1.756	.080	.4948	.580	-.00582
.195	25.96	.445	4.712	.695	1.726	.090	.4785	.590	-.01208
0.200	24.67	0.450	4.600	0.700	1.696	1.100	.4627		
.205	23.46	.455	4.492	.705	1.667	.110	.4472		
.210	22.34	.460	4.388	.710	1.639	.120	.4321		
.215	21.30	.465	4.287	.715	1.611	.130	.4175		
.220	20.33	.470	4.189	.720	1.584	.140	.4032		
0.225	19.42	0.475	4.094	0.725	1.557	1.150	.3891		
.230	18.57	.480	4.002	.730	1.531	.160	.3755		
.235	17.77	.485	3.913	.735	1.505	.170	.3621		
.240	17.03	.490	3.826	.740	1.480	.180	.3492		
.245	16.33	.495	3.742	.745	1.455	.190	.3364		

$$\frac{\cot x}{x} \text{ (continued)}$$

x	$\frac{\cot x}{x}$	x	$\frac{\cot x}{x}$	x	$\frac{\cot x}{x}$	x	$\frac{\cot x}{x}$	x	$\frac{\cot x}{x}$	x	$\frac{\cot x}{x}$
1.60	-.01825	2.00	-.2289	2.40	-.4549	2.80	-1.005	3.20	5.343	3.60	.5629
.61	-.02435	.01	-.2337	.41	-.4622	.81	-1.034	.21	4.545	.61	.5474
.62	-.03037	.02	-.2387	.42	-.4696	.82	-1.064	.22	3.953	.62	.5326
.63	-.03638	.03	-.2436	.43	-.4772	.83	-1.097	.23	3.492	.63	.5184
.64	-.04226	.04	-.2485	.44	-.4850	.84	-1.132	.24	3.126	.64	.5047
1.65	-.04812	2.05	-.2535	2.45	-.4929	2.85	-1.169	3.25	2.827	3.65	.4916
.66	-.05386	.06	-.2584	.46	-.5011	.86	-1.209	.26	2.579	.66	.4789
.67	-.05958	.07	-.2634	.47	-.5093	.87	-1.251	.27	2.369	.67	.4667
.68	-.06523	.08	-.2684	.48	-.5175	.88	-1.297	.28	2.189	.68	.4549
.69	-.07089	.09	-.2734	.49	-.5265	.89	-1.346	.29	2.033	.69	.4436
1.70	-.07641	2.10	-.2785	2.50	-.5354	2.90	-1.399	3.30	1.897	3.70	.4326
.71	-.08193	.11	-.2836	.51	-.5446	.91	-1.457	.31	1.777	.71	.4219
.72	-.08738	.12	-.2887	.52	-.5540	.92	-1.520	.32	1.670	.72	.4117
.73	-.09277	.13	-.2938	.53	-.5636	.93	-1.589	.33	1.575	.73	.4017
.74	-.09816	.14	-.2990	.54	-.5735	.94	-1.664	.34	1.489	.74	.3921
1.75	-.1035	2.15	-.3042	2.55	-.5836	2.95	-1.747	3.35	1.412	3.75	.3828
.76	-.1088	.16	-.3094	.56	-.5941	.96	-1.840	.36	1.341	.76	.3738
.77	-.1141	.17	-.3147	.57	-.6049	.97	-1.943	.37	1.277	.77	.3650
.78	-.1193	.18	-.3200	.58	-.6160	.98	-2.058	.38	1.217	.78	.3565
.79	-.1245	.19	-.3254	.59	-.6275	.99	-2.189	.39	1.163	.79	.3482
1.80	-.1296	2.20	-.3309	2.60	-.6396	3.00	-2.338	3.40	1.113	3.80	.3401
.81	-.1347	.21	-.3363	.61	-.6515	.01	-2.510	.41	1.066	.81	.3323
.82	-.1398	.22	-.3418	.62	-.6642	.02	-2.710	.42	1.023	.82	.3247
.83	-.1449	.23	-.3474	.63	-.6772	.03	-2.945	.43	.9827	.83	.3173
.84	-.1499	.24	-.3531	.64	-.6907	.04	-3.227	.44	.9450	.84	.3101
1.85	-.1550	2.25	-.3588	2.65	-.7048	3.05	-3.570	3.45	.9098	3.85	.3031
.86	-.1599	.26	-.3646	.66	-.7193	.06	-3.996	.46	.8768	.86	.2963
.87	-.1649	.27	-.3704	.67	-.7344	.07	-4.540	.47	.8457	.87	.2896
.88	-.1699	.28	-.3764	.68	-.7501	.08	-5.347	.48	.8164	.88	.2831
.89	-.1749	.29	-.3824	.69	-.7664	.09	-6.252	.49	.7888	.89	.2768
1.90	-.1798	2.30	-.3885	2.70	-.7835	3.10	-7.751	3.50	.7627	3.90	.2707
.91	-.1847	.31	-.3946	.71	-.8012	.11	-10.176	.51	.7380	.91	.2646
.92	-.1896	.32	-.4009	.72	-.8197	.12	-14.84	.52	.7146	.92	.2586
.93	-.1946	.33	-.4073	.73	-.8391	.13	-27.57	.53	.6923	.93	.2529
.94	-.1994	.34	-.4137	.74	-.8594	.14	-200.4	.54	.6711	.94	.2473
1.95	-.2044	2.35	-.4203	2.75	-.8807	3.15	37.74	3.55	.6509	3.95	.2418
.96	-.2092	.36	-.4269	.76	-.9029	.16	17.19	.56	.6317	.96	.2364
.97	-.2142	.37	-.4338	.77	-.9264	.17	11.10	.57	.6133	.97	.2311
.98	-.2190	.38	-.4406	.78	-.9510	.18	8.182	.58	.5958	.98	.2260
.99	-.2239	.39	-.4477	.79	-.9771	.19	6.473	.59	.5790	.99	.2209

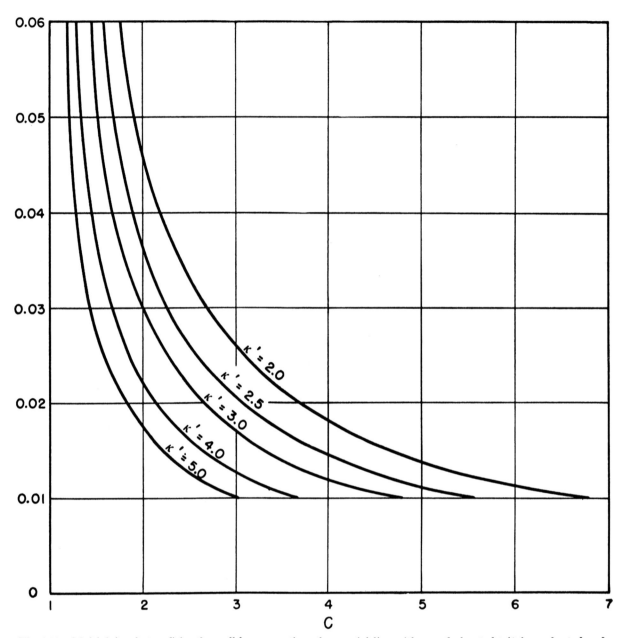

Fig. 2.31. Multiplying factor C for the wall-loss correction of a coaxial line with sample located $\lambda/4$ from shorted end.

VIII. Simplified Calculations for the Input Impedance Method

Calculation methods have been given assuming no wall losses and negligible losses in dielectric covers or quarter-wavelength terminating sections. The following equations and tables give correction terms for these losses and shortened calculations for use when x_0 is a multiple of a quarter wavelength. The values of Δx are assumed already corrected for loss in line-section 1.

1. *For sample at shorted end of line*:

$$\frac{\tan \beta_2 d_2}{\beta_2 d_2} = -\frac{\lambda_1}{2\pi d_2} \tan \frac{2\pi x_0}{\lambda_1}. \quad (2.68)$$

For coaxial line (TEM mode):

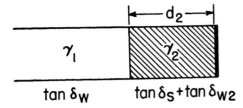

$$\tan \delta_s = \frac{\Delta x}{d_2} \frac{\beta_2 d_2 \left(1 + \tan^2 \frac{2\pi x_0}{\lambda_1}\right)}{\beta_2 d_2 (1 + \tan^2 \beta_2 d_2) - \tan \beta_2 d_2} \text{ (lossless line)}$$

$- \tan \delta_w$ (correction for wall loss in sample section of line)

$$- \frac{\tan \delta_w \tan \beta_2 d_2}{\beta_2 d_2 (1 + \tan^2 \beta_2 d_2) - \tan \beta_2 d_2} \text{ (correction}$$

for effect of $\tan \delta_w$ on reflection coefficient). (2.69)

For hollow wave guide (TE mode): †

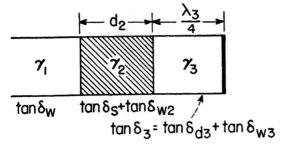

$$\tan \delta_s = \frac{\Delta x}{d_2} \frac{w - u \frac{1}{\epsilon_2'/\epsilon_0}}{w} \frac{\beta_2 d_2 \left(1 + \tan^2 \frac{2\pi x_0}{\lambda_1}\right)}{\beta_2 d_2 (1 + \tan^2 \beta_2 d_2) - \tan \beta_2 d_2}$$

(lossless line) (2.70)

$- \tan \delta_{w_2}$ (correction for wall loss in sample section of guide).

† Correction for effect of $\tan \delta_w$ on reflection coefficient neglected.

2. *For sample located a quarter wavelength from shorted end of line*:

$$\frac{\cot \beta_2 d}{\beta_2 d} = \frac{\lambda_1}{2\pi d_2} \tan \frac{2\pi x_0}{\lambda_1}. \quad (2.71)$$

For coaxial line (TEM mode):

$$\tan \delta_s = \frac{\Delta x}{d_2} \frac{\beta_2 d_2 \left(1 + \tan^2 \frac{2\pi x_0}{\lambda_1}\right)}{\beta_2 d_2 (1 + \cot^2 \beta_2 d_2) + \cot \beta_2 d_2} \text{ (lossless line)}$$

$- \tan \delta_w$ (correction for wall loss in sample section of line)

$$- \frac{\pi \tan \delta_3 \left(\frac{\epsilon_2' \epsilon_3'}{\epsilon_0 \epsilon_0}\right)^{1/2}}{2}$$

$$\times \frac{\left(\frac{1}{\epsilon_2'/\epsilon_0} + \tan^2 \frac{2\pi x_0}{\lambda_1}\right)}{\beta_2 d_2 (1 + \cot^2 \beta_2 d_2) + \cot \beta_2 d_2} \text{ (correction}$$

for dielectric plus wall loss in guide section 3)

$$+ \frac{\tan \delta_w \cot \beta_2 d_2}{\beta_2 d_2 (1 + \cot^2 \beta_2 d_2) + \cot \beta_2 d_2} \text{ (correction}$$

for effect of wall loss on reflection coefficient at sample boundary). (2.72)

The three correction factors involving $\tan \delta_w$ may be combined when section 3 is air line. Figure 2.31 shows a multiplying factor for $\tan \delta_w$ which gives the total correction as a function of sample length and dielectric constant. For hollow wave guide (TE mode): †

$$\tan \delta_s = \frac{\Delta x}{d_2} \frac{w - \frac{u}{\epsilon_2'/\epsilon_0}}{w} \frac{\beta_2 d_2 \left(1 + \tan^2 \frac{2\pi x_0}{\lambda_1}\right)}{\beta_2 d_2 (1 + \cot^2 \beta_2 d_2) + \cot \beta_2 d_2} \text{ (lossless}$$

line)

$- \tan \delta_{w_2}$ (correction for wall loss in sample section of guide)

$$- \frac{\pi}{2} \tan \delta_3 \left(\frac{\epsilon_3'}{\epsilon_2'}\right) \frac{[(\epsilon_2'/\epsilon_0) w - u]^{3/2}}{[(\epsilon_3'/\epsilon_0) w - u]^{1/2}} \frac{\frac{1}{(\epsilon_2'/\epsilon_0) w - u} + \tan^2 \frac{2\pi x_0}{\lambda_1}}{\beta_2 d_2 (1 + \cot^2 \beta_2 d_2) + \cot \beta_2 d_2}$$

(correction for dielectric plus wall loss in guide section 3). (2.73)

3. *Calculation of ϵ'/ϵ_0 and $\tan \delta_s$ for particular lengths of line sections with the losses in line-section 1 and in walls neglected.*

† Correction for effect of $\tan \delta_w$ on reflection coefficient neglected.

120 Dielectric Measuring Techniques

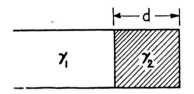

                        Coaxial line   *Hollow wave guide*

For $x_0/\lambda_1 = \frac{1}{4}$ and $m =$ number of quarter wavelengths in sample:

$$\frac{\epsilon_2'}{\epsilon_0} = \qquad \frac{m\lambda_1}{4d} \qquad\qquad \frac{1}{w}\left[\left(\frac{m\lambda_1}{4d}\right)^2 + u\right] \qquad (2.74)$$

$$\tan \delta_2 = \qquad \frac{\Delta x}{d\,\epsilon_2'/\epsilon_0} \qquad\qquad \frac{\Delta x}{dw\,\epsilon_2'/\epsilon_0} \qquad (2.75)$$

For $x_0/\lambda_1 = 0, \frac{1}{2}$, and $m =$ number of half wavelengths in sample:

$$\frac{\epsilon_2'}{\epsilon_0} = \qquad \frac{m\lambda_1}{2d} \qquad\qquad \frac{1}{w}\left[\left(\frac{m\lambda_1}{2d}\right)^2 + u\right] \qquad (2.76)$$

$$\tan \delta_2 = \qquad \frac{\Delta x}{d} \qquad\qquad \frac{\Delta x}{d}\,\frac{w - u\frac{\epsilon_0}{\epsilon_2'}}{w} \qquad (2.77)$$

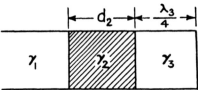

For $x_0/\lambda_1 = 0, \frac{1}{2}$ and $m =$ number of quarter wavelengths in sample:

$$\frac{\epsilon_2'}{\epsilon_0} = \qquad \left(\frac{m\lambda_1}{4d}\right)^2 \qquad\qquad \frac{1}{w}\left[\frac{m\lambda_1}{4d} + u\right] \qquad (2.78)$$

$$\tan \delta_2 = \qquad \frac{\dfrac{\Delta x}{d_2} - \left(\dfrac{\epsilon_3'}{\epsilon_2'}\right)^{1/2}\tan \delta_3}{1 - \dfrac{\pi^2}{16}\dfrac{\Delta x}{d_3}\tan \delta_3} \qquad\qquad \frac{\Delta x}{d_2}\,\frac{w - u\frac{\epsilon_0}{\epsilon_2'}}{w}\,\dagger \qquad (2.79)$$

For $x_0/\lambda_1 = \frac{1}{4}$, and $m =$ number of half wavelengths in sample:

$$\frac{\epsilon_2'}{\epsilon_0} = \qquad \left(\frac{m\lambda_1}{2d_2}\right)^2 \qquad\qquad \frac{1}{w}\left[\left(\frac{m\lambda_1}{2d}\right)^2 + u\right] \qquad (2.80)$$

$$\tan \delta_2 = \qquad \frac{\dfrac{\Delta x}{d\,\epsilon_2'/\epsilon_0} - \dfrac{1}{2}\left(\dfrac{\epsilon_3'}{\epsilon_2'}\right)^{1/2}\tan \delta_3}{1 - \dfrac{\pi^2}{16}\dfrac{\Delta x}{\epsilon_2'/\epsilon_0}\tan \delta_3} \qquad\qquad \frac{\Delta x}{d_2 w\,\epsilon_2'/\epsilon_0}\,\dagger \qquad (2.81)$$

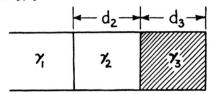

For $x_0/\lambda = \frac{1}{4}$, $d_2 = \lambda_2/4$, and $m =$ number of half wavelengths in sample:

$$\frac{\epsilon_3'}{\epsilon_0} = \qquad \left(\frac{m\lambda_1}{2d_3}\right)^2 \qquad\qquad \frac{1}{w}\left[\left(\frac{m\lambda_1}{2d}\right)^2 + u\right] \qquad (2.82)$$

$$\tan \delta_3 = \qquad \frac{\Delta x}{d_3\,\epsilon_2'/\epsilon_0} - \frac{\tan \delta_2}{2m}\left(\frac{\epsilon_2'}{\epsilon_1'}\right)^{1/2} \qquad\qquad \frac{\Delta x}{d_3 w\,\epsilon_2'/\epsilon_0} - \frac{w - \dfrac{u}{\epsilon_3'/\epsilon_0}}{2mw}\left(\frac{\epsilon_2'}{\epsilon_0}w - u\right)^{1/2}\tan \delta_2 \qquad (2.83)$$

† The loss in hollow wave guide, section 3, is neglected.

Coaxial line   *Hollow wave guide*

For $x_0/\lambda = \frac{1}{4}$, $d_2 = \lambda_2/2$, and $m =$ number of quarter wavelengths in sample:

$$\frac{\epsilon_3'}{\epsilon_0} = \left(\frac{m\lambda_1}{4d_3}\right)^2 \qquad \frac{1}{w}\left[\left(\frac{m\lambda_1}{4d_3}\right)^2 + u\right] \qquad (2.84)$$

$$\tan\delta_3 = \frac{\Delta x}{d_3\epsilon_3'/\epsilon_0} - \frac{2\tan\delta_2}{m}\left(\frac{\epsilon_2'}{\epsilon_3'}\right)^{1/2} \qquad \frac{\Delta x}{d_3 w\, \epsilon_3'/\epsilon_0} - \frac{w - \dfrac{u}{\epsilon_3'/\epsilon_0}}{m}\left(\dfrac{\dfrac{\epsilon_2'}{\epsilon_0}w - u}{\dfrac{\epsilon_3'}{\epsilon_0}w - u}\right)^{1/2} 2\tan\delta_2 \qquad (2.85)$$

IX. Two-Position Method

This method, which eliminates the use of hyperbolic charts, has been used especially by Surber and Crouch.[27] The sample is first terminated in a short circuit, then in a $\lambda/4$ section approximating an open circuit. For these two measurements

$$\frac{\tanh\gamma_2 d}{\gamma_2 d} = \frac{Z(0)_1/Z_1}{\gamma_1 d}, \qquad (2.86)$$

$$\frac{\coth\gamma_2 d}{\gamma_2 d} = \frac{Z(0)_2/Z_1}{\gamma_1 d}. \qquad (2.87)$$

The reciprocal product expresses γ_2 in terms of the measured quantities

$$\gamma_2^2 = \frac{\gamma_1^2}{\dfrac{Z(0)_1}{Z_1}\cdot\dfrac{Z(0)_2}{Z_1}}. \qquad (2.88)$$

Using M and ϕ to designate the magnitude and polar angle of each ratio $Z(0)/Z_1$, we obtain for the dielectric constant and loss:

$$\frac{\epsilon_2'}{\epsilon_0} = \frac{1}{w[1 + \tan^2(\phi_1 + \phi_2)]^{1/2} M_1 M_2} + \frac{u}{w}, \qquad (2.89)$$

$$\tan\delta = \tan(\phi_1 + \phi_2)\left[1 - \frac{u}{w\epsilon_2'/\epsilon_0}\right], \qquad (2.90)$$

where $u = (\lambda_1/\lambda_c)^2$ and $w = 1 + u$.

For medium loss materials Eqs. 2.89 and 2.90 can be written in terms of x_0 and Δx, provided that $\pi\,\Delta x/\lambda \ll \tan 2\pi x_0/\lambda$ and their product is small compared to unity.

$$\frac{\epsilon_2'}{\epsilon_0} = \frac{1}{-w\tan\dfrac{2\pi x_{0_1}}{\lambda_1}\tan\dfrac{2\pi x_{0_2}}{\lambda_1}} + \frac{u}{w}, \qquad (2.91)$$

$$\tan\delta = \left[1 - \frac{u}{w\epsilon_2'/\epsilon_0}\right]\left[\frac{\pi\,\Delta x_1}{\lambda_1}\left(\tan\frac{2\pi x_{0_1}}{\lambda_1} + \cot\frac{2\pi x_{0_1}}{\lambda_1}\right)\right.$$
$$\left. + \frac{\pi\,\Delta x_2}{\lambda_1}\left(\tan\frac{2\pi x_{0_2}}{\lambda_1} + \cot\frac{2\pi x_{0_2}}{\lambda_1}\right) - 2\delta_w - c\right], \qquad (2.92)$$

where δ_w is the wall loss in the section of line containing the sample and c is a correction term for losses in the quarter-wavelength section. For coaxial line the expression for c is

$$c = \frac{\pi\delta_3}{4}\left(\frac{\epsilon_3'}{\epsilon_0}\right)^{1/2}\left[\frac{1}{\dfrac{\epsilon_2'}{\epsilon_0}|M_2|} + |M_2|\right], \qquad (2.93)$$

[27] W. H. Surber, Jr., and G. E. Crouch, Jr., *J. Appl. Phys.* **19**, 1130 (1948). Their equations are given in rectangular form instead of the polar form shown here.

in which δ_3 and ϵ_3' are the total loss tangent and dielectric constant for the quarter-wavelength section, respectively, and $|M_2|$ is the magnitude of $\tan 2\pi x_{0_2}/\lambda_1$.

X. The Derivation of the Dielectric Field of the E_{010} Cavity

With the assumption that the coupling probes and metallic losses in the walls do not change the field distribution appreciably from the ideal case, Maxwell's equations for a completely filled cavity expressed in cylindrical co-ordinates (Z, r, θ) are as follows:

$$j\omega\mu H_\theta = \frac{\partial E_z}{\partial r} \qquad (2.94)$$

$$(\sigma + j\omega\epsilon')E_z = \frac{1}{r}\frac{\partial(rH_\theta)}{\partial r}, \qquad (2.95)$$

where H is the magnetic field intensity, E is the electric field intensity, μ, ϵ', σ are permeability, permittivity, and conductivity of the dielectric medium, and ω is the angular frequency of excitation.

The solutions of these equations are

$$H_\theta = AJ_1(kr)e^{j\omega t}, \qquad (2.96)$$

$$E_z = \frac{k}{\sigma + j\omega\epsilon'}\cdot AJ_0(kr)e^{j\omega t}, \qquad (2.97)$$

in which $k^2 = -j\omega\mu(\sigma + j\omega\epsilon')$ and J_0, J_1 are Bessel functions of the first kind.

With a sample in place the fields within the specimen are given by Eqs. 2.96 and 2.97 with the subscript 1 attached to the descriptive quantities μ, ϵ', σ, k, and β. In the air region, r does not vanish, so Bessel functions of the second kind are included in the solutions which have the form

$$H_\theta = [BJ_1(k_0 r) + CY_1(k_0 r)]e^{j\omega t} \qquad (2.98)$$

$$E_z = \frac{k_0}{j\omega\epsilon_0}[BJ_0(k_0 r) + CY_0(k_0 r)]e^{j\omega t}, \qquad (2.99)$$

in which $k_0 = \beta_0 = 2\pi/\lambda_0$ and Y_0 and Y_1 are Bessel functions of the second kind. The boundary conditions are that E_z vanishes at the cylindrical surface $r = a$ and that there is continuity of electric and magnetic fields at the specimen boundary $r = b$. Application of the boundary conditions Eqs. 2.96 to 2.99 results in three equations from which the amplitude constants can be eliminated to obtain

$$\frac{\epsilon'}{\epsilon_0} = \frac{\beta_1}{\beta_0}\frac{J_0(\beta_1 b)J_1(\beta_0 b)}{J_1(\beta_1 b)J_0(\beta_0 b)}\left[\frac{\dfrac{Y_0(\beta_0 a)}{J_0(\beta_0 a)} - \dfrac{Y_1(\beta_0 b)}{J_1(\beta_0 b)}}{\dfrac{Y_0(\beta_0 a)}{J_0(\beta_0 a)} - \dfrac{Y_0(\beta_0 b)}{J_0(\beta_0 b)}}\right]. \qquad (2.100)$$

In Eq. 2.100, $\beta_0 = 2\pi/\lambda_0$, λ_0 being the resonant-free space wavelength, and $\beta_1 = \beta_0(\epsilon'/\epsilon_0)^{1/2}$. For a loss-free system, Eq. 2.100 is exact for all values of a and b, but, since β_1 contains the unknown ϵ', the relation is not convenient for computation. As Jackson[28] and co-workers have shown, an approximate form of Eq. 2.100 is

$$\frac{\epsilon'}{\epsilon_0} = \left[1 - \frac{\beta_0 b}{8}\left(\frac{\epsilon'}{\epsilon_0} - 1\right)\right] \begin{bmatrix} \dfrac{Y_0(\beta_0 a)}{J_0(\beta_0 a)} - \dfrac{Y_1(\beta_0 b)}{J_1(\beta_0 b)} \\ \dfrac{Y_0(\beta_0 a)}{J_0(\beta_0 a)} - \dfrac{Y_0(\beta_0 b)}{J_0(\beta_0 b)} \end{bmatrix}, \quad (2.101)$$

and, on using the identity

$$Y_0(\beta_0 b)\cdot J_1(\beta_0 b) - Y_1 \beta_0 b \cdot J_0 \beta_0 b \equiv \frac{2}{\pi \beta_0 b} \quad (2.102)$$

and defining the function

$$F \equiv [Y_0(\beta_0 a)\cdot J_0(\beta_0 b) - Y_0(\beta_0 b)\cdot J_0(\beta_0 a)]\frac{\pi \beta_0 a}{2}, \quad (2.103)$$

it may be rewritten

$$\frac{\epsilon'}{\epsilon_0} = \left[1 - \frac{(\beta_0 b)^2}{8}\left(\frac{\epsilon'}{\epsilon_0}-1\right)\right]\left[1 + \frac{1}{F}\frac{a}{b}\frac{J_0(\beta_0 a)}{J_1(\beta_0 b)}\right]. \quad (2.104)$$

From Eq. 2.104 a solution is obtained explicitly in terms of a, b, and β_0 as

$$\frac{\epsilon'}{\epsilon_0} = 1 + \frac{\dfrac{a}{b}\cdot\dfrac{J_0(\beta_0 a)}{J_1(\beta_0 b)}}{F\left[1 + \dfrac{(\beta_0 b)^2}{8}\right] + \dfrac{(\beta_0 b)^2}{8}\cdot\dfrac{a}{b}\cdot\dfrac{J_0(\beta_0 a)}{J_1(\beta_0 b)}}. \quad (2.105)$$

A chart of F versus $(\beta_0 a)$ for various values of b/a up to 0.3 (ϵ'/ϵ_0 up to about 6) has been calculated.[28] For very thin samples a simpler form first given by Borgnis[29] results

$$\frac{\epsilon'}{\epsilon_0} = 1 + 0.539 \frac{a^2}{b^2}\frac{\Delta\nu}{\nu_0} \quad (2.106)$$

[28] F. Horner, T. A. Taylor, R. Dunsmuir, J. Lamb, and W. Jackson, *J. Inst. Elec. Eng.* (London) **93**, Part III, January, 1946, p. 53.

[29] F. Borgnis, *Physik. Z.* **43**, 284 (1942).

in which $\Delta\nu$ is the resonant-frequency change with sample in and out. This may be written in terms of sample and cavity volumes for comparison with TE modes

$$\frac{\epsilon'}{\epsilon_0} = 1 + 0.539 \frac{V}{v}\frac{\Delta\nu}{\nu_0}. \quad (2.107)$$

The expression for the loss tangent of the sample is given in terms of Q (of cavity with sample) and Q' (of the cavity with fictitious no-loss sample) as

$$\tan\delta = \frac{\dfrac{a^2}{b^2} + \nu^2\left(\dfrac{\epsilon'}{\epsilon_0}-1\right)}{\dfrac{\epsilon'}{\epsilon_0}\cdot\nu^2\left[1 + \dfrac{J_1^2(\beta_1 b)}{J_0^2(\beta_1 b)}\right]}\cdot\left[\frac{1}{Q} - \frac{1}{Q'}\right]. \quad (2.108)$$

The theoretical value of Q' is

$$Q' = \frac{1}{d}\frac{\dfrac{a^2}{b^2} + \nu^2\left(\dfrac{\epsilon'}{\epsilon_0}-1\right)}{\dfrac{a(a+l)}{b^2} + \nu^2\left(\dfrac{\epsilon'}{\epsilon_0}-1\right)}. \quad (2.109)$$

In practice an estimated value of Q' must be used, based on the ratio of experimental to theoretical values for an empty cavity at the same frequency. For very thin samples Eq. 2.108 reduces to

$$\tan\delta = \frac{0.269}{\epsilon'/\epsilon_0}\frac{a^2}{b^2}\left[\frac{1}{Q}-\frac{1}{Q'}\right]; \quad (2.110)$$

for comparison with TE modes, we can write

$$\frac{\epsilon''}{\epsilon_0} = 0.269 \frac{V}{v}\frac{\Delta\nu}{\nu_0}. \quad (2.111)$$

B · Permeability

By DAVID J. EPSTEIN

The macroscopic magnetic properties of matter may be described in terms of three field vectors: the magnetic induction **B**, the magnetic polarization **M**, and the magnetic field **H** (see IA, Sec. 2). In the rationalized mks system of units, the relation between these vectors is

$$\mathbf{B} = \mu_0(\mathbf{H} + \mathbf{M}) = \mu\mathbf{H}, \quad (1)$$

where $\mu_0 = 4\pi \times 10^{-7}$ [henry per meter] is the permeability of free space and μ is the permeability of the magnetic material. The magnitude of the relative permeability

$$\kappa_m = \mu/\mu_0 \quad (2)$$

may be used as a criterion to distinguish para-, dia-, and ferromagnetism. For diamagnetic materials κ_m will normally be slightly <1; for paramagnetic materials slightly >1; and for ferromagnetic materials it may be $\gg 1$, and range into the hundreds of thousands. In this discussion we shall be concerned only with the permeability of ferromagnetic materials.

The general relationship between **B** and **H** is by no means simple. In an anisotropic medium **B** and **H** may not have the same direction at a given point; if, furthermore, the material is inhomogeneous the field vectors will vary from one point to another. To effect a considerable simplification we shall confine our considerations to magnetic materials which are both homogeneous and isotropic, that is, materials for which the permeability will be a function of neither direction nor position.

Further complications in the definition of μ arise because **B** is a nonlinear, multivalued function of **H** for ferromagnetic materials. Moreover, since the magnetization process is time-dependent, μ will also be a function of frequency (or time). Thus Eq. 1 should more properly be written

$$\mathbf{B} = \mu(H, B, \omega)\mathbf{H}. \tag{3}$$

Definitions of Permeability in Static Fields

The fact that the μ of ferromagnetics is a function of several variables leads to the definition of different kinds of permeability. *The ASTM Standards on Magnetic Materials*[1] lists no fewer than ten different varieties of permeability, and to these we can add perhaps another half dozen that appear in the literature. Many

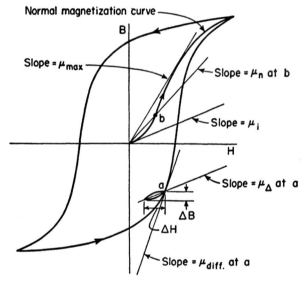

Fig. 1. Hysteresis loop, normal magnetization curve, and permeabilities.

of these are defined to fulfill rather specific practical needs and do not have particular significance for our present discussion. The terms which are of interest can be defined with the aid of the static characteristic of Fig. 1.

Of principal importance is the so-called *initial permeability* μ_i, defined as the slope of the normal magneti-

[1] American Society for Testing Materials, *ASTM Standards on Magnetic Materials*, Philadelphia, Pa., 1949, p. 11.

zation curve in the limit as the field H is reduced to zero. Also of interest is the *incremental permeability* μ_Δ, defined by the ratio $\Delta B / \Delta H$. In the limit as ΔH is reduced to the differential dH, the incremental permeability becomes the *reversible permeability* μ_{rev}. The initial permeability, it should be noted, is the reversible permeability at the origin. It is clear from Fig. 1 that μ_Δ and μ_{rev} will vary from point to point along both the hysteresis loop and the normal magnetization curve.

The tangent to either the loop or the normal curve at any point defines the *differential permeability* μ_{diff} at that point. This differs from the reversible permeability at the point by a quantity defined as the *irreversible permeability*; that is,

$$\mu_{\text{diff}} = \mu_{\text{rev}} + \mu_{\text{irrev}}. \tag{4}$$

For the demagnetized state μ_{irrev} becomes zero, and we have $\mu_{\text{diff}} = \mu_{\text{rev}} = \mu_i$.

The *normal permeability* μ_n is a quantity frequently given in data published by manufacturers of magnetic materials. It is defined by the slope of a line drawn from the origin of co-ordinates to a point on the normal magnetization curve. The largest value of this slope defines the maximum permeability μ_{max}.

Complex Permeability

The definitions introduced above, based as they are on magnetostatic considerations, result in permeabilities which are always real numbers. In extending the concept of permeability to frequency-dependent magnetization processes it is convenient to introduce the idea of a complex permeability (see IA, Eqs. 1.19 and 1.20). The use of a complex number allows us to account for the physical fact that in a sinusoidally varying magnetic field a phase angle θ arises between **B** and **H** due to energy losses associated with magnetic resonance and relaxation phenomena; that is, we must write

$$\mathbf{B}e^{j\omega t} = |\mu^*| \mathbf{H} e^{j(\omega t - \theta)} = (\mu' - j\mu'')\mathbf{H}e^{j\omega t}. \tag{5}$$

Such losses arise physically from the reorientation of the magnetic moments, and, it must be borne in mind, are quite distinct from the usual eddy-current and hysteresis loss. Hysteresis and eddy-current losses should be kept in their separate compartments, and the loss factor μ'' of the complex permeability should be used to designate situations in which: (a) the loss mechanism accompanies a resonance or relaxation process of nonmacroscopic origin; (b) vectors **B** and **H** are linearly related, that is, μ^* is a function of ω only. As noted above, the permeabilities which are independent of field strength are the initial and reversible permeabilities, and, hereafter, when we refer to complex permeability we shall be referring to these two quantities. When speaking of complex permeability we shall not employ any qualifying subscripts because the meaning will be clear from the context.

Although the idea of a complex permeability was suggested as long ago as 1913,[2] it has only since the early 1940's gained widespread use. Prior investigations into frequency-dependent magnetization processes were confined to metals, and experimental results were framed in terms of the quantities μ_R, variously called the resistive or outer permeability, and μ_L the inductive or inner permeability. The use of these quantities arose from the type of experiment performed in measuring the permeability; these terms are now outmoded in favor of μ^*.

To clarify the relation between μ_R, μ_L, and μ^*, let us consider a plane electromagnetic wave propagating in a ferromagnetic metal. To determine the magnetic properties of the material we shall measure the impedance seen by the wave. We recall from Sec. IA, Eq. 3.13, that the complex impedance is

$$Z = \left(\frac{\mu^*}{\epsilon^*}\right)^{1/2} = R + jX. \quad (6)$$

In a metal, the displacement current will be small compared to the conduction current, that is, $\epsilon' \ll \epsilon''$, and we may write

$$Z = \left(\frac{j\mu^*}{\epsilon''}\right)^{1/2} = R + jX. \quad (7)$$

Squaring this equation and equating real and imaginary parts yields

$$\mu' = 2RX\epsilon''$$
$$\mu'' = (R^2 - X^2)\epsilon'' \quad (8)$$

for the components of the complex permeability.

The resistive permeability is defined by equating the real part of the impedance to R in the following way:

$$R_e(Z) = R_e\left[\left(\frac{j\mu_R}{\epsilon''}\right)^{1/2}\right] = R. \quad (9)$$

Similarly, the inductive permeability is defined by equating the imaginary component of the impedance to X:

$$I_m(Z) = I_m\left[\left(\frac{j\mu_L}{\epsilon''}\right)^{1/2}\right] = X. \quad (10)$$

Utilizing the fact that $(j)^{1/2} = (1/2)^{1/2}(1 + j)$, we can carry through the operations indicated above to obtain

$$\mu_R = 2\epsilon'' R^2,$$
$$\mu_L = 2\epsilon'' X^2. \quad (11)$$

Thus μ_R is obtained from an experiment in which only the real part of the impedance is measured, whereas μ_L comes from a measurement of the reactive component.

[2] W. Arkadiew, *Physik. Z.* **14**, 928 (1913).

By combining Eqs. 8 and 11 we obtain the relations between μ_R, μ_L, and μ^*:

$$\mu' = (\mu_R \mu_L)^{1/2},$$
$$\mu'' = \tfrac{1}{2}(\mu_R - \mu_L),$$
$$\mu_R = |\mu^*| + \mu'',$$
$$\mu_L = |\mu^*| - \mu''. \quad (12)$$

These relations apply not only in our specific problem, but whenever we are dealing with materials in which the displacement current can be ignored relative to the conduction current, that is, whenever there is a strong skin effect. If this condition is not fulfilled there is no general relation between the complex permeability and μ_R and μ_L.[3]

It is clear from the foregoing that μ_R and μ_L are real quantities which usually differ in magnitude. They are so-called conjugate quantities, that is, if one parameter is known as a complete function of frequency, the other can be determined to within an arbitrary constant by application of the Kramers-Kronig relations.[4] The real and imaginary parts of the complex permeability are similar conjugates. Different symbols that have been used in the literature are summarized in Table 1.[5]

Table 1. Notations for various permeabilities

Name	Symbol	Other symbols			
Complex permeability	μ^*				
Real part of μ^*	μ'	μ_1	μ	μ	μ_R
Imaginary part of μ^*	μ''	μ_2	q'	ρ'	μ_I
Resistive permeability	μ_R	μ_k			
Inductive permeability	μ_L	μ_n			

Materials

Three distinct types of ferromagnetic materials are encountered in permeability measurements: metals, powdered metal cores, and ferromagnetic semiconductors. From the technological standpoint interest in the permeability of metals is restricted to frequencies below a few megacycles. The upper limit of frequency is set by eddy currents which not only are responsible for power loss, but also prevent the magnetic flux from completely penetrating the material, thereby reducing the effective permeability. Eddy-current effects diminish with the thickness of the lamination employed. To attain useful values of permeability in the megacycle region, metal tapes only a few mils thick must be used.

The fact that eddy-current shielding is so predominant in metals requires that a careful distinction be made between the effective permeability of the sample and the intrinsic permeability of the metal. The former will depend on the dimensions of the sample

[3] C. Kittel, *Phys. Rev.* **70**, 281 (1946).

[4] C. J. Gorter, *Paramagnetic Relaxation*, Elsevier Publishing Co., New York, 1947, p. 21.

[5] G. T. Rado, *Advances in Electronics*, Academic Press, New York, 1950, Vol. II, p. 251.

(Fig. 2). In order to find the intrinsic permeability, a skin-effect equation appropriate to the particular sample geometry must be solved. Extensive measurements of the permeability of metals beyond the frequency range of technical importance into the microwave region have been made in connection with investigations concerning the fundamental nature of the magnetization process.[6]

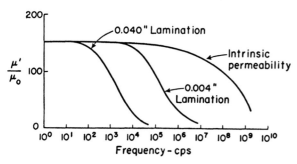

Fig. 2. Apparent reduction in the permeability of iron by eddy current shielding.

Powdered iron cores consist of finely divided metallic particles in an insulating matrix. The small particle size is effective in reducing eddy-current loss, but there is a considerable reduction in permeability below that of the pure metal because of dilution and nonmagnetic gaps between the particles. Permeabilities of only a hundred or less are realized in these cores even though the intrinsic permeability of the particles may range in the thousands. Powdered cores can be made with permeabilities that are substantially flat from direct current to several hundred megacycles (Fig. 3). They are usually preferred to metal tapes

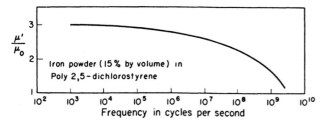

Fig. 3. Permeability of powdered iron as function of frequency.

as core materials for high Q coils. Metal tapes, despite their higher losses, find application as wide-band high-frequency transformers because of their higher permeability. Within recent years, metal tapes with rectangular hysteresis loops have acquired considerable importance as switching and memory devices in

[6] J. T. Allanson, *J. Inst. Elec. Engrs.* (London) *92*, Pt. III, 247 (1945).

moderate- and high-speed computer applications (see IIIC, Sec. 4).

The ferromagnetic semiconductors of practical importance at present are those mixed oxides of the transition metals which crystallize in the spinel structure. The natural crystal which is the prototype of this class of material is magnetite ($FeO \cdot Fe_2O_3$). A wide range of permeabilities can be obtained, depend-

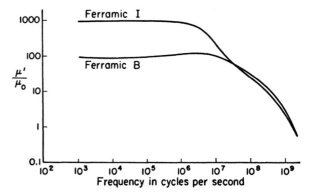

Fig. 4. Permeability as a function of frequency for two ferrites of different static μ.

ing on the composition (Fig. 4). As a rule, for materials having comparable values of the saturation moment, the range over which the permeability is substantially independent of frequency is inversely proportional to the static value of initial permeability.

Principles of Measurement

In order to determine μ^* we have to measure at least two independent experimental quantities which can be put into analytic expressions that permit of simultaneous solution for μ' and μ''. As noted previously, the resistive and reactive components of the sample impedance are, for example, two such experimental quantities. Other paired quantities which play similar roles are indicated in Table 2.

Table 2. Paired quantities for the determination of μ^*

Resistance	Reactance
Q	Resonant frequency
Decrement	Group velocity
Attenuation constant	Phase constant
Absorption	Reflection
Standing-wave ratio	Position of minimum

We cannot always determine μ^* by measuring only two experimental quantities since the properties of a magnetic material are characterized not only by μ^* but by ϵ^* as well (see IA, Sec. 1). Because these two quantities appear in product or ratio form it will be necessary, in general, to obtain four independent experimental quantities from which μ', μ'', ϵ', and ϵ'' can be calculated. Fortunately, from a practical standpoint, it is possible to disregard the role of ϵ^* whenever the sample can be located in a circuit position where the magnetic field energy is large compared to the electric field energy. Thus, at low or moderate frequencies samples are usually placed in coils, or solenoids. At higher frequencies the sample might be placed at the end of a short-circuited transmission line, or cavity, where the magnetic field strength is a maximum and the electric field strength is low. If we are dealing with magnetic metals at high frequency, the sample can be incorporated into the conducting system of the transmission line or cavity where it is subjected to weak electric fields.

Measurements in Static Fields

In static fields, the permeability may be determined by measuring the force exerted on the specimen by the applied field or by ballistic techniques. Ordinarily, this testing is done in a range of induction in which the permeability is a function of field strength. The value of initial permeability must, therefore, be obtained by extrapolation.

In the usual ballistic method (Fig. 5) the sample is a ring of a radial thickness small compared to the diameter. Reversal of current through the primary

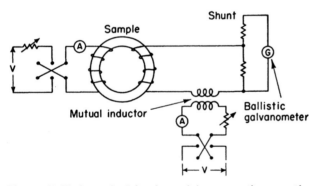

Fig. 5. Ballistic method for determining magnetic properties.

winding on the specimen deflects the ballistic galvanometer connected to the secondary. The galvanometer scale is calibrated by effecting a known current change in the primary circuit of a standard mutual inductor, the secondary winding of which is connected in series with the galvanometer circuit.

If the sample is available as a small rod or ellipsoid, measurements may be made by placing it in a homogeneous field established either by an iron-core electromagnet or an air-core solenoid (Fig. 6). The change in core induction, caused by a change in applied field, is determined with the aid of a ballistic galvanometer connected to a search coil wound around the specimen. In computing the permeability, suitable correction must be made for the demagnetizing field set up by the magnetic poles at the ends of the

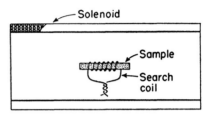

Fig. 6. Rod sample in air-core solenoid.

magnetized rod. The field within the sample, H_i, is not the applied field H_e but is

$$H_i = H_e - NM, \qquad (13)$$

where N, the demagnetizing factor, is dependent on the geometry of the specimen. The permeability defined as B_i/H_e is only an apparent one, the true permeability being B_i/H_i. We may write

$$\mu_{\text{true}} = K_c \mu_{\text{apparent}}, \qquad (14)$$

where K_c, the correction factor involving the demagnetizing constant, is given by

$$K_c = \frac{1-N}{1-N\dfrac{\mu_{\text{apparent}}}{\mu_0}}. \qquad (15)$$

Demagnetizing factors for ellipsoids can be calculated;[7] empirical factors for cylindrical rods of various length-to-diameter ratio have been tabulated by a number of authors.[8]

Where the sample shape permits, demagnetizing effects can be avoided by clamping the sample to a yoke of high-permeability material and of relatively large cross-sectional area (Fig. 7). A coil wound close to the sample is used for ballistic determination of the induction. The value of the magnetic field is determined with the aid of an air core connected across the arms of the yoke. There are a wide variety of yoke-type permeameters in use, differences among them being mainly in details of design.[9]

[7] J. A. Osborn, *Phys. Rev.* 67, 351 (1945).

[8] R. M. Bozorth and D. M. Chapin, *J. Appl. Phys.* 13, 320 (1942).

[9] R. M. Bozorth, *Ferromagnetism*, D. Van Nostrand Co., New York, 1951, p. 849.

A number of instruments have been built which allow a quasistatic curve of B vs. H to be traced automatically. The most sensitive of these is an apparatus due to Cioffi.[10] The problem in any B-H tracer lies in obtaining a signal proportional to B. If a ring specimen

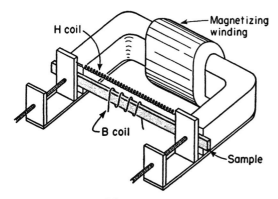

Fig. 7. Yoke-type permeameter.

is used, a signal proportional to H may readily be obtained as a voltage drop across a resistor in series with the magnetizing winding. The output of the secondary winding is a voltage proportional to $\dfrac{dB}{dt}$, and this must be integrated to obtain B. In Cioffi's instrument the required integration is obtained by the use of a feed-back amplifier system (Fig. 8). An input signal e_1

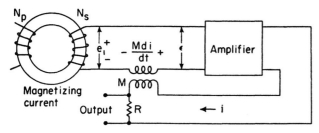

Fig. 8. Integration by feedback through a mutual inductor.

equal to the rate of change of flux linkages in the secondary winding, $N_s \dfrac{d\phi}{dt}$, is opposed by a voltage $M\left(\dfrac{di}{dt}\right)$ appearing across the secondary of a mutual inductor. The residual voltage

$$\epsilon = e_1 - M\frac{di}{dt} = N_s\frac{d\phi}{dt} - M\frac{di}{dt} \qquad (16)$$

drives a high-gain amplifier supplying the feed-back current i, which flows through the primary of the mutual inductor. If the gain is made sufficiently

[10] P. P. Cioffi, *Rev. Sci. Instr.* **21**, 624 (1950).

large, ϵ becomes virtually zero, and we have

$$N_s \frac{d\phi}{dt} = M \frac{di}{dt},$$

$$d\phi = \frac{M}{N_s} di. \qquad 17)$$

Thus a resistor placed in series with the current i will develop a voltage drop proportional to the change in induction in the specimen.

Measurement at Audio Frequencies

Accurate measurements of initial and reversible permeability can be performed in the audio-frequency range with the aid of some form of inductance bridge. Many different bridge arrangements are possible; we shall confine ourselves to the brief mention of a few of the more basic types.

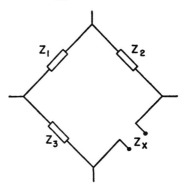

Fig. 9. Four-arm bridge.

For the general, four-arm bridge shown in Fig. 9, the balance condition is

$$Z_x = \frac{Z_2 Z_3}{Z_1}, \qquad (18)$$

where Z_x is the impedance of the unknown arm. In ordinary practice two arms of the bridge are kept fixed. Under these conditions we have two distinct situations for balance:

(I) $\quad |Z_x|\,\underline{|\phi_x} = A\,|Z_3|\,\underline{|\phi_3}, \qquad A = Z_2/Z_1,$

(II) $\quad |Z_x|\,\underline{|\phi_x} = \dfrac{B}{|Z_1|\,\underline{|\phi_1}}, \qquad B = Z_2 Z_3,$

where the ϕ's denote the phase angles of the various impedances, and where A is a fixed ratio and B a fixed product. Entirely equivalent with (I) is the case for which

$$|Z_x|\,\underline{|\phi_x} = A'\,|Z_2|\,\underline{|\phi_2}; \qquad A' = Z_3/Z_1. \quad (19)$$

In (I) the adjustable arm is adjacent to the arm containing the unknown; in (II) it is an opposite arm. If we confine ourselves to bridges with arms containing only the simplest impedance elements, the fixed terms A and B will be either purely real or purely imaginary. Practical considerations normally dictate that a real ratio be obtained by the use of resistive elements, and that an imaginary one be obtained by making one of the fixed arms purely resistive, the other purely capacitive.

Let us examine (I) in the light of these restrictions. If A is real it is clear that Z_3 must be an inductive arm, and we are led to the simple inductance bridge configuration shown in Table 3. Suppose we now make A imaginary, specifically let $A = j$; to fulfill our balance condition we must have $\phi_x = 90° + \phi_3$. Since ϕ_x is a positive angle less than 90°, ϕ_3 must be negative. In other words, Z_3 must contain negative reactance. This condition is realized in the Owen bridge.

Let us turn our attention to (II). We can have Z_2 and Z_3 both resistive, both capacitive, or we can have one resistive, the other capacitive. Thus the phase angle of B can be either 0, 180, or -90 degrees. For balance we must have $\phi_x = -\phi_1 + \underline{/B}$. This equality will be satisfied only for ϕ_1 negative and $\underline{/B} = 0$, that is, Z_1 must contain capacitive reactance and B must be a real number. The Maxwell-Wien bridge is the realization of this condition.

A great many modifications of these simple forms are possible, and some of the more important variations are shown in the lower half of Table 3. The equations for balance given in the table are only approximate. In order to obtain accurate results it is always necessary to make appropriate corrections for residual impedances: the resistance associated with capacitive and inductive elements, the inductance associated with resistive elements, stray capacity across resistors and inductors, etc. (see IIA, Sec. 1).

For precise measurements of μ' a toroidal sample of rectangular cross section is used. The value of μ'' can be determined only if

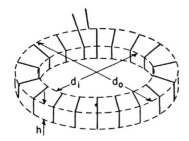

Fig. 10. Toroidal air core.

losses in the core material are comparable to copper losses in the winding. For semiconductor cores μ'' is frequently too small to be evaluated at the lower audio frequencies.

The inductance of a toroid with a nonmagnetic core is approximately

$$L = 0.0117 N^2 h \log_{10} \frac{d_0}{d_i} \quad \text{[microhenrys]}, \quad (20)$$

where N is the number of turns and the symbols h, d_0, and d_i are the coil dimensions (inches) (Fig. 10). This result is exact only for the ideal case, where the current can be treated as flowing in an infinitesimally thin sheet around the periphery of the core. The existence of discrete wires requires that a correction be made. The modified inductance is [11]

$$L_0 = L_0' - L_c = 0.0117 \left[N^2 H \log_{10} \frac{D_2}{D_1} - N(H + T)(A + 0.332) \right] \text{[microhenrys]}. \quad (21)$$

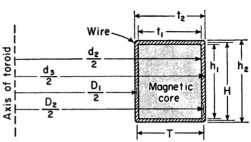

Fig. 11. Geometry for inductance calculation.

The symbols employed (Fig. 11) are defined below, dimensions being given in inches:

D_1 = effective inner diameter = $d_2 - t_1 - T$,

D_2 = effective outer diameter = $d_2 - t_1 + T$,

d_2 = outside diameter of unwound core,

t_1 = radial thickness of unwound core,

h_1 = height of unwound core,

dw_1 = over-all diameter of insulated wire,

d_3 = outside diameter of wound core,

t_2 = radial thickness of wound core,

h_2 = height of wound core,

dw_2 = diameter of bare wire,

H = effective height = $\dfrac{h_1 + dw + h_2 - dw}{2} = \dfrac{h_1 + h_2}{2}$

T = effective thickness = $\dfrac{t_1 + dw + t_2 - dw}{2} = \dfrac{t_1 + t_2}{2}$

$$A = \log_e \left(1.7452 \cdot \frac{2N\, dw_2}{\pi(D_1 + D_2)} \right)$$

To calculate the inductance of a toroid containing a magnetic material, it is necessary to know what fraction of the total flux is contained within the cross section of the core. There is no rigorous solution to this problem, and consequently we are forced to make certain assumptions about the flux distribution. We assume that the flux threading the core is μ'/μ_0 times the flux

[11] E. B. Rosa and F. W. Grover, *Bull. U. S. Bur. Standards* **8**, 122 (1912).

Table 3. Inductance bridges

Basic Types

Type	Balance Condition	Remarks
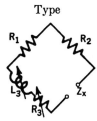 Simple inductance	$L_x = \dfrac{R_2}{R_1} L_3$ $R_x = \dfrac{R_2}{R_1} R_3$	In practice, adjustable elements are usually R_1 and R_2. This arrangement, while permitting use of decade resistor boxes, causes a "sliding zero" condition, i.e., balance condition for R_x is dependent on that for L_x.
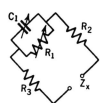 Maxwell-Wien	$L_x = C_1 R_2 R_3$ $R_x = \dfrac{R_2}{R_1} R_3$	Fixed condenser can be used and either R_2 or R_3 made variable. However, "sliding zero" condition results.
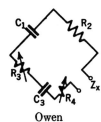 Owen	$L_x = C_1 R_2 R_3$ $R_x = \dfrac{C_1}{C_3} R_2 - R_4$	Independent balance for R_x and L_x achieved with variable resistors.

Modified Types

Type	Balance Condition	Remarks
Hay	$L_x = \dfrac{R_2 R_3 C_1}{1 + \omega^2 C_1^2 R_1^2}$ $R_x = \dfrac{R_1 R_2 R_3 \omega^2 C_1^2}{1 + \omega^2 C_1^2 R_1^2}$	Simple modification of Maxwell-Wien. Useful for high Q coils. Balance dependent on frequency.
Anderson	$L_x = CR_2\left[R\left(1+\dfrac{R_3}{R_1}\right)+R_3\right]$ $R_x = \dfrac{R_2 R_3}{R_1}$	Derivable from Maxwell-Wien by circuit transformation. Capable of high precision and wide range in measurement of L.
Series resonance	$L_x = \dfrac{1}{\omega^2 C_4}$ $R_x = \dfrac{R_2 R_3}{R_1}$	Modified simple inductance bridge. Useful for accurate measurements of loss.

within a nonmagnetic core of the same dimensions, and we further assume that the correction for departure from the idealized current sheet is independent of μ'. The new value of inductance thereby becomes

$$L_{tot} = 0.0117 \left[N^2 H \log_{10} \frac{D_2}{D_1} \left(\frac{A_{core}}{A_{coil}} \frac{\mu'}{\mu_0} + \frac{A_{coil} - A_{core}}{A_{coil}} \right) - N(H+T)(A+0.332) \right] \; [\mu h], \quad (22)$$

with A_{coil}, the coil area equal to HT, and $A_{core} = h_1 t_1$. The permeability of the material is obtained by solving Eq. 21:

$$\frac{\mu'}{\mu_0} = \frac{A_{coil}}{A_{core}} \left[\frac{L_{tot} + L_c}{L_0 + L_c} - \frac{A_{coil} - A_{core}}{A_{coil}} \right]. \quad (23)$$

The measured total resistance, R_{tot}, is made up of the sum of two terms, R_m and R_c. The former is associated with losses in the core material,

$$R_m = \frac{\mu''}{\mu_0} \omega L_0' \left(\frac{A_{core}}{A_{coil}} \right), \quad (24)$$

whereas the other arises from losses in the copper conductor. The magnetic loss is thus given by

$$\frac{\mu''}{\mu_0} = \frac{R_{tot} - R_c}{\omega L_0'} \left(\frac{A_{coil}}{A_{core}} \right). \quad (25)$$

The conductor resistance will be greater than the d-c resistance because of eddy-current losses in the copper.

When samples are available in the form of rods and bars, their permeability may be determined by measuring the change in inductance of a solenoid when the sample is introduced (Fig. 12). The permeability may

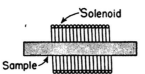

Fig. 12. Rod sample in short solenoid.

be expressed in terms of the inductance change ΔL, and the inductance L_e of the empty solenoid:

$$\frac{\mu'}{\mu_0} = 1 + K_{c_2} \frac{A_{coil}}{A_{core}} \frac{\Delta L}{L_e}. \quad (26)$$

K_{c_2} is a correction factor which takes into account sample demagnetizing effects, and its value must be found empirically.

Although K_{c_2} is a function of μ' and of the cross-sectional shape and area of the specimen, it is almost independent of sample length for rods greater than about three times the length of the solenoid. For a ¼-in. diameter rod placed in a solenoid ⅜ in. in diameter by 1 in. long, K_{c_2} has the following typical values:

K_{c_2}	μ'/μ_0
1	1
1.1	3
5	400

For values of $\mu' > 100$ the measured change in inductance becomes insensitive to the value of μ', and the accuracy of the method suffers. This situation arises because as the permeability of the rod is increased the field within the sample becomes governed more by the demagnetizing field than by the field applied externally. We can rewrite Eq. 15 in a form that demonstrates this point:

$$\frac{\mu_{app}}{\mu_0} = \frac{\mu_{true}/\mu_0}{1 + N\left(\frac{\mu_{true}}{\mu_0} - 1\right)}.$$

For the geometry we are using, N is, of course, a function of the permeability, but it is clear that μ_{app} increases far less rapidly than μ_{true}, and that in the limit of large values of μ_{true} the apparent permeability is governed by the demagnetizing factor.

If a capacitance rather than an inductance bridge is available, the inductance of either the toroid or the

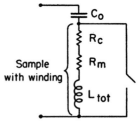

Fig. 13. Sample with series capacitor for measurement in capacitance bridge.

solenoid may be determined by measuring the change in capacitance, ΔC, produced in a series RLC circuit by short-circuiting the coil (Fig. 13). The coil inductance is given by

$$L_{tot} = \frac{\Delta C}{\omega^2 C_0 (C_0 + \Delta C)}, \quad (27)$$

where C_0 is the capacitance of the series condenser.

Radio-Frequency Measurements

The radio-frequency range is a transition region in which it is possible to employ lumped-circuit techniques, distributed-circuit techniques, or a combination of the two. By paying careful attention to details of construction, some of the audio frequency bridges, for example, the Maxwell-Wien bridge, can be made to operate up to frequencies of several megacycles, and coils similar to those employed at audio frequencies can be used. For the measurement of metals, bridges have been constructed in which the specimen is incorporated as the center conductor in a coaxial transmission line which is placed in one arm of the bridge. The other arms may contain lumped elements or may themselves be transmission lines.[12] Distributed systems may also be employed in meas-

[12] K. V. Kreielsheimer, *Ann. Physik* **17**, 293 (1933).

uring nonmetallic materials at radio frequencies. The sample, usually a toroidal ring, is placed at the end of a short-circuited coaxial line. The permeability of the specimen is determined from a measurement of the input impedance of the line. Such an impedance determination may be made by direct measurement with a radio-frequency impedance bridge. The line can also be resonated with external capacitance, and the input impedance determined by employing a frequency or susceptance variation technique (see IIA, Secs. 1 and 2).

In measuring coil inductance and resistance at radio frequencies, it is necessary to make corrections for the distributed capacity C_D of the coil (Fig. 14). The

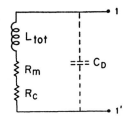

Fig. 14. Coil with distributed capacitance.

values of inductance and resistance measured at the terminals 1–1' are related to the true values as follows: [13]

$$R_{1-1'} \simeq (R_m + R_c)(1 + 2\omega^2 L_{tot} C_D),$$
$$L_{1-1'} \simeq L_{tot}(1 + \omega^2 L_{tot} C_D), \qquad (28)$$

the approximation being valid for operation well below the self-resonant frequency of the coil, that is, $\omega^2 < 1/L_{tot}C_D$.†

Above 10^5 cps, corrections for distributed capacity are difficult to make for coils wound directly on a

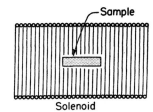

Fig. 15. Short rod sample in solenoid.

toroidal specimen. Preferably, one determines the change in inductance and resistance produced by introducing a sample into an air-core coil. The arrangement of a rod inserted into a short solenoid (previously described in connection with audio meas-

[13] V. E. Legg, *U. S. Bur. Standards Tech. J.* **15**, 39 (1936).

† It is evident that any additional stray capacity across terminals 1–1', e.g., capacity between bridge terminals, must also be corrected for.

urements) may be used. Other arrangements are possible: a small diameter rod may be placed in a relatively long solenoid,[14] as in Fig. 15; or demountable toroidal coils may be used.[15] Frequency or susceptance variation techniques provide a convenient method for measuring inductance changes obtained with such coil-core configurations.

For the susceptance variation methods (Fig. 16) the permeability of the core material is given in terms

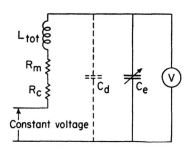

Fig. 16. Susceptance variation circuit.

of the total capacity C_{rt} required to resonate the coil with the core inserted, and the change in capacity, ΔC, needed to retune the coil when the specimen is removed. The specific form of the relation for μ' will depend upon the coil-sample geometry; for a rod in a short solenoid, we have

$$\frac{\mu'}{\mu_0} = K_{c_2} \left[\frac{A_{coil}}{A_{core}} \frac{\Delta C}{C_{ri}} + 1 \right]. \qquad (29)$$

The imaginary part of the permeability is obtained from measurements of the capacity bandwidth of the resonant circuit before and after sample insertion.

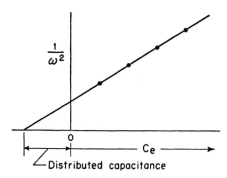

Fig. 17. Determination of distributed capacitance of coil.

The total capacity C_{ro} needed to resonate the empty coil is made up of the known circuit capacity placed across the coil plus the distributed capacity of the coil. The value of the latter capacity can be found by resonating the empty coil at several different

[14] Marcel Cogniat, *Cables and Transmission* **2**, 195 (1948).
[15] I. Epelboin, *L'onde électrique* **28**, 322 (1948).

frequencies to obtain a plot of external resonating capacity vs. $1/\omega^2$. The stray capacity can then be determined from an extrapolation of this plot (Fig. 17). It is normally assumed that introduction of the specimen does not alter either the value of distributed capacity or the dielectric losses in the coil insulation and supports.

Distributed Parameter Systems

Lumped parameter measuring systems have their upper limit of usefulness at about 100 Mc. However, it is frequently desirable to abandon lumped-circuit techniques in favor of distributed parameter methods at much lower frequencies. Not only does the stray capacity of the coil enter as an undesirably large correction factor, but the field within the specimen becomes nonuniform in a rather complicated way. This latter situation can become particularly serious in certain ferrite materials having large values of both μ^* and ϵ^*. The wavelength within these materials may become comparable to the dimensions of the sample, and an electrical resonance of uncertain mode may occur within the core material, giving rise to erroneous values of the measured permeability.[16]

Perhaps the simplest distributed-parameter measurement scheme is one in which a nonmetallic toroidal core is placed at the closed end of a coaxial line (Fig. 18).

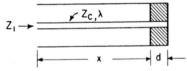

Fig. 18. Magnetic sample at end of coaxial line.

For a sample having a thickness d, short compared to the wavelength within the material, the complex permeability is

$$\frac{\mu^*}{\mu_0} \simeq \frac{\lambda}{2\pi d} \frac{\frac{Z_1}{Z_c} - j\tan\frac{2\pi x}{\lambda}}{\frac{Z_1}{Z_c}\tan\frac{2\pi x}{\lambda} + j}, \quad (30)$$

where Z_1 is the input impedance of the line, x is distance from the open end of the line to the front face of the sample, and Z_c and λ are respectively the characteristic impedance and the operating wavelength in the air-filled line. For samples having $\left|\frac{2\pi d}{\lambda}(\epsilon^*\mu^*)^{1/2}\right| \leq 0.3$

[16] F. G. Brockman, P. H. Dowling, and W. G. Steneck, *Phys. Rev.* **77**, 85 (1950).

the computational error arising from the use of Eq. 30 is less than 3 percent.

The input impedance of the line may be measured in any number of ways. At radio- and ultra-high frequencies, bridges and frequency- and susceptance-variation techniques can be used. At higher frequencies, it is more convenient to use a standing-wave detector or a line-length variation method (see IIA, 2).

In carrying out measurements on ferromagnetic metals, the specimen is made part of the conductor system. Early investigators employed open, two-wire transmission lines (Lecher wires), but in modern practice a completely shielded system, such as a coaxial line, is preferred, with the sample ordinarily being introduced as part of the center conductor. The diameter of the ferromagnetic conductor is ordinarily made large compared to the skin depth to simplify the calculations involved.

Until recently the high-frequency permeability of metals has been given in terms of μ_R and μ_L. Experimental techniques used for the determination of μ_R form a class distinct from those used to find μ_L. The resistive permeability is normally determined from some measurement involving attenuation (or cavity Q), whereas the inductive permeability is obtained from measurements of phase velocity (or resonant frequency of a cavity). Although a direct measurement of quantities such as phase velocity and attenuation are possible, more accurate results can be obtained by employing substitution methods in which the properties of a transmission line (or cavity) are measured, first with the ferromagnetic conductor in position and then with this conductor replaced by a nonferromagnetic one. To avoid substitution errors, great care must be exercised in mechanical construction.

Measurements of μ_R have been made by Hodsman et al.,[17] using the arrangement of Fig. 19. By means

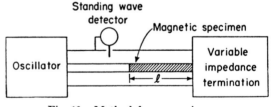

Fig. 19. Method for measuring μ_R.

of a standing-wave detector, the input impedance of a length of line containing the sample is measured for different values of the terminating impedance. When the results are plotted in the complex impedance plane

[17] G. F. Hodsman, G. Eichholz, and R. Millership, *Proc. Phys. Soc.* (London) **B62**, 377 (1949).

the measured points should fall on a circle of constant αl (Fig. 20). The value of α is found from the radius of the circle and the location of its center.

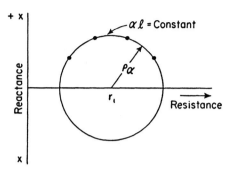

Fig. 20. Plot of impedance measurements in the complex plane.

Accurate measurements of μ_L are more difficult to make than corresponding measurements of μ_R on the same material. For one thing, it is clear from Eq. 12 that, at a given frequency, μ_R is always larger than μ_L. Iron, for example, has $\mu_R \simeq 90$ at 1000 Mc, whereas $\mu_L \simeq 15$. A further reason for the relative difficulty in measuring the inductive permeability is to be found in the fact that a given value of μ_L is less effective in changing the self-inductance of the center conductor than an equal value of μ_R is in changing the conductor resistance. Thus for equal values of μ_L and μ_R, there will be less change in phase velocity than in attenuation.

An accurate method of determining μ_L was developed by Glathart[18] and has been used successfully by other experimenters [19] (Fig. 21). Two similar cavi-

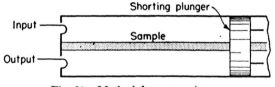

Fig. 21. Method for measuring μ_L.

ties are employed, one with a ferromagnetic center conductor, the other with a copper conductor. The transmission through each cavity is measured as a function of the position of the shorting plunger. The range of plunger motion is large enough to permit a series of resonance peaks to be traced out, and from these data the phase velocity in the respective lines is obtained. The value of μ_L can then be computed in terms of the reduced wavelength in the resonator containing the ferromagnetic specimen.

[18] J. L. Glathart, *Phys. Rev.* **55**, 833 (1939).
[19] R. Millership and F. V. Webster, *Proc. Phys. Soc. (London)* **B63**, 783 (1950).

Johnson and Rado [20] have employed an ingenious method for determining μ_R and μ_L simultaneously (Fig. 22). Though essentially one of substitution, their method avoids the need for any direct physical replace-

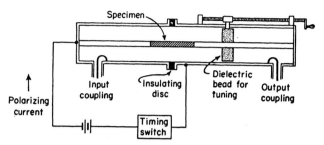

Fig. 22. Method for measurement of both μ_R and μ_L. (After Johnson and Rado.[20])

ment of one conductor by another. By passing a large d-c current through the ferromagnetic center conductor of a coaxial resonator, they are able to approach saturation in the specimen and thereby reduce its reversible permeability to unity. Measurements of the Q and the position of the dielectric tuning bead at resonance, made for both the demagnetized and magnetized states, yield data sufficient for a determination of μ_R and μ_L. Since polarizing currents of the order of 1000 amp are needed to produce saturation, it is necessary to apply this current as a pulse of several milliseconds duration to avoid heating effects.

Separation of ϵ^* and μ^*

So far we have restricted ourselves to experimental situations where it has been possible to disregard the role played by the electric field. In our previous treatment of a nonmetallic specimen placed at the short-circuited end of a transmission line (Fig. 18) we assumed the sample thickness d to be small compared to the wavelength within the specimen. Once this assumption breaks down, it is no longer possible to determine μ^* independently of ϵ^*.

In order to determine both μ^* and ϵ^*, the sample is measured at two positions in the line; in one position the field should be predominantly magnetic, whereas in the other the electric field should predominate. For simplicity of calculation the most convenient sample positions are those shown in Fig. 23. The sample is measured in the "short-circuit" position and also at a distance $\lambda/4$ away from the closed end of the line in the so-called open-circuit position. The properties of the specimen can be found in terms of the impedances measured at the front face of the sample for the two

[20] M. H. Johnson and G. T. Rado, *Phys. Rev.* **75**, 841 (1949).

positions. From IA, Eqs. 4.36 and 4.41, we have

$$Z_{SC}(0) = Z_2 \tanh \gamma_2 d,$$
$$Z_{OC}(0) = Z_2 \coth \gamma_2 d, \quad (31)$$

where Z_2 is the intrinsic impedance of the specimen and is equal to $(\mu^*/\epsilon^*)^{1/2}$, and γ_2, the propagation factor in the material, is equal to $j\omega(\epsilon^*\mu^*)^{1/2}$. Equation 31 may be manipulated into a form more suitable for computation, giving

$$Z_2 = [Z_{SC}(0)Z_{OC}(0)]^{1/2},$$
$$\gamma_2 d = \tanh^{-1}[Z_{SC}(0)/Z_{OC}(0)]^{1/2}. \quad (32)$$

We have two expressions for ϵ^* and μ^*, one containing their ratio, the other their product. With the aid of charts of $\tanh^{-1} Z$ [21] a simultaneous solution can be carried out to obtain both the complex permeability and permittivity (see also IIA, Sec. 2).

[21] A. E. Kennelly, *Chart Atlas of Complex Hyperbolic and Circular Functions*, Harvard University Press, Cambridge, Mass., 1924.

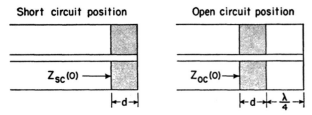

Fig. 23. Sample positions for determining μ^* and ϵ^*.

C · Microwave Spectroscopy

By MALCOLM W. P. STRANDBERG

The development of new techniques in the microwave region has opened up a new branch of spectroscopy extending roughly over the wavelength range from 3 mm to 30 cm. This microwave spectroscopy is concerned mainly with the absorption spectra of rotating molecules possessing permanent electric or magnetic dipole moments. The frequency regions and interests of various branches of spectroscopy are shown in Fig. 1.

Resolution

All branches of spectroscopy are essentially concerned with the resolution of their instruments which sets the accuracy with which a spectral line and its detailed structure can be measured. The term *resolution* used by the spectroscopist is identical with the Q of the electrical engineer ($\nu/\Delta\nu$). Any structure observed in a spectral line tests our quantum-me-

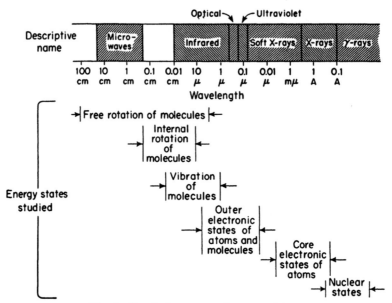

Fig. 1. Spectral ranges and energy states.

chanical description of molecules; it must be adequately explained by the theory, or the theory must be modified. Once explained, the fine structure is an exceedingly fruitful source of data to evaluate the parameters by which we describe atoms and molecules. An example of the great resolution achieved in microwave spectroscopy is given in Fig. 2. The absorption

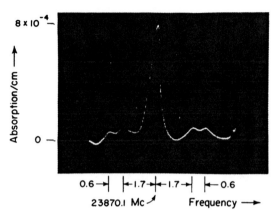

Fig. 2. $J = 3-$, $K = 3$-line of $N^{14}H_3$ inversion absorption (hyperfine structure due to N^{14} quadrupole moment).

shown is due to the ammonia molecule, which exhibits a hyperfine structure due to the quadrupole moment of the N^{14} nucleus. The satellites are separated from the main line by about 1 in 25,000 parts of the absorption frequency. The resolution here achieved is about 1 part in 250,000.

Instrumental limitations

Resolution in other branches of spectroscopy is determined by the dimensions of the slit, gratings, or prisms, and is limited by source intensity: the power incident on the detector must be above the noise level. The microwave region has no limitations imposed by the dispersion of instruments, since the sources (klystrons) emit monochromatic radiation. The radiation may be made monochromatic by electronic means to better than one part in 10^7 for periods of time convenient for measurements to be made (at 25,000 Mc per sec a source may be stabilized to 1 kc per sec for several minutes).

Natural limitations

Thus in the microwave region, the resolution is not limited by the instrument but by the intrinsic frequency width of the spectral line. There are many sources of this intrinsic *line breadth*. At microwave frequencies the essential factors are the Doppler width (of the order of 25 kc per sec) and pressure broadening. This broadening by the pressure of the gas or vapor is within the control of the experimenter. Since Doppler width sets an almost irreducible lower limit on line breadth, pressure widths the order of 100 kc per sec and larger are commonly used. The pressures for this width correspond to about 1 micron Hg pressure. Summarizing: a resolution of the order of 250,000 is readily achieved in the microwave region, whereas measuring accuracies are some ten times greater.

Phenomena Studied

Microwave spectroscopy is thus capable of high accuracy and resolution for the study of atomic or molecular energy states which have energy differences characteristic of microwave frequencies, and for which a permanent electric or magnetic moment exists. Energy differences of the order of the microwave frequencies (ca. 1 cm^{-1}) may arise from hyperfine structure in electronic states, as of hydrogen or cesium, for example; from fine structure in vibrational states as of ammonia; or from the structure of rotational energy states.

Data accurate to 1 part in 10^6 are meaningless unless related to a theory that includes effects caused by the perturbations of the energy states of atoms and molecules. For this reason microwave spectroscopy is essentially a study of such perturbations. Though the space available here is too limited to consider these effects in detail, it is well to mention at least some of them. For instance, in the study of rotational energy levels, account must be taken of the centrifugal distortion of the molecule. This effect, for light molecules such as water, contributes corrections of the rotational absorption frequencies of the molecule which are of the order of several percent. Similarly, molecules in different vibration states act essentially as different molecules because the moment of inertia has changed by measurable amounts. Finally, isotopic forms of the constituent nuclei give different absorption spectra. Figure 3 illustrates the situation on the linear molecule O=C=Se.

An example of smaller perturbations is the electronic correction of rotational energy levels. The rotational states are determined primarily by the three principal reciprocal moments of inertia. These reciprocal moments may be readily determined from the nuclear masses and internuclear distances, except for a correction due to the spatial distribution and lack of precise slipping of the electrons about their nuclear centers in the rotation of the molecule as a whole. Actually, the core electrons may be expected to slip quite accurately so that atomic masses may be used instead of

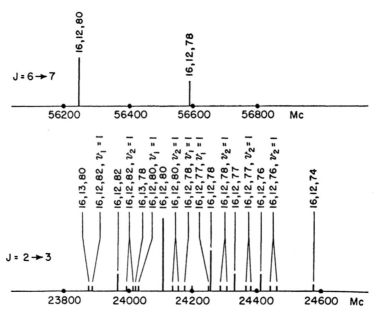

Fig. 3. Absorption spectrum of O=C=Se. (16, 12, 80: rotation line produced by molecule O^{16}=C^{12}=Se^{80} in zero vibration state. 16, 13, 80: line shifted due to replacement of C^{12} by C^{13}. **16, 12, 80**, $v_1 = 1$: line for molecule in first stretching vibration. 16, 12, 80, $v_2 = 1$: two lines for first bending vibration state.)

nuclear masses and a further correction, presumably due only to the valence electrons, will need to be made. Such a correction then would be only of the order of 1 part in 10^4 or 10^5, but it is still worthy of consideration.

One external perturbation which is readily studied with the high resolution afforded by microwave spectroscopy is the splitting of spectral lines by an electric field, the Stark effect. Perturbations of absorption frequencies of the order of several megacycles per second are obtained with fields of the order of 1000 volts per cm. The measurement of this frequency shift as a function of the electric field allows a measurement of the molecular electric dipole moment.

It will be recalled that a static electric field E interacts with the molecular electric dipole moment μ in such a way as to add to the molecular energy an amount (see IB, Sec. 3) $\Delta\mathcal{E} = -\mu \cdot E$. Actually, except for symmetric tops with degenerate inversion states, the perturbation is only second order, that is, it is proportional to $\mu^2 E^2$. Figure 4a gives the absorption of the linear molecules OCS for the transition from the first to the second rotation state ($J = 1 \rightarrow 2$) in the absence of an electric field. Figure 4b shows that this line splits into two components when a square-wave modulated electric field is applied to the OCS gas. When the field is zero the absorption as in Fig. 4a is observed. With the field the two components ($M = 0, 1$) appear.

When the electric field is known, the dipole moment may be evaluated for any pair of rotational states quite independent of the chemical purity of the sample.

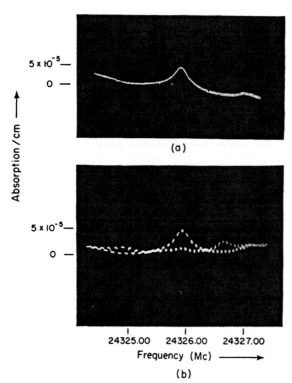

Fig. 4. (a) Absorption as function of frequency for $J = 1 \rightarrow J = 2$ rotation line of O=C=S. (b) Splitting of line by square-wave modulation of electric field (Stark effect).

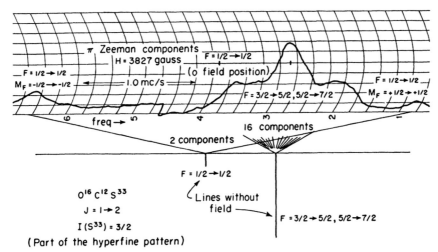

Fig. 5. Rotation lines $J = 1 \to 2$ of O=C=S molecule with and without magnetic field (Zeeman effect). ($I = 3/2$: nuclear angular momentum of S^{33}; $F = I + J$ = total angular momentum.)

Representative dipole moments so observed are given in Table 1.

Table 1. Electric dipole moments evaluated from microwave spectra

(In debyes)

CH_3F	2.07 ± 0.02	AsH_2D	0.22 ± 0.02
CH_3Cl	1.869 ± 0.010	AsF_3	2.815 ± 0.025
CH_3Br	1.797 ± 0.015	SbH_2D	0.116 ± 0.003
CH_3I	1.647 ± 0.014	$ClCN$	2.802 ± 0.020
CH_3CCH	0.75 ± 0.01	OCS	0.7085 ± 0.004
CH_3CN	3.92 ± 0.06	$OCSe$	0.754 ± 0.01
CH_3CF_3	2.321 ± 0.034	H_2CO	2.31 ± 0.04
H_3BCO	1.795 ± 0.01	H_2CCO	1.414 ± 0.01
SiF_3H	1.26 ± 0.01	H_2O	1.94 ± 0.10
CHF_3	1.64 ± 0.02	HDO	1.84 ± 0.01
POF_3	1.77 ± 0.02	HDS	1.02 ± 0.02
PH_2D	0.55 ± 0.01	N_2O	0.166 ± 0.002
PF_3	1.025 ± 0.009		

Static external magnetic fields may also be applied to the vapor under examination and result in Zeeman and associated effects. The rotation-induced magnetic moment of the molecule interacts with this static field to perturb the energy an amount $\Delta \varepsilon = -\mu \cdot B$. This interaction is again calculable in terms of molecular parameters. Hence the experimental observation of the magnetic line splitting allows the determination of molecular magnetic moments.

A typical Zeeman spectrum of the O=C=S molecule is shown in Fig. 5. This shows the splitting of an absorption line due to the molecule $O^{16}C^{12}S^{33}$. These data permit not only a determination of the molecular magnetic moment, but also of the nuclear magnetic moment of the S^{33} nucleus, since the nucleus is coupled to the molecular rotation by its electric quadruple moment. As for ammonia (see Fig. 2), this quadrupole coupling gives rise to a resolved hyperfine structure of the OCS^{33} absorption spectrum (Fig. 6). This splitting arises from the addition of the S^{33} nuclear spin I of 3/2, to the rotational angular momentum J, to give a total angular momentum $F = J + I, J + I - 1, \ldots,$

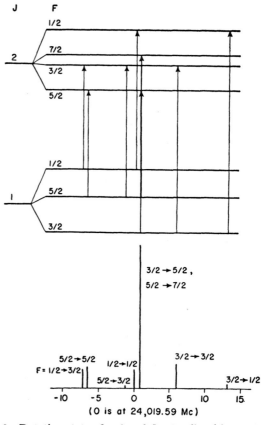

Fig. 6. Rotation states $J = 1$ and $J = 2$ split without external field (due to coupling of nuclear and molecular angular momenta by nuclear electric quadrupole interaction).

$J - I$. The pattern is unique for each value of nuclear spin I, and hence theoretical interpretation of the spectrum determines the value of I in this case as 3/2, just as Fig. 2 shows that the N^{14} has a nuclear spin I of 1.

Both the Zeeman and the Stark effects are important laboratory tools for the identification of rotational transitions.

The Spectroscope

A typical spectroscope is shown in Fig. 7. The klystron source sends microwave radiation through the sample cell to a crystal detector. The absorption is square wave-modulated by an electric field at a 6 kc per sec rate (Stark modulator). Only the modulated component of the gas absorption is amplified. The frequency of the klystron is varied in a sawtooth fashion (for example, ±5 Mc per sec). The much slower variation of the 6 kc per sec-amplifier output describes the absorption coefficient of the gas as the klystron sweeps through the resonant absorption frequency. The structure shown in Fig. 4a is due to a variation in the detecting crystal current of about ½ percent; a broad-band oscilloscope amplifier is used here. Small fluctuations in power output of the klystron with time or frequency are very objectionable in this direct method. However, if the signal shown in Fig. 4b is filtered for the 6 kc per sec fundamental,

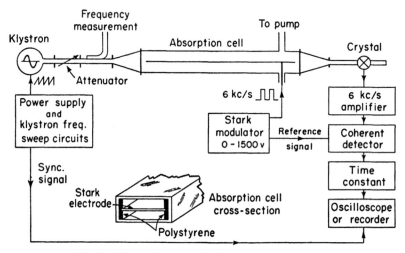

Fig. 7. Block diagram of microwave spectroscope.

only the absorption changes at this frequency comprise the amplifier output, and the effect of the klystron is eliminated. The result of such a filtering process is a much clearer picture of the structure of the absorption (Fig. 8).

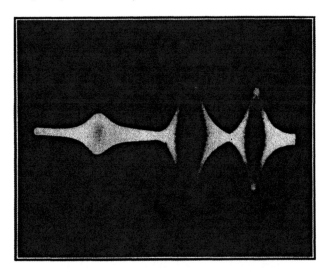

Fig. 8. Signal of Fig. 4b with fundamental of square wave remaining.

BIBLIOGRAPHY

Spectroscopy and the related theory are relatively old and sophisticated whereas experimental microwave spectroscopy is rather new. Thus papers relating the two in a general survey fashion are few. Two to be recommended are: W. Gordy, "Microwave Spectroscopy," *Revs. Mod. Phys.* **20**, 668 (1948), and D. K. Coles, "Microwave Spectroscopy," in *Advances in Electronics*, Vol. II, Academic Press, New York, 1950. The two volumes by G. Herzberg, *Spectra of Diatomic Molecules* and *Infrared and Raman Spectra of Polyatomic Molecules* (D. Van Nostrand Company, New York, 1935 and 1945) are excellent surveys of the theory of rotational spectra of molecules. The discussion of the quantum-mechanical form for absorption coefficient by J. H. Van Vleck and V. F. Weisskopf (*Revs. Mod. Phys.* **17**, 227 [1945]) is basic, though the presentation of R. Karplus and J. Schwinger (*Phys. Rev.* **73**, 1020 [1948]) is probably more elegant. Discussion of the appropriate Stark effect may be found in scattered papers, for example,

R. de L. Kronig, *Proc. Natl. Acad. Sci.* **12**, 488 (1926), W. G. Penney, *Phil. Mag.* **11**, 602 (1931), S. Golden and E. B. Wilson, Jr., *J. Chem. Phys.* **16**, 669 (1948); M. W. P. Strandberg, *ibid.*, **17**, 901 (1949), and M. W. P. Strandberg, T. Wentink, Jr., and R. L. Kyhl, *Phys. Rev.* **75**, 270 (1949). The Zeeman effect is discussed at length in J. R. Eshbach and M. W. P. Strandberg, *Phys. Rev.* **85**, 24 (1952); J. R. Eshbach, R. E. Hillger, and M. W. P. Strandberg, *ibid.*, **85**, 532 (1952). An example of an extensive examination of one molecule is given in H. R. Johnson and M. W. P. Strandberg, *J. Chem. Phys.* **20**, 687 (1952). Finally, there is a survey volume, *Microwave Spectroscopy*, by W. Gordy, W. V. Smith, and R. F. Trambarulo, published by John Wiley and Sons in 1953; and one by M. W. P. Strandberg, *Microwave Spectroscopy*, Methuen, London, 1954.

D · Magnetic Resonance

By FRANCIS BITTER

Although the magnetization of matter by a constant field has been described in detail for a considerable fraction of this century, the mechanisms involved in producing a change of magnetization have been obscure until quite recently. It is only during the last six or seven years that the theoretical and experimental investigations of magnetic resonance have shed much light on this fundamental phenomenon. We shall review some of the concepts concerning the rotational properties of particles, and some of the conclusions that are being arrived at as a result of studies of these properties.

The central fact in this discussion is that the particles in question have angular momenta. They are small gyroscopes, and an analysis of means available for their reorientation inevitably leads to the familiar peculiarities of the schoolboy's top. We shall first take up the classical and quantum-mechanical description of isolated gyroscopes to which constant or variable torques may be applied by means of fields. In the last half of this discussion we shall then take up the complications introduced when these gyroscopes interact with each other, as in actual matter.

Magnetic fields are due to charges in motion. A current I flowing around a circuit of area A has a magnetic moment **m** given by (see IA, Sec. 2)

$$|\mathbf{m}| = IA \quad [\text{amp-m}^2]. \qquad (1)$$

The energy $\mathcal{E}$ of a magnetic dipole **m** in a field **B** is

$$\mathcal{E} = -\mathbf{m} \cdot \mathbf{B} \quad [\text{joules}], \qquad (2)$$

if **B** is expressed in webers per m². The torque **T** exerted on the dipole by the field is

$$\mathbf{T} = [\mathbf{m} \times \mathbf{B}] \quad [\text{newton-m}]. \qquad (3)$$

A purely classical theory of the magnetic properties of matter is impossible because classical mechanics and electrodynamics together cannot account for the existence of atoms with permanent magnetic moments.

The first great contribution of the quantum theory to an understanding of the magnetic properties of matter was to show that the angular momentum of any mechanical system is quantized. The elementary particles with which we are concerned, electron, proton, and neutron, each has an angular momentum $\tfrac{1}{2}(h/2\pi)$ due to a spinning motion about its own axis. In addition, a group of these particles may have an angular momentum of any integral number of times $h/2\pi$ due to orbital motion about their common center of mass. These motions are so coupled together that any system of particles will have a resultant angular momentum $n\dfrac{h}{2\pi}$ or $\left(n+\dfrac{1}{2}\right)\dfrac{h}{2\pi}$, depending on whether the total number of particles is even or odd, n being some positive integer or zero.

Associated with this angular momentum is a magnetic moment. We shall return to a more precise quantum-mechanical formulation of what we mean by an atomic magnetic moment and how it is to be formulated. For the present it suffices to point out that there is, in general, a magnetic moment characteristic of each state of a given system, parallel or antiparallel to the resultant angular momentum. If **P**′ is the resultant angular momentum and **m** is the resultant magnetic moment, we may put

$$\mathbf{m} = \gamma \mathbf{P}'. \qquad (4)$$

The quantity γ is called the *gyromagnetic ratio*. For the orbital motion of a charge e having a mass m, $\gamma = e/2m$. For example, if the charge is moving in a circular orbit of radius r with angular velocity ω, we have $p' = mr^2 \cdot \omega$, $|\mathbf{m}| = I \cdot \pi r^2 = e\dfrac{\omega}{2\pi} \cdot \pi r^2 = \dfrac{er^2 \cdot \omega}{2}$, $\gamma = \dfrac{|m|}{p'} = \dfrac{e}{2m}$ (see IB, Eq. 2.47). The magnetic moment associated with the orbital motion of a particle having one unit of angular momentum $h/2\pi$ might therefore be expected to be

$$\mathbf{m}_B = \gamma p' = \frac{e}{2m}\frac{h}{2\pi} = \frac{eh}{4\pi m}. \quad (5)$$

This is, in fact, an important unit and is given the designation $\mathbf{m}_B$, or a *Bohr magneton* when m is the electronic mass (see IB, Eq. 2.48). Another common unit is the *nuclear magneton*, in which the mass of the proton is substituted for the mass of the electron; hence $\mathbf{m}_r = \frac{1}{1836}\mathbf{m}_B$. When orbital motions and spins of electrons and nucleons are coupled together, the gyromagnetic ratio and magnetic moment may have a wide range of values, and are not simple multiples of some basic unit, as is the angular momentum.

If the resultant angular momentum † of a system is $F\frac{h}{2\pi}$, quantum theory tells us that in a weak magnetic field the component of the angular momentum in the direction of the field is $m_F\frac{h}{2\pi}$, where *magnetic quantum number* m_F may have any of the values F, $F-1$, $F-2$, ..., $-F$. The ratio m_F/F may be thought of as the average value of $\cos\theta$, the angle between the angular momentum and the field. If $|\mathbf{m}|$ is the magnetic moment of the system, the magnetic energy is given by $\mathcal{E} = -\frac{m_F}{F}|\mathbf{m}|B$.

These considerations suffice for the computation of the equilibrium properties of simple paramagnetic systems. We now come to our first dynamical considerations. Newton's laws of motion state that the torque applied to any mechanical system is equal to the rate of change of its angular momentum:

$$\mathbf{T} = \frac{dp'}{dt}. \quad (6)$$

Substituting Eqs. 3 and 4, we get

$$\frac{d\mathbf{m}}{dt} = \gamma[\mathbf{m}\times\mathbf{B}]. \quad (7)$$

This is the classical equation of motion of the system. Let us apply it to the case of a dipole of fixed magnitude $|\mathbf{m}|$ in a constant field B_z parallel to the z-axis. Using polar angles to describe the orientation of the magnetic dipole, we have $m_x = |\mathbf{m}|\sin\theta\sin\phi$, $m_y = |\mathbf{m}|\sin\theta\cos\phi$, and $m_z = |\mathbf{m}|\cos\theta$ (Fig. 1). The

† The resultant angular momentum F of an atom is the vector sum of the electronic angular momentum J and the nuclear angular momentum I.

equation of motion (7) for each of the components of $\mathbf{m}$ becomes

$$\frac{dm_x}{dt} = |\mathbf{m}|\frac{d}{dt}(\sin\theta\sin\phi) = \gamma(m_y B_z - 0)$$
$$= \gamma|\mathbf{m}|\sin\theta\cos\phi B_z,$$
$$\frac{dm_y}{dt} = |\mathbf{m}|\frac{d}{dt}(\sin\theta\cos\phi) = \gamma(0 - m_x B_z) \quad (8)$$
$$= -\gamma|\mathbf{m}|\sin\theta\sin\phi B_z,$$
$$\frac{dm_z}{dt} = |\mathbf{m}|\frac{d}{dt}\cos\theta = \gamma(0 - 0) = 0.$$

The last of these equations states that θ must be constant, or that, according to Eq. 2, the energy of the

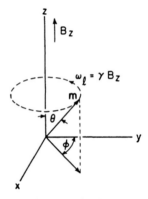

Fig. 1. Geometry used for derivation of Larmor precession.

system is constant. The magnetic forces are always at right angles to the direction of motion of electric charges and do no work. The first two equations both lead to the same result:

$$\frac{d\phi}{dt} \equiv \omega_L = \gamma B_z. \quad (9)$$

The magnetic moment of the system precesses with the angular velocity ω_L, called the *Larmor frequency*, around the field direction, maintaining a constant inclination to the field.

This is the classical, or Newtonian, result. It is also, if properly interpreted, the quantum-mechanical result. Any stationary state of a system is describable by a wave function, and the observable properties of the system are computed by means of operators which, acting on the wave function in question, yield the desired *expectation values*, or values to be expected in a measurement. It can be shown quite generally that for any rigid magnetic gyroscope the classical equation (7) is valid.

The full significance of Eq. 7 does not appear until we consider time-dependent fields, and the full significance of the quantum-mechanical treatment of the problem does not appear until we take up observations that cannot be described by the simple equation (7). We shall take up first some considerations involving isolated systems in sinusoidally varying fields and apply them to nuclear magnetic resonance problems, and then, in connection with paramagnetic resonance, consider some of the subtler points of the quantum-mechanical treatment.

The mathematical procedures from this point on are rather involved. We shall indicate the steps to be taken, but emphasize the assumptions made and conclusions drawn, rather than the derivations themselves. We have shown that a particle with a magnetic moment and an angular momentum will precess around a constant magnetic field at the Larmor frequency ω_L, given in Eq. 9. But any system having some natural frequency will respond selectively, or will "resonate," when driven by an external force at that frequency. We may therefore expect that, since the magnetic dipole precesses around the field, some sort of resonance phenomenon will occur if we add to the constant field a component at right angles, circulating in the same sense as the dipole, with the Larmor frequency. Mathematically, this experiment leads to equations in which we substitute for **B** into Eq. 7: $B_z = $ constant, $B_x = B_0 \sin \omega t$: $B_y = \pm B_0 \cos \omega t$. Thus three differential equations are obtained similar to Eq. 8, and the solutions will contain three arbitrary constants which specify the values of m_x, m_y, and m_z at the time $t = 0$. We shall not describe the solutions in detail because they are applicable only to the isolated atoms in an atomic beam experiment with which we are here not concerned.[1]

The next step in our analysis is to consider not one isolated dipole, but a group of dipoles in thermal equilibrium, and subjected to a constant field in the z-direction with a rotating component in the x,y-plane. Instead of dealing with the magnetic moment of a single particle, we shall deal with the sum of the magnetic moments of all the particles in a unit of volume. This is the magnetization **M**. If the particles did not interact with each other or with the other degrees of freedom involved in the thermal equilibrium of the system, the dynamic equation for **M** would be

$$\frac{d\mathbf{M}}{dt} = \gamma [\mathbf{M} \times \mathbf{B}]. \quad (10)$$

[1] For a complete discussion of this, and of the following problem, see I. I. Rabi, *Phys. Rev.* **51**, 652 (1937), and F. Bloch, *ibid.* **70**, 460 (1946).

To Eq. 10 we must add terms describing the interaction of the dipoles with each other. It is beyond our skill to do this explicitly. For each kind of problem we must make simplifying assumptions that still allow a description of observed phenomena.

The first case which we shall discuss is an assembly of nuclei having angular momenta and magnetic moments situated at the centers of diamagnetic atoms, that is, atoms whose resultant electronic angular momentum is zero. When such a substance is placed in a magnetic field, the field at the nucleus is simply the externally applied field with certain slight modifications due to the external electrons which we shall neglect for the present. Let us consider a definite problem, namely, the magnetic resonance of protons in a sample of water. The gyromagnetic ratio $\gamma = \dfrac{e}{2m} \left[\dfrac{\text{coul}}{\text{kg}} \right] = \dfrac{e}{2m} \times 10^{-4} \left[\dfrac{\text{cycles}}{\text{gauss}} \right]$ is for protons roughly 4.3 kc per gauss, so that in a field of 7000 gausses for example, the Larmor frequency is 30 Mc. The experimental arrangement will therefore have a sample of, for example, water, in a test tube, between the poles of a magnet. Around the test tube, with its axis at right angles to the field of the magnet, we wind a small coil which is to carry a high-frequency current, in the vicinity of 30 Mc (Fig. 2). The experiment now is to meas-

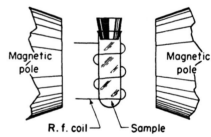

Fig. 2. Nuclear resonance experiment (schematic).

ure the impedance of the coil, or what amounts to the same thing, the complex permeability of the water core of the coil. We shall expect certain characteristic resonance phenomena when the driving frequency is equal to the Larmor frequency of the protons.

Two comments should be made about the experiment before we come to a discussion of the phenomena to be expected. The first is that we propose to use a linearly oscillating rather than a rotating field. The field of the coil may be thought of as the resultant of two fields rotating in opposite directions. Of these two, only that rotating in the same sense as the precession of the proton will be effective in producing resonance, and we may neglect the other. An oscillating field of amplitude B_0 is therefore, for present purposes, equivalent to a rotating field of amplitude $B_0/2$. The second point deals with the magnitude of the effects to be expected. The susceptibility of nuclei having a moment $|\mathbf{m}|$ is $N|\mathbf{m}|^2/3kT$, where N is the number of nuclei per unit volume (see IB, Eq. 3.9). An average solid or liquid containing nuclei having a moment of the order of magnitude of a few nuclear magnetons will have a susceptibility χ of the order of 10^{-10}. Nuclear resonance effects will therefore produce only very small changes in the voltage appearing across the r-f coil. It is therefore customary to make the coil part of a circuit in which the normal voltage across it is balanced out by some other equal and opposite voltage and in which only the change in voltage produced by the resonance effects is amplified and measured.

It is customary to speak of nuclei as holding their surrounding electrons by means of electrostatic forces. For the purposes of the present discussion it is perhaps better to think of the nuclei as existing at the bottom of a deep potential-energy well produced by the electrons. Although this electrostatic well is usually approximately spherical, it is not exactly so. In a crystal the crystallographic symmetry may be present. In a liquid, particularly a polar liquid or an ionic solution, there will be irregular fluctuations in the shape of the well. If the nucleus is not spherically symmetrical, we may expect torques of electric origin to act on the nucleus, and these must be considered in connection with the proposed modification of the fundamental equation of motion (Eq. 10). The electric quadrupole moment of a nucleus is a measure of its departure from electrical spherical symmetry. For protons, and indeed for all nuclei with a total nuclear angular momentum of $\frac{1}{2}(h/2\pi)$ for which only parallel and antiparallel orientations need be considered, the quadrupole interaction vanishes. We shall disregard electric quadrupole effects for the present.

This leaves us the problem of considering the torques exerted on nuclei by magnetic fields. It is important to remember that we are dealing with the field at a nucleus, rather than with some other average value which we customarily measure and consider. The nucleus is an exceedingly small probe which can tell us about local fields in molecules and in crystals in a most revealing manner. In fact, the theory of nuclear magnetic resonance and of the hyperfine structure of atomic energy levels has already progressed so far that a knowledge of the distribution of inhomogeneous fields over the volume of a nucleus has revealed to us something about the internal magnetic structure of some nuclei.[2]

[2] A further discussion of these effects in simple terms is to

The magnetic field at each nucleus has a mean or average value, and, in addition, performs excursions in magnitude and direction about this mean value in a more or less random fashion. These fluctuating fields are due to other nuclei, to paramagnetic impurities, or to other thermally agitated magnetic components of the surroundings. It is through this fluctuating component of the field at a nucleus that the magnetic moment interacts with its surroundings, and it is this interaction which produces the thermodynamic equilibrium with the other degrees of freedom in its surroundings. As a result of this interaction the protons in the sample of water we are considering will from time to time change their orientation in respect to the axis of the externally applied field.

Let us consider this process more quantitatively. Protons, with an angular momentum of $\frac{1}{2}(h/2\pi)$, may have one of two orientations in the applied field. The magnetic energies in these two states will be $\pm|\mathbf{m}|B$. Let us call the number of protons in the lower of these two states n_1, and the number in the upper, n_2. In the absence of the r-f field thermal equilibrium will be established, and we shall have $\dfrac{n_2}{n_1} = \exp - \dfrac{2|\mathbf{m}|B}{kT}$.

The number of protons per second making a transition from the lower to the upper state as a consequence of thermal interactions will be proportional to the number in the lower state. If the proportionality factor, which has the dimensions of a frequency, is c_1, we have for the rate of transfer from state 1 to 2, $r_{12} = c_1 n_1$. Similarly, we have for the number of transitions, thermally induced, from state 2 to 1, $r_{21} = c_2 n_2$. In thermal equilibrium these two rates must be equal, and therefore

$$\frac{c_1}{c_2} = \frac{n_2}{n_1} = \exp - \frac{2|\mathbf{m}|B}{kT} \simeq 1 - \frac{2|\mathbf{m}|B}{kT}. \quad (11)$$

The two excitation frequencies are almost equal, since $\dfrac{|\mathbf{m}|B}{kT}$ is a very small number (at room temperature in a field of 10,000 gausses, of the order of 10^{-5}). The mean excitation rate $c = \dfrac{c_1 + c_2}{2}$ is an important parameter in our considerations. Its reciprocal is the *mean lifetime* or *thermal relaxation time* τ of the protons in one of their magnetic sublevels; that is, $\dfrac{c_1 + c_2}{2} = C = 1/\tau$.

Let us now consider the radio-frequency field produced by the coil around the sample of water. A simple

be found, for example, in the author's textbook, *Nuclear Physics*, Addison-Wesley Press, 1950, pp. 127–136.

approach to the problem is to consider not the complex permeability of the water, but simply the resonance absorption of quanta of the proper energy $\mathcal{E} = 2|\mathbf{m}|B = h\nu$, or $2|\mathbf{m}|B = h\dfrac{\omega}{2\pi}$, $\omega = \dfrac{|\mathbf{m}|}{(\frac{1}{2})(h/2\pi)} B = \gamma B$. Note that this is the Larmor frequency, and that the gyromagnetic ratio is precisely that defined in Eq. 4. As previously stated, the quantum-mechanical analysis of the properties of a magnetic top gives the same result as the classical analysis. The *resonance absorption* may be thought of as inducing transitions from 1 to 2, or from 2 to 1. The rate at which these transitions are induced by the r-f field is again proportional to the number of atoms present. If we call the proportionality factor C, we have

$$r_{12} = c_1 n_1 + C n_1,$$
$$r_{21} = c_2 n_2 + C n_2. \qquad (12)$$

The proportionality factor C is itself proportional to the "intensity of the r-f radiation," or proportional to the square of the amplitude of the r-f field. From Eq. 12 and the relation $n_1 + n_2 = N$, the total number of protons present, we may solve for n_1 and n_2 explicitly. Under steady-state conditions we have $r_1 = r_2$, and consequently

$$n_1 = \dfrac{N}{2}\dfrac{c_2 + C}{c + C},$$
$$n_2 = \dfrac{N}{2}\dfrac{c_1 + C}{c + C}. \qquad (13)$$

For small r-f amplitudes, n_1 and n_2 have their equilibrium values. For large r-f amplitudes, the populations of the two states approach each other: $n_1 \to n_2 \to N/2$. The total energy per second absorbed by the protons from the r-f coil is simply the number of excess transitions from $1 \to 2$ over that from $2 \to 1$ per second multiplied by the energy change $2|\mathbf{m}|B$ of a transition:

$$W = 2|\mathbf{m}|B[C n_1 - C n_2] = 2|\mathbf{m}|BC\dfrac{N}{2}\left[\dfrac{c_2 - c_1}{c + C}\right].$$

Using Eq. 11, we find $c_2 - c_1 \simeq 2c_2 \dfrac{|\mathbf{m}|B}{kT} \simeq 2c \dfrac{|\mathbf{m}|B}{kT}$, and for the rate of absorption of energy, W,

$$W = 2\dfrac{(|\mathbf{m}|B)^2}{kT} N \dfrac{Cc}{c + C}$$
$$= 2\dfrac{(|\mathbf{m}|B)^2}{kT}\dfrac{1}{\tau}\dfrac{C}{\dfrac{1}{\tau} + C}. \qquad (14)$$

At low r-f amplitudes, or when $C \ll 1/\tau$, thermal equilibrium is not appreciably disturbed, and the expression Eq. 14 is proportional to C. At large r-f amplitudes, or for $C \gg 1/\tau$, the population of the two states approaches equality, and Eq. 14 becomes independent of C, and therefore independent of the r-f amplitude. These saturation effects may be used to determine the thermal relaxation time τ.

This discussion illustrates how we may justify the introduction of thermal interactions between dipoles by means of a thermal relaxation time, which is a measure of the time required to establish thermal equilibrium after an abrupt change in the applied field. In order to get a deeper insight we must make a more detailed theory. How wide is the resonance absorption line? What are the reactive as well as the resistive changes in the impedance of the coil? For an answer to these questions we must return to our fundamental equation (10).

F. Bloch, in the paper quoted above (footnote 1), has suggested that Eq. 10 be so modified that in the absence of an r-f field the magnetization will approach its equilibrium value exponentially with a time constant equal to the relaxation time. In a constant field B_z in the z-direction the z-component of the magnetization should approach its equilibrium value $M_0 = \chi_0 B_z$, and the x- and y-components should approach zero. The time constants for these two processes need not be the same. The resulting equations † are

$$\dfrac{dM_x}{dt} = \gamma[M_y B_z - M_z B_y] - \dfrac{M_x}{\tau_2}$$
$$\dfrac{dM_y}{dt} = \gamma[M_z B_x - M_x B_z] - \dfrac{M_y}{\tau_2} \qquad (15)$$
$$\dfrac{dM_z}{dt} = \gamma[M_x B_y - M_y B_x] - \dfrac{M_0 - M_z}{\tau_1}.$$

The expressions for the real and imaginary parts of the complex r-f susceptibility are, putting $\Delta\omega$ for $\omega_L - \omega$,

$$\chi' = \dfrac{\gamma \tau_2^2 \Delta\omega}{1 + (\tau_2 \Delta\omega)^2 + (\gamma B_0)^2 \tau_1 \tau_2} \chi_0 B_z, \qquad (16)$$

$$\chi'' = \dfrac{\gamma \tau_2}{1 + (\tau_2 \Delta\omega)^2 + (\gamma B_0)^2 \tau_1 \tau_2} \chi_0 B_z. \qquad (17)$$

We need not concern ourselves here with these rather formidable expressions beyond noting that, by observations on χ' and χ'', we can determine the Larmor frequency ω_L, and τ_1 and τ_2 experimentally. The physical interest in magnetic resonance phenomena from this point on lies in the significance of the values found for these parameters, and in the modifications in the

---

† For a discussion of the solutions of these equations, the reader is referred to Bloch's paper.

approach outlined above, required to account for various observations.

Two independent measurements are required to determine the parameters τ_1 and τ_2. These are the r-f amplitude required to produce saturation and the line width. In nuclear magnetic resonance phenomena in which the fluctuations in the magnetic field provide the only coupling of the nuclear magnetic moments with each other and with the other degrees of freedom of their surroundings, two aspects of the magnetic fluctuations are responsible. In our previous discussion we showed that only magnetic fields oscillating with the Larmor frequency could produce transitions from one orientation of the nucleus to another. In order to estimate the effectiveness of the fluctuating magnetic field in producing nuclear reorientations, we must make a Fourier analysis, and estimate the amplitudes of those components in the vicinity of the Larmor frequency. If these amplitudes are large, the thermal relaxation time τ_1 will be short. If there is little energy at the Larmor frequency, the thermal relaxation time will be long. In diamagnetic gases and liquids in which the fluctuating fields are due primarily to nuclear moments, long relaxation times are observed, often long compared to seconds. These relaxation times may be shortened by any means that increases the "magnetic noise" in the sample at the Larmor frequency. One means is to add paramagnetic ions, which, because of their large magnetic moments, increase the noise over a large frequency range. Another means is to increase the viscosity of the liquid. In liquids having low viscosity, changes in molecular configuration take place at a high rate compared to the few megacycles involved in the Larmor resonance. By increasing the viscosity, this rate is slowed down, the energy in the Larmor range of frequencies is increased, and the thermal relaxation time is consequently shortened.

The resonance line width, on the other hand, depends on the amount of energy in the magnetic noise spectrum at frequencies small compared to the Larmor frequency. The local fields at various nuclei will differ at any one instant by amounts ΔH, and if these differences persist for times long compared to the Larmor period, then the Larmor frequency itself will have a spread of the order of $\Delta \omega_L \simeq \gamma \Delta B$ for the various nuclei. Nuclear magnetic moments create fields of the order of 10 gausses at distances of a few angstroms, and this is the order of magnitude of the line widths observed in solids in which the nuclei change their orientations at a rate small compared to the Larmor frequency. In liquids, however, in which the nuclei are parts of molecules which change their configurations in times short compared to the Larmor periods, the internuclear fields are ineffective in producing line broadening. They average out to zero in a single period, and consequently much narrower resonance lines, actually of the order of milligausses, are observed in such liquids.[3]

As a consequence of the sharpness of the nuclear magnetic resonance lines, nuclear moments, or, rather, nuclear Larmor precession frequencies, may be accurately compared with each other. These frequencies depend on the nuclear gyromagnetic ratio, and on the field at the nucleus. It has been found that in some molecules, or in mixtures of several substances in each of which some one nucleus is present, the resonance condition for a given applied frequency is established for slightly different applied fields. When an atom or molecule is placed in a field B_z, the local field at the nucleus is not B_z, but may be slightly greater or less. The lines of magnetic induction may be bent slightly in toward the nucleus or out from the nucleus owing to the atomic electrons. These effects, due to diamagnetic shielding or to the incipient decoupling of the mutually canceling angular momenta in a diamagnetic molecule, make possible structural investigations of considerable interest.[4]

Sharp resonance lines are also of great importance in nuclear physics. In a mixture of nuclear isotopes in the same chemical environment we may assume that the local field is the same for all the isotopes. The ratios of the observed Larmor frequencies, which have, in some cases, been measured with an accuracy as great as a few parts in 10^7, give the ratio of the nuclear gyromagnetic ratios. These measurements have been most important, and promise to be even more so as they accumulate, in revealing features of internal nuclear structure.

Paramagnetic resonance is essentially similar to nuclear magnetic resonance except that the effects are due to electronic rather than nuclear magnetization. The electronic-gyromagnetic ratio of atoms is some thousands of times as large as the nuclear gyromagnetic ratio, and the Larmor frequencies in fields of the order of thousands of gausses lie in the microwave range. Resonance absorption measurements have been made on paramagnetic solids and on liquid solutions for several years, but the investigations of purely paramagnetic resonance phenomena, such as are discussed

[3] N. Bloembergen, E. M. Purcell, and R. V. Pound, *Phys. Rev. 73*, 679 (1948).

[4] For a further discussion, see, for example, W. C. Dickinson, *Phys. Rev. 81*, 717 (1951).

above, have been much less extensive than the nuclear resonance studies. An example of the kind of problem waiting to be solved is contained in some results obtained by W. C. Dickinson,[4] and more recently by G. W. Masters in the author's laboratory. The results are shown in Fig. 3. The line width of the proton

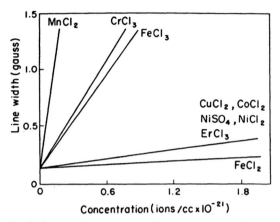

Fig. 3. Influence of paramagnetic ion concentration on line width of proton resonance in water.

resonance in aqueous solutions was observed as a function of the concentration of various paramagnetic ions. It is to be noticed that whereas relatively slight additions of $MnCl_2$ broaden the proton resonance considerably, large additions of other salts, for example, $FeCl_2$, produce no appreciable line broadening whatever. This indicates interesting differences in the properties of these ions, differences that might well be studied directly by means of paramagnetic resonance investigations.

Rather than attempt a review of paramagnetic studies that are in many ways similar to the nuclear resonance investigations already discussed in some detail,[5] we shall take up a new type of paramagnetic resonance experiment.[6] The experiment is performed on low-pressure mercury vapor in which some atoms are excited from the diamagnetic $^1S_0$ state to the paramagnetic $^3P_1$ state by the absorption of the ultraviolet resonance line having a wavelength of 2536 A. Polarized resonance radiation is used, so that certain magnetic sublevels of the excited state are selectively populated. For example, if we consider the even isotopes of mercury which have no nuclear moments to confuse the issue, circular polarized light of the proper sense illuminating the vapor in the direction of a magnetic field of a hundred gausses or so will selec-

[5] For such a review see the special issue of *Physica 17*, March-April, 1951, especially F. Bloch, "Nuclear Induction," p. 272.

[6] Described by J. Brossel and F. Bitter, *Phys. Rev.* **86**, 308 (1952).

tively excite atoms with the magnetic moment of the excited $^3P_1$ state parallel to the field. From this state atoms can radiate only by the emission of light having the same polarization as that which was absorbed. If now, in addition to the optical resonance absorption which put the atom in a particular excited state, we have a coil producing paramagnetic resonance in the excited state, the polarization of the light emitted by atoms that have been reorientated will have a different polarization. Paramagnetic resonance may be detected by the change produced in the polarization of the resonance radiation. Because in this experiment both optical and r-f resonance are used, it is called a *double resonance* experiment.

The complete elucidation of the results obtained requires us to use the fundamental quantum mechanics of particles and fields, and cannot be obtained by using Eq. 7. We are not concerned with the expectation values of the components of a magnetic moment, but rather with the probability of forced transitions from one to another magnetic sublevel of the excited state. These probabilities have been computed and predict the rather peculiar line shape actually found in this case, and shown in Fig. 4. The effects observed ex-

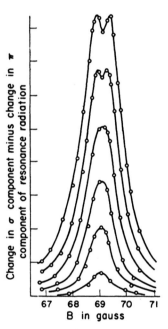

Fig. 4. Resonances observed in mercury isotopes having even mass number and zero nuclear spin.

perimentally were satisfactorily accounted for; they were obtained under conditions for which collisions presumably might be neglected. Although the depolarizing and quenching effects due to foreign gases, and of Hg-Hg collisions, have been extensively investigated, the availability of enriched isotopes and of this new double-resonance technique makes possible much more accurate and complete observations. Detailed investigations of collisions between atoms and molecules, and of the gradual broadening of energy levels as vapor pressures are increased and exchange effects become pronounced, should be most rewarding.

Finally, we come to magnetic resonance phenomena in ferromagnetic materials. The essential difference between ferro- and paramagnetic materials is that in the former *strong internal fields* tend to align the atomic moments even in the absence of externally applied fields. These internal fields are generally thousands of times as strong as the fields due to electromagnets. It is not meant here that the internal fields are magnetic. They are not. But they may be compared to magnetic fields in the sense that in both a given energy is required to produce a reorientation of an atomic magnet in their presence. In this sense the internal fields may be thought of as producing very large torques, and consequently also very much higher precession frequencies than were found in paramagnetic substances in normally available fields. These high frequencies fall in the infrared range, and have, to the author's knowledge, never been observed.

The resonance phenomena usually referred to as *ferromagnetic* do not involve the independent precession of individual atoms, but the precession of the magnetization of a macroscopic region as a whole. In this process the internal field remains parallel to the magnetization at all times, and therefore exerts no torque. There are, however, other torques than those due to the externally applied field to be considered, and it is these extra torques which give ferromagnetic resonance a certain distinctiveness.

Let us consider a ferromagnetic sphere spontaneously magnetized to an intensity M. We further suppose that this is a single cubic crystal of iron, for example. The energy of the crystal sphere will now depend on the orientation of the spontaneous magnetization with respect to the crystallographic axes. In iron this energy is a minimum when the magnetization is parallel to one of the cube edges, or perpendicular to either the (100), (010), or (001) planes. Work is therefore required to rotate the spontaneous magnetization away from one of these equilibrium orientations, and a torque must be exerted if work is done. In setting up an equation like Eq. 6 for a ferromagnetic substance, we must include in the expression for the torque not only [$\mathbf{M} \times \mathbf{B}$], but a more complex expression involving the orientation of M with respect to the crystallographic axes.

If, for any reason, our sample is stressed, as, for example, in a cold-worked or precipitation-hardened polycrystalline sample, the energy of a specimen will depend on the orientation of the magnetization with respect to the local principal stress axes. This effect will give rise to additional torques which must be included in Eq. 6.

Although some experiments have been made on ferromagnetic spheres, particularly of the semiconducting ferrites, this approach is generally not feasible because of the small penetration depth of high-frequency fields into metallic conductors. In such a case a flat sample is often used, with one surface forming a boundary of a microwave cavity. This brings into play additional torques due to the demagnetizing fields of strongly magnetized objects. For example, the energy of a magnetized ellipsoid is least when the spontaneous magnetization is parallel to the longest axis. In a magnetized thin sheet the energy is much greater when the magnetization is perpendicular to the plane of the sheet than when it is in the plane of the sheet. In general, therefore, the magnetization of a nonspherical ferromagnetic specimen will be subject to torques, depending on the orientation of the magnetization with respect to the symmetry axes of the sample, and these too must be included in Eq. 6. These aspects of ferromagnetic phenomena, which have been amply studied in static experiments, have been satisfactorily and quantitatively incorporated in the theory of ferromagnetic resonance.

There are, nevertheless, various points on which further clarification is desired; they offer a challenge to the theorist as well as to the experimentalist. An important point deals with the observed gyromagnetic ratio. Ferromagnetism is presumably due to electron spins, and we should therefore expect a gyromagnetic ratio of $2e/2mc$. Actually values near, but not exactly equal to, this quantity have been observed. This deviation may indicate that there is, in addition to the electron spins, a contribution to ferromagnetism from the orbital motion of electrons in atoms, but the last word has probably not yet been said on this subject.

Finally, the ferromagnetic relaxation times have to be interpreted. The experimental indication is that they are short, much shorter than a microsecond. No mechanism so far proposed seems adequate to account for the observations.[7]

[7] For further information the reader is referred to a clear account by C. Kittel, *J. phys. et radium* 12, 291 (1951), and to the *Reviews of Modern Physics*, Vol. 25, No. 1 (1953), which is devoted to an international conference on magnetism, held in September, 1952, at the University of Maryland under the auspices of the Office of Naval Research.

III · DIELECTRIC MATERIALS AND THEIR APPLICATIONS

A · Dielectric Materials

1 · Insulation Strength of High-Pressure Gases and of Vacuum †

By JOHN G. TRUMP

Solid, Liquid, Gaseous, and Vacuum Dielectrics

Of the four generic insulating media—solid, liquid, gases, and vacuum—the most extensive studies have been made on solid and liquid dielectrics. Their further improvement continues with emphasis on the realization of special properties such as chemical stability, low loss, high dielectric constant, and high temperature tolerance. Most of the past research on gaseous insulation has been directed at the fundamental processes of ionization and recombination at atmospheric or subatmospheric pressures.[1,2] Although measurements during the last 40 years on gases at high pressure have suggested their superiority over the denser media, utilization of the dielectric properties of compressed gases has awaited the modern demand for higher voltages and operating gradients. High vacuum, essential for the acceleration of elementary particles to high energy, is still in an early stage of scientific investigation.[3] A comparison of the relative dielectric strength in a uniform field of these four generic insulating media is shown in Fig. 1.1.

† Much of the work reported herein was performed in the High Voltage Research Laboratory of the Department of Electrical Engineering at M.I.T. with the support, in part, of the Office of Naval Research and of the Godfrey M. Hyams Trust.

[1] J. S. Townsend, *Electricity in Gases*, Clarendon Press, Oxford, 1915.
[2] L. B. Loeb, *Fundamental Processes of Electrical Discharge in Gases*, Wiley and Sons, New York, 1939.
[3] J. G. Trump and R. J. Van de Graaff, *J. Appl. Phys.* **18**, 327 (1947).

Fundamental processes in gases

In the passage of electric charge between electrodes immersed in a gas, the Townsend α process is the most prolific source of ionization. Townsend[4] showed that the number of ionizing collisions α by an electron of charge e per centimeter of its path in the direction of the electric field depends exponentially on the electric field intensity E, the energy $\mathcal{V}_i$ required to ionize the gas molecules, and the mean free path between collisions λ:

$$\alpha \simeq \frac{1}{\lambda} \exp\left(-\mathcal{V}_i/Ee\lambda\right).$$

Unless removed by attachment to a gas molecule, each free electron in an interelectrode space subject to a sufficiently high electric field is accompanied by an exponentially growing number of electrons produced by collisions as it moves the remaining distance X to the anode. This avalanche totaling $e^{\alpha x}$ electrons disappears into the anode with a velocity of the order of 10^7 cm per sec, and leaves behind a relatively stationary positive ion space charge to distort the electric field.[5]

Concurrently, certain γ processes renew the free electrons on the cathode side of the gap. These regenerative sources of ionization include photoelectric emission from the cathode and photoionization of the gas as the excited and ionized gas atoms radiate their energy, and secondary-electron emission from the cathode under bombardment by the positive ions

[4] J. S. Townsend, *Nature* **62**, 340 (1900); *Phil. Mag.* [6] **1**, 198 (1901); **3**, 557 (1902); **6**, 389, 598 (1903); **8**, 738 (1904).
[5] A. von Hippel and J. Franck, *Z. Physik* **57**, 696 (1929).

cleared from the gap by the electric field. As Townsend, Loeb, and others have shown, these ionization processes are distributed quite uniformly over the

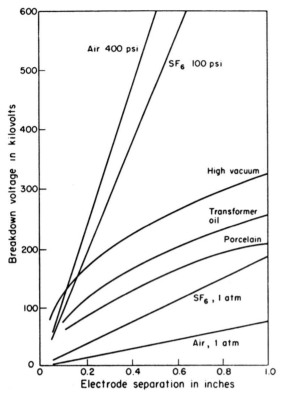

Fig. 1.1. Direct-current breakdown strength of typical solid, liquid, gaseous, and vacuum insulation in a uniform field.

section when the product of pressure and gap is well below 500 mm Hg × cm. Under these conditions, for an initial cathode emission current I_0 at $x = d$, the steady anode current is

$$I = I_0 \frac{e^{\alpha d}}{1 - \gamma(e^{\alpha d}-1)}.$$

At higher products of $p \times d$ there is an increasing tendency for the α ionization process to proceed as independent avalanches which attain progressively higher ion densities.[6] Superimposed on a steady current, the discrete and intense character of these electron pulses and their associated positive ion space charge accentuate the local field distortion. As the electric field intensity is raised, the heightened α and γ processes, aided by the increased field distortion, cause dielectric instability and the passage of a spark, arc, or glow discharge, depending on external circuit conditions. In the avalanche-streamer theory proposed

[6] L. B. Loeb and J. M. Meek, *J. Appl. Phys.* 11, 438 (1940).

by Loeb and Meek[7] and also by Raether[8] to account for long sparks in high-pressure gas, the condition for sparkover is reached when the local space-charge field produced by an advancing avalanche becomes comparable to the highest field strength on the electrode surface.

An increase in the density of the gas molecules reduces the mean free path between electron collisions and hence permits application of an increased voltage across the gap. It is convenient to refer to the measured quantity (gas pressure) rather than the more basic parameter (molecular density) since the two remain linearly related until close to the region of condensation. In a uniform field the sparkover voltage increases linearly with pressure up to about 10 atm, and then more slowly. This initial linearity between

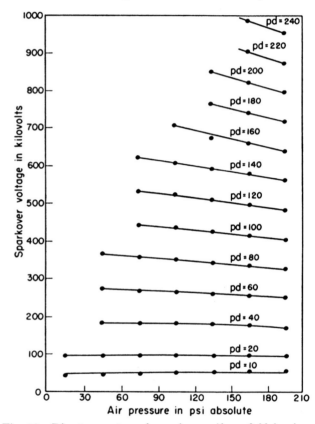

Fig. 1.2. Direct-current sparkover in a uniform field in air as a function of pressure (pounds per square inch) for various constant values of pressure × gap distance.

voltage and pressure suggests that the factors contributing toward electrical instability are not sensitive at moderate pressures to the total number of ionization events occurring between the electrodes. At

[7] L. B. Loeb and J. M. Meek, *The Mechanism of the Electric Spark*, Stanford University Press, 1941.

[8] H. Raether, *Z. Physik* 117, 375, 524 (1941).

higher pressures the cumulative effect of the increased ionization in the gap and the growing importance of new ionization mechanisms, particularly at the cathode, account for the gradual diminution in the rate of voltage increase with pressure.

This linearity of voltage strength with pressure is not predicted by Paschen's law, which suggests merely that the breakdown strength in a uniform field should remain constant if the product of $p \times d$ remains constant.[9] Figure 1.2 shows the relation between breakdown voltage in air as a function of its pressure at several constant values of $p \times d$. The changing slope of these lines indicates that new factors are introduced at the higher pressures which impair the ability of an electrode system to insulate a given voltage. Explanation for this phenomenon can be found in the increased photoionization of the gas at elevated pressures, in the diminished ability of the positive ion space charge to diffuse out of the active field, and in the significant contributions of the now highly stressed electrode surfaces, especially the cathode.

Electronegative gases

Since the investigations of Natterer [10] there has been a growing awareness that many compounds in the gaseous state, particularly those which contain chlorine, fluorine, bromine, oxygen, and other electronegative atoms, may have significantly higher insulating strength than air at the same pressure. The dielectric superiority of these molecules arises from their ability to remove free electrons by attachment. These electronegative molecules may also exert their influence by extracting a portion of the energy of incident electrons; these compounds are usually of higher molecular weight and complexity and thus offer increased opportunity for inelastic collisions.

The admixture of carbon tetrachloride (CCl_4) has long been known to increase markedly the dielectric strength of air, this contribution remaining about constant when air is raised to higher pressures.[11] Joliot [12] was first to use electronegative vapors in the insulation of high-voltage equipment when he introduced the vapor pressure of CCl_4 into a room containing an air-insulated Van de Graaff generator and reported its maximum voltage to be nearly doubled.[12] Pollock and Cooper, examining in a small test chamber over thirty gaseous compounds, selected Freon (CCl_2F_2) and sulfur hexafluoride (SF_6) as particularly promising.[13] These two gases have been extensively studied and over a wide pressure range exhibit nearly three times the dielectric strength of air at the same pressure. SF_6, discovered by Moissan and Lebeau in 1900 and long known for its unusual chemical stability, is particularly useful because of its high vapor pressure, 330 psi at 20°C.[14]

The pre-breakdown current and breakdown voltage characteristics of such electronegative gases are similar to those of permanent gases, but they occur at higher electric field strengths at any given gas pressure.[15, 16] Subjected to excitation and ionization in high-voltage applications the electronegative molecules may be dissociated into a variety of simpler structures including free chlorine and fluorine and thus deteriorate into significant amounts of toxic and corrosive products.[17] Schumb, Trump, and Priest [18] have shown that Freon and SF_6 maintain their high dielectric strength even under accelerated conditions of corona and sparkover and that effects of dissociation products can be minimized by the presence of suitable absorbers such as activated alumina and soda lime. Fluorocarbons such as C_3F_8 have been reported by Bashara to be nearly equal to SF_6 in dielectric strength but less subject to dissociation because of the relatively stronger carbon-fluorine bonds. When excessive ionization losses and low ambient temperatures are avoidable, such electronegative gases are practical and superior for gaseous insulation.

Figure 1.3 shows the insulating strength in a uniform field of a variety of gases from He to SF_6 as a function of gas pressure. These constant potential sparkover measurements by Kusko extend to higher total voltages and higher products of $p \times d$ than have been reported heretofore.[19] Lowest insulating performance is obtained with noble gases such as helium. Nitrogen, a stable and inert gas molecule, exhibits a dielectric strength slightly lower than that of air. The better

[9] F. Paschen, *Wied. Ann.* 37, 69 (1889).
[10] K. Natterer, *Ann. physik. Chem.* 38, 663 (1889).
[11] M. T. Rodine and R. G. Herb, *Phys. Rev.* 51, 508 (1937).
[12] F. Joliot, M. Feldenkrais, and A. Lazard, *Compt. rend.* 202, 291 (1936).

[13] H. C. Pollock and F. S. Cooper, *Phys. Rev.* 56, 170 (1939).
[14] W. C. Schumb, *Ind. Eng. Chem.* 39, 421 (1947).
[15] C. M. Hudson, L. E. Hoisington, and L. E. Royt, *Phys. Rev.* 52, 664 (1937).
[16] J. G. Trump, F. J. Safford, and R. W. Cloud, *Trans. Am Inst. Elec. Engrs.* 60, 132 (1941).
[17] D. Edelson, C. A. Bieling, and G. T. Kohman, "The Electrical Decomposition of Sulfur Hexafluoride." Paper presented to the Conference on Electrical Insulation, National Research Council, Lenox, Mass., Oct. 2, 1952.
[18] W. C. Schumb, J. G. Trump, and G. L. Priest, *Ind. Eng. Chem.* 41, 1348 (1949).
[19] A. Kusko, "A Study of the Electrical Behavior of Gases at High Pressures," Electrical Engineering Doctorate Thesis, Massachusetts Institute of Technology, 1951.

performance of CO_2 over N_2 may be accounted for by the electronegative oxygen atoms which serve to extract energy from incident electrons and to attach them, thus forming innocuous negative ions. When the pressure of CO_2 was increased to 600 psi, at which saturation occurs at 20°C, no discontinuity between the electric breakdown strength of gaseous CO_2 and liquid CO_2 was observed. This supports the work of

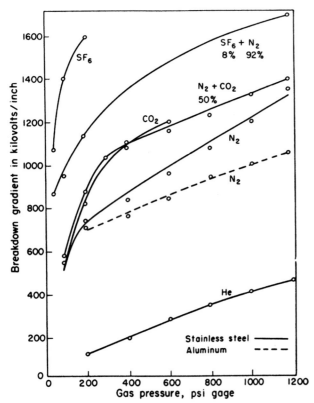

Fig. 1.3. Direct-current breakdown gradients in various gases (and mixtures) as function of pressure for uniform gap (0.5 inch). (After Kusko.[19])

Young, who demonstrated the smooth continuity in the dielectric strength of CO_2 from the gaseous to the liquid phase, using relatively low voltages with a more nearly ideal electrode arrangement.[20]

It is noteworthy that the dielectric strength of SF_6 at 100 psi was found to be equal to that of nitrogen at 1200 psi. Such an advantage in dielectric strength will not long be neglected in practical engineering applications. The value of adding a small percentage of SF_6 to a permanent gas is likewise illustrated.

Influence of electrodes on d-c breakdown in gases at high pressure

At elevated pressures the breakdown voltage may be observed to increase as much as 100 percent after

[20] D. R. Young, *J. Appl. Phys.* **21**, 222 (1950).

a number of interelectrode sparks of limited energy. Howell and others [21] have ascribed this improvement under controlled sparkover to the conditioning of the electrode surfaces by the reduction of high field emission areas such as might be caused by surface irregularities. If the surface roughness is reduced by highly buffing the electrodes, conditioning is accomplished with much less sparking.

Figure 1.4 is a photomicrograph of individual spark craters on well-buffed aluminum and stainless steel electrodes taken after conditioning by a considerable number of energy-limited sparks. Evidently, such conditioning may actually increase the surface irregularities while causing diminished electronic emissivity. Excessive sparking, particularly on aluminum, will finally reduce the voltage-insulating ability. The d-c breakdown values for gases reported in this paper are for highly conditioned electrodes, and the reported voltage is that value which could be maintained without sparkover for periods of 3 to 5 minutes. Breakdown values in gases have been found to vary widely among observers, depending on such factors as surface condition of the electrodes, the degree of conditioning, the criterion for breakdown, and the electrode area. If electrodes of much larger areas are used, conditioning by sparkover becomes more difficult, the stored energy becomes greater, and the permissible breakdown gradient may diminish to less than one-half the values herein reported.

Recently, a well-defined dependence of breakdown voltage on the electrode metal has been observed to begin at pressures of about 10 atm and to increase progressively with pressure.[22] Figure 1.5 shows the breakdown gradient between pairs of stainless steel and of aluminum electrodes as a function of pressure for a number of gases including air, CO_2, N_2, and their mixtures. The superior insulating strength of gaps between stainless steel electrodes at the higher pressures is evidently due to the lower contribution to the several γ ionization processes of this metal as compared with aluminum. Since these processes take place at or in the vicinity of the cathode electrode, this explanation suggests that only the cathode material should be an influential factor.

Figure 1.6 compares the d-c sparkover gradient in air for a uniform field electrode system in which the electrodes were made either of stainless steel or aluminum in all possible combinations. This work showed conclusively that the highest performance was ob-

[21] A. H. Howell, *Trans. Am. Inst. Elec. Engrs.* **58**, 193 (1939).
[22] J. G. Trump, R. W. Cloud, J. G. Mann, and E. P. Hanson, *Elec. Eng.* **69**, 961 (1950).

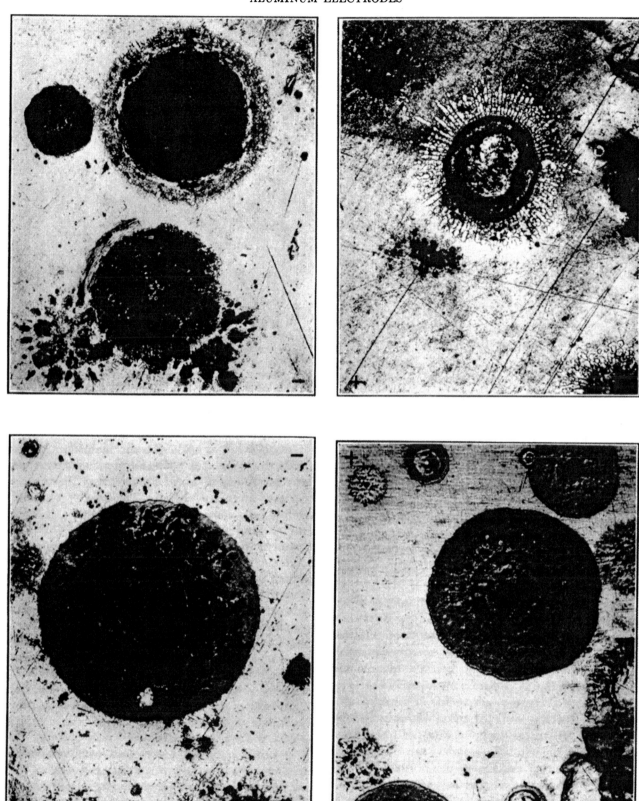

Fig. 1.4. Photomicrographs of spark craters on well-buffed aluminum and stainless steel electrodes after conditioning (magnification, 45 linear).

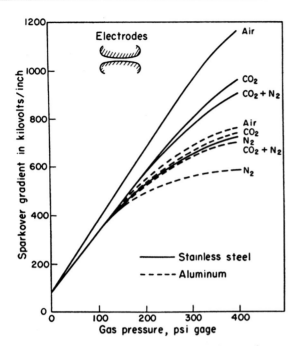

Fig. 1.5. Comparative insulating strength of several gases at high pressures for uniform fields between stainless steel and between aluminum electrodes.

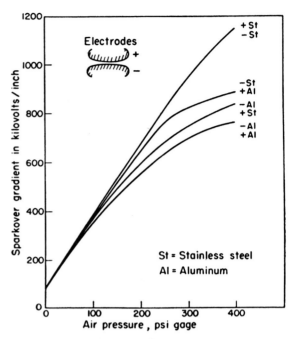

Fig. 1.6. Insulating strength in compressed air between electrodes of steel, aluminum, and of steel-aluminum.

tained with an all-stainless-steel system and the lowest with an all-aluminum system. Intermediate performance was observed for mixed electrodes, the stainless steel cathode being better. This work suggests that when a mixed electrode system is utilized in a gas gap an inferior anode material may result in the gradual impairment of the cathode by sputtering processes. There is evidence that the anode material is without much influence on breakdown potential of a mixed gap at high pressure when only small pre-discharge currents are permitted to flow.

As a consequence of higher gas pressure, higher gradients can be applied to the electrode surfaces in a uniform gaseous gap. Above 10 atm these gradients become able to contribute to the interelectrode current by high field electron emission from the cathode and by enhancing the photoelectric emission and the secondary-electron emission under positive ion bombardment. A cathode electrode metal characterized by high work function and low secondary electron emission should therefore be superior under such high gradient conditions. Although no a-c studies were made,[22] it is believed that in an a-c system the over-all performance at high pressures will be determined by the most inferior electrode material present in the gap.

[22] J. G. Trump, R. W. Cloud, J. G. Mann, and E. P. Hanson, *Elec. Eng.* **69**, 961 (1950).

Positive point-to-plane discharges in compressed gases

The high electric field distortion obtained by directing a metal point at a plane electrode immersed in a gas facilitates the flow of electric charge. This geometry may be useful when local conductivity of the insulating gas is desired as in the transfer of electric charge to or from the belt of a Van de Graaff generator or in the neutralization of static electricity.

For a negative point directed at a plane, corona current appears at a definite onset voltage and increases very rapidly with voltage until breakdown ensues. As the pressure is raised, there is a steady rise in the corona threshold voltage and in the voltage required for a given current. As before, increasing pressure reduces the mean free path and thus increases the voltage required to maintain the α coefficient at its prior value. Secondary changes due to diminished diffusivity of the positive ions and to increased photoionization of the gas can be observed, but modify the essential discharge characteristic only slightly.

In positive point-to-plane breakdown, however, the discharge phenomenon is dramatically altered. Vul was the first to show the existence of a critical gas pressure at which the maximum breakdown voltage between point and plane is realized.[23] The maximum

[23] I. M. Goldman and B. M. Vul, *Tech. Phys.* (U.S.S.R.) **1**, 497 (1935).

positive corona current decreases as the pressure is increased until at the critical pressure breakdown takes place at the onset corona voltage and little or no pre-breakdown current can be obtained. As the critical

Figure 1.8 shows the influence of voltage wave form on the positive point-to-plane characteristic. As would be expected, the a-c point-to-plane characteristic coincides almost exactly with the positive point-

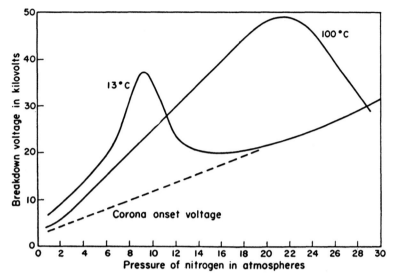

Fig. 1.7. Positive point-to-plane breakdown in nitrogen vs. pressure at 13° and 100°C (after Goldman and Vul [23]).

pressure is approached, the sparkover voltage becomes more erratic, and the spark discharge scrupulously avoids a direct path from point to plane. Above the critical pressure, sparkover occurs at lower voltages, and with no appreciable predischarge current.

This positive point-to-plane phenomenon constitutes a limitation to the use of very high pressures when excessively distorted electric fields are required. It seems associated with the formation, with increasing pressure, of a progressively tighter layer of positive ion space charge around the positive point. Vul demonstrated the space charge character of this phenomenon by showing that the critical pressure was shifted toward higher values when the temperature of the gas was raised from 13° to 100°C (Fig. 1.7). This temperature increase affects primarily the diffusivity of the positive ions and delays the formation of a tight positive ion sheath around the point. Howell [21] further elucidated the mechanism by showing the dependence of the critical pressure upon the nature of the gas. Whereas this critical pressure occurs at about 11 atm for air and nitrogen, it is found at 30 atm for helium and at intermediate values for other gases depending upon their ionic mobility. The positive point-to-plane phenomenon places a premium on the proper electrostatic design of gas-insulated high-voltage apparatus and introduces the need to use large radii of curvature on positive surfaces.

to-plane phenomenon. Impulse voltages, however, are progressively less dependent upon point polarity the shorter the pulse.

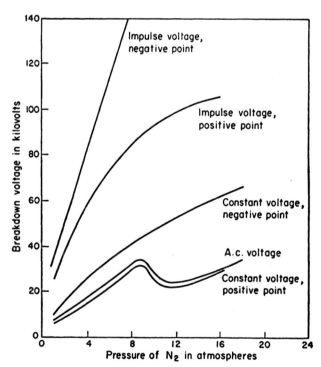

Fig. 1.8. Point-to-plane breakdown in nitrogen as a function of pressure, polarity and wave form. (After Goldman and Vul.[23])

Flashover of solid insulators in compressed gases

Solid insulation is required in liquid, gaseous, or vacuum insulated apparatus for the support of the high-voltage electrodes. Even in a uniform field the flashover voltage of such solid insulation is usually considerably lower than for the gas-filled gap of equal length. In the practical utilization of compressed gas insulation an effort must often be made to increase the

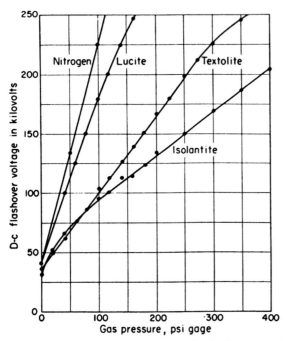

Fig. 1.9. Direct-current flashover strength of nitrogen and of corrugated cylinders of lucite, textolite, and isolantite in nitrogen at high pressures (uniform field, ½-inch gap).

surface flashover voltage of the solid insulator in order to match its volume breakdown strength or the insulating strength of the gaseous gap itself.

Experimental studies have been made of the d-c flashover strength of a variety of solid insulators in the form of short cylinders placed in a uniform field between plain steel electrodes.[24] It can be seen from Fig. 1.9 that the flashover voltage increases linearly with pressure at first and then more slowly. The flashover values increase nearly directly with the insulator length. Even for high-quality solid dielectrics the insulating performance is strongly dependent upon the insulator material, primarily because of the influence of end conditions, surface irregularities, and surface leakage. The highest performance was obtained with methyl methacrylate (Lucite) because of the ease with which ideal contact with the electrode could be made and because of its high resistivity and homogeneity. At the lower pressures Lucite attained

[24] J. G. Trump and J. Andrias, Trans. Am. Inst. Elec. Engrs. 60, 986 (1941).

very nearly the flashover strength which could be realized in the gas-filled N_2 gap alone.

The imminence of flashover along smooth insulator surfaces is often first revealed by a succession of faint gliding discharges originating at discrete points along the contact edge of the cathode electrode and insulator and proceeding along the insulator surface toward the opposite electrode. Further increase in voltage causes these surface gliding discharges to increase in brilliance and length and finally results in an intense flashover. Corrugations in the insulator surface, aside from increasing the length of the ultimate discharge path, exert an inhibiting action on such early discharge currents and thus generally more than compensate for the field distortion and increased field intensity which they introduce. The final flashover does not, in general, follow the contour of the corrugated insulator but glides over the tips of the corrugations and leaps across the gas gap between the corrugations.

Insulator corrugations were found to be of greatest value when inferior end conditions or wall defects were known to exist. In general, it appears desirable to terminate a stand-off insulator in its smallest diameter (i.e., at the depth of the corrugation). This has the effect of minimizing the ionization caused by small defects or crevices in the insulator-electrode contact. With this arrangement the end corrugation acquires a space charge which tends to repress the ionization at the end contact.

In distorted fields compressed-gas insulation does not improve the creepage strength along the solid insulator surfaces to the extent that would be expected from uniform field studies. This is probably related to the positive point-to-plane phenomenon which has been described. In the design of layer insulation for gas-insulated transformers, for example, it has been found desirable to use end clearances which are somewhat greater than would be required with a high-grade insulating oil.

Vacuum insulation

Vacuum insulation may be said to exist when the breakdown strength between metallic electrodes is independent of the pressure of the surrounding gas. The mean free path of the residual gas molecules must therefore be large compared with the interelectrode gap, that is, the contribution of gaseous ionization to the breakdown mechanism must be negligible. Though this condition is realized even at atmospheric pressure for gaps as small as a few wavelengths of sodium light, the pressure ordinarily satisfactory for the vacuum insulation of engineering devices ranges from about 10^{-4} to 10^{-6} mm of Hg or less.

In the voltage range below 20 kv, extensive studies [25,26] have shown that vacuum breakdown is initiated by electron emission from the cathode surfaces. At such low voltages, gradients as high as 5 million v per cm have been insulated at well-polished and outgassed cathode surfaces, and gradients at least an order of magnitude higher have been insulated at the anode.

With higher voltages, however, the maximum cathode gradient which can be insulated in a uniform field diminishes markedly. As shown in Fig. 1.10, using well-conditioned stainless-steel electrodes, the maximum cathode gradient for a d-c gap voltage of 100 kv is reduced to about 10^6 v per cm, and at 700 kv this gradient is reduced to less than 10^5 v per cm. The voltage which can be insulated across a 1-mm gap between vacuum-immersed electrodes shaped to produce a uniform field a few square centimeters in area is approximately as follows for polished but not highly buffed surfaces:

Material	Voltage (kilovolts)
Steel	122
Stainless steel	120
Nickel	96
Monel metal	60
Aluminum	41
Copper	37

When the surfaces are highly buffed, aluminum electrodes become superior to steel and stainless steel, though copper is only slightly improved.

It has now been established that breakdown in high vacuum at high voltages is not primarily dependent on high-field emission as was generally supposed, but involves processes which depend on the total voltage across the interelectrode gap.[3] The processes which contribute to the interchange of charged particles between cathode and anode include electron bombardment of the anode surface with subsequent emission of positive ions and photons, and ion bombardment of the cathode surface as well as photoelectric absorption with subsequent emission of electrons. These processes are evidently dependent upon the total voltage between the electrodes.

When the conditions are such that the interchange of charged particles and photons between the vacuum-insulated electrodes becomes cumulative, breakdown results. Thus an electron in the gap is accelerated toward the anode and on impact liberates A positive ions and C photons which travel to the cathode. On arrival, each ion will liberate B electrons, and each photon, D electrons. These emission coefficients are dependent primarily on the total gap voltage, the electric field at the metal surface, and on the nature and surface condition of the metal. When $AB + CD$ becomes greater than one, the interchange becomes unstable, secondary processes including thermionic emission develop, and complete breakdown as a metallic spark or arc ensues.

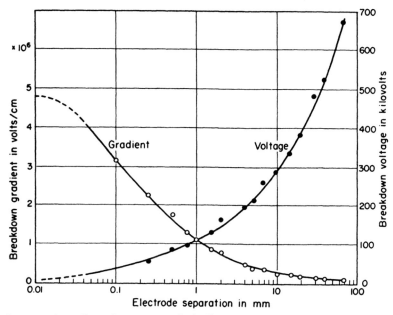

Fig. 1.10. Breakdown voltages and gradients between a 1-inch diameter stainless steel ball and a 2-inch diameter steel disk in vacuum.

[25] R. A. Millikan and C. C. Lauritsen, *Proc. Natl. Acad. Sci.* **14**, 45 (1928).

[26] R. H. Fowler and L. Nordheim, *Proc. Roy. Soc.* (*London*) A*119*, 173 (1928), and A*121*, 626 (1928).

The coefficient A has been measured as a function of electron energy for a parallel-plane steel gap. It increases sharply near the breakdown voltage to about 2×10^{-3}. For the AB particle mechanism to be important, the value of coefficient B must be of the order of several hundred near the breakdown voltage or some other particle process must be present. Recent measurements by Bourne[27] of the secondary-electron emission by various metals from aluminum to gold bombarded by positive ions ranging from hydrogen to mercury give evidence that the number of secondary electrons per bombarding ion is generally less than 20 and that there is complex dependency on the nature of the metal and on the bombarding ion and its energy.

Studies of the stable pre-breakdown current flowing between electrodes immersed in high vacuum [28] showed that the charge carried by electrons and negative ions to the anode was 300 to 1200 times greater than the corresponding positive ion current. An explanation for the observation that the insulating strength of a vacuum gap increases approximately as a square root of its length has been made by Cranberg.[29] He considers that breakdown is initiated when a critical amount of energy arrives at an electrode in the form of a clump of ionized atoms, the number of charges being proportional to the electric field intensity at the electrode of origin and the total energy of the clump therefore proportional to the square of the voltage divided by the gap.

High vacuum must be regarded as a basic insulating medium, free from many of the obscurities and limitations of material media and unique in that it is essential for the acceleration of elementary charged particles to moderate or high energies. Further fundamental and engineering research is needed to explain and minimize the processes which lead to its failure as a dielectric, particularly when high total voltages are applied.

Conclusions

1. In a uniform field, gases at high pressure can insulate higher total voltages than most solid or liquid dielectrics or high vacuum. Much of this advantage over the more viscous insulating media is lost when the electric field is seriously distorted.

2. Electronegative vapors such as CCl_2F_2 and C_3F_8 have nearly three times the voltage strength of common permanent gases, such as air or nitrogen at the same pressure in a uniform field. Dissociation of electronegative gases in the presence of ionization or sparkover and the formation of chemically active compounds are limitations which may be reduced by conservative design and the use of absorbing agents.

3. At high pressures, the limiting voltage in a uniform field is dependent on the electrode material, its surface smoothness, electrode conditioning, as well as the nature of the gas and the geometry of the gap.

4. For positive points directed at a plane, a critical gas pressure can be found above which the sparkover voltage diminishes and no significant pre-breakdown current flows.

5. The dielectric superiority of compressed gases is almost certain to be exploited to an increasing extent, particularly in devices in which high total voltages must be insulated such as electrostatic and electromagnetic sources of high energy X-rays and particles, high-voltage transformers and cables, and condensers.

6. High vacuum is a basic insulating medium free from many of the obscurities and limitations of material media and unique in that it is essential for the acceleration of elementary charged particles to moderate or high energies.

---

[27] H. C. Bourne, Jr., *Study of Theory of Voltage Breakdown in High Vacuum*, D.Sc. Thesis in Electrical Engineering, Massachusetts Institute of Technology, 1952.

[28] H. W. Anderson, *Elec. Eng.* 54, 1315 (1935).

[29] L. Cranberg, *The Initiation of Electrical Breakdown in Vacuum*, AEC Report LADC-1059.

2 · Liquid Dielectrics

By JOHN D. PIPER

This chapter describes properties and uses of insulating liquids, particularly the properties as related to conduction phenomena. Liquid dielectrics serve mainly as impregnants for high-voltage cables and capacitors, and as filling compounds for transformers, circuit breakers, and terminals. In some applications the liquid has additional important functions, for example, as a heat-transfer agent in transformers or an arc quencher in circuit breakers. Petroleum oils are most commonly used as liquid dielectrics. Synthetic hydrocarbons and halogenated aromatic hydrocarbons are also extensively employed. For high-temperature applications, the silicone oils, fluorinated hydrocarbons, and Fluorochemicals † show considerable promise; silicone oils served during the Second World War as spark-plug-filling compounds. Certain vegetable oils and esters are used in diminishing quantities.

General properties

The three important electric properties of liquids are conductivity, dielectric constant, and electric strength. Table 2.1 indicates that conductivity and electric

Table 2.1. Dielectric properties of some liquid dielectrics at 25°C

Liquid	Conductivity, mho/centimeter	Dielectric Constant	Dielectric Strength, kilovolts (ASTM D-887, 100 mils)
Cable oils in			
solid cables (nine, vintage 1930)	10^{-13} to 10^{-15}	2.3 to 2.6	29 to >35
oil-filled cables	10^{-14} to 10^{-16}	2.3 to 2.6	29 to >35
oil-pressure cables	10^{-14} to 10^{-16}	2.3 to 2.6	29 to >35
gas-pressure cables	10^{-14} to 10^{-16}	2.3 to 2.6	29 to >35
Transformer fluids			
Oil	10^{-14} to 10^{-16}	2.2 to 2.3	26 to 30
Askarels	10^{-13} to 10^{-14}	4.8	30 to 35
Silicones	10^{-14}	2.8	25 to 30
Fluorochemicals	10^{-14} to 10^{-16}	1.85	37 to 40

strength scatter appreciably since they are sensitive to impurities and other influences, whereas the dielectric constant appears as a characteristic of the material proper.

In contrast to the relatively similar electric characteristics, other physical and chemical properties of these dielectrics, for example, the viscosity, may differ markedly (Fig. 2.1). Even among cable-impregnating compounds, the viscosities vary widely; the impregnant for oil-filled cables is nearly as fluid as transformer oil, whereas some gas-pressure cables are impregnated with an extremely viscous material. Dielectric fluids differ markedly also in their flash points and in thermal stability. The silicone oils are remarkably stable but burn readily when ignited. The chlorinated and fluorinated liquids are noninflammable in the usual sense of the word, but of lower thermal stability.

† Fluorinated chemicals produced by Minnesota Mining and Manufacturing Co., St. Paul, Minn.

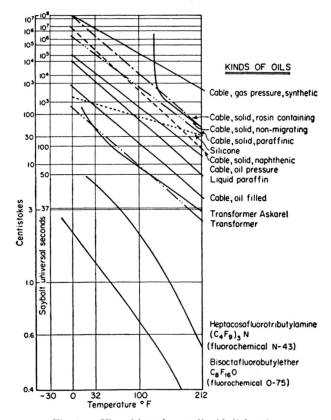

Fig. 2.1. Viscosities of some liquid dielectrics.

Choice of liquid

In selecting a liquid dielectric for a particular application, not only the properties of the liquid count but also other factors like saving of space, influence of the environment, cost, and industrial experience. By replacing oil with askarel impregnant, for example, the size of a capacitor can be reduced by about 30 percent.[1] Often choice is made on the basis of chemical stability, and occasionally unusual physical properties. Thus the manufacturer of highly volatile fluorochemicals suggests that users take advantage of the latent heat of vaporization for heat dissipation.[2] Illustrating the importance of environment, Table 2.2 shows the dielectric effect of various materials on a halogenated aromatic hydrocarbon. To cite an example of the impact of industrial experience, the Detroit Edison Company has had over ten years of successful experience with a gas-pressure cable impregnated with a very high viscosity oil. Impregnants of lower viscosity might function equally satisfactorily, but until in-

[1] F. M. Clark, *Am. Soc. Testing Materials Special Tech. Bull.* 95.

[2] Minnesota Mining and Manufacturing Co., 3M Company Fluorochemicals, Electrical Properties, *Bulletin*, New Products Division, St. Paul, Minn.

dustrial experience has been acquired the company would be reluctant to change impregnants. Finally, the factor of low cost keeps petroleum liquids in a dominant position.

Table 2.2. Power factor contamination produced by transformer materials aged in askarel at 100°C for 96 hours

Material	Askarel Power Factor, percent	Dielectric Strength, kilovolts
None	1.0	35
Black varnished cloth	85	42
Copper	1.5	40
Pressboard	2.0	37
Manila paper	1.5	39
Bakelite	1.6	41
Shellac	6.0	36
Iron	5.0	39
Synthetic rubber	70.0	39

Environment factors

Liquids are rarely the sole dielectric in commercial installations. In transformers the amount of liquid is usually great as compared with solid dielectrics; the electric stress is usually low; and the liquid functions mainly as a heat-transfer medium. In service, the transformer fluids are heated while in contact with other insulating materials and metals. In cables and capacitors the liquid also comes in contact with such materials as copper and paper, lead and its alloys, aluminum, steel, varnished cambric, polyethylene, and other synthetics. Add to this list the constituents of the atmosphere, and it becomes obvious that any study of the behavior of liquid dielectrics must consider the effects of its environment.

Failures

Much of what we know about liquid dielectrics has been learned directly or indirectly because of equipment failures. For example, cables break down or transformers become so filled with sludge that heat dissipation is impeded. After a failure, the "experts" assemble and try to determine why it occurred. Too often, all they can agree upon is that a hole is burned in the insulation. A manifestation of impending trouble may be the evidence of electrical discharge in the insulation or an increase in power factor. The latter phenomenon will receive principal attention in the remainder of this discussion.

Conduction Phenomena

Effect of voltage gradient and temperature

Most investigators have found that the conductivity increases with the voltage gradient both in highly purified liquids [3,4] and in commercial impregnants (Fig. 2.2). Conductivity also increases with increasing temperature; but the dielectric constant diminishes

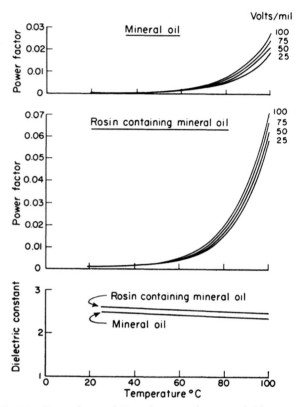

Fig. 2.2. Dependence of 60-cycle power factor on field strength and temperature for two cable coils.

because, as the liquid expands, fewer molecules are contained between the plates of the capacitor. A major reason for the conductivity increase is the decrease in viscosity, hence an increase in the mobility of the ions. Whitehead [5] states that between 30° and 60°C the product of viscosity and conductivity is nearly constant. However, the author [6] found that increasing the temperature from 20° to 80°C increases the conductivity considerably more than can be accounted for by the change in viscosity (Fig. 2.3). The product of the viscosity and the conductivity is inversely proportional to the reciprocal of the absolute tem-

[3] E. B. Baker and H. A. Boltz, *Phys. Rev. 51*, 275 (1937).
[4] R. W. Dornte, *Ind. Eng. Chem. 32*, 1529 (1940).
[5] J. B. Whitehead, *Impregnated Paper Insulation*, John Wiley and Sons, New York, 1935, p. 67.
[6] Unpublished data.

Liquid Dielectrics

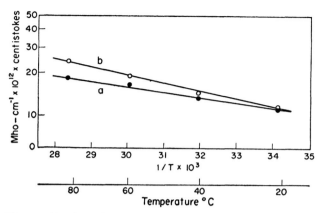

Fig. 2.3. Product of viscosity and a-c conductivity versus $1/T$. (a) Circuit-breaker oil and (b) transformer oil, both deteriorated from service.

perature. The conductivity of two unused mineral oils and a silicone oil varies exponentially with $1/T$, as shown in Fig. 2.4.

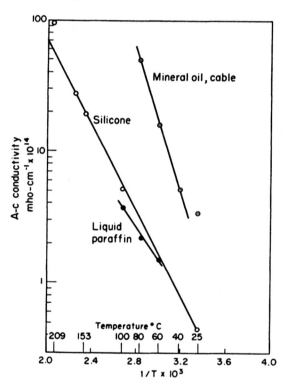

Fig. 2.4. Conductivities of some liquid dielectrics before use.

Causes of increased conductivity

The principal causes of conductivity increase of liquid dielectrics or insulation containing them are contamination and chemical and electrochemical reactions with the environment. Table 2.3 gives an example of contamination by handling. Transformer oils are transferred through flexible hoses. The poorest hose tested caused severe contamination, as evidenced by the increasing power factor, and even the best one had a noticeable contaminating effect.

Table 2.3. Contamination of transformer oil by contact with oil hoses as shown by power factor

	Oil in Contact with Hose at Room Temperature			Oil in Contact with Hose at 65°C			
	After 1 day	After 4 days	After 7 days	After 1 day	After 2 days	After 4 days	After 7 days
Poorest hose tested	0.0274	0.0573	0.0812	0.1023	......	......	......
Best hose tested	0.0006	0.0005	0.0006	......	0.0008	0.0010	0.0015
Oil alone	0.0002	0.0002	0.0002	0.0003	......	0.0002	......

Mutual reactions between the materials in dielectric equipment are well illustrated by extensive studies on the reaction among aluminum, paper, and askarels in capacitors. Figure 2.5 [7] shows the rapid rate of failure of capacitors not containing stabilizers [8] (curves a and

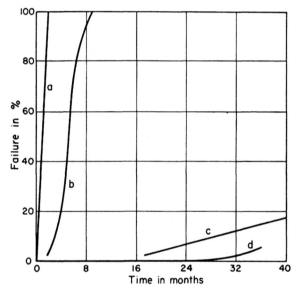

Fig. 2.5. Cumulative failure distribution curves illustrating type of reaction between askarels, paper, and aluminum-foil electrodes.[7] (a) Linen paper, 65°C; (b) Kraft paper, 65°C; (c) linen paper, room temperature; (d) stabilized samples, 65°C.

b) and the remarkable increase in life effected with them (curve d).

Reactions between dielectric liquids and oxygen

[7] D. A. McLean, L. Egerton, G. T. Kohman, and M. Brotherton, *Ind. Eng. Chem.* **34**, 101 (1942).

[8] For a comparison of the effectiveness of various stabilizers see, for example, L. J. Berberich and R. Friedman, *Ind. Eng. Chem.* **40**, 117 (1948).

have long been a source of trouble. Some of the means used for following the extent of oxidation have been: determination of absorption of oxygen, acid number, active hydrogen, peroxide, metal soaps, color change, sludge, interfacial tension, and hydrophils. Balsbaugh and co-workers [9,10] investigated the rela-

Davenport [15] discovered evidence for the presence of strong acids as well as weak ones in oxidized insulating oil.

In addition to purely chemical reactions in liquid dielectrics, electrochemical reactions take place under the influence of electric discharge in the gaseous and

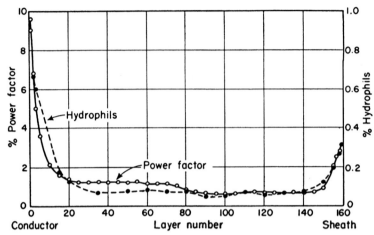

Fig. 2.6. General correspondence of radical hydrophil and power-factor curves for 132-kv oil-filled cable after six years in service.

tionship between amount of oxygen absorbed and the conductivity, the dielectric constant, and color. Wyatt [11,12] and co-workers found marked similarity between many curves representing the hydrophil content of oils taken from various layers of cable insulation and the power factors of the respective layers of oil-impregnated paper (Fig. 2.6). An outstanding contribution of Balsbaugh's [13] experiments was the observation that for both oil and oil-impregnated paper, subjected to oxidation in the presence of copper catalysts, the conductivity in the early stages of oxidation increased more when the supply of oxygen was limited than when it was unlimited. The effect of copper salts in causing high power factors in a seriously deteriorated oil-filled cable had been pointed out by Hirshfeld, Meyer, and Wyatt.[14] Evans and

liquid phases. The general aspects of these reactions have been thoroughly reviewed by Glockler and Lind.[16] Important to the behavior of dielectrics is the formation of gaseous ions, gases, and condensation products. As illustrated for benzene [17] (Fig. 2.7), a great variety of gaseous ions are formed even when

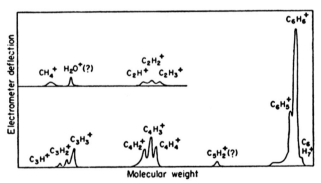

Fig. 2.7. Gaseous ions formed by electrical discharge in benzene.[17]

the discharge takes place in compounds of simple structure. When hydrocarbon oils are subjected to electrical discharge or electron bombardment under

[9] J. C. Balsbaugh, A. G. Assaf, and W. W. Pendleton, *Ind. Eng. Chem.* 33, 1321 (1941).

[10] J. C. Balsbaugh and J. L. Oncley, *Ind. Eng. Chem.* 31, 318 (1939).

[11] K. S. Wyatt, E. W. Spring, C. H. Fellows, *Trans. Am. Inst. Elec. Engrs.* 52, 1035 (1933).

[12] K. S. Wyatt, Deterioration of Insulating Oils through Interaction with Cable Metals, paper presented before International Conference on Large Electric Systems, Paris, 1937.

[13] J. C. Balsbaugh and A. G. Assaf, *Trans. Am. Inst. Elec. Engrs.* 62, 311 (1943).

[14] C. F. Hirshfeld, Meyer, and Wyatt, "Study of Mechanism of Cable Deterioration," 4th Progress Report, 1930, p. 44, printed but not published by Association of Edison Illuminating Companies.

[15] R. N. Evans and J. E. Davenport, *Ind. Eng. Chem.*, Anal. Ed. 9, 321 (1937).

[16] G. Glockler and S. C. Lind, *The Electrochemistry of Gases and Other Dielectrics*, John Wiley and Sons, New York, 1939, especially Chapter VII.

[17] E. G. Linder, *Phys. Rev.* 41, 149 (1932).

reduced pressures, hydrogen and low-molecular hydrocarbon gases evolve. If the gases are not removed, the rate of evolution lessens (Fig. 2.8).[18] The volume of gases evolved depends upon the structure, volatility, and viscosity of the oil [19] and the partial pressure of gases in contact with it,[20] to cite only a few of the

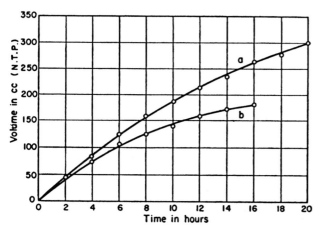

Fig. 2.8. Effect of partial pressure of hydrogen on rate of hydrogen evolution from cable compound subjected to electron bombardment.[18] (a) Gas pumped off; (b) gas not pumped off.

variables. In general, paraffinic hydrocarbons evolve the greatest amounts of gases and aromatics the least, as judged by the overall results of various investigators.

Electrical discharge in dielectric liquids usually produces more polymerized or condensed hydrocarbon than gas, as illustrated in Fig. 2.9 for decalin.[21] Cables taken from service often contain a material known as "wax" or Cable X which is insoluble in known solvents and is probably some three-dimensional hydrocarbon polymer. Apparently the polymer formed from paraffinic compounds is more insoluble than that from aromatic compounds. More polymer seems to be formed from the aromatic compounds, but instead of separating from the oil as a "wax" much of it remains in colloidal suspension. The author believes that the kind of polymer formed depends also upon the intensity of the electrical discharge, the insoluble products ranging anywhere from light-colored dielectric films to carbonaceous material.

Instead of evolving gases electrical discharge re-

[18] C. S. Schoepfle and L. H. Connell, *Ind. Eng. Chem.* **21**, 529 (1929).

[19] J. Sticher and D. E. F. Thomas, *Trans. Am. Inst. Elec. Engrs.* **58**, 709 (1939).

[20] G. W. Nederbragt, *Inst. Elec. Eng.* **79**, 282 (1936).

[21] J. Sticher and J. D. Piper, *Ind. Eng. Chem.* **33**, 1567 (1941).

actions in hydrocarbons may absorb gases.[16] Most bombardment studies on liquid dielectrics have been carried out at pressures considerably less than atmospheric. Such studies invariably show considerable hydrogen evolution but may not represent actual service conditions. Many years ago the Detroit Edison

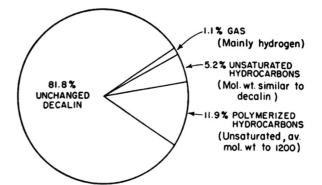

Fig. 2.9. Products from nine 4-hour bombardments of decalin.[20] (Composition is expressed in percent by weight of original decalin.)

Company carried out an extensive investigation to determine the composition of gases evolved from operating cables. From laboratory studies it had been anticipated that large volumes of hydrogen would be found. Usually the gas was simply air from which a part of the oxygen had been removed [22] (Table 2.4).

Table 2.4. Composition of gases collected from operating cables, solid type

(Analysis of samples from gas collectors)

Number of gas collectors—9
Number of samples—131
Total amount of gas collected—6831.4 cc from March, 1927, to September, 1928

	Average Percent by Volume	Maximum Percent	Minimum Percent
CO_2	0.17	7.0	0
O_2	14.79	21.0	4.9
H_2	0.58	5.3	0
CO	0.18	1.8	0
Methane	0	0	0
Ethane	0	0	0
N_2	84.28	94.8	76.8
	100.00		

In general, the oxygen in the environment of bombarded liquid dielectrics seems to combine with the polymer. Until recently, little thought was given to

[22] Hirshfeld and Meyer, "Study of the Mechanism of Cable Deterioration," printed but not published by Association of Edison Illuminating Companies, 1928, p. 70.

the possibility that nitrogen might react in like manner. However, the General Cable Corporation found that all the hydrocarbon oils that they had subjected to electrical discharge in the presence of nitrogen near or above atmospheric pressure, reacted with the nitrogen which unites with the polymer. In the writer's

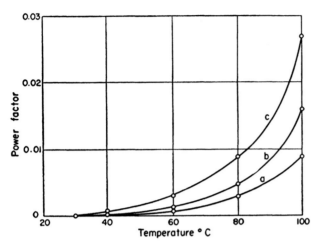

Fig. 2.10. Effect of corona discharge on the power factor of a cable compound:[23] (a) before bombardment; (b) bombarded 6½ hours; (c) bombarded 13 hours.

opinion this is one of the most significant findings in the field of practical liquid dielectrics in recent years. Apparently this phenomenon has also been taking place, unrecognized, in solid-type cables where the electrical discharge usually takes place only when the cables are operating at pressures less than atmospheric. After the results of General Cable Corporation's study had been revealed, the writer uncovered a little-known report [23] which showed the empirical formula of a particular sample of Cable X to be $C_{17}H_{30}ON$. The percentage of nitrogen found in that

Table 2.5. Fixation of nitrogen in hydrocarbon polymers formed by corona discharge

Material	Percent Nitrogen	Reference
Cable X, empirical formula $C_{17}H_{30}ON$	5.3	Fellows, unpublished data, 1927
Eight cable oils examined in 1931. Not subjected to discharge.	0.006 to 0.053	Fleiger, unpublished data
Insoluble polymer formed from cable compounds subjected to corona discharge under nitrogen pressure above one atmosphere	Up to 25	General Cable Corp., unpublished data

[23] C. H. Fellows, Unpublished report of The Detroit Edison Company, 1927.

sample of Cable X (Table 2.5) was a hundred times higher than that found in unbombarded cable oils analyzed a few years later by Fleiger, although much less than that found by the General Cable Corporation in the insoluble polymer or "wax" from some of the oils they studied.

Electrical discharge in the gas space in the environment of liquid dielectrics generally causes a more or less permanent increase in the conductivity of the dielectric [19] (Fig. 2.10). Application of a d-c potential causes a drop in the conductivity of bombarded decalin, and removal of the d-c field restores the conductivity to its original value (Fig. 2.11). Whitehead [5] observed a similar behavior in other liquid dielectrics and ascribed it to space charges. The material causing the high conductivity in the bombarded decalin was a polymer essentially hydrocarbon in nature; it proved insoluble in the base oil after it had been separated from lower-molecular polymers which apparently had kept it in stable dispersion.[21]

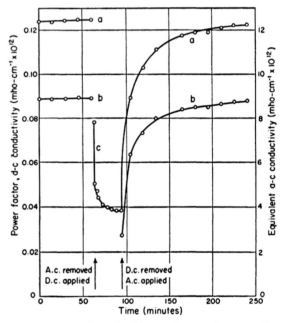

Fig. 2.11. Effect of one-half-hour direct voltage application on power factor and conductivity of bombarded decalin.[23] (a) Power factor, 200 volts, 60 cycles; (b) equivalent a-c conductivity, 200 volts, 60 cycles; (c) d-c conductivity. (Electrode spacing, 0.010 inch; 50°C.)

Although the conductivity increases markedly when hydrocarbons are subjected to electrical discharge in the absence of air, oxygen usually accelerates the process. The General Cable Corporation experiments showed that nitrogen or carbon dioxide likewise causes marked increases.

Materials producing increased conductivity

Because the correlation between the results of chemical and electrical tests on deteriorated insulating oil left much to be desired, the author and colleagues carried out additional studies to determine the kinds of materials that increase the conductivities of insulating oils and oil-impregnated paper appreciably. Compounds that could conceivably be formed in hydrocarbon oils and impregnated paper were purified, added to oils of low conductivity, and the conductivity increase was measured.

Soluble members of the carboxylic acid series had only a small effect on the power factor of liquid paraffin,[25] but the lower molecular ones had a large effect on the power factor of impregnated cable paper.[26] Formic acid, which was sparingly soluble, raised markedly the power factor of the oil, and this effect increased as the temperature was lowered. Similarly phenol[27] exerted little influence when in solution, but a marked influence when it came out of solution on cooling. By microscopic examination it was found that the mixture of phenol and liquid paraffin first formed an emul-

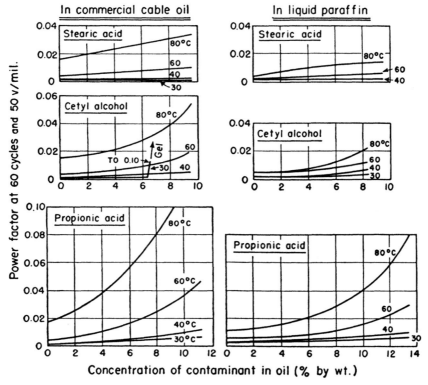

Fig. 2.12. Difference in effect of soluble oxidation products on power factor of commercial cable oil and liquid paraffin.[24]

Figure 2.12 shows the effect of oxidation products on the power factor of liquid paraffin and a commercial cable oil. When present in the low concentrations corresponding to deteriorated oils the effect was small in the paraffin but appreciably larger in the cable oil.[24] By adding decalin to increase the fluidity a marked increase in power factor occurred in the commercial cable oil, but none in liquid paraffin (Fig. 2.13). Liquid paraffin was therefore used as the base oil for most of the remaining studies with the following results:

[24] Research Department Staff, The Detroit Edison Company, "Some Further Results of Cable Research," AEIC Report, 1933–34, pp. 87 and 95.

sion upon cooling and later crystallized (Fig. 2.14). The high power factor resulted when the mixture existed as an emulsion. Emulsions of castor oil and liquid paraffin did not cause high power factors when the two phases separated on cooling. When, however, methyl *n*-amyl ketone, which itself had little effect, was added to the mixture an appreciable increase in power factor took place as the mixture became heterogeneous with decreasing temperature. The resulting

[25] J. D. Piper, D. E. F. Thomas, and C. C. Smith, *Ind. Eng. Chem.* **28**, 309 (1936).

[26] J. D. Piper, D. E. F. Thomas, and C. C. Smith, *Ind. Eng. Chem.* **28**, 843 (1936).

[27] J. D. Piper, C. C. Smith, N. A. Kerstein, and A. G. Fleiger, *Ind. Eng. Chem.* **32**, 1510 (1940).

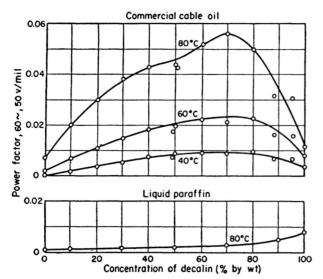

Fig. 2.13. Conducting materials in commercial cable oil revealed by dilution with decalin.[24]

conductivities of the heterogeneous mixtures were much greater than could be ascribed to a Maxwell-Wagner mechanism. Lead soaps (Fig. 2.15) and

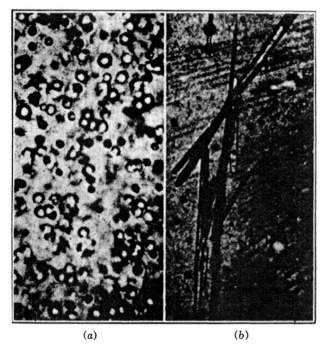

Fig. 2.14. Microscopic appearance of a system containing 5.11 percent phenol (×85): (a) shortly after the system became heterogenous; (b) later.

copper soaps caused high power factors at temperatures appreciably above those at which their solutions became visibly heterogeneous upon cooling.[28] In cer-

[28] J. D. Piper, A. G. Fleiger, C. C. Smith, and N. A. Kerstein, *Ind. Eng. Chem.* **31**, 307 (1939).

tain of these solutions gels formed near the point of heterogeneity. At temperatures sufficiently above the point of heterogeneity, the conductivity was surprisingly low. In one instance, during the preparation of a solution of a copper soap in a sealed tube in the absence of air, the blue solution changed to a reddish color. This mixture had a high power factor. Presumably the cupric soap reacted with the oil to form a cuprous compound. Cuprous compounds were found to be exceedingly insoluble in oil. Efforts to prepare stable solutions or dispersions of known cuprous compounds failed. Another class of compounds investigated were nitrogen bases and their salts.[29] The bases

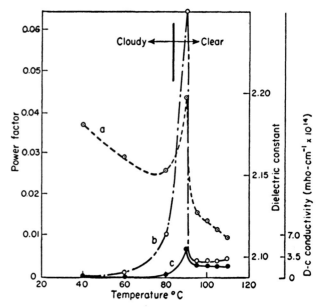

Fig. 2.15. Dielectric properties during cooling of liquid paraffin containing 0.15 percent lead myristate:[28] (a) dielectric constant; (b) power factor; (c) d-c conductivity.

by themselves were no more effective in causing high power factors than carboxy acids. When both were added, the results were additive up to some initial concentration. Beyond this point salts evidently formed, causing considerably greater power factor increases. This concentration was great as compared to that of nitrogen compounds found in commercial insulating oils, but not to that of oils bombarded in nitrogen.

Sulfur compounds investigated included lauryl sulfonic acid [30] representing soluble sulfonic acids that might be formed in insulating oils. This was the only oil-soluble material investigated that produced high

[29] J. D. Piper, A. G. Fleiger, C. C. Smith, and N. A. Kerstein, *Ind. Eng. Chem.* **34**, 1505 (1942).

[30] J. D. Piper, P. Treend, and K. S. Bevis, *Ind. Eng. Chem.* **40**, 323 (1948).

conductivities. Those conductivities, however, were much lower than those of colloidal solutions of gilsonite, an asphaltic material that is used as electrical insulation (Fig. 2.16). The 60-cycle conductivities of paper impregnated with the solutions behaved in reverse order and was much greater for lauryl sulfonic acid than for gilsonite.

ions by impregnating dry paper with the solutions, keeping oil-paper ratios low, and comparing the low frequency conductivities. Electrolytically dissociable materials cause markedly higher conductivities in impregnated paper than in oil. The reverse is true for colloidal materials. The principle, although not exhaustively tested, is at least true for asphalt and a

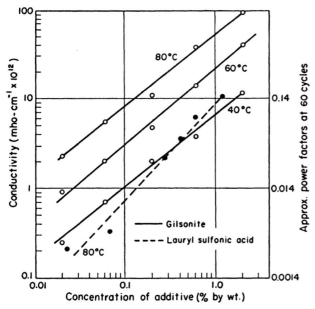

Fig. 2.16. Conductivity of liquid paraffin solutions of gilsonite and of lauryl sulfonic acid.[30]

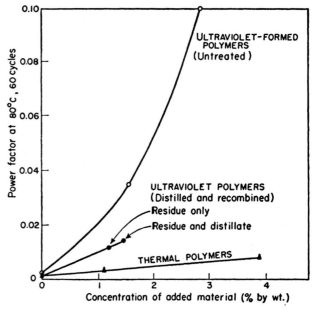

Fig. 2.17. Power factor of decalin containing indene polymers,[20] formed thermally or by ultraviolet irradiation.

Numerous hydrocarbons containing somewhat dissociable hydrogen [21] were also investigated but none caused significant increases in conductivity. Hydrocarbon polymers [21] likewise had little effect except those formed by irradiation with ultraviolet light (Fig. 2.17).

Conclusions

Although the increased conductivity of deteriorated oils may in some instances be caused by the formation of compounds that are electrolytically dissociable, such as sulfonic acids and salts of nitrogen bases, usually it is caused primarily by the formation of colloidal ions. A feature common to such colloids is a constituent that by itself is insoluble in the oil. Examples of such materials are "soluble" sludge, asphalt, cuprous soaps or oxides, and condensation products that result from electrical discharge reactions involving oxygen, nitrogen, and sulfur compounds as well as hydrocarbons. The materials that cause high conductivities primarily through electrolytic dissociation may be differentiated from those that cause high conductivities primarily through the formation of colloidal

lead soap, representing colloidal materials, and for lauryl sulfonic acid and several carboxylic acids, representing electrolytically dissociable materials.

Nature of ions

In addition to conductometric techniques, Gemant has introduced electropotential and radioactive-tracer methods to determine the nature of ions in dielectric liquids. The electropotential technique, as in aqueous solutions, consists of selecting an electrode that is reversible to the ion under study and determining the potential between two such electrodes placed in solutions of different concentrations. The conductivities of the solutions in the two half cells serve as a measure of the active concentrations of the ions. For a reversible electrode the potential of the cell, in millivolts, is

$$E = 118(1 - t_I) \log \frac{\sigma_1}{\sigma_2},$$

where t_I = transference number of the ion examined, and

$\frac{\sigma_1}{\sigma_2}$ = the conductivity ratio between the two half cells.

In tests for hydrogen-ion concentration Gemant used a glass electrode and found a linear dependence of the potential on the conductivity ratio for solutions of p-toluene sulfonic acid in dioxane and xylene, trichloroacetic acid in dioxane, and picric acid in xylene.[31]

To investigate solutions of the low dielectric constants of insulating oils he found it necessary[32] to depart from a strict concentration cell and to use a solution of lauryl sulfonic acid in xylene in one cell and a more dilute solution of the same acid in transformer oil in the other. The plot of potential vs. conductivity ratio was also linear in this case (Fig. 2.18),

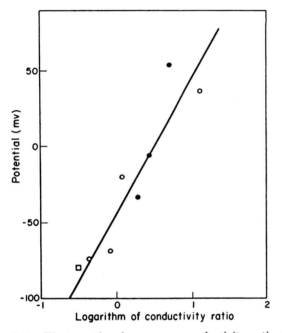

Fig. 2.18. Electromotive force versus conductivity ratio for hydrocarbon oil containing lauryl sulfonic acid.[32]

demonstrating that the cell behaves like a concentration cell and that the conductivity of lauryl sulfonic acid in mineral oil results from electrolytic dissociation into hydrogen ion and anion. In contrast, there was no correlation between the potential and the conductivity ratio for solutions of asphalt or for oxidized transformer oil. Gemant concluded[32] that for those systems "the predominant mechanism of ion generation is different from the usual acidic dissociation."

In tests for negative ions silver electrodes coated with silver propionate and silver myristate were used.[33] These electrodes reacted essentially reversibly, not only with their respective anions, but also with aliphatic anions in general. Using such electrodes in high-conductivity systems[34] containing an amine, an aliphatic acid, and a phenol in transformer oil, Gemant concluded that the negative ions consist of phenol-solvated anions of the aliphatic acids. Similar tests with oxidized insulating oils lead to the conclusion that the negative ions in oxidized oils may also be solvated aliphatic ions. To the writer, the data indicate that the aliphatic ions are indeed associated with some other material but that this material need not necessarily be solvated molecules; it could be colloidal material of the kinds previously discussed.

In his search for organic cations in insulating oils Gemant has investigated amines and quinones. Up to the present these investigations have been confined to conductometric tests, since reversible electrodes for the cations have not yet been found. The combinations of amines and acids in transformer oil proved to be of low conductivity, confirming the writer's experience with liquid paraffin.[29] When, however, a phenol was added[34] a marked increase in conductivity resulted (Table 2.6). The addition of phenol to a base-acid-transformer-oil system at 25°C acts similarly to the addition of more base and acid to the writer's base-acid-liquid paraffin system at 80°C. The ionic mobilities were nearly the same in both cases because the viscosity of the liquid paraffin at 80°C was almost identical with that of transformer oil at 25°C. On the basis of equal concentrations by weight of total additive the conductivities of the amine-acid systems, although high as compared with oil alone, were not nearly as great as those of the lauryl sulfonic acid systems (bottom of Table 2.6), and these in turn were less than those of the gilsonite systems.

One of Gemant's systems containing an amine and an acid gave very high conductivities in low concentration and without the addition of a phenol. This system, triethanolamine-oleic acid in transformer oil, appeared to be homogeneous to the naked eye and had conductivities independent of frequency. The triethanolamine, however, was practically insoluble alone, and had a rather limited solubility in the presence of the acid. Thus the system fits the observation previously discussed, that systems in which one component is insoluble, even under conditions in which they appear to be homogeneous, often have high conductivities.

In a preliminary report on quinones (1952 Conference on Electrical Insulation National Research Council) Gemant showed that systems containing a hydroquinone, upon being mildly oxidized, produce high conductivities. An ionic structure, intermediate be-

---

[31] A. Gemant, J. Chem. Phys. 12, 79 (1944).
[32] A. Gemant, J. Chem. Phys. 13, 146 (1945).
[33] A. Gemant, J. Electrochem. Soc. 94, 160 (1948).
[34] A. Gemant, J. Chem. Phys. 14, 424 (1946).

tween the hydroquinoid and the quinoid, is believed to form, this structure acting as a cation much as an amine does. Further work has to insure that the hydroquinone is not undergoing condensation reactions to form tarry matter that might account for the high conductivity.

Table 2.6. Conductivities of solutions in hydrocarbon oils

Concentrations, mole/liter			Percent (total)	Conductivity, mho/cm	Solvent	Temp.
Tributyl Amine	Oleic Acid	o-Cresol				
0.05	0.05	....		0.017×10^{-12}	Transformer oil	25°C
0.05	....	0.05		0.029	" "	"
....	0.05	0.05		0.021	" "	"
0.05	0.05	0.05	3.4	0.9	" "	"
Tributyl Amine	Myristic Acid	α-Naphthol				
0.05	0.05	....		0.014	" "	"
0.05	....	0.05		0.079	" "	"
....	0.05	0.05		0.055	" "	"
0.05	0.05	0.05	3.3	4.0	" "	"
Tributyl Amine	Acetic Acid					
0.037 *	0.037 *			0.022	Liquid paraffin	80°C
0.22	0.22		3.2	1.8	" "	"
Tributyl Amine	Lauric Acid					
0.025 *	0.025 *			0.015	" "	"
0.15	0.15		3.4	0.2	" "	"
Piperidine						
0.085 *	0.085 *			0.07	" "	"
0.20	0.20		3.4	11.2	" "	"
Lauryl Sulfonic Acid						
			3.3 †	70 †	" "	"
			3.3 ‡	63 ‡	Transformer oil	25°C
Gilsonite			3.3 †	260 †	Liquid paraffin	80°C

* Up to these concentrations the effect of the amine and acid on both the conductivity and the polarization were simply additive; beyond these concentrations both effects were markedly enhanced.
† Extrapolated from Fig. 2.16.
‡ Extrapolated from Fig. 2 of ref. 32.

The nature of the negative ions produced by an aliphatic acid in hydrocarbon oil can be studied by using radioactive tracers. From diffusion rate determinations on $0.005\,N$ tridecanoic acid, identified in the carboxyl group by C^{14}, the diameter of the acid in transformer oil was found by Gemant[35] to be 7 A, corresponding to molecular dispersion. As previously discussed, such solutions have very low conductivities even in high concentrations. When an amine and a phenol were added, 0.3 and $0.4\,N$, respectively, the diameter of the tridecanoate complex was doubled.

Tagged tridecanoic acid reacts with various metals under room conditions to form films on the metal surfaces.[36] Addition of an amine increased the rate of this reaction greatly; further addition of a phenol had little effect. Oxidation of the original oil also increased the rate of film formation, probably indicating that oxygen bases are formed, related to quinones.

A third radioactive tracer technique has been used by Gemant[37] on suspensions of particle size in the micron range (silver myristate, potassium tridecanoate, potassium acid tridecanoate, magnesium oxide, and graphite in transformer oil). To these suspensions were added the ionically conducting acid-amine-phenol system previously discussed which caused a marked decrease in zeta-potential. Like other investigators, he found that the suspended particles were usually negative in sign, but sometimes positive. Acids caused negatively charged particles to become positive. Some years ago this writer,[21] using a cardioid ultramicroscope and a cell in which a field was applied, concluded that, although suspended material can often be perceived to move in a high-conductivity oil, the high conductivity is but little related to the movement of such large particles. Rather, it seems to be related to the presence of colloidal material that is either of too small a particle size or with optical properties so near those of the oil as to render discrete particles indiscernible. The radioactive-tracer techniques may permit a closer investigation of materials previously described as *soluble sludge*, asphalt, and products from bombardment, thus bridging the gap between coarse suspensions and associated molecules.

By persistent and painstaking effort our knowledge of the mechanism of conduction in deteriorated petroleum oils is being enlarged. Although much work has been done in the application of synthetics, the mechanism of conduction in synthetic dielectric liquids has been scarcely explored. "The harvest is ready, but the laborers are few."

[35] A. Gemant, *J. Phys. & Colloid Chem.* **54**, 569 (1950).
[36] A. Gemant, *J. Electrochem. Soc.* **99**, 279 (1952).
[37] A. Gemant, *J. Phys. & Colloid Chem.* **56**, 238 (1952).

3 · Plastics as Dielectrics

By WARREN F. BUSSE

Many thousand plastics have been made and described by the organic chemist, and the dielectric properties of several hundred of them have been measured in the Laboratory for Insulation Research at the Massachusetts Institute of Technology (see Part V). Some thirty-five types have attained enough commercial importance or general interest to be listed in the *Plastics Properties Chart for 1951* of the Plastics Catalog Corporation. This discussion will cover some of the newer plastic dielectrics that are now coming onto the market, and the relation of mechanical and dielectric properties to molecular structure.

As background information we shall review briefly some of the chemical structures present in various types of plastics, and relations between the mechanical and dielectric properties of the plastics and those of the small molecules from which they are derived. The value of a dielectric material often depends as much on its mechanical as on its dielectric properties.

Properties of Small Molecules

Plastics are made by combining large numbers of small molecules into fewer big ones. Each of the small molecules, and of the atoms and groups in them, is in a constant state of motion and vibration due to thermal agitation. Collisions occur at rates of the order of 10^{12} per sec. The Brownian motion of a molecule as a whole causes it to diffuse in a random path through a liquid. The speed of this diffusion depends on the size of the molecule and the resistance encountered from other molecules, that is, on the viscosity of the surrounding medium.

For a molecule of a radius 2×10^{-8} cm in a liquid of 0.01-poise viscosity, the average time required to diffuse one diameter is $\sim 10^{-11}$ sec. Because of the random nature of the process, the square of the distance traversed in diffusion is proportional to the elapsed time. Hence, 10 diameters (4×10^{-7} cm) will require 100 times as long (10^{-9} sec), and the traversal of macroscopic distances of a few millimeters demands hours. A similar microdiffusion of segments of long molecules plays an important part in the elasticity of rubber and the toughness of plastics.

Some of the small molecules used to make plastics are hydrocarbons and, when quite pure, have very low loss factors and dielectric constants only slightly greater than the square of their indices of refraction (see IA, Eq. 3.24). Symmetrically substituted hydrocarbons such as carbon tetrachloride or paradichlorobenzene also have low loss factors, but their dielectric constants are considerably higher than calculated from the index of refraction for visible light. This is due to infrared absorption bands. Substituted hydrocarbons such as phenols, anilines, and alcohols contain polar groups which give the molecules a permanent dipole moment. Such molecules tend to rotate when placed in an alternating field. The Debye theory accounts in a first approximation for the change in dielectric constant and dissipation factor with frequency and with viscosity. The peak of the loss factor maximum (plotted as a function of frequency) is a measure of the rotational relaxation time of these polar molecules (see IB, Eq. 4.28).

Since the relaxation time depends on the viscosity of the surrounding medium we would expect changes in dielectric constants and dissipation factors on changing the temperature, similar to those we get on changing the frequency.

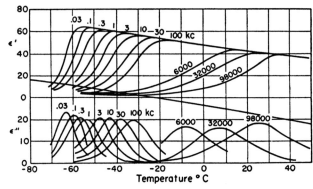

Fig. 3.1. Temperature dependence of dielectric constant and loss factor for glycerol.[1]

Figure 3.1 shows the effect of temperature and frequency on the dielectric constant and the loss factor of glycerol as determined by Morgan and Yager[1] and

[1] S. O. Morgan and W. A. Yager, *Ind. Eng. Chem.* **32**, 1519 (1940).

Mizushima.[2] For any given frequency the dielectric constant of the liquid is low at low temperature, where the viscosity is high. As the temperature is raised, the viscosity decreases, and a point is reached where the molecules begin to follow the field changes and the dielectric constant thus rises. In this temperature range the dissipation factor traverses a maximum. The higher the frequency, the higher the temperature at which the dielectric constant begins to increase. An extreme example of the effect of hindered rotation of polar molecules is water when freezing; the dielectric constant drops from about 80 to around 4.[3]

It should be noted that the dipole molecules show only a slight preferential orientation in the field direction (see IB, Sec. 3). Debye[4] pointed out that, at a field intensity of 1 volt per cm, the effect equals the alignment of one dipole molecule in 5,000,000 in field direction; this fractional orientation would suffice to produce a dielectric constant of 80. Thus the molecules of the water undergo their thermal vibration and Brownian motion in practically the same random way with or without the field. Even when the resistance to rotation is increased enormously, as when water is changed to ice, a slight amount of orientation still remains. The dielectric constant of ice shows a change with frequency in the megacycle range just as glycerol at lower temperatures. The dissipation factor[5] on lowering the temperature of ice from $-2°C$ to $-15°C$ decreased from 26×10^{-4} to 2×10^{-4} in the X-band region (3.2 cm).

Of importance in liquids is the effect of the dielectric constant on the ionization and d-c conductivity when ionic impurities are present. Water is an outstanding example, since it allows almost complete ionization of many dissolved salts. The problem recurs in certain sulfones such as tetramethylene sulfone (sulfolane) or substituted sulfolanes. These liquids have dielectric constants of 15 to 20 or higher,[6] but it is very difficult to get them sufficiently pure for use in condensers because almost any impurity is ionized and leads to fairly high d-c conductivity.

Chemistry of Plastics

When small molecules link to form the macromolecules of plastics, many different types of structures may result. We shall consider here only two extremes.

The linear (fishline) type

Most thermoplastic resins approximate a structure, in which several thousand atoms are tied together by strong valence bonds in one direction, whereas the macromolecules are only a few atoms thick at right angles to this direction. A molecule of polyisobutylene, for example, with a molecular weight of 250,000 would have a carbon chain about 8000 carbon atoms long but, on the average, only two carbon atoms thick. Thus its proportions correspond about to those of a piece of fishline a yard long. This illustration is not quite to the point, for the fishline will fall limp and dead, whereas the thermal bombardment of adjacent atoms will make the molecule of polyisobutylene wiggle and squirm in a continuous ramdom motion. A better model might therefore be an angleworm $\frac{1}{10}$ in. in diameter and perhaps 40 ft long. The chances for tangling, stretching, and flowing in a mass of such structures are obvious and play a decisive part in the properties of plastics.

The crosslinked (fishnet) type

The thermosetting resins and cured rubbers, in contrast, form a three-dimensional fishnet type. Instead of being long, thin, and flexible, the molecules here resemble an irregular three-dimensional fishnet, and their properties depend very much on the size and distribution of the mesh links. For very large meshes with only a few crosslinks, the properties might be almost like those of the fishline types. An example is uncured (natural or synthetic) rubber, with its high elongation and recovery. On the other hand, very many crosslinks make the material hard and stiff and perhaps brittle, as exemplified by hard rubber or phenolformaldehyde resin.

The formation of phenolformaldehyde resin is illustrated in Fig. 3.2.[7] Resins of similar structures can be made with aniline or melamine or urea, and formaldehyde. The degree of elongation or the toughness of the product depends on the distribution of crosslinks and the number of chain atoms between the links. Since these are old, well-known resins, we will not discuss them further.

The fishline type of molecule includes a wide variety of chemical structures. Figure 3.3 shows a few vinyl-type polymers. *Polyethylene* is made by linking large numbers of ethylene molecules in a chain. This is the simplest vinyl-type resin and one of the

[2] S. Mizushima, *Bull. Chem. Soc., Japan*, 1, 83, 145, 163 (1926).
[3] *International Critical Tables*, E. W. Washburn, Editor, McGraw-Hill Book Co., New York, 1929, Vol. VI, p. 78.
[4] C. P. Smyth, *Dielectric Constant and Molecular Structure*, The Chemical Catalog Co., New York, p. 38, 1931.
[5] W. A. Cumming. *J. Appl. Phys.* 23, 768 (1952).
[6] W. F. Busse, U. S. Patent 2540748, Feb. 6, 1951.

[7] J. J. P. Staudinger, *British Plastics* 25, 160 (1952).

most important. We would predict that it has excellent electrical properties. The surprising fact is that the mechanical properties are also much better

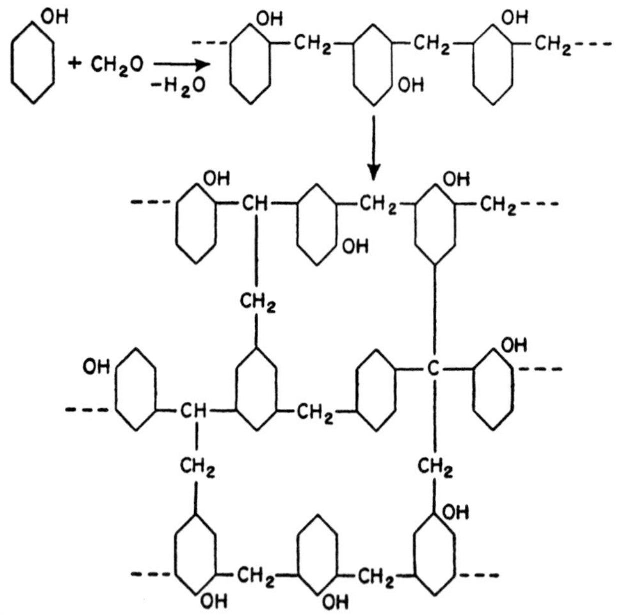

Fig. 3.2. Formation of phenol-formaldehyde polymer.[7]

than one might expect from looking at this chain molecule. *Polyvinylchloride* is important commercially; it also permits a comparison of the rotation of polar groups in an electrical field with the distortion

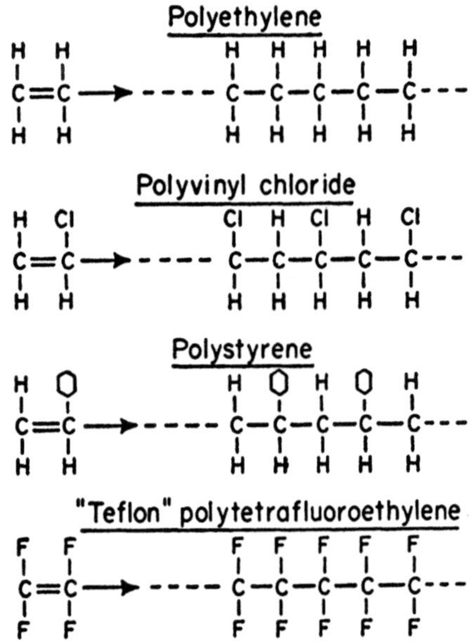

Fig. 3.3. Typical linear polymers.

of the molecules in a mechanical field. *Teflon*, the polytetrafluoroethylene resin, is in a class by itself, as will be discussed later. *Polystyrene*, although excellent electrically, is hard and brittle at room temperature and has a relatively low softening point of about 85°C.

If we replace the phenyl group of polystyrene by a carbazole group (Fig. 3.4), *polyvinylcarbazole*, a plastic with a much higher softening point (>150°C), results. However, the polyvinylcarbazole is even more

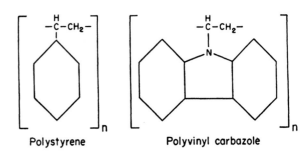

Fig. 3.4. Effect of substitution of phenyl by carbazole group.

brittle than polystyrene, and this drawback, together with very high cost, has prevented its wide use. One of its promising applications is as an impregnant for paper capacitors. The condensers are evacuated, impregnated with the monomer, and heated under pressure until polymerization is complete. Such capacitors will stand up longer at high temperatures than oil-impregnated paper capacitors (Fig. 3.5).[8]

Figure 3.6 gives a longer list of vinyl compounds with brief indications of their nature.[7] The compounds cover a wide spectrum of properties, all the way from elastic materials to hard plastics and fiber-forming compounds.

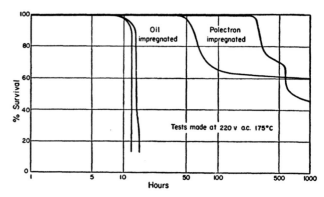

Fig. 3.5. Life tests of condensers impregnated with oil and polyvinyl-carbazole.[8]

Some other types of linear polymers are presented in Fig. 3.7. The *polyesters* of the first line comprise a large family of products made by combining various dibasic acids with various glycols. When the acid is

[8] W. F. Busse, J. M. Lambert, C. McKinley, and H. R. Davidson, *Ind. Eng. Chem.* **40**, **2271** (1948).

terephthalic acid, that is, benzene with two carboxyl groups at opposite ends, and the alcohol is ethylene glycol the product is *Dacron* polyester fiber, also available in the form of thin films under the du Pont

X	Y	Nature of polymer	Type of polymer
H	H	Crystalline	Polythene
CH_3	CH_3	Rubbery	Polyisobutene
H	$O.OC.CH_3$	Plastic	Polyvinyl acetate
H	$CO.O.CH_3$	Rubbery	Polymethyl acrylate
CH_3	$CO.O.CH_3$	Plastic	Polymethyl methacrylate
CL	Cl	Fibre-forming	Polyvinylidene chloride
H	Cl	Plastic	Polyvinyl chloride
H	C_6H_5	Plastic	Polystyrene
H	CN	Fibre-forming	Polyacrylonitrile

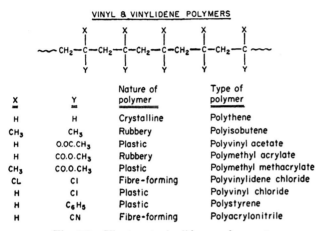

Fig. 3.6. Vinyl and vinylidene polymers.[7]

tradename *Mylar.*† Contrary to what we might expect from the structure, this plastic has very good electrical properties in the power frequency range (see below).

The *polyamides* in Fig. 3.7 include the whole family of *nylons.* When a 6-carbon diamine is combined with a 6-carbon dibasic acid the result is 66 nylon.

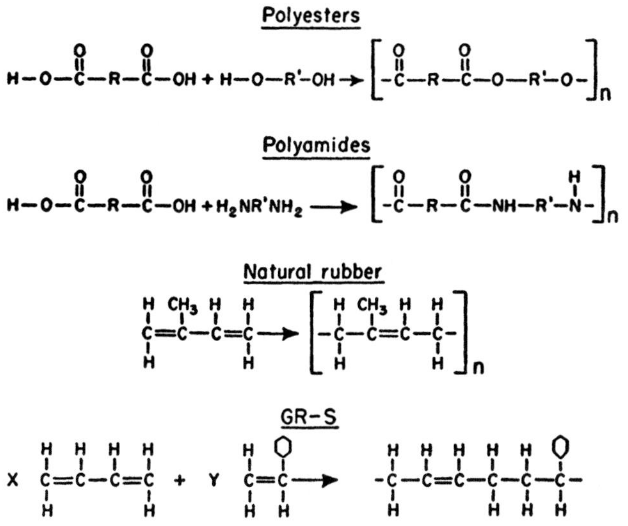

Fig. 3.7. Polymers (polyester, for example, Mylar; polyamide, for example, nylon).

If a 10-carbon dibasic acid is used with a 6-carbon diamine we have the 610 nylon, etc. Although these plastics do not have the best electrical properties, particularly in the presence of moisture, their mechanical properties are so good that they have an important place in jacket material for wire insulation.

The *natural rubber* shown on the third line differs from all the previous compounds in that it is unsaturated. This double bond allows reaction with sulfur and accelerators to form crosslinks and three-dimentional structures. If the CH_3 group is replaced by Cl, the product is neoprene, the first commercial synthetic rubber produced in this country.

The last line in Fig. 3.7 represents a copolymer of butadiene and styrene. The amount of each component can be varied over wide ranges. Materials with high proportions of styrene are particularly useful for electrical insulation.

Three other types of polymers which cannot be omitted from a general discussion of dielectrics are shown in Fig. 3.8. Cellulose and cellulose acetate are

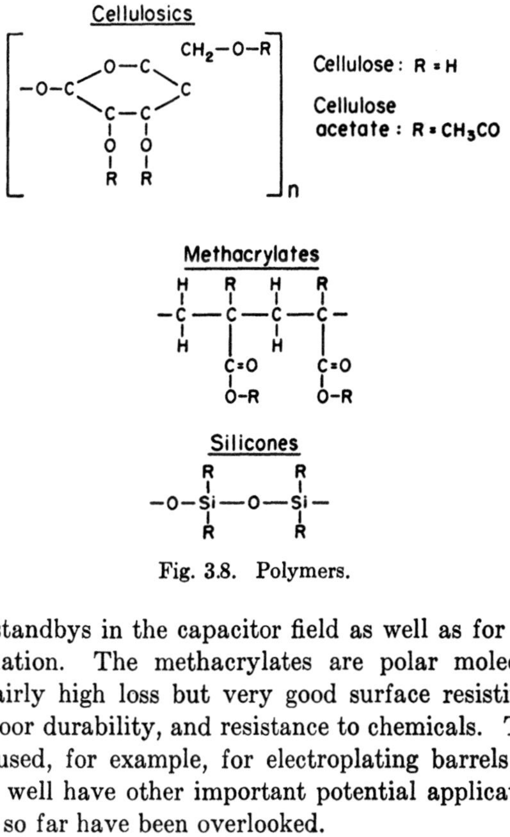

Fig. 3.8. Polymers.

old standbys in the capacitor field as well as for wire insulation. The methacrylates are polar molecules of fairly high loss but very good surface resistivity, outdoor durability, and resistance to chemicals. They are used, for example, for electroplating barrels and may well have other important potential applications that so far have been overlooked.

The silicone resins, while not outstanding from the electrical standpoint (loss factor = 0.001 to 0.01) and of low tensile strengths (600 to 1200 psi), have a number of valuable properties. They are stable and useful over a wide temperature range. Polmanteer et al.[9]

† A similar product is put out by the British under the tradename Terylene.

[9] K. E. Polmanteer, P. C. Servais, and G. M. Konkle, *Ind. Eng. Chem.* *44,* 1576 (1952).

have described types that are flexible down to −112°F. Other silicones [10] will withstand prolonged exposures to temperatures up to 500°F, and short exposures to even higher temperatures. In addition to this wide-temperature range, the resins are generally resistant to acids, alkalies, and mineral oils, of low moisture adsorption, and relatively unaffected by corona discharge or ozone. Flashovers do not produce carbon.

A wide variety of silicone oils and resins can be obtained. Rochow [11] describes the chemistry behind these preparations in some detail. Interesting variations in physical and electrical properties result by suitable variation in preparation and structure of the resin. Alkyl and aryl silicones have been synthesized as well as copolymers of the two. By varying the proportions of the different groups in the copolymers a wide range of properties can be obtained. Rochow describes a methyl silicone of power factor 0.0045 at 56°C, whereas a similar methylphenyl silicone (1.00 methyl groups and 0.80 phenyl groups per silicon atom) has a power factor of 0.0005 at 100°C.

Mechanical and Electrical Properties of Selected Plastics

Rubber

Let us consider the mechanical properties of rubber, since it shows, in a pronounced way, certain properties that are less apparent in other dielectrics.

If we take a rubber band, stretch and then release it, the happenings on a macroscopic scale permit some fundamental conclusions about the molecular structure and rearrangement.[12]

(1) In a band stretched, say 500 percent, the average distance between the atoms along the length of the band must have increased sixfold. Released, it returns to the original distance without significant change in density. Since the force fields between atoms fall off rapidly with a high power of the distance between the atoms (see IB, Sec. 3), the distances along primary valence bonds cannot possibly be increased sixfold without destroying the bond. Hence the deformation of the rubber must involve largely a rearrangement of the atoms, a change in the molecular shape, without greatly increasing the length of any valence bonds. This is possible if the molecule is a long flexible kinky chain, say, of the proportions of the 40-foot angleworm discussed above.

[10] D. C. R. Miller, *Can. Chem. Process Ind.* **33**, 764, 858 (1949).

[11] E. G. Rochow, *An Introduction to the Chemistry of the Silicones*, John Wiley and Sons, Inc., New York, 1951, 2nd Ed., p. 78.

[12] W. F. Busse, *J. Phys. Chem.* **36**, 2862 (1932).

(2) With these large deformations, the unraveling segments of the kinky chain molecules must move relatively large distances past other molecules with little friction. This implies that the chains have weak secondary valence forces. Hence, for small deformations, rubber may be considered as a low-viscosity liquid in which diffusion of segments of the chains can occur in micro-microseconds.

(3) There is a limit to the deformation of the band. The molecules do not continue to slip past each other, as they do, for example, in drawing taffy. This implies a large-scale structure, produced by a few crosslinks between the chains of the order of one crosslink per 100 chain carbon atoms. We can also get the effect of such crosslinks from temporary mechanical entanglements, such as occur in unmilled crude rubber, or from polar groups that give strong van der Waals attractions between chains, as in polyvinyl chloride, or from crystallization, as in polyethylene.

(4) The fourth factor in this simple experiment, and the one of the greatest technical importance, is that the rubber, when the deforming force is reduced, returns most of the energy used to stretch it. This implies some mechanism for storing free energy.

The recovery cannot depend on storing up potential energy, for the rubber, when stretched, actually heats up, and it cools on retracting, as can easily be demonstrated with the rubber band. About twenty years ago it was first recognized that the straightening of a long tangled chain molecule would decrease its entropy (its disorder), and this would increase its free energy. When the deforming force is released, the thermal vibrations of adjacent atoms and segments of molecules will tend to make the molecule resume a random shape, and thus shorten its length and increase its entropy. This is the basis of the kinetic theory of rubber elasticity.[13]

(5) Another factor which may be noted in this experiment with the rubber band is that it loses its transparency and becomes opaque on stretching. This observation led Katz [14] to suspect there was a new phase produced in the rubber on stretching, and to examine it with X-rays. Figure 3.9 shows the amorphous X-ray pattern of unstretched rubber and the highly ordered, crystalline pattern of stretched rubber. These crystallites act as reinforcing pigments and temporary crosslinking agents which greatly improve the physical properties of the rubber. If uncured rubber is held at temperatures of 10° to −20°C for long times, it will also crystallize, but the crystals

[13] L. R. G. Treolar, *The Physics of Rubber Elasticity*, The Clarendon Press, Oxford, 1949, pp. 7 ff.

[14] J. R. Katz, *Kolloid Z.* **36**, 300 (1925).

will be in random order. The rubber will then be hard and boardy. It should be noted that in the absence of the orienting and restraining effects of an external force, the rubber must be cooled to a much lower temperature to crystallize.

to hold the molecules of the amorphous regions together in a rigid structure; the rubber becomes brittle and "glassy."

If we plot the volume (or density) of rubber as a function of temperature, we observe a change in slope

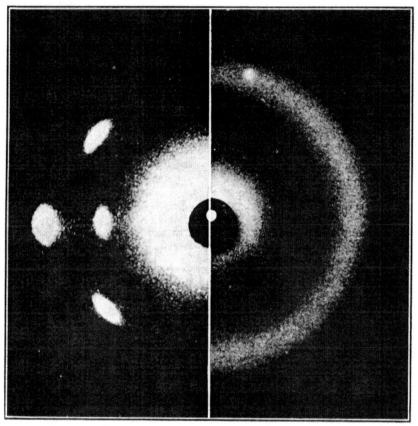

Fig. 3.9. X-ray pattern of stretched and unstretched rubber.

Even under the best conditions, however, only a part of the rubber will become ordered in crystallites, and these will be separated by amorphous regions. If the temperature is further reduced to about $-80°C$, the thermal vibrations will become so small that the weak van der Waals forces between the chains are sufficient

at the temperature where the rubber becomes brittle or glassy; this change is called a second-order transition.

Table 3.1. Second-order transition temperatures

Material	°C	Material	°C
Polyisobutylene	-74	Polyvinylidene chloride	-17
Natural rubber	-73		
GR-S rubber	-61	Polymethyl acrylate	3
Hycar or rubber	-23	Polymethyl methacrylate	57–68
Cellulose acetate, B-96	68.6	Polystyrene	81
Polyvinyl acetate	28	Polyvinyl carbazole	84
Polyvinyl alcohol	85	Nylon	47
		Ebonite	80–85

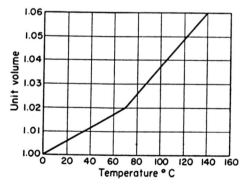

Fig. 3.10. Volume-temperature relations for polymethylmethacrylate.

Figure 3.10 shows such a curve [15] for polymethyl methacrylate, and Table 3.1 gives the second-order transition temperatures [15] for a number of polymers of in-

[15] H. Mark and A. V. Tobolsky, *High Polymers*, Vol. II, *Physical Chemistry of High Polymeric Systems*, Interscience Publishers, New York, 1950, pp. 346 and 347.

terest as dielectrics. All flexible materials have low second-order transition temperatures whereas hard, stiff materials have high second-order transition temperatures. In general, polymers tend to be:

(1) Hard or brittle if the second-order transition is above room temperature.

(2) Hard and tough if the second-order transition is below room temperature, but the crystalline melting point is above room temperature.

(3) Flexible if the crystalline melting point is below room temperature and the molecular weight high enough to make the macroscopic viscosity very high.

Let us see how important these factors are in some of the other dielectrics.

Polyethylene

The very simple, uniform hydrocarbon structure of Fig. 3.3 suggests some properties of this plastic. The low van der Waals forces around this chain should result in a low brittleness temperature; it actually lies near $-70°C$. Furthermore, a compound of high molecular weight with such a regular structure should be quite crystalline. Dole et al.[16] report that polyethylene is 40 to 60 percent crystalline at room temperature. The crystallinity decreases at higher temperature and becomes zero at around 105°C. This crystallinity at room temperature makes the polyethylene inert to many of the solvents and chemicals which would be expected to attack hydrocarbons. The fact that it is saturated makes it resistant to chemical attack also.

A number of the observed properties of polyethylene are not entirely consistent with the structure shown. We would expect no dipole moment, hence zero electrical loss factor. Commercial samples have a low, but not zero loss, particularly after the plastic has been milled. The Laboratory for Insulation Research,

[16] M. Dole, et al., *J. Chem. Phys.* **20**, 781 (1952).

Massachusetts Institute of Technology,[17] and others[18] have measured polyethylene after various amounts of milling, with results similar to those of curves 1 to 3 of Fig. 3.11. (Jackson and Forsyth[19] found the power factor peak at about 10^8 cycles.) Obviously the milling produces dipole groups, probably through oxidation, which cause loss peaks near 10^9 cycles. Antioxidants are usually added to protect the polyethylene during fabricating operations where minimum loss is needed; curve 4 results.

Polyethylene must be protected against exposure to sunlight. After 6 months to one year of exposure, unprotected polyethylene will acquire enough dipole groups, not only to raise its power factor, but also to raise its brittle point from $-70°C$ to ambient temperatures. Although it is well known from the work of the Bell Telephone Laboratories[20] that the addition of a few percent well-dispersed carbon black to the polyethylene will act as an efficient light screen, and

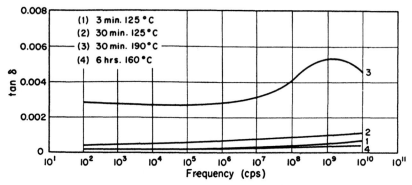

Fig. 3.11. Loss factors of polyethylene after various amounts of milling. Curves 1–3[16]; curve 4, milling with antioxidant.

prevent this type of deterioration for many years, we still see far too many television lead-in wires insulated with polyethylene not adequately protected from light. These will not give the service expected.

Hypalon chlorosulfonated polyethylene

If the regularity of the polyethylene structure is broken by chlorine atoms substituted on the chain at random, the crystallinity should decrease and the material become more flexible. Eventually, when enough chlorine is added, the dipole attractions of the chlorine atoms should make the material hard and stiff. Figure 3.12 shows that this is actually the case. At around 30 percent chlorine, the stiffness goes

[17] Tables of Dielectric Materials, Vol. II, NDRC Contract OEMsr-191, Report 8, Laboratory for Insulation Research, Massachusetts Institute of Technology, June, 1945.
[18] J. G. Powles and W. G. Oakes, *Nature* **157**, 840 (1946).
[19] W. Reddish, *Chemistry and Industry*, 1951, p. 1077.
[20] V. T. Wallder et al., *Ind. Eng. Chem.* **42**, 2320 (1950).

through a minimum, and a soft and elastic product results. Oakes and Richards [21] found that the peak relaxation time at 16°C varies from about 10^{-8} sec for unchlorinated polyethylene to around 1 sec for a sample containing about 50 percent chlorine.

Chlorinated polyethylenes are thermoplastics like the vinyl and vinylidene chlorides. However, if about one $-SO_2Cl$ group is also introduced into the chain for every 90 to 100 carbon atoms, the product can be cured to a nonthermoplastic material by forming crosslinks with metal oxide such as MgO, PbO, or

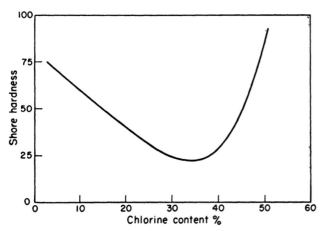

Fig. 3.12. Hardness vs. chlorine content of chlorinated polyethylene.

diamines, or certain vulcanization accelerators. This chlorosulfonated polyethylene has recently been put on the market by du Pont under the trademark Hypalon.

Hypalon can be cured to give rubbers with mechanical properties comparable to those of other synthetic rubbers, and with a number of other properties that make it of special interest as a dielectric. One of these is its almost complete ozone resistance, which suggests its use for high-voltage insulation. Table 3.2 shows the relative times required for ozone cracks

Table 3.2. **Stability of elastomers to ozone exposure**

Elastomer	Time to Crack *
Natural rubber	3 min.
GR-S	3 min.
Neoprene	30 min.
GR-I	14 hours
Hypalon	>100 hours

* Exposed to 130 ppm ozone.

to develop when cured high gum compounds of various elastomers were bent double and exposed to 130 ppm of ozone.

[21] W. G. Oakes and R. B. Richards, *Trans. Faraday Soc. 42A*, 197 (1946).

The dielectric properties of Hypalon, like those of other elastomers, depend very much on the compounding agents. Data for a typical compound (Hypalon 100 parts, Staybelite 2.5, PbO 40, Captax 3) and for a polyvinyl composition containing 30 parts tricresyl phosphate are given in Table 3.3.

Table 3.3. **Dielectric properties of Hypalon chlorosulfonated polyethylene and polyvinyl chloride**

	100 cps	1 kc	100 kc
Hypalon			
κ'	6.19	6.04	5.10
tan δ	0.014	0.020	0.098
Polyvinyl chloride [21] (30 pts. plasticizer)			
κ'	7.0	5.9	4.0
tan δ	0.082	0.12	0.13

The d-c resistivity of the Hypalon is over 10^{13} ohm cm, high enough for all low-frequency applications. The material does not require carbon black to give tough stocks. Mixtures with GRN can be made which have good oil resistance as well as good abrasion resistance and electrical properties.

The development of other Hypalon chlorosulfonated polyethylenes containing different ratios of chlorine and sulfur offers some interesting possibilities. Experimental batches with only about 5 percent chlorine and 1 percent sulfur have approached the electrical properties of polyethylene, yet because they can be cured they will be useful at temperatures above 105°C where polyethylene melts.

Polyvinyl chloride

Polyvinyl chloride is old and well established as a dielectric, but still being studied to check theories of the relations between the response to electrical and mechanical deformations. Some early experiments attempting to relate temperature and plasticizer content and dielectric properties to the hardness of polyvinyl chloride containing 0, 30, and 60 parts tricresyl phosphate are shown in Fig. 3.13.[22] Measurements were made at 15, 60, and 1000 cps, and it was found that a 66-fold frequency change at one temperature shifted the position of the dipole peaks by about as much as at 10°C change in temperature at a constant frequency. The "hardness" or Young's modulus E of these same samples was calculated from the deformation d produced by hemispherical indentors of diameter D and load L, using Hertz's formula: [23]

$$E = 0.795 \frac{L}{d^{3/2} D^{1/2}}.$$

[22] J. M. Davies, R. F. Miller, and W. F. Busse, *J. Am. Chem. Soc. 63*, 361 (1941).
[23] L. Larrick, *Rubber Chem. and Tech. 13*, 969 (1940).

Although the loss maxima do not occur at a constant hardness, the data suggest the existence of a relation between hardness and dielectric relaxation time.

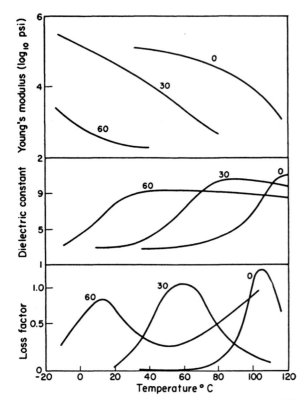

Fig. 3.13. Effect of temperature on properties of stabilized polyvinyl chloride–tricresyl phosphate systems (0, 30, 60% TCP).[21,22]

Many people found it difficult to accept this view when it was first proposed. Although the dielectric loss maximum is determined by a rate process,[24] it was argued that the hardness was determined by a potential energy function hence could not be related to dielectric loss. We know today that in these high polymers the hardness actually depends on the rate at which the molecules move to readjust themselves to the applied force. The indentation hardness changes with the time the load is applied. Thus hardness is, to some degree at least, a rate measurement.

Further work on this subject was done by Fuoss and Kirkwood,[25] who attempted to develop a theory relating the dielectric loss with a simple relaxation mechanism. Fuoss also showed that under some conditions there were two peaks in the loss factor curve. Alfrey[26] developed an expression for the compliance as a function of time (or frequency) and molecular weight.

Nielsen, Buchdahl, and Levreault[27] have made a direct comparison of the a-c electrical properties and the dynamic mechanical properties of polyvinyl chloride containing various plasticizers over a range of temperatures. A torsion pendulum was used to determine the dynamic shear modulus and dissipation factor. Although they ignored the important effects of frequency on mechanical properties, their results illustrate clearly the similarity of the mechanical and electrical effects (Fig. 3.14). Ferry and Fitz-

[24] W. Kauzmann, Revs. Mod. Phys. 14, 12 (1942).
[25] R. M. Fuoss and J. G. Kirkwood, J. Am. Chem. Soc. 63, 385 (1941).
[26] T. Alfrey, J. Chem. Phys. 12, 374 (1944).
[27] L. E. Nielsen, R. Buchdahl, and R. Levreault, J. Appl. Phys. 21, 607 (1950).

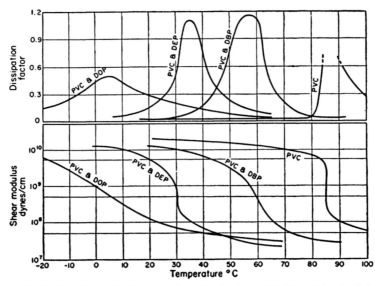

Fig. 3.14. Change of mechanical and electrical properties of polyvinyl chloride with plasticizer and temperature.[27]

gerald [28] measured the dielectric constant and dissipation factor and the mechanical stiffness and loss factor of plasticized polyvinyl chloride at a wide range of frequencies. The results show that the units active in the two processes are closely related.

One exception to this relation should be noted. Reddish [19] found the 60-cycle loss peak of methacrylate to be ca. 50°C whereas the materials softened at around 100°C. This was attributed to the fact that the dipoles were on side chains, rather than on the main chain.

Fluorocarbon resins

The need for dielectrics for high-temperature service caused a great interest in the fluorocarbons.[29, 30, 31] Teflon tetrafluoroethylene resins and "Kel-F" chlorotrifluoroethylene resins [32] are the two outstanding examples of this class. They are not new, but recent commercial developments (larger production, lower price, new ways of handling) lend new interest to them.

A comparison of some physical and electrical properties of these materials shows where each has its greatest usefulness (Table 3.4). The replacement of one fluorine atom in C_2F_4 with chlorine gives a polymer of higher rigidity, strength, and better molding properties. It also leads to poorer electrical properties and a somewhat lower maximum use temperature. The symmetrical structure of polytetrafluoroethylene makes it quite crystalline; the individual molecules are quite stiff. A first-order transition occurs near 20°C, and the crystals melt at 327°C to a very high viscosity gel.

Recent tests show the viscosity of Teflon to be around 10^{11} poises at 380°C. This fact, together with the weakness of the gel to shearing forces, and the long time required for the sheared surfaces to sinter together, makes the fabrication problems with Teflon rather difficult. However, progress is being made. It is possible, for example, to extrude a paste of particles of Teflon mixed with a lubricant such as cetane on a wire, and then bake the wire to vaporize the cetane and sinter the Teflon to a solid, coherent coating. Another process which offers promise is the calendering of Teflon on wires and even on rectangular conductors. Some wire is being coated by a ram extrusion process, but at slow rates. Experiments are under way to increase operating speeds for all these processes.

Mylar polyester film

A new plastic approaching commercial scale manufacture is Mylar polyester film; it is similar in composition to Dacron polyester fibers. A few physical properties are shown in Table 3.5, together with corresponding properties of cellulose acetate.[33] The differences in softening point, strength, and elongation are notable.

The dielectric properties of Mylar are shown in Figs. 3.15 and 3.16. The dissipation factor is fairly high in the megacycle range at room temperature, but drops off at higher and lower frequencies. At power frequencies, the dissipation factor is very low and decreases as the temperature is increased. The volume resistivity is over 10^{15} ohm-cm at 100°C.

The high softening point enables Mylar to be used at temperatures above the operating limit of paper insulation. It also has good resistance to weathering

Table 3.4. Physical properties of fluorocarbon resins

	Kel F [32]	Teflon
Molding qualities	Excellent	....
Tensile strength, psi	5700	1800
Elongation, %	30	110
Modulus of elasticity	1.9×10^5	0.6×10^5
Compressive strength, psi	30,000–80,000	1700
Maximum continuous-use temperature, °F	390	500
Dielectric constant, 60–10^6 cycles	2.3–2.8	2.0
Dissipation factor, 60 cycles	0.01	<0.0002
10^3 cycles	0.02	<0.0002
10^6 cycles	0.01	<0.0002

Table 3.5. Properties of Mylar polyester film

	Mylar	Cellulose Acetate
Tensile strength (psi)	25,000	11,000
Break elongation, %	130	20
Bending modulus (psi)	500,000	350,000
Softening point (°C)	240–245	60–110
Dielectric strength (v/mil)	5,000	3,000
Volume resistivity (ohm-cm)	4×10^{15}	10^{10}
Surface resistivity (ohms—100% RH)	4.8×10^{11}	....

[28] E. R. Fitzgerald and J. D. Ferry, *J. Colloid Sci.* **8**, 1 (1953).
[29] "Fluorocarbons—Face, Fame and Fortune," *Modern Plastics* **30**, 79 (1952).
[30] B. H. Maddock and W. M. Land, *Modern Plastics* **30**, 79 (1952).
[31] "Versatile Fluorine Plastics," *Chem. Eng. News* **30**, 2688 (1952).
[32] See M. W. Kellog Co., *Technical Bulletin*, Jan. 12, 1949.
[33] R. C. Krueger and A. B. Ness, "The Electrical, Physical, and Chemical Properties of 'Mylar' Polyester Film," presented at the meeting of the American Institute of Electrical Engineers, New York, January, 1952.

and even soil burial. These properties suggest that it may be of interest for motor and transformer insulation at power frequencies, and perhaps even for very high frequency insulation where subject to weathering.

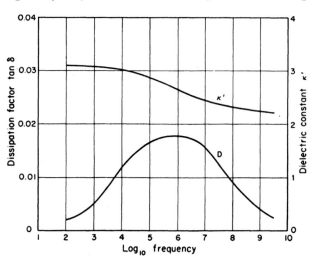

Fig. 3.15. Dielectric properties of Mylar polyester resin.

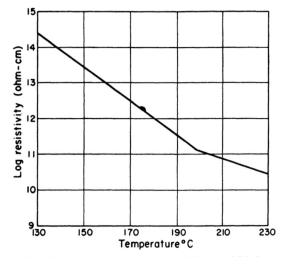

Fig. 3.16. Effect of temperature on resistivity of Mylar polyester resin.

Miscellaneous Dielectric Properties

Two more dielectric properties are of practical importance and challenge those who are trying to understand and improve dielectrics.

Corona breakdown

Practically all types of organic and most inorganic dielectrics, including mica, will break down eventually if they are tested under corona conditions where they are subject to ionic bombardment.[34] Figure 3.17 shows typical data. This type of breakdown can be prevented by avoiding gas ionization, as by immersing the electrode and dielectric in oil.

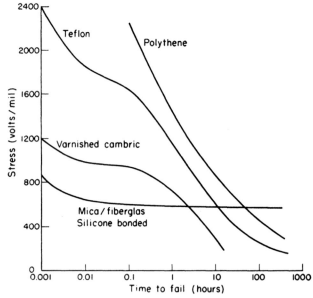

Fig. 3.17. Failure of typical dielectrics under corona discharge.[31]

Conducting plastics

Conducting plastics may, at first sight, seem out of place here, for conductivity usually is the last thing wanted in a dielectric. However, the property of high conductivity is important in a number of cases. Rubber is sometimes made conducting by adding acetylene black to prevent the accumulation of static, as in airplane tires. In this way also the rubber pads used in operating rooms are made conducting to avoid sparks that might ignite the ether in a patient's lungs. Markites[35] have received considerable publicity as conducting plastics, but there have been few, if any, reports of practical applications.

A somewhat different use of conducting plastics is indicated in the Sylvania Electric's development of "Panalescent light" which concerns emission of light from phosphors in an alternating potential field.[36] Here the light must come out through one of the electrodes, so the dielectric must not only have a conducting surface, but it must also be transparent. A glass has been developed with a surface resistivity of 1 to 10 ohms per square. A transparent plastic dielectric of equal surface conductivity would have considerable advantage in weight, safety, and ease of fabrication.

---

[34] C. G. Brodhun, *Phys. Rev.* **86**, 653 (1952).

[35] M. A. Coler et al., *Proc. Inst. Radio Engrs.* **38**, 117 (1950).

[36] E. C. Payne, E. L. Mager, and C. W. Jerome, *Illum. Eng.* **45**, 688 (1950); *Sylvania Techn.* **4**, 2-5 (1951).

Another place where conducting, transparent plastics are needed is in airplane canopies and other glazing uses. Here two orders of conductivity would be useful. A very small conductivity would prevent the build-up of high static charges and corona that affect radar and other electronic equipment. Conductivities of a much higher order could be used to defrost windshields or canopies by electric heating. The development of such transparent, conducting plastics stands today as a challenge to the plastics industry.

4 · Ceramics

By HANS THURNAUER

The field of ceramic dielectrics comprises a wide range of materials which cannot be treated with justice in a short outline. It is possible to give only a bird's-eye view as to classification and properties. At present there does not exist one complete textbook on the subject. For further information, the books in references 1, 2, 3, 4, 9, 21, and 24 might be consulted. We have attempted to select this bibliography in such a way that the most important developments of recent years are highlighted. Data quoted in our tables and graphs are characteristic values, but may vary within rather wide limits from one product to another. Table 4.1 summarizes the chemical composition of the minerals to which this survey refers.

Table 4.1. Mineral compositions mentioned in this survey, used either as raw materials or constituents of fired ceramic bodies

Kaolinite (main constituent mineral of clay or Kaolin)	$(OH)_4Al_2Si_2O_5$
Flint	SiO_2
Feldspar	$K_xNa_{1-x}[Al \cdot Si_3]O_8$
Mullite	$Al_6 \cdot Si_2O_{13}$
Cordierite	$Mg_2Al_2Si_5O_{15}$
Corundum	Al_2O_3
Talc or steatite	$(OH)_2 \cdot Mg_3 \cdot Si_4O_{10}$
Clinoenstatite	$MgSiO_3$
Forsterite	Mg_2SiO_4
Zircon	$ZrSiO_4$
Spodumene	$Li_2Al_2Si_4O_{12}$
Wollastonite	$CaSiO_3$
Rutile	TiO_2
Barium titanate	$BaTiO_3$
Sodium tantalate	$NaTaO_3$
Sodium columbate	$NaNbO_3$
Cadmium columbate	$Cd_2Nb_2O_7$
Magnetite	Fe_3O_4
Spinel (mineral)	$MgAl_2O_4$
Ferrite spinel (typical)	$Ni_xZn_{1-x}Fe_2O_4$

Definition

The word *ceramic* is derived from the Greek word *keramos*, meaning "a potter," "potter's clay," or "pottery," and is related to the older Sanskrit root, "to burn." The definition generally given for a ceramic is "a product obtained through the action of fire upon an earthy material." The definition is sufficiently broad to include not only structural products, such as refractories and building materials, but also glass, enameled ware, abrasives, cements, electrical and thermal insulation.[1,2]

This discussion will deal chiefly with ceramic dielectrics which are molded from earthy, inorganic materials and permanently hardened by a firing or sintering process. It will not include ceramic products molded in the viscous liquid state, while hot; in other words, glass or glass-bonded products are excluded. It is recognized, however, that these products play an increasingly important role as dielectric materials.

Classification

Ceramic dielectrics may be conveniently classified in four groups: (1) materials with a dielectric constant below 12; (2) with a dielectric constant above 12; (3) with piezoelectric and ferroelectric properties; (4) with ferromagnetic properties. We shall discuss compositions within each group and compare their mechanical and dielectric properties.

(1) Materials with a dielectric constant below 12

The ceramics of this group are at present the most important ones for our economy. About one-half of the total whiteware production in the United States is devoted to the production of low- and high-tension

[1] Hewitt Wilson, *Ceramics; Clay Technology*, McGraw-Hill Book Co., New York, 1927.
[2] F. H. Norton, *Elements of Ceramics*, Addison-Wesley Press, Cambridge, Mass., 1952.

electrical insulators. The other half comprises the manufacture of dinnerware, artware, and chemical porcelain.[3,4]

(a) Low- and high-tension electrical porcelain

Conventional electrical porcelain is used for both low- and high-tension insulation. It consists of a mixture of the minerals clay-flint-feldspar. Clay or kaolin, hydrous aluminum silicates of varying purity and of the general formula $[(OH)_4Al_2Si_2O_5]$, give plasticity to the ceramic mixture. Flint, which is silica (SiO_2), acts as an inert material that helps to maintain the shape of the formed article during firing; and feldspar, an alkali-aluminum silicate $[K_xNa_{1-x}(AlSi_3)O_8]$, serves as flux. The quality of the final product and its degree of vitrification depend upon the fineness of grinding, the method of forming, and the firing temperature. The fired product contains two main crystalline phases: mullite crystals $(Al_6 \cdot Si_2 \cdot O_{13})$ and undissolved quartz (SiO_2) crystals.

[3] Rexford Newcomb, Jr., *Ceramic Whitewares*, Pitman Publishing Corp., 1947; E. Rosenthal, *Porcelain and Other Ceramic Insulating Materials*, Chapman and Hall, London, 1944.

[4] H. Salmang, *Die Keramik*, Springer, Berlin, 1951.

These crystals are imbedded in a continuous glassy phase, originating from the feldspar. By varying the proportions of the three main ingredients, it is possible to emphasize certain properties, such as heat-shock resistance, dielectric strength, or mechanical strength (Fig. 4.1). Some of the properties of typical electrical porcelain bodies are given in Table 4.2. This table should also be consulted for comparing the properties of ceramic materials discussed later on. For high tension insulation, water absorption must be zero in order to insure high dielectric and mechanical strength. This is not as critical for low tension insulators, although the degree of vitrification serves as a good indicator of quality.

The glaze on high tension insulators has important functions. If the glaze is properly compounded, that is, of an expansion coefficient slightly lower than the porcelain, the latter is under compression and the mechanical strength of the insulator improved by as much as 30 percent. Such a glaze also improves the electrical resistance of insulators under humid conditions.

To reduce corona on high-voltage insulators and eliminate interference with electronic devices, high-

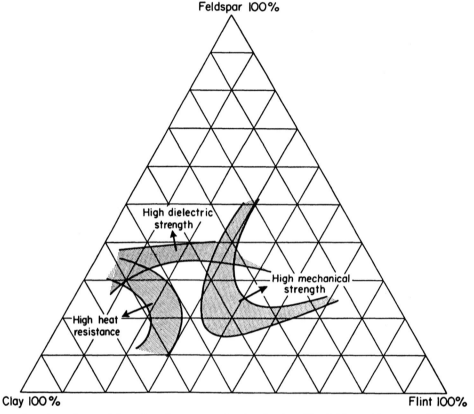

Fig. 4.1. Location of porcelains of various properties in the clay-flint-feldspar diagram. (H. Salmang.[4])

Ceramics

Table 4.2. Typical physical properties of ceramic dielectrics

Vitrified products

	1	2	3	4	5	6	7
Material→	High-Voltage Porcelain	Alumina Porcelain	Steatite	Forsterite	Zircon Porcelain	Lithia Porcelain	Titania, Titanate Ceramics
Typical Applications→ Properties ↓	Power Line Insulation	Sparkplug Cores, Thermocouple Insulation, Protection Tubes	High-Frequency Insulation, Electrical Appliance Insulation	High-Frequency Insulation, Ceramic-to-Metal Seals	Sparkplug Cores, High Voltage-High Temperature Insulation	Temperature-Stable Inductances, Heat Resistant Insulation	Ceramic Capacitors, Piezoelectric Ceramics
Specific gravity (g/cc)	2.3-2.5	3.1-3.9	2.5-2.7	2.7-2.9	3.5-3.8	2.34	3.5-5.5
Water absorption (%)	0.0	0.0	0.0	0.0	0.0	0.0	0.0
Coefficient of linear thermal expansion/°C (20-700)	$5.0-6.8 \times 10^{-6}$	$5.5-8.1 \times 10^{-6}$	$8.6-10.5 \times 10^{-6}$	11×10^{-6}	$3.5-5.5 \times 10^{-6}$	1×10^{-6}	$7.0-10.0 \times 10^{-6}$
Safe operating temperature (°C)	1000	1350-1500	1000-1100	1000-1100	1000-1200	1000	....
Thermal conductivity (cal/cm²/cm/sec/°C)	0.002-0.005	0.007-0.05	0.005-0.006	0.005-0.010	0.010-0.015	....	0.008-0.01
Tensile strength (psi)	3000-8000	8000-30,000	8000-10,000	8000-10,000	10,000-15,000	....	4000-10,000
Compressive strength (psi)	25,000-50,000	80,000-250,000	65,000-130,000	60,000-100,000	80,000-150,000	60,000	40,000-120,000
Flexural strength (psi)	9000-15,000	20,000-45,000	16,000-24,000	18,000-20,000	20,000-35,000	8000	10,000-22,000
Impact strength (ft-lb; ½-in. rod)	0.2-0.3	0.5-0.7	0.3-0.4	0.03-0.04	0.4-0.5	0.3	0.3-0.5
Modulus of elasticity (psi)	$7-14 \times 10^6$	$15-52 \times 10^6$	$13-15 \times 10^6$	$13-15 \times 10^6$	$20-30 \times 10^6$	....	$10-15 \times 10^6$
Thermal shock resistance	Moderately good	Excellent	Moderate	Poor	Good	Excellent	Poor
Dielectric strength (v/mil; ¼-in. thick specimen)	250-400	250-400	200-350	200-300	250-350	200-300	50-300
Resistivity (ohm/cm³) at room temperature	$10^{12}-10^{14}$	$10^{14}-10^{15}$	$10^{13}-10^{15}$	$10^{13}-10^{15}$	$10^{13}-10^{15}$	....	10^8-10^{15}
Te-value (°C)	200-500	500-800	450-1000	above 1000	700-900	....	200-400
Power factor at 1 Mc	0.006-0.010	0.001-0.002	0.0008-0.0035	0.0003	0.0006-0.0020	0.05	0.0002-0.050
Dielectric constant	6.0-7.0	8-9	5.5-7.5	6.2	8.0-9.0	5.6	15-10,000
L-Grade (JAN Spec. T-10)	L-2	L-2-L-5	L-3-L-5	L-6	L-4	L-3	....

Semivitreous and refractory products

	8	9	10	11
Material→	Low-Voltage Porcelain	Cordierite Refractories	Alumina, Aluminum Silicate Refractories	Massive Fired Talc, Pyrophyllite
Typical Applications→ Properties ↓	Switch Bases, Low-Voltage Wire Holders, Light Receptacles	Resistor Supports, Burner Tips, Heat Insulation, Arc Chambers	Vacuum Spacers, High-Temperature Insulation	High-Frequency Insulation, Vacuum Tube Spacers, Ceramic Models
Specific gravity (g/cc)	2.2-2.4	1.6-2.1	2.2-2.4	2.3-2.8
Water absorption (%)	0.5-2.0	5.0-15.0	10.0-20.0	1.0-3.0
Coefficient of linear thermal expansion/°C (20-700)	$5.0-6.5 \times 10^{-6}$	$2.5-3.0 \times 10^{-6}$	$5.0-7.0 \times 10^{-6}$	11.5×10^{-6}
Safe operating temperature (°C)	900	1250	1300-1700	1200
Thermal conductivity (cal/cm²/cm/sec/°C)	0.004-0.005	0.003-0.004	0.004-0.005	0.003-0.005
Tensile strength (psi)	1500-2500	1000-3500	700-3000	2500
Compressive strength (psi)	25,000-50,000	20,000-45,000	15,000-60,000	20,000-30,000
Flexural strength (psi)	3500-6000	1500-7000	1500-6000	7000-9000
Impact strength (ft-lb; ½-in. rod)	0.2-0.3	0.2-0.25	0.17-0.25	0.2-0.3
Modulus of elasticity (psi)	$7-10 \times 10^6$	$2-5 \times 10^6$	$2-5 \times 10^6$	$4-5 \times 10^6$
Thermal shock resistance	Moderate	Excellent	Excellent	Good
Dielectric strength (v/mil; ¼-in. thick specimen)	40-100	40-100	40-100	80-100
Resistivity (ohm/cm³) at room temperature	$10^{12}-10^{14}$	$10^{12}-10^{14}$	$10^{12}-10^{14}$	$10^{12}-10^{15}$
Te-value (°C)	300-400	400-700	400-700	600-900
Power factor at 1 Mc	0.010-0.020	0.004-0.010	0.0002-0.010	0.0008-0.010
Dielectric constant	6.0-7.0	4.5-5.5	4.5-6.5	5.0-6.0
L-Grade (JAN Spec. T-10)	....	....	....	....

voltage suspension insulators are frequently equipped with semiconducting glazes or coatings over areas which are in contact with metallic conductors. These types of glazes can be recognized by their mat appearance and gun-metal color. They contain semiconducting oxides, such as iron oxide or partially reduced titanium dioxide. Similar glazes are also applied to eliminate corona and reduce high-altitude flashover on transformer bushings or lead-through insulators in general. These coatings are of special importance for airborne radio and radar equipment. They are designed with a surface resistivity of about 3×10^9 ohms per square.[5]

(b) High-temperature porcelains

Prime considerations in selecting ceramics for insulators and supports of electrical heating elements

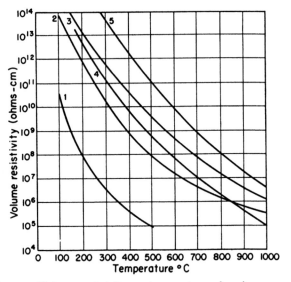

Fig. 4.2. Volume resistivity vs. temperature of various ceramics: (1) high-voltage ceramics; (2) general-purpose steatite; (3) low-loss steatite; (4) zircon porcelain; (5) sintered alumina.

or resistor wires are freedom from deformation under mechanical load, high insulation resistance at operating temperatures, and resistance to thermal shock. Conventional low- or high-voltage porcelains, as described under 1(a), are not satisfactory because they contain a glassy phase which may cause deformation of the insulator and low insulation resistance at elevated temperatures. More or less porous ceramics are therefore selected with only small amounts or no glassy binders. Materials of low thermal expansion and good heat conductivity are preferred, because of their superior heat-shock characteristics. The outstanding ceramic in this group contains as the predominant crystalline phase, cordierite ($Mg_2Al_2Si_5O_{15}$). Cordierite crystals develop during the firing of ceramic compositions consisting of mixtures of aluminum silicates (like clay) and magnesium silicates (such as talc). The mineral cordierite has a very low thermal expansion, approaching that of fused quartz. Steatite ceramics are also frequently used because of their high insulation resistance at high temperatures (Fig. 4.2).

(c) Insulators for sparkplugs

Insulators used in sparkplugs for high-compression automotive and special aircraft engines must withstand most severe working conditions. They must remain effective at temperatures of 2000°F, when thermal as well as electrical stresses are high. Literally millions of dollars have been spent to develop improved sparkplug insulators and methods to produce them.[6,7] The best compositions contain corundum (Al_2O_3) either in wholly crystalline form or glass-bonded. The presence of a small amount of glassy phase makes it possible to control the recrystallization of the corundum crystals during the firing and cooling period and helps the development of small crystals of uniform size. Some sparkplug compositions contain small amounts of coloring oxides, especially chromium oxide, which produces a pink color when in solid solution in corundum.

Progress in forming sparkplug bodies has kept step with body development. Besides conventional methods of extrusion and turning, two techniques have been developed by the two leading producers that are worth mentioning because they probably are the most advanced ones of forming ceramic dielectrics.

One method produces, by hot-compressing or injection-molding, complete insulator shapes. Thermosetting or thermoplastic resins are added to the ceramic composition and the organic binder burned out during the firing process. The other method uses an isostatic process of molding the ceramic powder, contained in a rubber mold, in a hydraulic medium. The advantage of this latter method is the forming of a ceramic blank of uniform density, which does not deform during the firing process and has uniform dielectric properties.

[6] H. B. Bartlett and K. Schwartzwalder, *Bull. Am. Ceramic Soc.* **28**, 462 (1949).

[7] Karl Schwartzwalder, *Bull. Am. Ceramic Soc.* **28**, 459 (1949).

[5] W. W. Pendleton, *Am. Inst. Elec. Engrs.* (August 7, 1947), Paper 47-236.

(d) Vitrified materials for high-frequency insulation

In selecting or developing a ceramic material for use at high frequencies these properties have to be considered: low dielectric loss, high dielectric strength, and high resistivity under all operating conditions. The material must be vitreous, dimensionally stable, of good mechanical strength, and capable of being formed and shaped to close dimensional tolerances. To meet this range of special requirements, a number of low-loss-type ceramic materials are produced commercially.

The Armed Services have set up specifications for these ceramics,[8] which have served during and since the Second World War as a guide for the improvements made in low-loss low-dielectric-constant materials (Table 4.3). The requirements in these specifi-

Table 4.3. Insulating materials, ceramic radio class L, JAN-I-10

Properties	Requirements					
1 Porosity:	No penetration of liquid under 10,000 psi pressure					
2 Flexural strength:	Not less than 3000 psi					
3 Resistance to thermal change:						
Grade	A			B		
	20 cycles			5 cycles		
Requirements	Boiling water to ice water					
4 Dielectric strength:	Not less than 180 rms volts per mil					
5 Loss factor	(Power factor X dielectric constant)					
Grade	L-1	L-2	L-3	L-4	L-5	L-6
Loss factor	<0.150	<0.070	<0.035	<0.016	<0.008	<0.004
Dielectric constant:	Not over 12 after immersion in water for 48 hours					

cations call for certain flexural strength, dielectric strength, porosity, thermal shock resistance, power factor, and dielectric constant. Qualification approval and grading (L-1 to L-6) are based on the loss factor of the material. The ceramics are further graded as "A" or "B," according to their ability to withstand thermal shock.

Much of the research work on high-frequency ceramics in recent years has been directed toward the development of lower dielectric-loss materials to meet L-6 requirements.

Steatite. Low-loss steatite is the most widely used ceramic in electronic applications. Among the low-loss materials, it is unsurpassed for economy in manufacture, especially in shapes which can be produced by either automatic dry pressing or extrusion methods. It can be made to close tolerances and has good mechanical strength. This advantage in manufacturing

[8] JAN-I-10, "Insulating Materials," Ceramic Radio, Class L.

stems from the basic raw material, *talc* or *steatite* of the composition $(OH)_2 \cdot Mg_3 \cdot Si_4O_{10}$, a soft mineral which is readily available and does not cause excessive wear of dies and tools. Typical low-loss steatite compositions consist of talc with clay and alkaline-earth oxides as fluxes. The material, after forming, is fired to a temperature of 1300° to 1400°C, depending on composition. Fired steatite bodies consist essentially of closely knit crystals of magnesium metasilicate ($MgSiO_3$), in the form of *clinoenstatite* or one of its polymorphous phases. The bond between the crystals consists of a glass high in alkaline-earth oxides. To obtain plasticity and strength before firing, temporary organic binders, such as waxes, gums, dextrine, or polyvinyl alcohol, are added. Steatite bodies fall under Class L-3 to L-5 of JAN-I-10 Specifications.

Forsterite. Where lower dielectric losses are required, forsterite ceramics are used. The main mineral constituent in the fired ceramic is magnesium orthosilicate (Mg_2SiO_4). The high melting point of pure forsterite (1890°C) serves to advantage for making basic refractory products. Forsterite refractories should not be confused with vitrified Forsterite ceramics for high-frequency applications.

Forsterite ceramics for dielectric purposes are produced from compositions containing talc to which magnesium oxide has been added to satisfy the stoichiometric composition Mg_2SiO_4. Suitable fluxes are again alkaline-earth oxides, and a ceramic body, consisting chiefly of forsterite crystals, is developed and matures at about 1350°C. Complete conversion of MgO and $MgSiO_3$ to Mg_2SiO_4 can be ascertained by the X-ray diffraction pattern of the fired body.

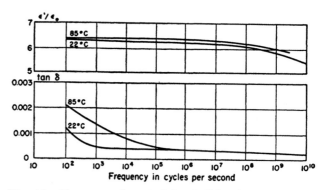

Fig. 4.3. Frequency characteristic of dielectric constant and power factor for Forsterite ceramic 243. (Laboratory for Insulation Research, M.I.T.)

Forsterite ceramics have excellent dielectric properties at elevated temperatures (Fig. 4.3). The Te value (the temperature at which the resistivity is 10^6 ohm-cm) lies above 1000°C. The linear thermal expansion of forsterite ceramics is practically a straight line between room temperature and 1000°C. The coefficient of expansion is approximately 11×10^{-6} per °C, that is, rather high for a ceramic material. This is a disadvantage as far as thermal shock resistance is concerned; forsterite ceramics are quite sensitive to heat shock. It is a decided advantage, however, in cases where vacuum- or pressure-tight metal-ceramic seals are required. Certain metal alloys of the nickel-iron type match these ceramics in thermal expansion, and by using these two materials, strain-free vacuum-tight seals can be made (Fig. 4.4).

Techniques for producing such seals have been worked out. One method, first suggested by Bondley (General Electric Company), applies titanium metal as the bonding medium between the metal and ceramic, as titanium hydride in powder form, and decomposes it to the metal by firing in vacuo.[9] Hydrides of zirconium, thorium, and tantalum have also been used successfully. In another method a mixture of iron and molybdenum or iron and manganese powders serves as bonding medium and is fused to the ceramic in a hydrogen atmosphere. The conductive coating can be electroplated and is suitable for brazing, silver or soft-soldering. The production of vacuum-tight seals between metals and ceramics is by no means simple, and variations in sizes and shapes require special techniques. Extreme cleanliness is essential, and the metal and ceramic joints must be carefully ground to accurate dimensions to leave very small but well-defined clearances. Doolittle (Machlett Laboratories,

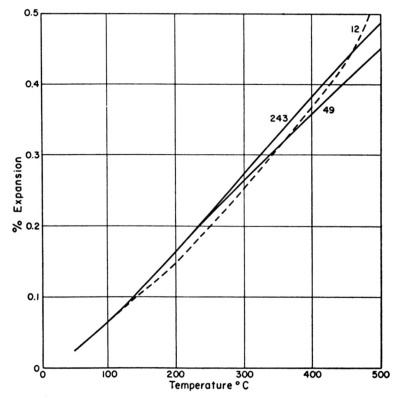

Fig. 4.4. Linear thermal expansion of Corning G-12 glass, Forsterite ceramic 243, and Ni-Fe No. 49 of Carpenter Steel Company.

Inc.) investigated the merits of the two basic processes, the molybdenum and the hydride process, and came to these conclusions:

(1) For small assemblies not required to insulate high voltages at high frequencies, the molybdenum-metallizing process is preferred. Quantity set-ups are feasible for fairly rapid production.

(2) The vacuum-hydride process is preferable for larger assemblies because

(a) The brazing cycle is faster, consisting of fewer operations.
(b) Parts are more thoroughly outgassed, with no gas pockets remaining.
(c) Controlled atmospheres are eliminated.
(d) High chrome alloy metal parts may be used.

[9] Walter H. Kohl, *Material Technology for Electron Tubes*, Reinhold Publishing Corp., New York, Chapters 15 and 16.

Zircon. Zircon porcelain compositions consist of the mineral zircon ($ZrO_2 \cdot SiO_2$) mixed with clay and fluxes to promote vitrification. The fired material has good mechanical strength and, as its outstanding property, a low coefficient of thermal expansion. Among the low-loss vitrified ceramics, zircon bodies rank high because of their good heat-shock characteristic.[10] The best methods of forming zircon bodies follow those developed for wet-process porcelain, because zircon compositions have a sufficient amount of clay content to be considered plastic bodies. They can be formed by dry pressing and extrusion, similar to steatite, but require tungsten carbide tools, because of the highly abrasive nature of zircon sand. Zircon bodies are generally rated as Grade L-4 dielectrics.

Alumina. The outstanding properties of alumina ceramics, which make these materials useful for spark-plug insulators, are also useful for high-frequency insulation. Special compositions have been developed which show low dielectric loss and qualify alumina ceramics as Grade L-5. The excellent mechanical strength of alumina ceramics (for example, flexural strength and impact strength are about twice that of steatite) is of advantage for insulators exposed to mechanical shock or where small cross sections are required. Typical uses are hermetically sealed terminals for transformers, capacitor housings, relays, switches, or motors. These types of seals can be made by fusing a metallic coating of silver, mixed with a small amount of glass flux, to the ceramic surface. Such coating is suitable for making a soft solder metal-ceramic seal. Soft solder has sufficient flexibility so that the matching of expansion between metal and ceramic is not as important as for the hard-solder seals, mentioned under Forsterite.

Cordierite. As mentioned before [see 1(b)], ceramics with low thermal expansion of the cordierite type are being used as high-temperature porcelains for supporting electrical heating elements. For such applications, porous materials suffice; a vitrified type, suitable for high-frequency insulation, presents certain manufacturing difficulties because the firing range (the temperature range between the temperature at which zero water absorption can be attained and that at which the material will deform during firing) is very narrow.[10,11] The Signal Corps has sponsored a research project on vitrified cordierite bodies. Certain compositions have been developed of zero absorption, a loss factor of Grade L-3, and a firing range sufficiently broad to permit commercial production.

Lithia porcelain. An interesting material is lithia porcelain. The predominant crystalline phase is *beta spodumene* ($Li_2Al_2Si_4O_{12}$). This type of material can be so compounded that the thermal expansion is zero or even negative. It has promising possibilities for use as a temperature-stable dielectric.[12]

Wollastonite. One of the more recently developed ceramics in the group of low-loss dielectrics is a material which is based on the mineral wollastonite (calcium silicate, $CaSiO_3$). The dielectric losses are sufficiently low to rate this material as L-6. Vitrification occurs at 1200° to 1250°C, considerably lower than for steatite, zircon, and similar bodies. The mechanical strength is good, and this material may find some uses in high-frequency applications.

(e) Refractory-type ceramics for electronic use

In electronic applications, refractory ceramic materials serve as supports of metallic parts in vacuum tubes. The temperature requirements for such insulators vary greatly, but are especially severe where the insulator serves to protect the cathode heater from the nickel sleeve. At high temperatures, in vacuo, the limiting factor for many refractories is not the melting point but the instability in contact with other substances (Table 4.4), and the rate of volatilization.[13]

Table 4.4. Temperature (°C) at which stability ceases in a surface-to-surface contact between the materials indicated [13]

	W	Mo	ThO_2	ZrO_2	MgO	BeO
C	1500^8	1500^8	2000^4	1600^4	1800^8	2300^2
BeO	2000^2	1900^4	2100^4	1900^2	1800^2	
MgO	2000^2	1600^4	2200^4	2000^4		
ZrO_2	1600^4	2200^4	2200^4			
ThO_2	2200^4	$1900^4(?)$				
Mo	2000^8					

Superscripts indicate time in minutes.
Pressure range 0.1 to 0.5 micron.

Pure aluminum oxide insulators, sintered at about 1800°C, of porosity between 10 to 20 percent, are widely used for insulating and supporting the cathode heater in vacuum tubes. They are also satisfactory for vacuum-tube spacer insulators.

Insulators machined from block talc are used where a high degree of dimensional accuracy is required.

---

[10] E. H. Smoke and J. H. Koenig, *Ceramic Age*, Feb. (1951).
[11] R. J. Beals and R. L. Cook, *J. Am. Ceramic Soc.* 35, 53 (1952).
[12] R. E. Stark and B. H. Dilks, *Materials and Methods*, 35, 98 (1952).
[13] P. D. Johnson, *J. Am. Ceramic Soc.* 33, 168 (1950).

Machined insulators are substantially free from distortion during firing and can be made to a high degree of precision without grinding after firing. These insulators can be readily degassed because they are free from glassy inclusions that might form gas pockets.

(2) Materials with dielectric constant above 12

This group, also called high dielectric constant ceramics, is based primarily on the mineral rutile (TiO_2) and on titanates (combinations of titanium dioxide with other oxides). Attention has recently been given also to other compositions, free of TiO_2. By changing compositions, it is possible to make a large variety of ceramic dielectrics with well-defined dielectric properties, to suit special requirements. This variety of compositions and properties makes the field of ceramic capacitor dielectrics interesting, but also rather confusing. The Radio-Electronics-Television Manufacturers Association has attempted to classify these materials into main Class I and Class II dielectrics [14] (Table 4.5).

Table 4.5. **Classification of ceramic dielectrics for capacitors**

	Class I	Class II
K range (κ' at room temp.)	6 to 500	500 to 10,000
Temperature coefficient of capacitance	P120 to N5600	....
Power factor	0.04 to 0.4%	0.4 to 3%
Insulation resistance	10,000 to 100,000 megohms	10,000 to 100,000 megohms
Maximum capacitance decrease (-55 to $+85°C$)	0 to 40%	15 to 80%
Maximum capacitance increase (-55 to $+85°C$)	0 to 40%	0 to 25%
Aging Characteristics of κ	None	Approx. 4% per decade time

(a) Class I dielectrics for capacitor use

Class I dielectrics are suited for resonant circuits or other applications where high Q and stability of κ' are essential. The temperature coefficient of dielectric constant of these dielectrics is a function of crystalline structure of the dielectric and can be predetermined for a number of compositions. Capacitance-temperature characteristics from $+200$ to -5600 ppm per °C are available in materials of dielectric constants between 15 to 500 at room temperature (Fig. 4.5). By controlling the temperature coefficient of κ', these di-

[14] RETMA Standard REC-107-A, Ceramic Dielectric Capacitors, MIL-C-11015A.

electrics can be used as temperature compensators in electronic circuits. Circuit elements such as coils, transformers, resistors, and tube capacitances change with temperature. If these elements are trimmed with a capacitor of opposite drift, the circuit tends to maintain the same frequency characteristic at all operating temperatures. When a change of κ' with temperature

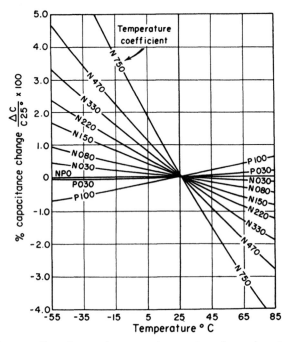

Fig. 4.5. Capacitance change vs. temperature for various temperature-compensating ceramic capacitors.

is not desirable, dielectrics with no drift (zero temperature coefficient) over a wide temperature range may be used.

In order to obtain the best dielectric properties, careful control of manufacturing processes is essential. Minute impurities may cause slight reduction of TiO_2 accompanied by a disastrous drop in Q value and insulation resistance. For the same reason a careful control of firing conditions is imperative.

(b) Class II dielectrics for capacitor use

The growing demand for miniaturization has helped the development of ceramic capacitors for other than temperature-compensating applications. General-purpose capacitors of very small dimensions can be made within a capacity range of 5 $\mu\mu f$ to 0.1 μf. They are useful where changes in capacity with operating temperature are not too critical, for example, as by-pass and coupling capacitors. Materials with dielectric constants between 500 and 6000 at room temperature are available. Most of these compositions are based on barium titanate (BaO·TiO₂). Barium titanate, a ferroelectric material,[15] exists in several crystal modifications. Above 120°C the cubic form exists, which is nonferroelectric. Between +5° and +120°C the crystal habit is tetragonal and ferroelectric. Below +5°C, the material changes to orthorhombic, and near −70°C to trigonal, but remains ferroelectric. Ferroelectric properties are characterized by high dielectric permeability, ferroelectric hysteresis, polarization saturation, piezo effect, and a Curie temperature [15] (see IB, Sec. 5). For capacitor applications only the high dielectric constant is desired; the other properties must be suppressed as much as possible. Fortunately, by various oxide additions to barium titanate or by changing the ratio of BaO to TiO₂ from the stoichiometric ratio, one arrives at compositions which are useful capacitor dielectrics. It is possible to shift the Curie temperature and to decrease its effect on the change of κ' with temperature. In other words, it is possible to make ceramic compositions of fairly even temperature-capacitance characteristics.

Barium titanate ceramics show a definite "aging" effect of dielectric constant and power factor. The aging generally follows a function $\kappa' = \kappa_0' - m \log t$, where κ' is the dielectric constant at time t, κ_0' the initial dielectric constant measured at a point in time arbitrarily designated "one," and m designates the rate of decay in κ'.[16] Unfortunately, m, the measure of change, increases with the initial dielectric constant.

The mechanism of "aging" of barium titanate ceramics is not yet well understood. It depends on a number of variables, such as composition, previous heat treatment, and manufacturing methods in general. Investigations are under way to obtain a better picture of aging and to find steps either to control or to eliminate it.

A group of crystals with structures similar to that of BaTiO₃ show similar dielectric behavior.[17] The compounds NaCbO₃, KCbO₃, NaTaO₃, KTaO₃, LaGaO₃, and LaFeO₃ show high dielectric constants, which increase with temperature; some show a break as a result of polymorphic transitions. Of special interest are the columbates of sodium and cadmium.[18] They are made from mixtures of sodium carbonate, cadmium oxide, and columbium oxide, using conventional ceramic methods. Columbate capacitors may meet the need for small-size by-pass capacitors capable of operating at 250°C for use in compact military equipment.

(c) Barium titanate capacitors as nonlinear circuit elements

Ferroelectric properties, undesirable for capacitors, can be very useful for nonlinear circuit applications. The nonlinearity of the dielectric constant with applied field strength suggests its use in dielectric amplifiers [19] (see IIIC, Sec. 3). The large temperature dependence of the dielectric constant, especially near the

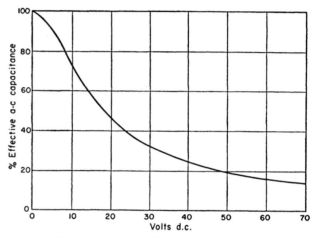

Fig. 4.6. Change of a-c capacitance vs. d-c polarizing voltage for BaTiO₃-SrTiO₃ ceramic; κ' at room temperature = 8000.

Curie point, makes this application a rather difficult problem. Efforts are being made at present to develop capacitors which show relatively small changes in capacity with temperature, but are highly affected by variations in applied field strength (Fig. 4.6).

(3) Titanate ceramics with piezoelectric and ferroelectric properties

Dielectric materials in general are known to be electrostrictive, that is, a mechanical deformation takes place proportional to the square of an applied electric field. This effect is extremely small in ordinary dielectrics and has no inverse, that is, mechanical deformation does not produce polarization (see IB, Sec. 5). Titanate ceramics, on the other hand, possess true piezoelectric properties,[15] characterized by a direct and inverse response of a material to an electric field and to mechanical forces. Barium titanate can be given a permanent and strong piezoelectric response by treatment with an electric polarizing field for a short period of time.[19] The material to be polarized is

---

[15] A. von Hippel, *Revs. Mod. Phys.* **22**, 221 (1950); *Z. Physik* **133**, 158 (1952).

[16] R. G. Graf, *Ceramic Age*, Dec. (1951), p. 16.

[17] B. T. Matthias, *Phys. Rev.* **75**, 1771 (1949).

[18] Chandler Wentworth, "Columbate Dielectrics," RCA Report LB-825; also E. Wainer and C. Wentworth, *J. Am. Ceramic Soc.* **35**, 207 (1952).

[19] S. Roberts, *Phys. Rev.* **71**, 890 (1947).

heated above its Curie temperature, an electric polarizing field is applied, and the material is cooled under applied voltage. The piezoelectric response of prepolarized barium titanate drops to about 80 percent of its initial value after a short period but remains practically constant thereafter.

The strength of a piezoelectric effect can be expressed by the piezoelectric coupling coefficient.[20, 21] A coupling coefficient of 100 percent means that all dielectric polarization would appear as an elastic stress, or, inversely, that an applied external mechanical force would be converted into electric voltage. Prepolarized titanates have coupling coefficients up to 50 percent, which makes them extremely useful as transducers (Table 4.6). Rochelle salt has a coupling

Table 4.6. Piezoelectric data on common electromechanical transducer materials [20]

	Mode	d (10^{-12} m/v)	κ'	g (v·m/Newton)	k
Rochelle salt (34°C)	Lateral	165	200	0.093	0.54
Quartz	Parallel	2.3	4.5	0.058	0.10
	Lateral	2.3	4.5	0.058	0.10
$NH_4H_2PO_4$ (ADP)	Lateral	24	15.3	0.177	0.28
Barium titanate (crystal)	Parallel	190	1700	0.0125	0.46
$BaTiO_3$ ceramic (prepolarized)	Lateral	78	1700	0.0052	0.19

d = Piezoelectric modulus, indicates mechanical strain per applied electrical field strength.
κ' = Relative dielectric constant.
g = Voltage output coefficient, indicates piezoelectric field strength per applied mechanical stress.
k = Piezoelectric coupling coefficient.

coefficient of 54 percent but poor mechanical and thermal stability. Quartz has a coupling coefficient of only 10 percent.

The advantages of barium titanate ceramics over other piezoelectric materials are obvious. They show good chemical stability, and in this respect are superior to organic or water-soluble crystals. They can be formed to any desired shape most suitable for a particular application. It is possible to focus and concentrate the energy generated in the ceramic element to the work area. Elements can be made in the form of bowls, tubes, plates, disks, or a multitude of elements can be set in any suitable arrangement. Because of the high coupling coefficient of barium titanate, tuning to resonance is not as critical as with quartz transducers. The piezoelectric properties are maintained nearly to the Curie temperature of the material; operating temperatures may be as high as 70°C.[21]

In order to raise operating temperatures and promote stability, it is desirable to raise the Curie temperature above that of pure barium titanate; this can be accomplished by the addition of a small amount of lead titanate. Pure lead titanate has a Curie temperature of 490°C.

Introduction of lead titanate into the barium titanate lattice has the advantage of obtaining a remanent polarization that is more permanent. The remanent polarization of an ordinary barium titanate ceramic can be removed by a negative field of 5000 volts per cm, but with 4 percent lead titanate, it is only slightly diminished up to a negative voltage of 25,000 volts per cm. It is completely restored by cycling to 25,000 volts per cm.

Ultrasonic transducers of barium titanate are being used for a large number of technical applications, as for the emulsification of oil-soluble and water-soluble liquids, stimulation or destruction of bacteria, transformation of crystal structures, mixing of powders and paints, homogenization of milk, and agglomeration of smoke and dust particles.[22] Other applications are microphones, pick-ups, accelerometers used for measuring high-frequency shock and vibration, and strain gauges [23] (see also IIIC, Sec. 2).

(4) Ferromagnetic ceramics

Nonmetallic ferromagnetic materials were introduced in this country immediately after the Second World War, when it became known that they had found practical application in Europe during the war. The "soft" ferromagnetic materials now being used extensively as magnetic core materials are generally known as ferrites. They are compounds of various metal oxides, sintered into hard, dense materials according to standard ceramic practices.

Magnetite (Fe_3O_4) is the earliest known magnetic substance, and a number of related chemical compounds have long been known to possess magnetic properties. The most notable are compounds of copper, magnesium, manganese, nickel, and zinc. Synthetic ferrites were first patented by G. Hilpert in Germany in 1909 and proposed as core materials for high-frequency applications. Various investigators in Europe and Japan worked on ferrites and developed ferrites with both soft and hard (permanent) magnetic properties. Snoek [24] in Holland, one of the pioneers in ferrite development, carried out extensive studies

---

[20] H. Jaffe, *Ind. Eng. Chem.* **42**, 264 (1950).
[21] W. P. Mason, *Piezoelectric Crystals and Their Application to Ultrasonics*, D. Van Nostrand Co., 1950.
[22] Brush Development Co., Bulletin "The Brush Hydroelectric Generator," 1949.
[23] Gulton Manufacturing Corp., Metuchen, N. J., Bull. A403, "Glennite Self-generating Accelerometer."
[24] J. L. Snoek, *New Developments in Ferromagnetic Materials*, Elsevier Pub. Co., New York, 1947.

to correlate magnetic and dielectric properties with chemical composition and crystal structure.

Ferrites are known to have crystalline structures similar to that of the mineral spinel ($MgO \cdot Al_2O_3$). Two variations of the spinel structure exist, normal spinel and inverse spinel, depending on the arrangements of the atoms in the crystal lattice. Correspondingly, there are normal and inverse ferrites. Only inverse ferrites are ferromagnetic. Typical inverse ferrites are those of nickel, cobalt, or manganese. Zinc ferrite crystallizes in the normal spinel structure, and is nonmagnetic.

The ferromagnetic spinels are magnetostrictive, some positive, others negative. Since the different ferrites can form homogeneous mixed crystals, it is possible to make a variety of materials with varying degrees of magnetostriction, permeability, hysteresis, etc.[25] Magnetostriction can be minimized by mixing ferrites of positive and negative magnetostriction, and thus high permeability materials are obtained. By adding nonmagnetic zinc ferrite to magnetic ferrites, the Curie temperature of materials can be shifted at will, and the electrical resistivity raised. Further variations can be introduced by changing firing conditions, which affect crystal sizes and state of oxidation.

One of the most important advantages of nonmetallic ferrites over metallic magnetic materials is their high electrical resistance, which keeps the eddy-current losses low. This characteristic is of special importance at higher frequencies, since eddy-current losses increase as the square of the frequency. The ferrites have small hysteresis losses and high permeability compared with powdered iron core material. Compositions and manufacturing procedures can be varied to effect wide variation of properties. It is possible, for instance, to change the shape of the hysteresis loop. Of special interest recently has been the development of ferrites with square loop characteristics for cores in memory devices of electronic computers (see IIIC, Sec. 4).

Ferrites are not considered to be substitutes for iron cores in power transformers since their saturation magnetization is appreciably lower than that of iron, but they are extremely useful in many applications, especially at high frequencies.

[25] See, for example, F. G. Brockman, *Elec. Eng.* 68, 1077 (1949); J. J. Went and E. W. Gorter, *Philips Tech. Rev. 13*, 181 (1952); E. Albers-Schoenberg, *Ceramic Age 55–56*, 14 (October, 1950). For further information on ferromagnetics, see also the companion volume, *Dielectrics and Waves*.

B · Dielectrics in Equipments

1 · Dielectrics in Power and Distribution Equipment†

By LEO J. BERBERICH

There is no piece of electrical equipment that does not depend on insulation in some form to maintain the flow of current in the proper channels. Field experience indicates that a high percentage of all failures in electrical apparatus is caused by electrical insulation failure, thus proving the extreme importance of the field of electrical insulation and dielectrics to the electrical industry. This is particularly true of the power equipment field, where insulation failures in the large and heavy units may become quite costly, but, more important, may affect the continued reliable flow of electrical energy to our cities, industries, and farms.

† The author wishes to acknowledge the assistance and suggestions of a number of his associates in the Westinghouse Electric Corporation in the preparation of this chapter.

This discussion deals with the insulation problems of power and distribution equipment. By power and distribution equipment we mean transformers, motors, generators, switchgear, power capacitors, high-voltage cables, etc., used in the central station industry. We will begin with the classification of insulating materials needed in specifying them for various types of equipment. Next, important insulation components and systems used in transformers, power capacitors, switchgear, and rotating equipment will be reviewed. Finally, factors influencing the deterioration of insulating materials will be discussed. Power cables will not be included in this brief presentation since this subject is treated elsewhere (see IIIB, Secs. 4 and 5).

Insulation Classification

The system of insulation classification used at present in the United States is described in A.I.E.E. Standard No. 1.[1] This Standard recognizes that the principal determining factor in establishing temperature limits on various types of electrical equipment is the thermal endurance of the insulating materials employed. It also recognizes that insulation does not fail immediately when a certain critical temperature is reached. The chemical deterioration is gradual and causes a similar gradual decrease in the mechanical and electrical strength properties. Time and temperature co-operate in producing such changes, for example, embrittlement and cracking of organic materials, aided by vibration and severe mechanical forces. Cellulosic products like paper and cotton cloth lose water, carbon dioxide, and carbon monoxide on being heated at high temperatures, and finally only a powdery carbonaceous residue is left, with no mechanical strength.

Mechanical and electrical strength often cannot be correlated directly or quantitatively. A material may still possess considerable electrical strength even though the mechanical strength has been reduced to low values; this is true especially in the absence of vibration or other mechanical forces which might disturb the physical integrity of the insulation. A decrease in mechanical strength, however, does usually indicate that deterioration has taken place and is, therefore, often used as a test criterion.

Insulation deterioration may be treated as a chemical rate phenomenon.[2] Deterioration studies have taken various forms; some have been made on materials as such,[3,4] others on systems of materials as used in electrical machinery.[5,6,7] (See also IIIA, Sec. 2.) In the latter case, the effectiveness of the physical support enters and the severity of the mechanical forces. These forces may have an electrical origin or

[1] A.I.E.E. Standard No. 1, "General Principles upon Which Temperature Limits Are Based in the Rating of Electrical Machines and Equipment," American Institute of Electrical Engineers, New York, N. Y.

[2] T. W. Dakin, *Trans. Am. Inst. Elec. Engrs.* 67, Pt. 1, 113 (1948).

[3] T. A. Kauppi and G. L. Moses, *Trans. Am. Inst. Elec. Engrs.* 64, 90 (1945).

[4] H. C. Stewart, L. C. Whitman, and A. L. Scheideler, *Trans. Am. Inst. Elec. Engrs.* 72, Pt. 3, 267 (1953).

[5] G. Grant, III, T. A. Kauppi, G. L. Moses, and G. P. Gibson, *Trans. Am. Inst. Elec. Engrs.* 68, Pt. 2, 1133 (1949).

[6] K. N. Mathes, *Trans. Am. Inst. Elec. Engrs.* 71, Pt. 3, 254 (1952).

[7] G. A. Cypher and R. Harrington, *Trans. Am. Inst. Elec. Engrs.* 71, Pt. 3, 251 (1952).

Table 1.1. Limiting temperatures for various insulation classes

For Class O material	90°C
For Class A material	105°
For Class B material	130°
For Class H material	180°
For Class C material	No limit selected

Table 1.2. Classification of insulating materials

Class	Description of Material
O	Class O insulation consists of cotton, silk, paper, and similar organic materials when neither impregnated * nor immersed in a liquid dielectric.
A	Class A insulation consists of (1) cotton, silk, paper, and similar organic materials when either impregnated * or immersed in a liquid dielectric; (2) molded and laminated materials with cellulose filler, phenolic resins, and other resins of similar properties; (3) films and sheets of cellulose acetate and other cellulose derivatives of similar properties; and (4) varnishes (enamel) as applied to conductors.
B	Class B insulation consists of mica, asbestos, glass fiber, and similar inorganic materials in built-up form with organic binding substances. A small proportion of Class A materials may be used for structural purposes only.†
H	Class H insulation consists of (1) mica, asbestos, fiberglass, and similar inorganic materials in built-up form with binding substances composed of silicone compounds, or materials with equivalent properties; (2) silicone compounds in rubbery or resinous forms, or materials with equivalent properties. A minute proportion of organic films or paper materials may be used only where essential for structural purposes during manufacture.†
C	Class C insulation consists entirely of mica, porcelain, glass, quartz, and similar inorganic materials.

* An insulation is considered to be "impregnated" when a suitable substance replaces the air between its fibers, even if this substance does not completely fill the spaces between the insulated conductors. The impregnating substance, in order to be considered suitable, must have good insulating properties; must entirely cover the fibers and render them adherent to each other and to the conductor; must not produce interstices within itself as a consequence of evaporation of the solvent or through any other cause; must not flow during the operation of the machine at full working load or at the temperature limit specified; and must not unduly deteriorate under prolonged action of heat.

† The electrical and mechanical properties of the insulated winding must not be impaired by the application of the hottest-spot temperature permitted for the specific insulation class. The word "impaired" is here used in the sense of causing any change which could disqualify the insulating material for continuously performing its intended function, whether creepage spacing, mechanical support, or dielectric barrier action.

be produced by the thermal expansion and contraction of the insulation as the equipment goes through temperature cycles. Since thermal expansion and contraction may reach appreciable values in large machines, somewhat lower temperature operating conditions are usually imposed on them.

The length of life desired for a given insulation system is largely a question of economics. The choice of materials must be associated with cost, reasonable maintenance, the size of the equipment, and reasonable life. The insulation classification system now used in this country attempts to take these considerations into account to some degree. Evaluation of materials on a thermal and electrical basis and of insulation systems as they are used in electrical equipment, on a functional basis, is only now beginning. It will undoubtedly require a number of years before sufficient information will be obtained for a more rational basis of insulation classification than we now have.

From the results of experience with equipment in service, and of laboratory tests on various materials, limiting temperatures sometimes called "hottest-spot" temperatures have been assigned to the various insulation classes (Table 1.1).

The various classes are broadly defined in Table 1.2.

An examination of Table 1.2 reveals that this system of classification leaves much to be desired. New materials are difficult to classify, and the definitions are frequently vague. For example, in the definition for Class B insulation, it is stated that a small proportion of Class A materials may be used for structural purposes only. But how much Class A material can be used in any given case? It is hoped that a system of insulation classification based on thermal and electrical stability will prove more rational and more useful.

Dielectrics in Transformers

Transformers are in the "front lines" as far as attack by lightning and other high-voltage surges are concerned; they must safely withstand impulse voltages of peak values many times the steady 60-cycle operating voltage. Improvements in recent years in transformer insulation and in the performance of lightning arresters have resulted in much greater freedom from failure due to lightning and other surges. This, in turn, has increased the reliability of electric service to a remarkable degree.

Omitting the small specialty-type transformers of radio, television, and allied industries, we may distinguish between distribution and power transformers. Distribution transformers usually have voltage ratings of 120 to 240 volts on the secondary to about 25 kv on the primary, and kva ratings of 1.5 to about 500 kva. They distribute power to the ultimate user. Power transformers may have voltage ratings of 15 to 330 kv and higher, and power ratings up to and above 300,000 kva. Most distribution and power transformers are oil immersed. In recent years, however, dry-type transformers of both types at ratings up to and including 15 kv and 10,000 kva are being built in increasing numbers.

Transformers are constructed according to two designs, the *core form* and the *shell form*. In the former, the magnetic circuit forms a core through the coil structure; and in the shell form the magnetic circuit is designed as a shell about the coils. Since most manufacturers use the core form in the small and intermediate sizes, this discussion will concern itself largely with the core-form design.

Components of transformer insulations

In general, all transformers require: (1) conductor or turn-to-turn insulation, (2) layer-to-layer or coil-to-coil insulation, (3) low-voltage coil-to-ground insulation, (4) high-voltage coil-to-low-voltage coil insulation, and (5) high-voltage coil-to-ground insulation. These various types of insulation are illustrated in a schematic drawing of the center phase of a three-phase transformer showing a sectional view of the coil structure (Fig. 1.1). The insulation problem is usually simpler when the low-voltage coils are placed next to the magnetic iron, surrounded by the high-voltage coils, as in Fig. 1.1, which applies either to the liquid-filled or to the dry-type designs of medium kva ratings.

The low-voltage coil-to-ground and the high-to-low-voltage coil insulations usually consist of solid tubes combined with liquid- or gas-filled spaces. These spaces facilitate the removal of heat from the core and coil structure through convection of the medium, and also add to the total insulation strength.

The turn-to-turn insulation is generally placed directly on the conductor, as organic enamel on the round wire in smaller distribution types, or as paper or glass tapes wrapped on the rectangular conductors of the larger distribution and power transformers. The material chosen and its treatment vary with the insulation temperature class and whether it is a liquid-filled or a dry-type design. This is also true of the layer-to-layer, coil-to-coil and coil-to-ground insulation, which may vary from kraft paper in the smaller

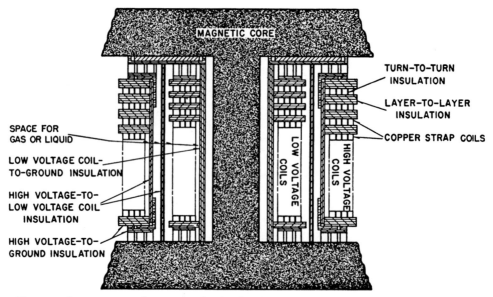

Fig. 1.1. Core-type transformer showing insulated parts for one phase of a three-phase unit.

units to relatively thick radial spacers made of heavy pressboard paper, glass fabric, paper-filled plastic laminates, or porcelain.

Fig. 1.2. Core and coil structure of a three-phase Class H dry-type transformer rated 1250 kva, 13.8 kv to 480 volts.

Some idea of the form these various types of insulations take in a Class H dry-type design is given in Fig. 1.2.† Unfortunately, the low-voltage coils are obscured in this picture by the high-voltage coil structure and the insulating tubing. The turn insulation in this Class H design is glass-taped directly onto the conductor and then treated with a silicone varnish. The layer or intercoil insulation consists of porcelain spacers. The corrugated porcelain insulators at the upper and lower ends of the stack of high-voltage coils serve as insulation from the coils-to-ground and as physical support for the coils. The barrier tubes between high- and low-voltage coils, the upper and lower ends of which can be seen, are, in general, tubes of silicone-treated mica, glass fabric, silicone rubber, or a combination of these materials.

Class B dry-type designs are similar to Class H, except that the varnishes used in the treatment of the conductor or turn insulation, as well as in the treatment of the barrier tubes, are usually phenolic or phenolic-alkyd types rather than silicones. Few, if any, dry-type Class A power or distribution transformers are built in this country, but in them the turn insulation might be paper or organic enamel; the layer insulation paper, pressboard, or a plastic laminate; and ground insulation also pressboard or a plastic laminate. These materials are dried and treated with conventional organic varnishes to prevent the absorption of moisture and other contaminants.

Figure 1.3 shows one phase of a three-phase, core-

† All the photographic illustrations in this section are by courtesy of Westinghouse.

type oil-insulated transformer for a moderately high voltage rating. The cut-away view at the top and bottom reveals the barrier tubes and coil structure. The conductor or turn-to-turn insulation in this transformer consists of manila or kraft paper taped upon

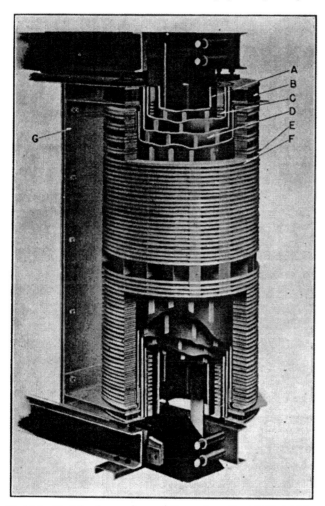

Fig. 1.3. Sketch of one phase of three-phase liquid-filled high-voltage transformer. *A* and *D*. Barrier tubes. *B*. High-voltage coil-to-ground insulation. *C*. Low-voltage coils. *E*. Coil-to-coil spacers. *F*. High-voltage coils. *G*. Phase-to-phase insulation.

the conductor, the layer or coil-to-coil insulation usually consists of pressboard spacers. The barrier tubes between the low-voltage coils and ground and between low- and high-voltage coils, as well as the spacer strips between tubes, are also usually made of pressboard paper. The purpose of the spacer strips is to allow the free circulation of oil in and around the core and coil structure. The ground insulation consists of paper-base phenolic laminates and pressboard. The pressboard and other cellulosic insulation is very carefully dried by application of heat and vacuum and then impregnated with oil or askarel, depending on the type of liquid used. The various shapes required and the extent to which pressboard is used between coils and between coil groups of a high-voltage transformer are illustrated in Fig. 1.4, which shows a section through a coil group of a high-voltage shell-form transformer.

The general arrangement of coils, leads, and barrier tubes in an oil-filled power transformer can be seen in Fig. 1.5. The coil leads and the leads to the bush-

Fig. 1.4. Cross section through a coil group of oil-insulated shell-type high-voltage power transformer, showing use of pressboard.

ings are supported by maple wood or pressboard strips. The radiators at the extreme right and left of Fig. 1.5 permit cooling of the oil. The addition of fans at the extreme right permits an increase in kva rating.

Oil-impregnated cellulose

Of all the materials mentioned, oil-impregnated cellulose in the form of paper or pressboard comes closest to meeting the requirements for liquid-filled transformer insulation. In Fig. 1.6,[8] the breakdown strength of oil-impregnated pressboard is compared with that of oil alone, as a function of time of voltage application ranging from a few tenths of microseconds to one minute. The test impulse waves used for the range up to several hundred microseconds had linear rise time and exponential decay. The values plotted

[8] P. L. Bellaschi and W. L. Teague, *Trans. Am. Inst. Elec. Engrs.* **56**, 164 (1937).

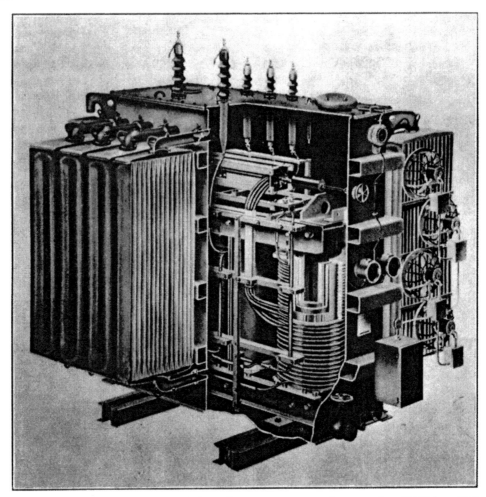

Fig. 1.5. Sectionalized view of an oil-insulated three-phase core-type transformer rating 10,000 kva, 69 kv to 6.9 kv.

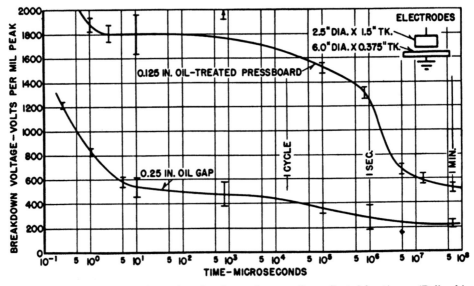

Fig. 1.6. Dielectric strengths of oil-treated pressboard and transformer oil as affected by time. (Bellaschi and Teague.[8])

were obtained from breakdown near the crest of the wave. The longer-time data, from 10^5 microseconds to one minute, refer to 60-cycle voltages applied for varying lengths of time. The data illustrate the striking barrier action of the network of cellulose fibers in the pressboard; the impregnated pressboard has approximately six times the dielectric strength of a free oil layer of equal thickness. The fact that the breakdown strength under impulse conditions is considerably higher than at long-time 60-cycle application is a desirable protection against lightning and other surges.

The lack of thermal stability at elevated temperatures limits the use of cellulose to continous operation up to 105°C, the maximum continuous temperature for Class A insulation. In spite of some deterioration of cellulose above 105°C, the A.S.A. operating guides [9] for oil-insulated transformers permit higher than Class A temperatures for short periods of time during overloads, some sacrifice of life expectancy being fully recognized.

Cellulose absorbs moisture very readily from the atmosphere. Data on the effect of moisture on electrical properties of power transformer insulation are given by Teague and McWhirter,[10] and the moisture problem in oil-treated cellulosic insulation is extensively reviewed by Clark.[11] The importance of keeping cellulosic insulation free of moisture during its life in a transformer cannot be overemphasized.

Transformer oil

The function of the petroleum oil, besides providing dielectric strength and insulation, is to cool the transformer by circulating through the core and coil structure. This means it must be freely flowing at all outdoor operating temperatures, which may reach −40°C or lower in some parts of the country. Petroleum oil is subject to oxidation when exposed to oxygen at elevated temperatures, but in most present-day transformer designs access to oxygen is limited. In oxidizing, the principal products formed [12,13] are peroxides, water, organic acids, and sludge. The first three may affect the rate of cellulose deterioration, and the acids together with water may cause corrosion of the metal parts of the transformer. Sludge, which is a high-molecular-weight reaction product of oxidized petroleum hydrocarbons, can impair the heat transfer by precipitating out of oil solution and coating the core and coil structure as well as the tank walls. In extreme cases, it may block the passages through the core and coil structure, and thus further impede heat transfer.

Considerable progress has been made in mitigating oil oxidation by use of (1) sealed transformers, (2) nitrogen in the gas space of large power transformers,[13] (3) activated clay or alumina treaters, and (4) oxidation inhibitors.[12,14,15] So far, *inhibitors* have not come into general use in power transformers because they are sealed against oxygen. However, most manufacturers use inhibited oils in the distribution types, where the units may be pole-mounted for years without attention. That inhibitors are quite effective in the presence of oxygen is illustrated by the results of laboratory oxidation tests [12] (Fig. 1.7) (modified Sligh test [16]). The effect of inhibitors in transformers is usually not so pronounced, however, because even in a distribution transformer the access of oxygen is limited. Oil B is a moderately refined conventional transformer oil, oil A a more highly refined oil of commerce.

When petroleum hydrocarbon oils are subjected to an arc discharge, extensive breakdown of the hydrocarbon results. Of the products formed on arcing, hydrogen and the gaseous hydrocarbons may lead to explosion and fire hazards. It is therefore not general practice to install oil-insulated transformers inside buildings or other hazardous locations unless placed within specially constructed vaults supplied with adequate ventilation to the outside. Dry-type and askarel-filled transformers have come into extensive use where fire and explosion hazards may exist.

Askarel is the generic name of a number of synthetic chlorinated aromatic hydrocarbons developed for use in transformers and paper capacitors, where fireproof liquids are desired.[17] The askarels are more stable to oxidation than petroleum oils, and under normal transformers operating conditions do not form acids or sludge. Under arcing conditions they do not give off flammable or explosive gases. Unfortunately, however, they do give off hydrochloric acid gas when

[9] "American Standard Guides for Operation of Transformers, Regulators and Reactors," C57.32–C57.36, American Standards Association, New York, N. Y.

[10] W. L. Teague and J. H. McWhirter, *Trans. Am. Inst. Elec. Engrs.* **71**, Pt. 3, 743 (1952).

[11] F. M. Clark, *Ind. Eng. Chem.* **44**, 887 (1952).

[12] L. J. Berberich, "Oxidation Inhibitors in Electrical Insulating Oils," *ASTM Bull.* **149**, 65 (1947).

[13] J. G. Ford, *Elec. Eng.* **67**, 1066 (1948).

[14] E. D. Treanor and E. L. Raab, *Trans. Am. Inst. Elec. Engrs.* **70**, Pt. 2, 1414 (1951).

[15] H. Halperin and H. A. Adler, *Trans. Am. Inst. Elec. Engrs.* **70**, Pt. 2, 1205 (1951).

[16] L. L. Davis, B. H. Lincoln, G. D. Byrkitt, and W. A. Jones, *Ind. Eng. Chem.* **33**, 339 (1941).

[17] F. M. Clark, *Ind. Eng. Chem.* **29**, 698 (1937).

subjected to arcing, which not only is toxic to human beings but also attacks the cellulosic insulation. The effect on cellulose, however, has been partially corrected by the use of a *scavenger*.

Tin tetraphenyl [18] and similar compounds capable of absorbing hydrochloric acid are used in this application. For heavy arcs, however, the scavenger cannot absorb all the hydrochloric acid formed, and thus some hazard still remains. For these reasons the popularity of dry-type transformers has increased in recent years for use inside buildings. It should also be

carbons [20] developed during the Second World War. The dielectric and other properties of some recently developed fluorocarbon liquids are described by Bashara.[21] Although these liquids are inert, stable, and completely fireproof, the ones presently available are still too low-boiling and too high in price to replace either petroleum oil or askarel directly. A commercial network transformer,[22] however, has been developed recently which makes use of a relatively low-boiling fluorocarbon liquid by spraying it on the core and coil structure and utilizing its heat of vaporiza-

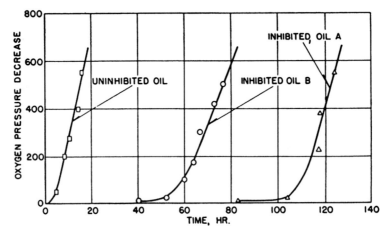

Fig. 1.7. Oxygen absorption by modified Sligh test at 120°C (copper catalyst). (Mahncke.[12])

noted that transformer askarel, a eutectic mixture of trichlorobenzene and chlorinated diphenyl, has high solvent power for most varnishes and some insulating materials. Extreme care must therefore be exercised in the selection of varnishes and other materials used in askarel-filled transformers.

Askarel has not been used in the higher-voltage transformers because the impulse strength of askarel-impregnated pressboard is appreciably less than that of oil-impregnated pressboard. Recent test results [19] throw some light on the reasons for this behavior, but further elucidation is required. The askarel by itself shows a higher impulse strength and 60-cycle strength than the oil. In spite of this excellent dielectric strength at 60 cycles, it deteriorates rapidly at low-voltage stresses at high frequencies (megacycles), liberating hydrochloric acid and carbon. This phenomenon occurs also with other halogenated aromatic hydrocarbons.

There is as yet no perfect all-purpose liquid for use in transformers. Some progress has been made with the announcement of completely fluorinated hydro-

[18] F. M. Clark, *Chem. Eng. News* 25, 2976 (1947).
[19] T. W. Dakin and C. N. Works, *Trans. Am. Inst. Elec. Engrs.* 71, Pt. 1, 321 (1952).

tion. In this manner, the heat generated by the transformer can be dissipated to the atmosphere with much less liquid than is normally used in a conventional liquid-filled transformer. The fluorocarbon liquids undoubtedly will find more use in electrical applications as equipment is designed utilizing their special properties and as their price is reduced with increased production.

Summary

The requirements for insulating materials in transformers may be summarized as follows:

(1) Good electrical properties to withstand not only the rated 60-cycle voltages, but also impulse voltages of peak values many times the normal operating voltage.

(2) Good mechanical properties, particularly strength and toughness, to withstand fabrication and handling during manufacture and the electromagnetic

[20] "Symposium on Fluorine Compounds," *Ind. Eng. Chem.* 37, 235 (1947).
[21] N. M. Bashara, *Trans. Am. Inst. Elec. Eng.* 72, Pt. 1, 79 (Technical Paper 53-135) (1953).
[22] "Transformer Cooling by Evaporation," *Westinghouse Engineer* 11, 166 (1951).

forces during normal operation, overloads, and short circuits.

(3) Good thermal stability to maintain both the electrical and the mechanical properties at a high level during the life of the transformer, which may be 20 years or more.

The most important materials used in liquid-filled transformers are oil-impregnated paper, pressboard, and petroleum oil, in dry-type transformers, air or nitrogen, fiberglass, glass, porcelain, organic and silicone varnishes. The silicones or other Class H insulating materials have found appreciable use only in sealed dry-type transformers. Liquid-filled transformers are still all Class A because no suitable low-cost substitutes for petroleum oil and oil-impregnated cellulose have been found.

Power Capacitor Dielectrics

Most industrial power requirements are reactive in nature and, unless compensated, a lagging current is drawn from the power lines. This means additional generating capacity and extra copper in the transmission circuits. Since capacitors take a leading current on an a-c circuit, they supply leading kva, referred to as kilovars, which can be used to compensate for lagging kva of industrial power loads.

By far the greatest proportion of power capacitors are used in 60-cycle service. The voltage ratings of single units vary from 230 to 13,800 volts in approximately ten steps. For higher voltages, an appropriate number of units are connected in series. The kvar ratings vary from 0.5 to 10 for special-purpose units and from 7.5 to 25 kvar for general-purpose units. A liquid-impregnated paper dielectric is generally employed in all capacitors of the power type.

Some power capacitors are used also in higher frequency applications, usually for power factor correction in high-frequency heaters and induction heating furnaces, limited to an upper frequency of about 12,000 cycles. Since the kvar rating of a capacitor increases linearly with the frequency and the dielectric losses increase at an even greater rate, the point is soon reached where it becomes difficult to dissipate the heat generated within the capacitor at the higher frequencies. Therefore, at frequencies of about 180 to 12,000 cycles, water cooling is usually employed in units specially designed for this service. It is possible, of course, to go to higher frequencies even without water cooling when special low-loss dielectric materials are employed. The dielectric losses definitely determine the upper frequency limit for the liquid-impregnated paper dielectric commonly employed in all power capacitors. Ratings up to 300 kvar in single units are common for this type.

Some power capacitors are used also in d-c or modified d-c service. Typical of d-c application is capacitative filtering in a power rectifier circuit. Typical of modified d-c or energy storage service are (1) surge generators for producing test-voltage surges similar to lightning surges; (2) energy storage welding; and (3) high-intensity flash X-ray and light photography. Voltage ratings for d-c capacitors may go to 100 kv. In general, capacitors for d-c service can be operated at higher voltage stresses since the d-c dielectric losses are usually much lower than a-c losses. Power capacitor design and applications are treated at length by Marbury.[23]

Components of power capacitors

Most power capacitors are made by rolling thin aluminum foils interleaved with multiple sheets of special kraft cellulose paper onto a mandrel. After a sufficient number of turns are wound, the combination of metal foil and paper is removed from the mandrel and pressed into a flat oval shape. The capacitor section so formed is then assembled within an insulating box, and the whole assembly inserted in a metal case (Fig. 1.8). Connections are made to the foils by inserting thin metal strips or tabs at the upper ends of the capacitor sections. Then the sections are connected in various series or parallel combinations by soldered flexible leads.

Next, the capacitor is placed in a vacuum oven for several days to remove air and moisture and then completely impregnated with a liquid while vacuum is maintained. Before impregnation, the liquid is usually treated with an absorptive or Fuller's earth type of clay to remove the last traces of ionic impurities. Finally, the capacitor case is hermetically sealed by soldering, welding, or other means.

Although synthetic films, such as the recently developed polyethylene terephthalate, find application in some types of capacitors,[24] in the power capacitor field paper still reigns supreme. This is partly a question of economics, but, more important, paper after impregnation offers many of the desirable properties [25]

[23] R. E. Marbury, *Power Capacitors*, McGraw-Hill Book Co., New York, 1949.
[24] M. C. Wooley, G. T. Kohman, and W. McMahon, *Trans. Am. Inst. Elec. Engrs.* 72, Pt. 1, 33 (1953).
[25] H. H. Race, R. J. Hemphill, and H. S. Endicott, *Gen. Elec. Rev.* 43, 492 (1940).

required for operation at high-voltage stresses. The early paper-capacitor papers were made largely of linen paper derived from flax pulp. During the last twenty years, specially refined kraft (wood pulp) paper has almost completely replaced linen paper in this application.

Capacitor paper is usually made on a standard paper machine; and rolls up to 90 in. wide are calendered to produce a smooth dense sheet, and then slit into the widths required. The thickness ranges from about 0.0002 to 0.001 in. The problem of manufacturing such thin papers relatively free of holes, conducting particles, and other defects, is one of the most difficult in the paper-making art. Brooks[26] in a statistical analysis on thin papers showed the pronounced effect of conducting particles on the electric breakdown strength. It is because of these ever-present flaws or particles that two, three, or more sheets of paper have to be used between pairs of aluminum foils. Considerable progress has been made by the paper manufacturers in recent years not only in making thinner papers but also in reducing the number of flaws. High dielectric strength is particularly important when it is considered that 60-cycle power capacitors are usually operated at continuously applied voltage stresses up to about 400 volts per mil, whereas in rotating machinery, transformers, and power cables the stresses range only from 25 to 100 volts per mil. On d-c applications, stresses even of 1000 volts per mil are common.

Another important characteristic is freedom from soluble impurities which affect the power factor and dielectric losses of the finished capacitor. Figure 1.9 illustrates the progress made in this respect during the

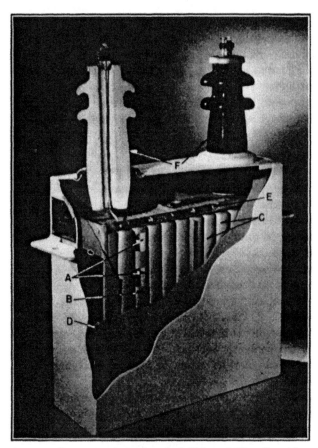

Fig. 1.8. Cut-away view of typical 15-kva, 2400-volt outdoor power capacitor. *A.* Aluminum foil. *B.* Paper. *C.* Capacitor sections. *D.* Pressboard box liner. *E.* Discharge resistors. *F.* Bushing.

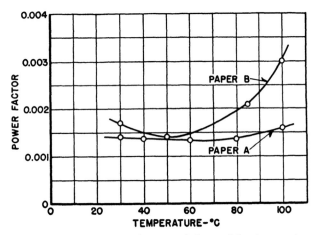

Fig. 1.9. Apparent power factor as influenced by temperature for dry kraft paper.

last decade. Paper *B* is typical of the earlier papers, paper *A* represents the improved papers now available. It is the power factor of the paper, together with that of the impregnating liquid, which determines the overall loss in the capacitor. The improvement shown has made it possible to increase the maximum kva rating of 60-cycle units from 15 to 25 kva and at the same time to reduce the cost per kva.

Paper, furthermore, has suitable mechanical strength for easy winding of the capacitor sections, and the dielectric constant (approximately 6.5 when the fiber is impregnated with a suitable liquid) provides a very satisfactory over-all dielectric constant. Two undesirable properties are lack of high thermal stability and sensitivity to moisture.[27]

The cellulose molecule breaks down on heating to form water, carbon dioxide, and carbon monoxide, and finally a carbonaceous residue. The drying temperature must therefore be held within the range of 100°

[26] H. Brooks, *Trans. Am. Inst. Elec. Engrs.* **66**, 1137 (1947).

[27] E. W. Greenfield, *J. Franklin Inst.* **222**, 345 (1936).

to 135°C because the decomposition rate of cellulose becomes excessive at higher temperatures. Water is a product of this decomposition, and unfortunately this water of constitution is difficult to distinguish from the water adsorbed by the paper.

Application of vacuum markedly accelerates the rate of drying (Fig. 1.10). The gases removed from these

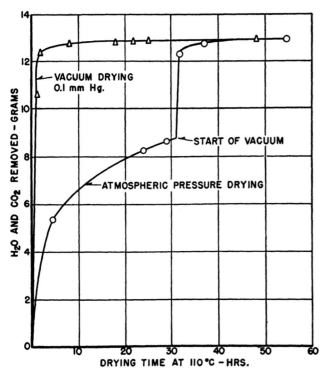

Fig. 1.10. Water and carbon dioxide removed at 110°C with and without vacuum from capacitors containing 215 grams of paper.

capacitors at 110°C were largely moisture vapors, but they did contain appreciable amounts of carbon dioxide, indicating that some cellulose was being decomposed. The progress of the drying reflects in the power factor of the capacitor unit (Fig. 1.11). The power factor reaches a constant value after about 80 hours, but the last traces of moisture are not yet removed, as the gas evolution curve indicates. The bone-dry condition is approached in approximately 120 hours in our example. For obtaining optimum life, capacitors are therefore usually dried beyond the point where the electrical properties, such as power factor or resistivity, reach constant values.

Impregnating Liquid. The early paper power capacitors used petroleum oil ranging in viscosity from the viscous types used to impregnate solid-core paper cables to the relatively freely flowing transformer oils. Even after excellent processing to eliminate gas voids from the structure the maximum operating voltage was only about 200 volts per mil. Attempts to raise the operating stress resulted in gaseous discharges and rapid decomposition. The decomposition of hydrocarbons when subjected to gaseous electric discharge was investigated extensively by Schoepfle and Fellows,[28] Berberich,[29] and others. The products are

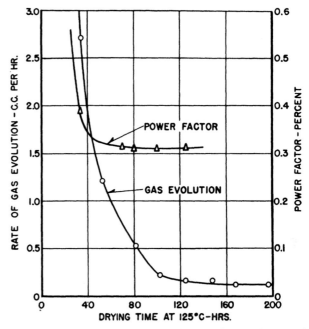

Fig. 1.11. Rate of gas evolution and power factor of a capacitor (39 grams of paper, 125°C, 0.01 mm Hg).

largely hydrogen, some hydrocarbon gases and liquids, and a solid wax-like relatively insoluble material of undetermined composition (X-wax) (see IIIA, Sec. 2).

Petroleum oil has another limitation for this application. Its dielectric constant is only about 2.2 as compared to ~6.5 for the cellulose fiber. This results, not only in lower over-all capacitance per unit volume of dielectric, but also in an undesirable stress distribution between oil and cellulose fiber. The oil films in series with the paper fibers are subjected to approximately three times the stress with a consequent enhancement of the discharge effects just described.

These undesirable features of petroleum oil resulted in the development of askarel as a capacitor impregnant [17] in the early 1930's. The particular askarel used most extensively is pentachlorodiphenyl ($\kappa' \sim 5.0$ at 25°C and 60 cycles). This higher dielectric con-

[28] C. S. Schoepfle and C. H. Fellows, *Ind. Eng. Chem.* **23**, 1396 (1931).

[29] L. J. Berberich, *Ind. Eng. Chem.* **30**, 280 (1938).

stant, and the better stress distribution between paper fiber and liquid, resulted in a greater than 50 percent reduction in size for the same kva rating (Fig. 1.12). Even though the cost of the askarel, known by such various trade names as 1254 Aroclor, Inerteen, and Pyranol, is more than five times that of the oil formerly used, an over-all cost reduction was effected because not only less liquid was required per kva, but also less paper, aluminum foil, steel, and other construction materials. Askarel-impregnated capacitors

Fig. 1.12. Comparison of askarel-impregnated (left) and oil-impregnated (right) capacitor (10 kva, 2400 volts, 60 cycles).

are operated at a-c stresses of the order of 400 volts per mil. This development is a striking example of how a synthetic material, even though higher in cost than the natural material it replaces, may lead to a marked engineering advance.

Pentachlorodiphenyl is a flameproof liquid of the density 1.54 at 25°C, viscosity of 2800 Saybolt seconds at 100°F, and an ASTM pour point (approximate freezing point) of 10°C. When it freezes, dipole rotation ceases, and the dielectric constant falls from a value of about 5.0 at 25°C to about 3.2 at −30°C. This drop, combined with a somewhat smaller decrease in dielectric constant of the polar cellulose, results in considerable decrease in capacitance of an askarel-impregnated capacitor at low temperatures. Since, however, the dielectric losses at low temperatures are also high, a rapid rise in temperature results which eventually restores the higher capacitance values. The variation of capacitance and power factor with temperature and frequency [23,30] and the properties of pentachlorodiphenyl and other diphenyls with both higher and lower chlorine contents are described extensively in the literature.[31-34] A chlorinated aromatic hydrocarbon liquid developed relatively recently is discussed by Warner.[35]

The chlorinated aromatic hydrocarbons, particularly the chlorinated diphenyl types, have a long life on 60-cycle alternating current, even though they are operated at high stresses in power capacitors. However, when high temperatures and high d-c stresses are combined, the life is relatively short. Attention was first called to this fact, and to the possibility of improving the life by means of additives, by McLean[36] and co-workers. The exact mechanism of the reaction which takes place is still not definitely known, but it is obvious that it must be electrolytic in nature, since it does not occur with temperature alone or to any appreciable extent with a-c stresses. In general, it is believed that the chlorinated hydrocarbon is decomposed, liberating both active hydrogen and chlorine. It is possible that chlorine reacts with the aluminum of the electrodes to form aluminum chloride, a catalyst known to promote decomposition of chlorinated hydrocarbons. It is also possible that hydrogen plays a leading role in this deterioration process, since there is evidence that it can take place even in petroleum oils, but at a much slower rate. Fortunately, it has been found that by adding small amounts of certain compounds to the chlorinated hydrocarbons the life of capacitors can be prolonged to a marked degree. Figure 1.13 shows the effectiveness of anthraquinone, azobenzene, and benzil.[37]

Considerable work has been done in the field of stabilizers,[37-39] and many of the effective materials

[30] L. J. Berberich, C. V. Fields, and R. E. Marbury, *Trans. Am. Inst. Elec. Engrs.* 63, 1173 (1944); *Proc. Inst. Radio Engrs.* 33, 389 (1945).

[31] W. Jackson, *Proc. Royal Soc. (London)* 153A, 158 (1935).

[32] S. O. Morgan, *Trans. Am. Electrochem. Soc.* 65, 109 (1934).

[33] A. H. White and S. O. Morgan, *J. Franklin Inst.* 216, 635 (1933).

[34] "The Aroclors—Physical Properties and Suggested Applications," *Tech. Bull.* P-115, Monsanto Chemical Co., St. Louis, Mo., 1950.

[35] A. J. Warner, *Trans. Am. Inst. Elec. Engrs.* 71, Pt. 1, 330 (1952).

[36] D. A. McLean, L. Egerton, G. T. Kohman, and M. Brotherton, *Ind. Eng. Chem.* 34, 101 (1942).

[37] L. J. Berberich and R. Friedman, *Ind. Eng. Chem.* 40, 117 (1948).

[38] D. A. McLean and L. Egerton, *Ind. Eng. Chem.* 37, 73 (1945).

[39] D. A. McLean, L. Egerton, and C. C. Houtz, *Ind. Eng. Chem.* 38, 1110 (1946).

are known to be capable of reacting with hydrogen or forming complexes with aluminum chloride. Compounds, such as tin tetraphenyl, which very readily react with HCl, however, are almost completely ineffective.[37] This fact, together with recent evidence that straight aromatic hydrocarbon additions have some stabilizing action, indicates that hydrogen may play a leading role in the deterioration process.

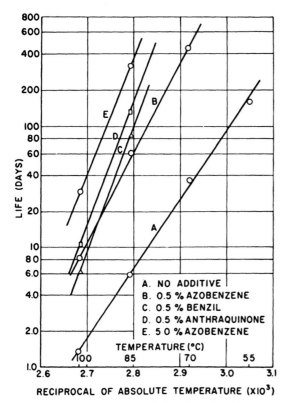

Fig. 1.13. Direct-current life of chlorinated diphenyl (kraft paper) capacitors (1000 volts per mil).

Summary

In power capacitors, almost universally, liquid-impregnated kraft paper is used as the dielectric. Besides reasonable cost, paper possesses a very desirable combination of physical and electrical properties for this application. Careful drying of the paper and complete impregnation are essential. Early capacitors were impregnated with petroleum oil, but oils have been almost completely replaced by chlorinated diphenyl (askarel) in the last twenty years. Its higher dielectric constant and the better electric stress distribution between paper and askarel has achieved more than 50 percent reduction in capacitor size. Oil is superior to chlorinated diphenyl only in capacitance constancy at low temperatures and life on direct current at high temperatures. These deficiencies of chlorinated diphenyl have been remedied, however, to some extent by reducing the freezing point and adding stabilizers.

Dielectrics in Switchgear

The term *switchgear* refers in general to the switching and interrupting devices and associated control equipment of electrical circuits. If the switch automatically interrupts the circuit when some critical current or voltage rating is exceeded, it is called a *circuit breaker*. Alternating-current circuits are considerably less difficult to interrupt than d-c circuits, because the current goes through zero each half cycle. In general, a-c circuit interruption requires first the substitution of an arc for part of the metallic circuit and then its deionization at the moment of current zero, so that the arc will not reignite.

Low-voltage designs, in general, use synthetic resin moldings to carry the metallic parts. In the low-current types, the molding compounds usually consist of phenolic resins with fillers ranging from wood-flour to asbestos fibers. The latter filler is used where moderately high temperatures are encountered. For still higher temperatures ceramic parts are used. In some types of low-voltage switches and circuit breakers, the entire interrupting mechanism is encased in a synthetic resin or ceramic case. When the arc is likely to come in contact with the molded parts, melamine, urea, or special alkyd resins are used because they have greater arc resistance than the phenolics.

Circuit breakers for higher voltages and current fall into two classes: (1) air circuit breakers, generally applied inside buildings, and (2) oil circuit breakers, generally applied outdoors. Air circuit breakers may be further subdivided into types operating in air at atmospheric pressure and those employing compressed air to interrupt the arc. The former are used at voltages up to about 13.8 kv and may have interrupting capacities up to one-half million kva. In the compressed air types voltage ratings go to 69 kv, with interrupting capacities as high as 3.5 million kva. Oil circuit breakers are built for distribution circuit voltages as low as 2.3 kv, and up to the highest transmission voltages of 330 kv, and interrupting capacities range from 50,000 to 25 million kva. The important insulation problems in all these designs are (1) the support of the stationary and moving contacts within an enclosed structure, and (2) selection of materials which will aid in the interruption of the arc.

Air circuit breakers

Figure 1.14 shows the contact and arc-chute assembly for a typical magnetic breaker for atmospheric pressure (2.3 to 13.8 kv). When the moving contact separates, the magnetic effect pushes the arc upward to arc horns in series with a magnetic blow-out coil. The field from this coil forces the arc into a stack of insulating rectangular splitter plates. These plates, as the side view indicates, have a slit. By staggering the position of the slits in the various plates the arc

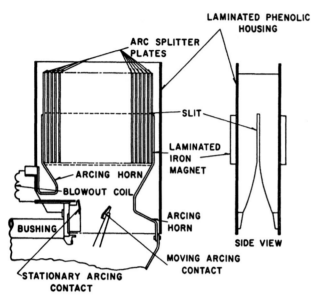

Fig. 1.14. Arc chute and contact assemblies for magnetic type of air breaker.

is forced to take a zigzag path, thus increasing its length and contact with the plates, which makes quick deionization and interruption possible. The supports for the contacts are usually resin-treated-paper bushings or rods. The operating rod (not shown) for the moving contact is usually made of phenolic laminated wood, which combines good mechanical and electrical strength properties. The arc-chute housing may be fabricated of vulcanized fiber or a paper-base phenolic laminate.

The lower-voltage breakers, 600 volts and below, often employ steel for splitter plates because they are low in cost and produce good magnetic and cooling action; the higher-voltage breakers operate with insulating plates (usually asbestos board or refractory ceramics). The refractory has to withstand the heat shock of the arc and surface fusion. Special ceramics with high melting ingredients, such as the zircon porcelains, perform very well in this application. They possess good thermal conductivity, a reasonably low thermal coefficient of expansion, and are made somewhat porous to lessen the heat shock when the hot arc suddenly comes into contact with the material. Zircon porcelains are finding an increasing number of higher temperature applications [40,41] (see IIIA, Sec. 4).

The contact and arc-chute assemblies for a typical cross-blast compressed-air breaker are shown in Fig. 1.15. Again the contact supports, arc-chute housing, and air-blast tube are usually made of paper-base laminated phenolic tubes or rods, and the operating rod of phenolic-laminated wood. Laminated wood has a higher tensile strength than either paper-base or

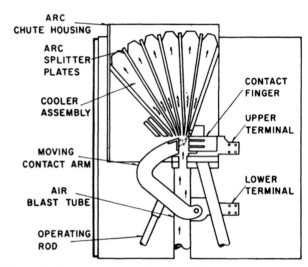

Fig. 1.15. Compressed air blast circuit breaker arc chute and contact assemblies.

cloth-base laminated material; stronger laminates can be made with rayon or glass cloth instead of paper or cotton cloth, but the cost is appreciably higher than that of wood laminates.

The arc-chute assembly consists of alternately placed diverging arc-splitter plates and cooler assemblies. Ceramics are not used here as arc-splitter plates, since even the best become conducting on the surface under intimate contact with the arc produced by the compressed air blast. Hard (vulcanized) fiber, a cellulosic product made by treating pulped cotton rags with zinc chloride, performs well in this application. The hot arc breaks down cellulose molecules, and the water vapor, carbon dioxide, and carbon monoxide produced not only help to extinguish the arc, but also protect the plate through the formation of a gaseous, insulating surface layer. The small amount

---

[40] E. W. Lindsay and L. J. Berberich, *Trans. Am. Inst. Elec. Engrs.* 67, Pt. 1, 734 (1948).

[41] R. Russell, Jr., and L. J. Berberich, *Electronics* 17, No. 5, 136 (1944).

of carbon formed in this thermal breakdown process is blown away, thus removing the danger of surface conduction. The cooler assemblies, consisting of hard-fiber plates and perforated metal plates (not shown), are designed to give maximum turbulence and cooling by contact with the hot arc gases.

When the contacts start to open in this type of breaker, a blast of air blows the arc into the splitter and cooler assemblies. The mixture of the hot arc gases with the gases released by the hard-fiber splitters, together with the turbulent cooling effect produced by the compressed air blast passing through the cooler assemblies, deionize the arc and prevent reignition after a current zero. The hard fiber is subject to some erosion, but its life is surprisingly long, and few other materials can match its performance in this application. Polytetrafluorethylene (Teflon) is one of the very few synthetic materials of less erosion that match the performance of hard fiber. Its considerably higher price, however, limits its use to special applications.

Oil circuit breakers

Circuit breakers of the highest voltage and kva ratings are of the oil-filled type. Besides the oil and the high-voltage bushings, phenolic laminates and vulcanized fiber play roles similar to those in the air breakers (Fig. 1.16). The lift rod for the moving contact must withstand the line-to-ground voltage as well as the mechanical stresses induced by quick operation (laminated wood). The stationary contact and part of the arc-interrupting assembly are housed in a paper-base laminated phenolic tube. The arc-interrupting assembly may take various forms in order to reduce arcing time and energy liberation and thus diminish oil deterioration, burning of contacts, and gas pressure rise. The device shown in Fig. 1.16 is a De-ion-grid, a structure of hard-fiber plates enclosing the contact gaps. Magnetic types have interleaved iron plates to force the arc into specially designed pockets containing oil. Other designs utilize self-generated gas pressure to drive oil into the arc. The turbulent action of the gases evolved from the oil and the hard fiber are effectively utilized in deionizing and extinguishing the arc.[42,43]

Circuit-breaker oil

The oil used in circuit breakers in this country possesses the same characteristics and is interchangeable with that used in transformers. In addition to the usual functions of insulation and heat dissipation it assists in the arc interruption.

The temperatures at the center of the arc may reach values as high as 5000°C. The composition of the arc gases of petroleum oil generated in an experimental breaker is given in Table 1.3. Since the gases help to

Table 1.3. Composition of arc gases; plain-break type of breaker

Hydrogen	82.9%
Methane	9.1
Ethylene	4.2
Ethane	3.6
Propane	0.2

extinguish the arc, a liquid which generates large quantities of gas per unit of arc energy is desired and petroleum oil performs well in this respect. No liquid with better properties than oil has been found for this application. Dibutyl sebacate closely approaches oil in gas evolution (Fig. 1.17), but this higher-priced synthetic liquid has no special advantage. The askarels generate copious quantities of HCl gas, which is toxic and corrosive to metals.

The amount of carbon produced per liter of gas generated at N.T.P. varies from about 0.65 to 0.68 gram

[42] E. W. Kimbark, *Power System Stability—Power Circuit Breakers and Protective Relays*, Vol. II, John Wiley and Sons, New York, 1950.

[43] H. Pender and W. A. Del Mar (Editors), Electric Power Section of *Electrical Engineers Handbook*, 4th Ed., John Wiley and Sons, New York, 1949.

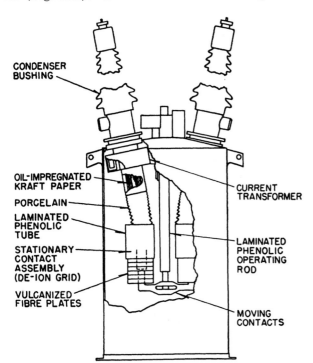

Fig. 1.16. Schematic diagram of high-voltage power circuit breaker, showing insulated parts.

for conventional transformer oil. Carbon particles reduce the breakdown strength if they remain suspended in oil in appreciable quantities; under normal circumstances, however, the carbon agglomerates and settles to the bottom of the tank. Moisture speeds

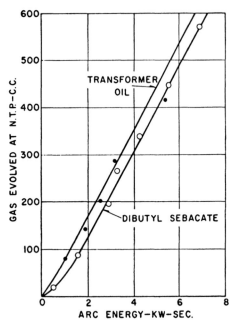

Fig. 1.17. Gas evolved in plain break interrupter.

agglomeration and settling, but unfortunately no use can be made of this for obvious reasons. Carbon may cause difficulty by depositing on creepage surfaces, thus lowering the flashover voltage strength. In general, breakers are designed with vertical internal creepage surfaces to minimize this effect. Circuit-breaker oils rarely need be changed because of excessive oxidation, but may require treatment when an excessive amount of carbon develops in applications subject to frequent interrupting service.

Breaker bushings

Often, particularly in oil circuit breakers, the stationary contacts are supported from the bushings (Fig. 1.16), rating from 2.5 to 330 kv and higher. The lower-voltage bushings may consist of solid cylinders of porcelain, and shellac or other resin-treated paper wrapped upon the current-carrying stud. Many of the higher-voltage bushings, from about 69 kv and up, however, are filled with oil; constructional details vary with the manufacturer. Some use a system of coaxial porcelain or treated paper cylinders with the space between them filled with oil. Others use the "condenser" type of bushing. Here paper is wound on the stud, and metal foils are wrapped on at intervals throughout the diameter, in such a way that the capacitance between successive foils is constant. This insures relatively uniform voltage distribution and thus higher over-all dielectric strength.

The early condenser bushings were made by wrapping shellac or synthetic resin-treated paper and resin-treated foils on a stud. This combination was coated with a special varnish and placed inside a porcelain case for outdoor use. The space between the paper-covered stud and porcelain case was filled with a specially formulated asphaltic compound, which increased the over-all dielectric strength and prevented entry of moisture. Such bushings are still made, but one manufacturer has been supplying oil-impregnated paper bushings over the last ten-year period for all higher-voltage outdoor applications.[44] A typical outdoor bushing of the oil-impregnated type is shown in Fig. 1.18. The stepped-shape structure of metal foil and

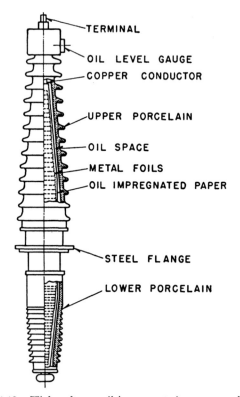

Fig. 1.18. High-voltage, oil-impregnated, paper-condenser bushing.

paper results in maximum dielectric strength at the steel flange by which the bushing is supported on the top of the tank. Corrugated porcelain castings, above and below the flange, seal against moisture and increase the flashover strength, particularly on the upper

[44] H. J. Lingal, H. L. Cole and T. R. Watts, *Trans. Am. Inst. Elec. Engrs.* 62, 269 (1943).

end exposed to the weather. The same type of bushing is used in transformers.

Summary

Switchgear employs a variety of dielectric materials, as phenolic, melamine, urea, and alkyd molding compounds, paper- and wood-laminated phenolics, oil- and resin-treated paper, ordinary and special ceramics, hard (vulcanized) fiber, and petroleum oil. The last two assist in extinguishing the arc by gas evolution. The melamines, alkyds, and ureas are chosen for arc resistance, but the ureas have the disadvantage of poor dimensional stability. Petroleum oil is the best liquid arc-interrupting medium thus far, even though it yields some carbon. The fluorocarbon liquids produce less carbon on arcing than petroleum oils, but also less gas, and that produced may be corrosive. Bushings for the highest-voltage ratings are insulated with oil-impregnated paper.

Dielectrics in Rotating Machinery

Rotating equipment may be divided into low-voltage and high-voltage categories, with a dividing line at 6600 volts. Because of the difficulty of insulating higher voltages, very little rotating equipment is built in this country above 22 kv. A few specially designed generators in England [45] were rated at 33 kv. The greater portion of the equipment used in the generation of power is rated from 11 to 14.4 kv. The kva or horsepower rating may range from a very small fraction of a horsepower for a razor motor to 250,000 kva, and perhaps higher, for some of our large central station generators.

The smaller, low-voltage equipment may be insulated with materials of Class A, B, or H (see Table 1.1). The higher-voltage and larger central station generating equipment is universally insulated with Class B insulation. The major ground insulation in this case is provided by mica in some form. Class H insulation has not yet come into use in the large machines because the effects of expansion and contraction over the larger temperature range make this impractical. In the smaller machines, considerable progress has been made in recent years in reducing size for a given rating by the use of silicones and other Class H materials, especially in totally enclosed types. In large machines, important improvements have been made by replacing asphalt and shellac for the bonding of mica flakes with synthetic solventless resins. In general, the insulation problems for armature windings are more severe than for field winding, since the voltages are higher in the armature. This brief treatment of the subject will therefore be limited to armature insulation (for a more detailed discussion, see footnotes 46 and 47).

Components of armature insulation

All armature coils require (1) conductor insulation, (2) turn insulation, and (3) ground insulation. In the smaller Class A machines, the conductor insulation is baked-on enamel, cotton covering, or combinations of both; in the smaller Class B machines, fiberglass coated with an organic varnish; in Class H designs a silicone varnish is substituted for the organic varnish. In the smaller low-voltage machines, a U-shaped slot liner serves as ground insulation (fish paper in Class A, paper-backed mica or fiberglass-backed in Classes B or H). After winding, the A and B types are impregnated with thermosetting synthetic organic varnishes, the H types with silicone varnishes. In the medium-size lower voltage machines, form-wound coils are used (Fig. 1.19).

High-voltage generator insulation

Insulating large high-voltage generators is a more complex problem. Figure 1.20 shows cross sections of the top and bottom coil sides in a stator slot. The coil consists of three turns, each subdivided into eight strands to increase the flexibility and reduce eddy currents. Insulation must be applied to the strands, to the turns, and to the coil as a whole. The strand insulation, mica tape or fiberglass, prevents flow of eddy currents between strands. The conductor or turn insulation is usually mica tape. The ground insulation of the coil consists of mica flakes bonded by resins. The outer surfaces of the slot portions of the coil are treated with a conducting varnish to produce good contact between coil surface and slot, thus eliminating corona discharges. Also the end turns of the coils are treated over a short distance beyond the slot edges with a high-resistance varnish to reduce the electric stress and possibility of corona. The conductive compounds are made by incorporating specially treated carbon or carbonaceous pigments in a high-grade organic varnish.[48]

---

[45] C. A. Parsons and J. Rosen, *J. Inst. Elec. Engrs. (London)* 67, 1065 (1929).

[46] G. L. Moses, *Electrical Insulation—Its Application to Shipboard Electrical Power Equipment*, McGraw-Hill Book Co., New York, 1951.

[47] D. F. Miner, *Insulation of Electrical Apparatus*, McGraw-Hill Book Co., New York, 1941.

[48] C. F. Hill, L. J. Berberich, and J. S. Askey, "Corona Blackout Achieved in High Voltage Machines," *Westinghouse Engineer* 1, No. 2, 43 (1941).

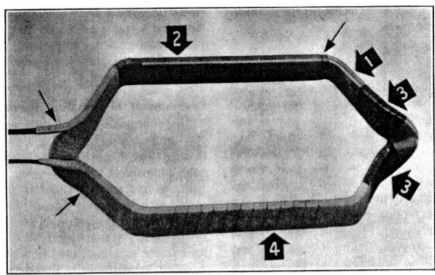

Fig. 1.19. A typical Class A low-voltage stator coil for medium-sized generator. 1. Double cotton-covered conductors. 2. Fish paper-backed mica wrapper. 3. Varnished cambric tape. 4. Cotton binder tape.

Mica is used because it possesses a unique combination of electrical, mechanical, and thermal properties not yet equaled by any of the many synthetics. Unfortunately, mica is not available in strong continuous sheets; tapes must be fabricated from flakes ranging in area from a fraction of a square inch to several square inches. Recently, progress has been made in the development of mica in continuous form. This mica paper [49,50] is made of very small flakes on a modified paper machine. It requires a bonding resin and fibrous backing material to give it sufficient strength for taping applications, just as the mica products fabricated from larger flakes. Voltage endurance and other properties do not yet appear to equal those for the large flake mica products, but improvements will undoubtedly be made. At the present time, most of the high-voltage insulation is still made with muscovite mica flakes, mostly from India.

The early high-voltage insulations were bonded with shellac, later replaced by more flexible asphaltic compounds as the coils became larger. These asphaltic compounds, thermoplastic in nature, caused insulation migration on large turbine generators where the long coils may be subjected to severe temperature cycles. After a number of cycles, sufficient cumulative motion may take place to disturb the physical integrity of the ground insulation and cause electrical failure. This type of failure occurs only rarely and has been eliminated by a thermosetting resin-bonded insulation developed within the last few years.[51] Insulation movements can reportedly be minimized also for

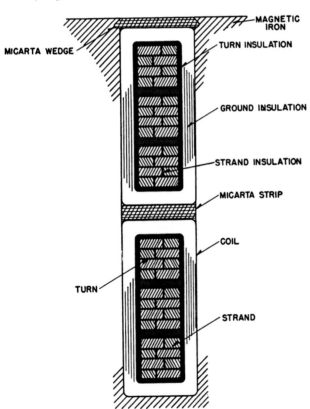

Fig. 1.20. Cross-sectional view, showing typical construction of stator coil.

[49] R. L. Griffith and E. R. Younglove, *Elec. Eng.* **71**, 463 (1952).

[50] E. A. Kern, H. A. Letteron, and P. L. Staats, *Gen. Elec. Rev.* **55**, No. 3, 54 (1952).

[51] C. M. Laffoon, C. F. Hill, G. L. Moses, and L. J. Berberich, *Trans. Am. Inst. Elec. Engrs.* **70**, Pt. 1, 721 (1951).

asphaltic material,[52] but the details of this development are not available.

The new insulation (trade name Thermalastic) consists basically of a continuous mica tape impregnated with a solventless heat-reactive thermosetting synthetic resin. A finishing tape of fiberglass is applied over the mica tape, then the coils are placed in a vacuum oven to remove moisture and air, and vacuum-

new insulation showed a small initial displacement, then no further movement took place, whereas the displacement for the asphalt-bonded insulation increased even after 900 cycles (Fig. 1.23). The displacement was registered on dial gauges just beyond the edges of the slots. Visually, in asphalt-bonded coils separations in the surface tape as wide as one inch were observed, whereas in the new synthetic resin-bonded

Fig. 1.21. Half coils for turbine generator. (Workman shown applying conducting varnish. *A.* Coil end turns. *B.* Stress grading high-resistance varnish. *C.* Conducting varnish.)

Fig. 1.22. Coils wound into stator of hydrogen-cooled generator. *A.* Phase coil connections. *B.* Coil slot in stator bore. *C.* Arch-bound coil end turns.

impregnated with a solventless heat-reactive resin which penetrates and fills the voids in the whole insulation wall. Next, the coils are placed in special mobile presses in an oven. The resin cures to a solid without the elimination of any by-products and a well-filled coil with the slot portions molded to slot size results (Fig. 1.21). Figure 1.22 shows the coils after they are wound into the stator of a hydrogen-cooled turbine generator.

In order to compare the behavior of the new insulation with that of the asphalt-bonded mica tape during temperature cycling, tests were made on full-length coils in a simulated section of stator (108 in. long). The temperature was carried through repetitive cycles of 90 minutes each from about 40° to 120°C. The

[52] C. E. Kilbourne, Discussion of paper of Ref. 51, p. 726.

coils no visible disturbance of the surface of the coils occurred.

The reason for this superior performance of the new insulation on temperature cycles is that asphalt merely flows, whereas the thermosetting crosslinked resin exercises a restoring force. The elastic recovery at 100°C for this resin was found to be about 95 percent. Many generators employing this new insulation have been built, and no case of insulation migration has been observed in the field.

In addition to mechanical improvement, the new insulation provides considerably lower dielectric loss than the asphalt-bonded type and increased dielectric strength (Fig. 1.24). This is due to the greater degree of filling of the voids made possible with the solventless resin. Also the voids do not reappear in service, since flow is eliminated. This insulation will become available on many other types of rotating machine windings in the near future. This development is one example of how the properties of insulation can be improved through the joint efforts of research, development, and application engineers.

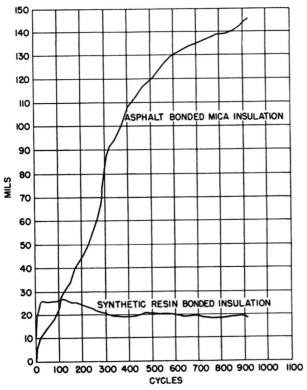

Fig. 1.23. Permanent migration of old and new insulation in thermal cycling test.

Summary

The maintenance of good mechanical as well as electrical properties is extremely important to the reliable operation of machines. The importance of good mechanical properties of the insulation is emphasized by the expansion and contraction during temperature cycles in large machines. These effects become so severe at the high temperatures in central-station generators of large physical size that no such machines have yet been built capable of operating at Class H temperatures. Good mechanical properties are no less important in smaller machines subject to vibration and sudden changes in magnetic forces. In addition, good high voltage and thermal endurance are required.

A great variety of materials meet the requirements of Class A, B, and H in the smaller low-voltage machines, as paper, cotton fabric, glass fabric, mica, and combinations of these with natural and synthetic resinous binders. For higher temperature applications, silicone resins, silicone rubber, and, to a lesser extent, the fluorocarbon resins are being used in steadily increasing amounts. For many of the smaller and all the large high-voltage machines, mica still constitutes the primary insulation. In the bonding of mica flakes, the replacement of thermoplastics by solventless thermosetting resins has resulted in a marked improvement in this type of insulation.

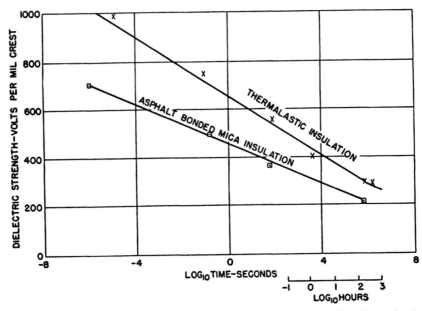

Fig. 1.24. Dielectric strength and voltage endurance behavior of the old and new insulations.

Principal Deteriorating Influences †

Effects of heat and oxygen

Oxidation is responsible for most of the deterioration usually observed in transformer oils; it has also a marked effect on the physical properties of both natural and synthetic rubbers.[53] In other materials,

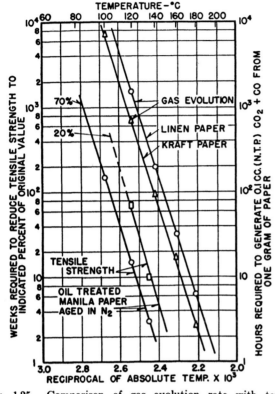

Fig. 1.25. Comparison of gas evolution rate with tensile strength reduction for paper. (Gas evolution data from Murphy.[55])

such as cellulose, asbestos, and mica, oxidation plays only a minor role in comparison to thermal degradation (pyrolysis). For example, muscovite mica loses its water of crystallization gradually in the region of 600° to 700°C and is finally reduced to a powder. The same thing happens to chrysotile asbestos between 300° to 400°C.

Since cellulosic papers have found so many important applications in this field, their rate of deterioration with temperature has been carefully studied. The deterioration of most materials follows the chemical rate laws.[54] When the logarithm of life, usually indicated by changes in some physical or mechanical property,

† See also IIIA, Sec. 2.

[53] J. R. Shelton and W. L. Cox, *Ind. Eng. Chem.* **45**, 392 (1953).

[54] See, for example, T. W. Dakin, *Trans. Am. Inst. Elec. Engrs.* **67**, Pt. 1, 113 (1948).

is plotted against the reciprocal of the absolute temperature, a straight line results (Fig. 1.25). The curves at the right of Fig. 1.25 are due to Murphy[55] and demonstrate that the tensile-strength reduction is caused by the thermal breakdown of the cellulose molecule. These test data were obtained in the absence of oxygen. The rate of deterioration of rope paper, a backing material for mica tape, is more rapid when sealed in nitrogen than when exposed to air (Fig. 1.26). In the sealed tube, water and other prod-

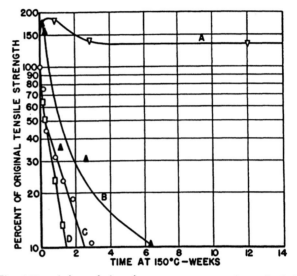

Fig. 1.26. Aging of American rope paper and acrylonitrile (Orlon) at 150°C (*A*, Orlon in N_2; *B*, Orlon in air; *C*, paper in air; *D*, paper in N_2).

ucts of deterioration are trapped, and it has been shown that moisture, in particular, has a pronounced effect on cellulose deterioration.[56] The synthetic acrylonitrile fiber *Orlon*, on the other hand, is practically unaffected by 14 weeks of aging in nitrogen at 150°C, but in air a pronounced reduction in tensile strength is observed. This indicates that for Orlon, in contrast to cellulose, oxidation plays a major part in temperature aging.

Significant benefits in life may be derived by sealing a phenolic-impregnated asbestos Class B insulation system in nitrogen.[57,58] The favorable aging behavior of some materials in nitrogen has prompted the construction of equipment sealed in tanks filled with this gas.

[55] E. J. Murphy, *Trans. Electrochem. Soc.* **83**, 161 (1943).

[56] F. M. Clark, *Trans. Am. Inst. Elec. Engrs.* **60**, 778 (1941); **61**, 742 (1942).

[57] H. C. Stewart and L. C. Whitman, *Trans. Am. Inst. Elec. Engrs.* **70**, Pt. 1, 436 (1951).

[58] H. C. Stewart, L. C. Whitman, and A. L. Scheideler, *Trans. Am. Inst. Elec. Engrs.* **72**, Pt. 3, 267 (1953).

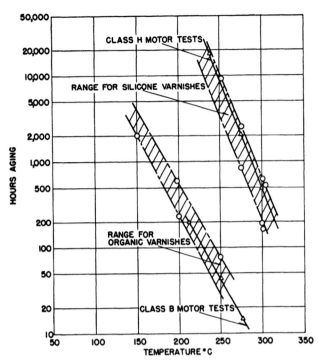

Fig. 1.27. Thermal endurance of silicone varnishes compared with organic varnishes. (Dow Corning and Westinghouse data.)

tric-strength test. The varnish sample was considered to have failed when the dielectric strength deteriorated to 50 percent of the original value. Organic varnishes investigated included oil-modified alkyds, phenolics, and combinations of both. The silicones were normal impregnating varnishes.

In addition to these tests on varnish samples, an investigation was made on Class B and Class H insulated motors in which the resistance to moisture, as determined by insulation resistance measurements after humidification, was used as the criterion.[59] The motor test results fall into the same range as those on the varnish samples applied to glass cloth. The data of Fig. 1.27 show that the silicone varnishes have about 100°C advantage over the conventional organic varnishes for the same thermal life. Thus, the markedly superior thermal stability of Class H insulation, now based largely on silicones, is indicated. Fluorocarbon liquids and solids, such as Teflon and a fluorocarbon rubber, promise to have even higher thermal stability than the silicones. However, these materials will not be used in large volumes until better handling techniques have been developed, and their high cost is reduced.

Figure 1.27 gives some statistical data on the thermal endurance of silicone varnishes as compared with organic varnishes. The samples for these tests consisted

Effects of corona

Corona in air produces ozone (O_3) and the several oxides of nitrogen. Ozone, as a powerful oxidizing

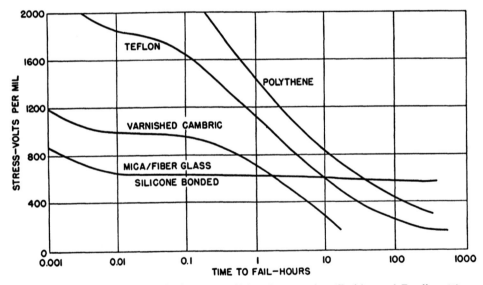

Fig. 1.28. Effect of corona discharges on dielectric strength. (Perkins and Brodhun.[61])

of 7-mil heat-cleaned glass-fiber cloth, treated with two applications of each of the varnishes investigated. The aging tests were carried out in the circulating air ovens, and samples removed periodically for a dielec-

agent, attacks chemically most organic materials. The oxides of nitrogen in the presence of moisture

[59] G. Grant, T. A. Kauppi, G. L. Moses, and G. P. Gibson, *Trans. Am. Inst. Elec. Engrs.* **68**, Pt. 2, 1133 (1949).

lead to the formation of nitric and nitrous acids which will corrode metals and embrittle cellulose and other organic materials. Besides these chemical effects, corona discharges can also have a physical effect. The high-velocity electrons and ions can erode organic materials, and even glass and ceramics. A comprehensive study of corona discharges on polyethylene [60] demonstrated that discharges in voids with limited access to oxygen cause a slow erosion; eventually semicarbonized channels form, and breakdown follows.

Instructive data were obtained by Perkins and Brodhun in air by applying 0.25 in. diameter electrodes to films of the various materials listed [61] (Fig. 1.28). Discharges occurred at the edges of the electrodes, increasing in severity with voltage. Obviously polyethylene, Teflon, and varnish cambric are strongly affected by the discharges; the dielectric strength decreased markedly with time. The glass-fiber mica combination, however, is substantially unaffected. This is one of the reasons why mica is used in high-voltage machine insulation. The effect on the organic materials persisted when tests were made in an inert atmosphere, but disappeared under oil which eliminated the discharges.

Summary

Most all the commonly used organic insulating materials are subject to deterioration in normal use. Of the deteriorating influences normally encountered, the

[60] J. H. Mason, *Proc. Inst. Elec. Engrs.* (London) **98**, Pt. 1, 44 (1951).

[61] J. R. Perkins and C. G. Brodhun, "Teflon-Polytetrafluoroethylene Resin as an Electrical Insulator," *Annual Report of the National Research Council Conference on Electrical Insulation*, p. 31, 1949.

effects of heat and of corona discharges are the most important.

Materials differ widely in their thermal stability. The rate of deterioration of some, like rubber and petroleum oils, is strongly increased by the presence of oxygen; others, like mica, asbestos, and to a lesser extent cellulose, are practically unaffected by oxygen. Cellulosic materials deteriorate more rapidly in a nitrogen-filled container than in open air. This is true because the products of deterioration accelerate further deterioration in a sealed tube where they cannot escape. Other materials, however, show favorable life characteristics in nitrogen. Therefore, dry-type as well as oil-impregnated apparatus sealed in nitrogen appears to be destined for an important future. The present Class H insulation is largely designed around the silicones; silicone varnishes have about 100°C thermal advantage over conventional organic varnishes.

Corona discharges have a physical as well as a chemical effect on most organic materials. The physical effect, apparently erosion of high-velocity ions, is quite pronounced in most varnishes, oils, and resins. It is somewhat surprising that Teflon is attacked almost as readily as its hydrocarbon analogue, polyethylene. Mica, glass, and other inorganic materials are not seriously affected unless the discharge is of sufficient intensity to produce high temperatures. In apparatus where voids and the corona discharges cannot be eliminated, the more resistant inorganic materials must be used. The chemical effects of corona discharges are important where air is available to form ozone and the oxides of nitrogen. Every attempt should be made to design apparatus so that both the internal and external discharges are held to a minimum.

2 · Dielectrics in Electronic Equipments[†]

By ARTHUR J. WARNER

Modern electronic equipments, in all their complexity, are very dependent on the nature and reliability of the dielectric materials used in their construction. When such dielectrics are discussed, it immediately

[†] It is a pleasure to acknowledge the contributions of my co-workers at Federal Telecommunication Laboratories, especially F. A. Muller, H. G. Nordlin, and G. R. Leef, of our Dielectrics Laboratory, and the support of the Signal Corps Engineering Laboratory, Fort Monmouth, New Jersey, under whose sponsorship some of the work reported herein was con-

becomes apparent that we are faced with an almost limitless field, since it would be practically impossible to think of any dielectric material that is not, or has not been, used in such equipments, ranging from gases, through liquids to solids, and from the earliest, high-quality naturally occurring insulator, amber, to the

ducted. We are also grateful to the American Society for Testing Materials for permission to reproduce Figs. 2.5, 2.6, and 2.7 from the *ASTM Bulletin*, April, 1950.

more recent fluorine-containing plastics. In addition, we must consider that the materials will be required to operate from direct current to the highest frequency ranges (at least 24,000 Mc), from zero percent relative humidity to direct moisture condensation, and from very low to very high temperatures.

It is not surprising, therefore, that perhaps no single material has actually been studied over all possible combinations of environmental conditions, and this fact gives particular emphasis to the great need for continued research into dielectrics and the theories governing their behavior. In this discussion we propose to review certain typical materials and applications to demonstrate how such materials are applied in practice, and to show how further insight into their nature and performance may be obtained by more fundamental studies.

Dielectric materials may be divided, for purposes of this discussion, into two general classes, and the end-use requirement will necessitate a choice from one or the other of these groups. They are (1) the polar, or high electrical loss materials, and (2) the nonpolar, or low electrical loss materials.

The first group includes a wide range of materials whose electrical characteristics may approach those of the low-loss class, as well as products having partially conducting properties. Their application is limited essentially to low-frequency applications (from direct current to a few hundred kilocycles).

The second group comprises the so-called high-frequency, or high-quality, insulating materials. As examples of nonpolar, or low-loss materials, we may cite polystyrene, polyethylene, polytetrafluoroethylene, quartz, nitrogen, and carbon tetrachloride, and for the polar type the phenolic plastics, plasticized cellulose acetate, nylon, monochlorotrifluoroethylene, castor oil, etc. (see IIIA, Sec. 3). In many cases, physical property considerations may necessitate the employment of a material for high-frequency use which, strictly considered, is polar.

In this discussion we shall deal in some detail with examples from the two groups, treating each one in a somewhat different manner in an attempt to bring out various points of interest.

High Loss Materials as Wire Insulants

Large quantities of wires and cables are used in connecting together the various components of electronic equipments, and these, with the major exception of the low-loss, high-frequency cables, utilize large quantities of high-loss dielectrics.

Originally the acceptable insulation for wires, particularly for what is termed *hook-up wire*, was either a rubber tape applied to the bare conductor longitudinally and vulcanized under pressure in an autoclave, or a textile lapping impregnated with a synthetic lacquer such as cellulose acetate. The chief disadvantage of the rubber tape at that time was the difficulty of identifying one wire from another in complicated runs, which often led to the necessity of applying a colored textile wrapping over the rubber insulation; this represented an additional and therefore costly procedure. On the other hand, the straight textile insulated wires were highly susceptible to moisture absorption (with a consequent drop in the insulation resistance) and also to deterioration by biological forces.

With the development of synthetic resins of the thermoplastic type applied by a relatively simple extrusion operation, a new series of wires appeared. The first samples were disappointing in a number of ways. Their ability to withstand high applied voltages over long periods of time was variable and poor; initially acceptable insulation resistance rapidly became intolerably low with immersion in water; wires, which at first were extremely flexible, hardened and even became brittle after heat aging; and even the promise of bright colors suitable for color coding faded as did the colors. It would be a long story indeed to follow the detailed investigations which went into the study of vinyl insulation and led to the present, more extensive, series of compounds especially designed for specific applications. We shall discuss here a few case histories as indicative of the problems encountered and the solutions developed.

Vinyl compounds for wire insulation comprise, in their simplest form, the base resin, a plasticizer, and a stabilizer; in addition, lubricants, coloring agents, and fillers may be present. The base resin is either the polymer of vinyl chloride itself, or a copolymer of this material with some other suitable vinyl monomer such as vinyl acetate, the latter being present in a relatively minor amount.

It has been found that, in order to prepare a compound having acceptable flexibility at room temperature without adversely affecting its ability to withstand temperatures of the order of 85°C, the amount of plasticizer to be used should be between 35 and 50 percent of the total weight of the finished compound. The exact amount to be used for a given flexibility is a complex problem having to do primarily with the molecular configuration of the particular plasticizer under consideration.

Several years ago it was found that certain vinyl compounds showed great differences in electrical performance, particularly in respect to insulation resistance after immersion in water; dielectric measurements played an important role in elucidating this

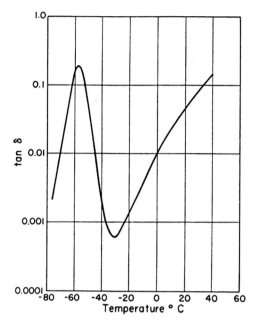

Fig. 2.1. Dissipation factor of dinonylphthalate as function of temperature.

problem. Thus a coil of vinyl-insulated wire of poor electrical properties, when tested for insulation resistance after immersion in water and with an applied direct voltage, showed the normal decrease with time over the first few minutes, followed by a period of a slower rate of decrease eventually going to a minimum value. With the voltage still applied, the insulation resistance then started to increase, reaching a more or less equilibrium value after some 24 hours. The final insulation resistance was higher than the one-minute electrification value. From these results it was postulated that there must be ionizable material present in the insulation and that the initial drop in resistance was due to the moisture absorption of the vinyl material, with the consequent formation of conducting paths. With time a second process occurred, namely, the removal of the ions by transfer to the water and "plating out" on the electrodes. The final value of insulation resistance, therefore, was that of the purified insulation.

This prompted a detailed examination of the electrical properties of the plasticizers used for wire insulations, which showed that large differences existed not only between the material obtained from different suppliers, but also from drum to drum of nominally the same material.

The dissipation factor of all liquid, monomeric polar plasticizers used in vinyl compounds, when measured as a function of temperature at a fixed frequency, gives the typical curve of Fig. 2.1. The peak at the lower temperature is due to the dipolar loss of the material itself; the increase beyond the minimum, as the temperature is raised, represents the contribution of the ionizable impurities. Purification will often decrease the value of the latter portion of the curve, without affecting either height or position of the dissipation factor maximum itself.

Results for the polymeric plasticizer, polypropylene glycol sebacate,

$$-CO[-O-CH_2-CH-COO(CH_2)_8-COO-]_n-CH_2-,$$
$$\mid$$
$$CH_3$$

are shown in Figs. 2.2 and 2.3. Paraplex G-25 is the trade name of the commercial material, and C-1012 is a specially purified plasticizer of the same molecular weight and chemical composition. The lower dissipation factor of C-1012 is reflected in a corresponding improvement in the d-c resistivity of some 1.5 decades.

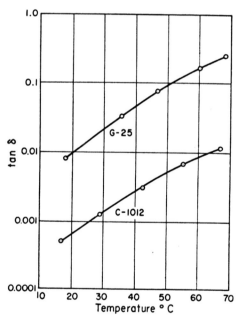

Fig. 2.2 Effect of purifications on dielectric loss of polypropylene glycol sebacate (400 cycles).

In addition to ionizable materials that may be present in the original resin or plasticizer, impurities may be introduced during processing. The base resin is unstable at the high temperatures used for mixing and extrusion, and to minimize this it is conventional to add a so-called stabilizer. For purposes of this discussion, and admitting the subject to be more complex, we can say that a stabilizer acts to combine with any traces of decomposition products formed during processing or in use (usually assumed to be hydrochloric acid), before these decomposition products can exert a catalytic effect on the degradation process itself. Some materials are more effective in this respect than others (see IIIA, Sec. 3 and IIIB, Sec. 1).

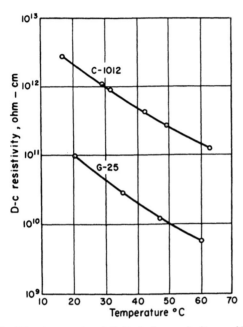

Fig. 2.3. Direct-current resistivity before and after purification.

Figure 2.4 shows the dissipation-factor results obtained on IN-102, a vinyl compound, as originally prepared. This compound had an undesirable insulation resistance. Changing the stabilizer system to an improved one conferring better thermal stability gave improved properties (curve A), and substituting the improved C-1012 plasticizer for the original G-25 in addition improved the situation still more (curve B).

No significant change in either the position or the magnitude of the dissipation factor maximum has occurred, although a major improvement has been made in reducing the electrical losses due to ionic impurities, especially at the higher temperatures. Wires insulated with a vinyl composition of the loss characteristic of curve B have a high insulation resistance over a wide range of conditions and are very stable with time.

The study of plasticizers and plasticized vinyl compounds by electrical methods has thrown considerable light on the whole subject of improved wire insulations. Among perhaps the more interesting aspects of this work is the attempt to correlate electrical and mechanical properties and, in particular, to determine how the low-temperature behavior is related to plasticizer structure. For many years the chief guide to low-temperature plasticizer work was the viscosity theory originally put forward by Leilich [1] and later extended, particularly by Jones.[2] In effect, this theory stated that a plasticized vinyl compound could be considered as a mixture of two liquids of different viscosities and that the final properties were dependent primarily on the viscosity of the liquid plasticizer itself since the same resin was used in each case. Jones, in particular,

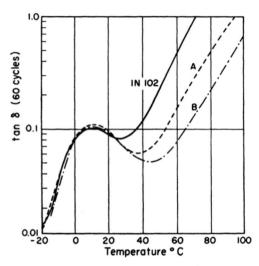

Fig. 2.4. Improvement of thermal stability by change of plasticizer.

pointed out that apart from the absolute viscosity, the slope of the viscosity-temperature curve was also important. For a certain class of plasticizers, particularly those whose compatibility for the resin was high, this hypothesis was useful, but the practical compounder knew of many materials that deviated sharply from these simple concepts. In particular, the discovery of the so-called polymeric plasticizers [3] dealt a sharp blow to this theory.

In contrast to the usual *monomeric* plasticizers of low molecular weight, *polymeric* plasticizers are materials of high molecular weight (greater than 1000)

[1] K. Leilich, *Kolloid. Z.* **99**, 107 (1942).
[2] H. Jones, "Plasticizers in Rubbers and Plastics," a paper read before a joint meeting of the Plastics Industry and the Institution of the Rubber Industry, November 27, 1944, the Engineers Club, Manchester, England; *Trans. Inst. Rubber Ind.* **21**, 298 (1946); H. Jones and E. Chadwick, *J. Oil Colour Chemists Assoc.* **30**, 199 (1947); see also R. F. Boyer, *J. Appl. Phys.* **22**, 723 (1951).
[3] R. F. Boyer, *J. Appl. Phys.* **20**, 540 (1949).

formed by the condensation or polymerization of simpler molecules. Among examples of the former are dioctyl phthalate $(C_8H_{17}COO)_2C_6H_4$, and tricresyl phosphate $(CH_3C_6H_4)_3PO_4$, and of the latter, polypropylene glycol sebacate (G-25),

—CO[—O—CH$_2$—CH—COO(CH$_2$)$_8$—COO—]$_n$—CH$_2$—,
|
CH$_3$

and butadiene-acrylonitrile copolymer (Paracril).

As discussed earlier, polar liquids have a well-defined maximum of their dissipation factor occurring at a definite temperature for any specific frequency of measurement. Figure 2.5 shows this situation for several common plasticizers. The base vinyl resin, including the stabilizer, shows a similar distinct peak in its dissipation factor (Fig. 2.6), and for this discussion will be considered to behave like a liquid. From a molecular standpoint, the temperature of the dissipation maximum at a fixed frequency may be taken as a measure of the temperature at which the internal viscosity of the various materials being measured is the same, since it represents the temperature at which the dipoles present in the material have the same relaxation time.

Should the base resin and the plasticizer be compatible, the position of the dissipation factor maximum (T_m) can be calculated as follows:

$$\frac{1}{T_m(\text{calc})} = \frac{a}{a+b}\left(\frac{1}{T_p}\right) + \frac{b}{a+b}\left(\frac{1}{T_b}\right),$$

where T_p is the temperature of the dissipation factor maximum for the plasticizer, T_b that for the base resin, and a and b are the volumes of the plasticizer and base resin, respectively. All the temperatures are expressed, of course, on the absolute scale.

Experiments in which the T_m value was actually measured show that the foregoing expression holds reasonably well for those plasticizers known from experience to be "good," and, surprisingly perhaps, including the polymeric plasticizers. However, differences do exist between the T_m value as calculated and the T_m observed, and such differences are particularly marked for the less efficient plasticizers. These divergencies suggested the use of this deviation as a measure of the lack of strict compatibility of the resin and the plasticizer, and the so-called k value, defined as

$$k = \frac{T_b - T_m(\text{obs})}{T_b - T_m(\text{calc})}$$

was developed.

Table 2.1 shows various plasticizers with their T_p and "k" values listed, from which it will be seen that plasticizers with "k" values markedly lower than 1.0

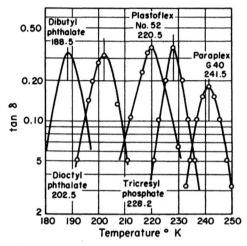

Fig. 2.5. Temperature dependence of dielectric loss for various plasticizers.

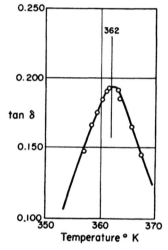

Fig. 2.6. Dissipation factor peak of plastic with stabilizer (0.8 part stabilizer, 2.4 parts lubricant, 46.3 parts vinylite).

Table 2.1. Plasticized polyvinylchloroacetate

Plasticizer	T_p	T_m (obs)	T_m (calc)	"k"	Brittle Point
G-25	238°C	302°C	304.8°C	1.05	256°C
AP-53	233°	301°	304.0°	1.04	253°
Trioctyl phosphate	175°	258°	262.4°	1.04	221°
Methyl pentachlorostearate	228°	304°	304.4°	1.00	273°
Tricresyl phosphate	228°	303°	303.1°	1.00	273°
Dioctyl sebacate	175°	258°	256.6°	0.98	217°
Dibutoxyethyl sebacate	181°	269°	263.9°	0.94	217°
Di-n-octyl phthalate	193°	279°	273.5°	0.93	239°
Di(2-ethylhexyl)-phthalate	202.5°	286°	280.6°	0.93	241°
Plastoflex No. 52	220.5°	301°	293.2°	0.88	263°
Dibutyl phthalate	188.5°	284°	273.6°	0.88	247°
Di(2-ethylhexyl)-tetrahydrophthalate	197.5°	288°	276.4°	0.85	241°
Plastoflex No. 50	205.5°	299°	284.6°	0.80	261°
Plastoflex No. 55	191.5°	303°	273.2°	0.66	249°

are those known to the practical compounder of vinyl compounds for their tendency to sweat out in use, a practical manifestation of a lack of compatibility. The uniqueness of the "k" value, and the use to which it can be put in practice, is detailed in a previous article.[4]

It is possible to predict the low-temperature behavior of plasticized polyvinyl compounds as shown in Fig. 2.7, where the so-called flex temperature [5] is plotted as a function of the observed T_m.

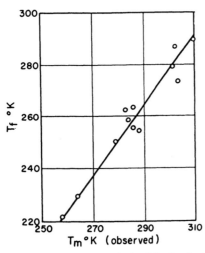

Fig. 2.7. Flex temperature (T_f) of plasticized polyvinyl chloroacetate as function of T_m. (T_f data from Reed and Connor.[5])

The straight-line relationship indicates that it is now possible to design a compound having a specific temperature flexibility from a knowledge of the electrical properties of its components, provided of course that the desired value is within the capability of the ingredients themselves.

In more recent years the demands on the performance of insulated wires for use in electronic equipments have become increasingly severe. Equipments are being designed in which almost all the space within the cabinet is utilized and for which complete hermetic sealing is required. This requirement has of necessity meant an over-all increase in operating temperatures and even higher local "hot-spots," and poses new problems in insulation materials. Typical among the problems recently examined was that presented by a special hermetically sealed power supply unit where electrical failure was occurring between the pins of a connector after many hours of operation. This connector had, attached externally thereto, cables running to other equipments. Detailed examination showed that under sealed and operating conditions temperatures as high as 90°C were reached, with local "hot-spots" going even higher, and that the back of the plug, inside the unit, was sticky to the touch. Measurements showed that the insulation resistance between pins of the connector was low, particularly as the temperature was raised, and that it was this low resistance that ultimately caused failure. An examination of new connectors showed no immediate reason for this low insulation resistance. However, under the specifications to which this equipment was built, various components were required to be "fungus proofed," including the connector in question; and this was achieved by conventional techniques using a commercially available and approved lacquer. The hook-up wires were approved SRIR types with vinyl insulation. To summarize: the cause of failure was found to be the volatilization of residual solvents from the lacquer, together with plasticizer from the vinyl insulation and the deposition thereof on the back of the connector. The connector was the coldest part of the equipment, cooled by the external wires attached thereto. The anti-fungus lacquer on the back of the connector served as a convenient recipient for these volatile ingredients. It absorbed them and caused a serious drop in insulation resistance, followed by electrolysis, corrosion of the pins, and finally electrical breakdown.

The adoption of a prolonged baking cycle before sealing would have satisfactorily solved most of the problem, but the high temperatures measured gave cause for worry that the vinyl insulation itself would not be satisfactory over a long period of time. It was necessary, therefore, to develop a wire insulation suitable for high-temperature use which preferably should contain no plasticizer. Nylon-jacketed vinyl wires were considered, but the susceptibility of nylon to oxidation at elevated temperatures, with consequent development of brittleness, as well as the permeability of the nylon to plasticizer from the vinyl, made this approach unfavorable. What was needed was a new insulation, capable of operation up to 150°C, but possessing basically the properties of a vinyl resin. Such a material potentially is polymonochlorotrifluoroethylene, whose molecular structure is that of vinyl chloride in which all the hydrogens have been replaced by fluorine.

Figure 2.8 shows the electrical properties of this material in one of its commercial forms (Kel-F). The temperature of the dissipation factor maximum occurs at about 270°K compared to 362°K for the base vinyl resin. This means that Kel-F should possess adequate low-temperature flexibility without plasticization, in-

[4] A. J. Warner, *ASTM Bull. 165*, 53 (1950).
[5] M. C. Reed and L. Connor, *Ind. Eng. Chem.* 40, 1414 (1948).

asmuch as 270°K is below the T_m value for most of the plasticized vinyls; this is found to be true in practice. After considerable experimentation, methods were developed for the extrusion of Kel-F directly

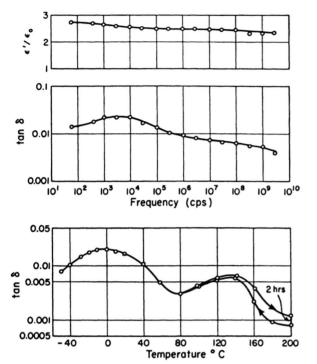

Fig. 2.8. Dielectric characteristics of Kel-F (60 cycles; upper two curves at 25°C).

onto wire, resulting in a high-insulation, abrasion-resistant coating useful for operation up to 135°C, thus solving the problem under discussion.

Low-Loss Materials

The most important dielectrics for modern electronic equipments operating over a very wide range of frequencies are perhaps the low electrical loss or nonpolar materials; however, because of limitations of space, only a few selected examples will be discussed. As representative of this group, we shall deal with polytetrafluoroethylene (Teflon), polyethylene, and polystyrene, giving some of the most recent data on these materials. All three possess the basic properties requisite for use in electronic equipments, namely, low and essentially constant dissipation factor, a low dielectric constant, and a high intrinsic dielectric strength. Physically they differ widely: polytetrafluoroethylene is waxlike, capable of withstanding temperatures from −100°C to +250°C, and is difficult to fabricate; polyethylene is semiflexible, melts fairly sharply at 105°C, has a useful range of temperature operation from −60°C to +85°C, and is subject to oxidation if unprotected; and polystyrene is hard and rigid, optically clear, has only a moderate temperature range of operation, but is cheap, abundant, and relatively easy to process.

Polytetrafluoroethylene

On a first examination, polytetrafluoroethylene would seem to be the ideal dielectric (Fig. 2.9). Its dielectric constant is 2.0 over the frequency range shown, and its volume resistivity is of the order of 10^{19} ohm-cm at 25°C. In addition, it is noninflammable and completely inert to most chemicals. Apart from its price, which has been steadily declining since its introduction, what militates against its being the ideal dielectric? In the first place, it has a high coefficient of expansion (Fig. 2.10), a factor which must be watched when one designs precision components for a potential operating range of 350°C. Second, it has a tendency to creep under load, with the initial deformation occurring fairly rapidly, as shown in Fig. 2.11. Finally, and most important, it has erratic dielectric strength and degrades in the presence of ionization when a structure insulated with the material is operated above its corona-starting level. Since it is impossible to fuse the material properly, most com-

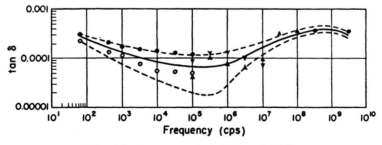

Fig. 2.9. Loss tangent of Teflon at 24°C.

quencies are perhaps the low electrical loss or nonpolar materials; however, because of limitations of space, only a few selected examples will be discussed. As representative of this group, we shall deal with mercially formed articles are really sintered, and the basic properties are thereby lost to some extent. For example, it has been found impossible to prepare wires insulated with polytetrafluoroethylene in thin walls showing the low-loss characteristics expected. When perfectly dry, the insulation approaches the values for solid material, but on exposure to high

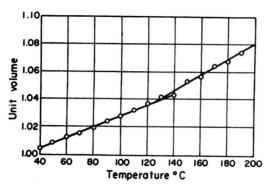

Fig. 2.10. Thermal expansion of Teflon (from 40° to 128°C: 39.2×10^{-5} per °C; 128° to 200°C: 52.7×10^{-5} per °C).

relative humidity the insulation picks up moisture, with a consequent impairment of the electrical properties. One of the biggest problems still facing the

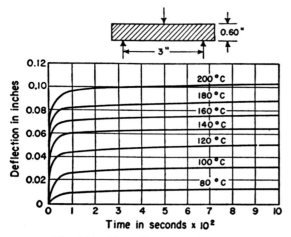

Fig. 2.11. Heat distortion of Teflon.

materials engineer, therefore, is how to fabricate polytetrafluoroethylene and preserve its inherent electrical properties.

Polyethylene

Polyethylene is very familiar as the primary insulation in use today for the majority of high-frequency cables as well as for other communications applications. Its relative ease of handling, its availability, its low electrical losses, high softening point, and low moisture absorption have made the use of the material very attractive. Although introduced into production in this country as recently as 1943, it was not long before its detailed behavior caused speculation as to the exact structure of the material. Even though evidence to the contrary is now available, it is still common to see the structure of polyethylene depicted as a straight polymethylene chain: CH_3—CH_2—CH_2—$(CH_2)_n$—CH_2—CH_2—CH_3.

Evidence against this simple structure was quickly forthcoming. On the basis of known polymethylenes it was calculated that the observed melting point of polyethylene, 105°C, was lower than it should have been by some 20°. It had been observed quite early that when polyethylene was heated and subjected to mechanical shearing in the presence of air, such as would occur naturally during the processing and extruding operations, there was a sharp impairment of the high-frequency properties of the material and that there was formation of hydroperoxide groups. A more detailed study of the dissipation factor as a function of frequency showed a small, but quite definite, peak occurring at frequencies between 100 and 1000 Mc. Bryant,[6] by the use particularly of infrared measurements, showed that instead of polyethylene's having a straight-chain structure, it was quite highly branched, and the individual branches could be fifteen to twenty carbon atoms in length, terminated with methyl groups. The susceptibility of polyethylene to oxidation at elevated temperatures, which has led to the addition of antioxidants to the material, is believed due to this highly complicated branched structure which possesses weak links along its chain where such branches occur. It cannot be ignored, either, that the method of preparation, namely, polymerization in the presence of elementary oxygen, probably builds in oxygen-containing carbonyl groups, and that these carbonyl groups are the cause of the high-frequency losses observed. Even when stabilized against degradation by heat, polyethylene is still subject to deterioration when exposed to ultraviolet light, especially in the region 2900 to 3300 A, and this degradation leads rapidly to embrittlement and loss of electrical characteristics. Despite all the work that has been done on this subject in the past few years, finely divided carbon black, properly distributed, is still the only effective stabilizer against light deterioration available.

When considering polyethylene as used in electronic equipments, therefore, we must consider that we are using a compounded material, and most of the values obtained and reported are on such a mixture. Some work, however, has been done on the unstabilized material, enough to highlight the differences that exist between the properties of the material per se and those found for the fabricated articles.

The dielectric strength of polyethylene has been

[6] W. M. D. Bryant, *J. Polymer Sci.* **2**, 547 (1947).

measured at a frequency of 50 cps and at direct current by a number of investigators,[7] and the results as a function of temperature are shown in Fig. 2.12. It will be seen that the a-c values at room temperature are appreciably lower than the d-c ones, although at the low temperatures there is not so much difference.

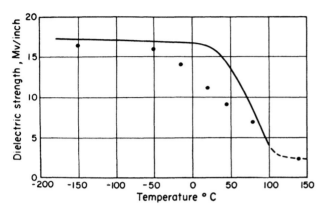

Fig. 2.12. Dielectric strength of polyethylene film (1.5 microns; — d-c values; • 50-cycle values).

The effect of antioxidant addition to polyethylene on the d-c breakdown was studied by Oakes,[8] and Table 2.2 shows the results he obtained. Under normal con-

Table 2.2. Effect of antioxidant addition [8]

Percentage of antioxidant	0.00	0.01	0.10	1.00
Number of specimens tested	15	19	16	15
Average dielectric strength, volts/mil	17,300	16,300	15,800	14,600

Measurements made with direct current, 300 volts/second.

ditions on an extruded polyethylene cable, it is difficult to get values within a decade of this intrinsic value; and it is believed that the chief cause of this

Table 2.3. Volume resistivity of polyethylene with temperature

	Volume Resistivity (ohm-cm)	
	25°C	54°C
Stabilized polyethylene	$>1.2 \times 10^{19}$	4.8×10^{15}
	$>1.1 \times 10^{19}$	4.1×10^{15}
Unstabilized polyethylene	1.7×10^{18}	2.2×10^{17}
	$>1.1 \times 10^{19}$	2.0×10^{17}

[7] W. G. Oakes, *J. Inst. Elec. Engrs.* (London) *95*, Pt. 1, 36 (1948).

[8] W. G. Oakes, *Proc. Inst. Elec. Engrs.* (London) *96*, Pt. 1, 37 (1949); A. E. W. Austen and H. Pelzer, *J. Inst. Elec. Engrs.* (London) *93*, Pt. 1, 525 (1946); D. W. Bird and H. Pelzer, *Proc. Inst. Elec. Engrs.* (London) *96*, Pt. 1, 44 (1949).

is the dissolved air in the material. The difference between stabilized and unstabilized polyethylene can also be observed in precise measurements of their volume-resistivity relationship (Table 2.3).

We have not discussed the effect of the crystalline-amorphous nature of polyethylene on its behavior from the mechanical standpoint, but this has undoubtedly contributed to some of the puzzling phenomena observed in the past, and is the subject of continued extensive study.[9] Despite its problems and deficiencies, polyethylene remains one of the most useful dielectrics yet developed.

Polystyrene

One of the oldest synthetic, low-electrical-loss materials known is polystyrene, a dielectric which has been under study for a great many years, but about which much concerning its detailed behavior is yet to be uncovered. This polymer, the high-molecular-weight derivative of vinyl benzene, $[C_6H_5CH\text{—}CH_2\text{—}]_n$ (where n is usually greater than 500), has long been used in a multitude of applications in electronic equipments, where rigidity and ability to insulate conducting members even at the highest frequencies are prerequisites.

Despite its apparent simplicity of molecular configuration, the electrical losses in polystyrene are not as low as might be anticipated, and much research has been done to elucidate its fine structure. We would like to present here some of the more recent work in this field to illustrate the progress made. At this time we can say that probably the electrical properties of polystyrene are better known than those of any other plastic material.

It has been known for some time that, despite its hydrocarbon nature, polystyrene not only has a finite moisture vapor transmission, but also actually absorbs water. The small amount of water absorbed has an effect on the values obtained for dissipation factor, and is undoubtedly partly the reason for some of the discrepancies reported in the literature. Careful measurements on specially conditioned samples show that the effect of moisture is reproducible, and, as one would expect, has the greater effect the lower the frequency of measurement. Thus, at 1 kc the dissipation factor of polystyrene increases approximately one decade when the relative humidity is changed from 0 to 90 percent. An approximately linear relationship ex-

[9] C. W. Bunn, *Trans. Faraday Soc. 35*, 482 (1939); *Proc. Roy. Soc.* (London) A*180*, 82 (1942); C. W. Bunn and T. C. Alcock, *Trans. Faraday Soc. 41*, 317 (1945).

ists between the logarithm of dissipation factor and the percent relative humidity. At 3000 Mc, increasing the relative humidity from 0 to 90 percent causes an increase in dissipation factor of 0.00031. Commercial polystyrenes still vary somewhat in their electrical losses, values at 3000 Mc varying from 0.00007 to 0.00022, such differences undoubtedly being due to differences in manufacturing techniques.

The temperature coefficient of the dielectric constant of polystyrene at 1 kc is -0.00027 per °C between $-20°$ and $+50°C$, and the corresponding temperature coefficient of dissipation factor is $+0.000002$ per °C between $+25°$ and $70°C$. At temperatures between $-60°$ and $0°C$ the dissipation factor is nearly constant, and above $80°C$ it increases sharply with increasing temperature. It is possible to use this type of electrical measurement to arrive at the second-order transition temperature of the material, and, with the data reported above, a value of $+84°C$ for this parameter was recorded, in good agreement with the generally accepted value for this material.

In discussing polyethylene it was pointed out that foreign, nonhydrocarbon, groups were found in the polymer, and in commercial polystyrene a similar situation exists. Determination of the oxygen content of various polymers shows values that are quite high, of the order of one-half of one percent. It has been shown by Reinhart et al. at the National Bureau of Standards,[10] using mass spectrometric methods, that when polystyrene is exposed to 100°C in vacuo, the initial products of decomposition are mostly water, carbon dioxide, various alcohols, and hydrogen. Such products probably derive from the decomposition of thermolabile oxygen-containing groups in the polymer formed during polymerization or storage. It is undoubtedly true that many of the observed fine-structure results on polystyrene, both mechanical and electrical, are caused by such foreign groups, and we have yet to see a pure polystyrene.[11]

Effect of Ionizing Radiation

A problem of recent importance in the application of dielectrics to electronic equipments is the effect of ionizing radiation, and some of the results of studies made in this field will be of interest in this discussion.

It has been known for many years that certain insulating materials, when exposed to X-rays or gamma-

[10] B. G. Achhammer. M. J. Reiney, and F. W. Reinhart, *J. Res. Natl. Bur. Standards* **47**, 116 (1951).

[11] For additional information on polystyrene, see R. H. Boundy, R. F. Boyer. and S. M. Stoesser, "Styrene," *Am. Chem. Soc. Monograph* 115. Reinhold, New York, 1952.

rays, show a deterioration in their insulating properties, and that some recovery occurs after removal of the ionizing source. Several materials have been studied under various conditions, in particular polystyrene, and as a result not only has a better understanding of the effects of irradiation been obtained but the mechanism of conduction has been elucidated. In the first place, it has been demonstrated that no permanent degradation of polystyrene occurs as a result of gamma irradiation for total dosages as high as 10^6 roentgens and at dosage rates as high as 1.5×10^5 roentgens per hour. Most of the work, however, has been carried out at lower radiation intensities, using cobalt-60 as a source material.

By the use of specially developed test equipment and of very thin samples stacked in parallel-plate capacitor fashion, it has been possible to record volume resistivities as high as 10^{21} ohm-cm, such values being obtained, for example, on special polystyrene samples after prolonged electrification and on perfectly dry samples. The experiments during irradiation were carried out in a good vacuum to avoid ionization of the air under the applied voltage.

Figure 2.13 shows the result when a sample of polystyrene is irradiated at a dosage rate of 13 roentgens

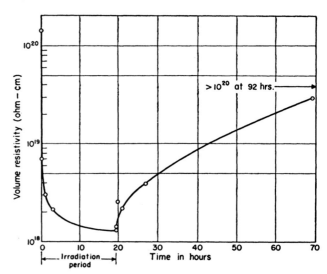

Fig. 2.13. Volume resistivity of polystyrene, during and after irradiation.

per hour, and the irradiating source removed after 19 hours of exposure. The initial, sharp drop in resistivity decreases after the first 2 hours, and becomes almost nil after 15 hours. The recovery after irradiation is rapid at first, slowing down later, and not reaching the original value until some 70 hours after the ionizing source has been removed. Similar curves have been obtained under various other conditions of irradiation, but it is believed that the one shown illustrates basically the phenomena observed.

As a result of work on a number of different insulating materials, an expression has been derived relating the conductivity σ with the intensity of irradiation:

$$\sigma = \sigma_b (I/I_b)^n,$$

where σ = conductivity in ohm^{-1} cm^{-1},
σ_b = normal conductivity (at background irradiation rate),
I_b = background intensity of irradiation (10^{-5} r − hr^{-1}),
I = intensity of gamma irradiation in roentgen − hour^{-1},
n = a constant.

	$\log \sigma_b$	n
Polystyrene	−22.8	0.71
Teflon	−22.6	0.62
Polyethylene + antioxidant	−21.3	0.66
Styroflex	−21.0	0.51
Polyethylene	−19.6	0.48

It is evident from the expression above that the uppermost limit of resistivity is set by the background radiation which exists at all times.

3 · Dielectrics in Capacitors

By LOUIS KAHN

All of us concerned in the design, manufacture, and application of capacitors have often wished for an ideal dielectric to be used in an ideal capacitor, a capacitor that would be smaller, operate at higher temperature, have a higher insulation resistance, a lower power factor, longer life, and be available at a lower price.

When pinned down to actual specifications, the applications engineer will probably describe some type for a specific application of his immediate interest. Then it becomes apparent that an insulation resistance of 50 megohm per microfarad is more than satisfactory for a line by-pass capacitor, but 10^6 megohm per microfarad is about the bare minimum for another application. A power factor of 20 percent is not detrimental in a cathode by-pass capacitor of an audio-frequency amplifier, but a power factor of 0.01 percent is required when a larger mica capacitor is to be used for some other purpose. Obviously, a comprehensive discussion of capacitor dielectrics would have to describe all dielectrics presently available and their recommended uses.

Table 3.1, for example, lists capacitors now used specifically for d-c applications. These are, in addition to the paper capacitors, electrolytic capacitors with aluminum or tantalum, ceramic capacitors with a large variety of ceramic bodies, and film dielectric capacitors.

Voltage Ratings

At the present time, in the United States, the voltage rating of capacitors is based entirely on certain accelerated life tests during which the capacitor is subjected, in general, to the maximum operating temperature and some multiple of the rated voltage. The test is based on the assumption that the Fifth Power Law holds for impregnated paper capacitors: life (hours) = $K(\mathcal{U}_R/\mathcal{U}_T)^5$, where $\mathcal{U}_R$ is the rated and $\mathcal{U}_T$ the test voltage. This is an empirical law; actually we find that the life of capacitors over the operating temperature may vary inversely with the fourth to the seventh power of the voltage. The standard 250-hour test at twice the working voltage predicts a minimum life of one year at rated voltage. This method of rating, still relatively new, was first incorporated in the American War Standards of 1942. In Europe this method of capacitor rating is not accepted; instead, the working voltage is based on an infinite life.

The whole problem of voltage rating must be reconsidered. The application of capacitors in present-day equipment is, in general, not limited to long-time

Table 3.1. Capacitor types for d-c applications

Gases or Vacuum	Film	Stacked
Paper with Impregnants	Polystyrene	Mica
Castor oil	Mylar	Glazed or enamel
Chlorinated diphenyls (Aroclor 1254, Aroclor 1254 with T.C.B.)	Teflon	*Electrolytic Capacitors*
Chlorinated naphthalene	Polyacetates	Tantalum Aluminum
Diallyl phthalates	Polyethylene	Wet Wet
Mineral oil	Diaplex	Dry Dry
Mineral wax	*Ceramic*	Foil Foil
Opal wax	Temperature compensated	Pellet
Polyisobutylene	Tubular	
Polyesters	Disk	
Polyvinylcarbazole	Blocking	
	High voltage	

operation. Many applications require extremely short-time operations; in some cases, the total operating life of the equipment may be of the order of minutes only. On the other hand, there are applications in which a minimum life of twenty years must be anticipated. In addition, the problem of reliability must be considered, which, to a certain extent, is a function of voltage rating.

Paper Capacitors

At the present time, practically all capacitors made in the United States use a kraft paper ranging in thickness from 0.18 to 3 mils. The paper is made from a highly purified pulp of a density ranging from 0.8 to 1.1. An analysis of a typical high-quality paper is given in Table 3.2.

Table 3.2. Analysis of a typically high-quality paper

Thickness	0.30 mil (average)
Apparent density	1.010 (dry)
Moisture	5.6%
Base weight	4.73#/ream 24 × 36 (500) BD
Coverage	91,147 sq in./lb
Porosity	0.7 cc/15 sec (average)
Conducting paths	1.0/sq ft (average)
Ash	0.271%
pH	7.0
Chlorides	4–6 ppm
Sulfates	no detectable amount
Dielectric strength	1227 v/mil, dry
Resistivity of water extract	283,000 ohm-cm, at 25°C
Power factor	30°C 0.150%, 100°C 0.214% at 60 cps
Dielectric constant	2.23

In an effort to make better paper for capacitor application, the paper manufacturers have gone to extreme care in the selection and treatment of their raw pulp. The use of centrifuges, known as centrifinors, in the paper industry for the removal of metallic or other particles of densities greater than water is not unusual, and even ionic demineralizing equipment for water treatment has been used. In the last ten years, the paper manufacturers have also given more thought to the various constituents of paper other than cellulose in order to determine the effects of materials such as lignins and pentosans on the electrical properties. As a result a great improvement in the quality of the paper has been achieved, particularly in reducing the power factor, not only of the paper itself, but also of the finished capacitor.

For many years, it was not unusual to find that finished capacitors, made with paper of a dry power factor of ca. 0.15 percent and an impregnant of ca. 0.05 percent, had power factors of ca. 0.4 to 0.5 percent. Today, the power factor of the impregnated capacitors has been reduced to 0.2 percent, indicating that certain undesirable materials have been eliminated from the paper, since the impregnants are the same as formerly.

Capacitance, power factor, and insulation resistance

The permissible values of these characteristics and of their variations are determined by the application. For d-c capacitors the power factor as such is of no importance, but it serves as a check on materials and processing methods. The correlation between capacitor life and insulation resistance or power factor as such is extremely poor, unless these values are extremely low or high. A more reliable evaluation can be based on the *variations* of capacitance, power factor, and insulation resistance as function of temperature, frequency, voltage, and time.

Impregnating materials for paper capacitors

The characteristics listed in Table 3.3 must be evaluated, not only for the liquid proper, but also with

Table 3.3. Characteristics of impregnated materials for paper capacitors

(1) Dielectric constant
(2) Power factor
(3) Insulation resistance
(4) Viscosity
(5) Pour points
(6) The volume coefficients of expansion
(7) Specific gravity
(8) Chemical stability
(9) Decomposition products in an electric arc
(10) Toxicity

respect to performance in the finished capacitor. Compatibility with other materials used in the processing

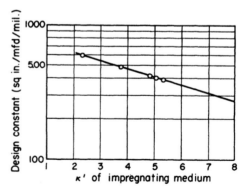

Fig. 3.1. Design constant of capacitors as function of the dielectric constant of impregnant.

and construction of the capacitor, the economics of capacitor design, effects on personnel during the manufacturing process, long-time stability, effects of nuclear radiation, and various other considerations enter (see IIIB, Secs. 1 and 2).

Figure 3.1 shows the dependence of the *design constant* of a capacitor (square inches per microfarad per mil of dielectric thickness) on the dielectric constant

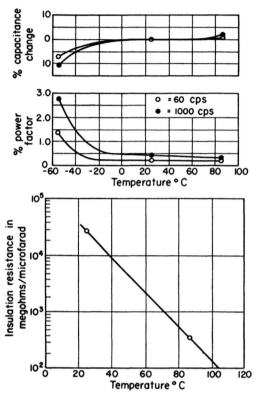

Fig. 3.2. Temperature characteristics of mineral-oil-impregnated capacitors.

of the impregnating medium. (The points on the graph indicate dry paper (~2.25), mineral oil or wax, a polyester resin, castor oil, chlorinated diphenyl, and chlorinated naphthalene.)

Table 3.4 lists the most important electrical properties of presently used impregnants and the capacitor characteristics that can be expected with them as impregnants.

In Table 3.5 and Figs. 3.2 and 3.3 the characteristics of capacitors impregnated with mineral oil and with

Table 3.5. Comparison of two types of capacitors

	Mineral Oil	Chlorinated Diphenyl
Volume	1	0.667
Weight	1	1
Insulation resistance (repacitance) at 25°C	4000 to 22,000	8000 to 12,000 megohm/μf
Insulation resistance at 85°C	125 to 400	75 to 250
P.F., 60 cps, 25°C	0.20%	0.20%
P.F., 1000 cps, 25°C	0.3%	0.35%
Capacitance at −55°C as compared to 25°C	94%	73% at 60 cps / 55% at 1 Mc/sec

chlorinated diphenyl are compared at various temperatures. Obviously, when space is not the most critical factor, most circuit designers will specify the mineral-oil type.

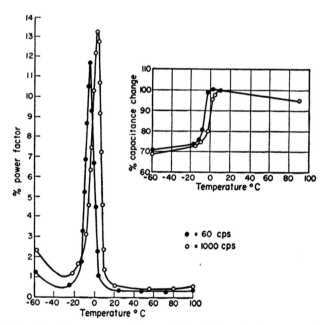

Fig. 3.3. Temperature characteristics of Aroclor-impregnated capacitors (Aroclor 1254).

Table 3.4. Impregnant and capacitor characteristics for paper capacitors

	Impregnant *			Capacitor		
Material	κ'	P.F., %	ρ, ohm-cm	μf/mil	P.F., %	I.R., megohm/μf
Mineral oil	2.23	0.03	10^{14}	600	0.20	20,000
Mineral wax	2.2	0.05	10^{15}	600	0.35	10,000
Petroleum jelly	2.2	0.05	10^{12}	600	0.35	10,000
Polyisobutylene	2.2	0.03	10^{15}	600	0.20	25,000
Silicone liquids	2.6	0.05	10^{16}	560	0.35	22,000
Castor oil	4.7	0.08	10^{12}	455	0.5	2,000
Chlorinated diphenyl	4.9	0.05	10^{13}	430	0.2	15,000
Chlorinated naphthalene	5.2	0.2	10^{11}	390	0.4	5,000

* At 100°C, except chlorinated diphenyls at 65°C and chlorinated naphthalene at 125°C.

Other Types of Capacitors

Film capacitors

These capacitors are relatively new because only recently plastic films have become available in thickness and quality that make their use in capacitors feasible (Table 3.6). Other materials have been used to a

Table 3.6. Film capacitors

Material	κ'	P.F.	Maximum Temperature	25°C I.R., megohm μf	Minimum Thickness
Polystyrene	2.6	0.0001	85°C	10^{14}	0.0004
Teflon	2.2	0.0001	250°	10^{14}	0.00025
Polyethylene	2.2	0.0001	90°	10^{14}	0.001
Mylar *	2.8	0.001	150°	10^{13}	0.000025

* Material now being evaluated.

limited extent. In general, film-type capacitors are used without impregnation or liquid fill. At higher

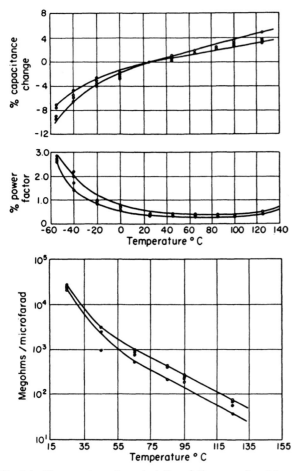

Fig. 3.4. Temperature characteristics of film capacitor (styrene copolymer; power factor at 1000 cycles).

voltages, a liquid fill is desirable to reduce corona and increase voltage flashover limits. The proper choice of a filling material is difficult, however, especially for operation at high temperature, since the advantage of the very high surface resistivity of most film materials may be lost. Typical temperature characteristics for a styrene copolymer are shown in Fig. 3.4.

Ceramic capacitors

In general use for less than 15 years in this country, the small-size, low-cost, and special capacitance characteristics of this type have earned it a definite place in electronic applications. Ceramic capacitors can be divided into two broad groups: temperature-compensating types, and high dielectric-constant types for by-passing functions only.

The first ceramic bodies used for capacitors had dielectric constants ranging from 4 to 90, and with temperature coefficients of capacitance ranging from about 50 ppm positive to 750 ppm negative. The capacitance-temperature curves were straight, and the materials had excellent resistivity and power-factor properties. The recent high dielectric-constant materials have values that may approach 6000 for commercial bodies and experimental values up to 100,000. Resistivities are fairly low, and power factors high. Moreover, allowable voltage working stresses are low (of the order of 30 to 50 volts per mil), and the materials more fragile than the common ceramic bodies (see also IIIA, Sec. 4).

Because of the simplicity of the flat-type ceramic capacitor and its low cost, coupled with very small size as compared to most tubular or mica capacitors, ceramic capacitors have displaced the foregoing types for most low-capacitance (0.01 μf and lower) applications wherever the large changes in capacitance with temperature and the high power factors can be tolerated.

Electrolytic capacitors

Electrolytic capacitors operate with an insulating film formed at the anode. Except for voltage ratings below 6 volts, this film is formed prior to the winding of the capacitor roll. An electrolyte is introduced completely impregnating the paper spacer, which serves primarily as storage for the electrolyte. The electrolyte produces and also reforms the dielectric film in case it is ruptured or reduced in thickness or density. Since the film thickness is very small, and large effective surface areas can be obtained by roughening the anode, the capacitance values per unit volume are high. Maximum operating and surge voltages cannot be exceeded without causing failure of the unit.

Two electrode materials are in use today for electrolytic capacitors, aluminum and tantalum. At present, aluminum capacitors have working voltages up to 525 volts and surge voltages of 575 volts; tantalum capacitors have working voltages of 150 volts and surge voltages of 200 volts. These are ratings for single units only. By series connection higher voltage ratings can be obtained, but the size advantage is reduced because the voltage distribution across the individual elements of a series-connected assembly is not uniform. Therefore, a resistor network must be used across the capacitors to insure an equal voltage distribution, or the number of series elements must be increased to reduce the maximum possible voltage across a single element.

Commercial tantalum capacitors are relatively new. Tantalum oxide, the dielectric of the capacitor, is extremely refractory, and tantalum itself relatively inert, thus permitting the use of electrolytes that cannot serve with aluminum. Capacitors of very low leakages can be obtained, and the maximum and minimum operating temperatures extended.

Mica capacitors

This type has been in use since the start of the electrical industry and will continue to be used because of its unusual properties. The property of mica to split into thin sheets of uniform thickness, its high dielectric constant (7 to 8), and low power factor, plus the fact that mica requires only mechanical handling, make it extremely satisfactory for capacitors. A major disadvantage is the limited size of individual mica sheets and the high cost of preparation, thus limiting its use to low capacitance values.

Glass-ribbon and enamel capacitors

Both these types have been developed in the last ten years and are more or less in the same class as mica capacitors.

The use of glass as a dielectric is not new; in fact, glass is about the oldest capacitor dielectric. However, the development of glass ribbons 1 mil thick permits the manufacture of capacitors comparable to mica. Finished glass capacitors have temperature coefficients very close to 140 ppm per °C, whereas the design and construction of mica capacitors affect their temperature coefficient considerably. (Values as high as 100 ppm per °C might be found unless special precautions are taken; hence six different classes of mica capacitors are listed.)

The other type of capacitor in the mica class is one using a ceramic glaze or enamel as the dielectric. This type was developed by du Pont under Signal Corps contract in an effort to mass-produce capacitors. Essentially, alternate layers of glaze and silver are sprayed and the entire unit fused into a single block. Commercial quantities of both types may be available soon.

4 · Rubber and Plastics in Cables

By JOHN T. BLAKE

The last dozen years have seen important changes in the wire and cable industry. Synthetic rubbers and synthetic resins have been introduced as insulating and jacket components, a field previously dominated almost exclusively by natural rubber. Oil-impregnated paper and varnished cambric are used as insulation to some extent, but only in comparatively small quantities.

For many decades natural rubber compounds were the principal solid flexible insulations on wires and cables. The only basic improvement in this material was the invention of deproteinized rubber for low water absorption.[1,2,3] Gradual improvements in rubber insulation were made through new filler combinations, vulcanization techniques, and the development of powerful antioxidants. Rubber insulation resulted, better adapted to specific applications and of longer service life. By 1940 natural rubber insulation had reached maturity and further improvements could be expected only on a minor scale.

In contrast, the field of synthetic materials is still in its infancy. Synthetic rubbers and resins can be developed more nearly to specifications, and the future gives promise of improvements as striking as those of recent years.

Service Requirements

In considering these new materials and comparing them with natural rubber, it is important to keep in mind the features which are of real importance in

[1] C. R. Boggs and J. T. Blake, *Ind. Eng. Chem.* 18, 224 (1926).

[2] C. R. Boggs and J. T. Blake, *Ind. Eng. Chem.* 28, 1198 (1936).

[3] U. S. Patent 1,997,355, April 7, 1925; Canadian Patent 277,204.

terms of satisfactory service. The purchaser of a wire or a cable desires a product which can be installed without damage and which will withstand the mechanical hazards of its use. He requires suitable electrical properties. Of most importance, these various qualities should be stable over a long period of time under the particular conditions of service. This means that the product should have good resistance to oxidation, heat, light, ozone, water, oil, chemicals, microorganisms, and any other hazards to which the material may be exposed.

It is well known that the war forced the rapid development of synthetic rubber. Neoprene had been used to a minor extent since about 1934, but was a high-priced specialty. Synthetic rubber insulations dominate since 1943. During the change-over period both the manufacturer and the user were faced with new standards in the manufacturing techniques as well as in the properties of the final product. Since these standards were somewhat different from the previous ones, some apprehension was felt that inferior products were being introduced. It has turned out that these synthetic insulations have not only been satisfactory, but are actually more desirable in many respects than those made from natural rubber.

example, at 200 percent elongation) is a rough measure of toughness. Tear, abrasion, and compression resistance are also useful in evaluating this property. One measure of toughness is the extent to which an insulation resists deformation at various temperatures. Obviously we would like the properties of an insulation to be independent of temperature, but this ideal is never reached.

The electrical properties of an insulation must be suitable for the particular use. This usually means that the dielectric constant and the power factor should be low, and the insulation resistance and dielectric strength should be high. Table 4.1 shows typical values for various insulations.

Table 4.1. Electrical properties of various cable insulations

	Natural Rubber	GR-S	Butyl Rubber	Neoprene	PVC	Polyethylene
Dielectric constant	2.6–5	2.7–5	3–5	high	5–10	2.3
Power factor	0.8–4%	0.9–4%	1–4%	high	7–13%	0.04%
Dielectric strength						
volts/mil—a.c.	600	600	400	low	800	1,200
volts/mil—d.c.	1,500	1,500	1,000	low	2,000	3,000
Dielectric resistance						
Mok *	5,000 to 40,000	5,000 to 40,000	10,000 to 80,000	1	500 to 4,000	100,000+

* Mok: $R/\log(D/d)$, where R is resistance in megohms per 1000 ft, D the outer, and d the inner diameter of the cable insulation.

Physical and Electrical Properties

The physical properties required for wire and cable insulations depend on the particular service conditions. Tensile strength and elongation are commonly used as criteria of quality, although, as a measure of practical serviceability, tensile strength is in itself of little significance. Elongation of 100 percent would be adequate for practically all uses, although most specifications demand much higher values. The physical property that we demand in an electrical insulation is toughness, so that it may withstand installation and service hazards. It is extremely difficult to define such a property. The modulus of elasticity (for

Compounding Ingredients for Rubber Insulation

Rubber is never used by itself as an insulation. It is compounded with various fillers, which in the inorganic field may be silica, silicates, whiting, clay, talc, carbon blacks, etc. In addition, a rubber compound may contain reclaimed rubbers, asphaltic, and other hydrocarbon fillers, vulcanized vegetable oils, softeners, processing aids, protective waxes, organic and inorganic accelerators, accelerator activators (like zinc oxide and stearic acid), and vulcanizing agents such as sulfur, selenium, tellurium, and complex organic compounds. Table 4.2 shows the formulation of a typical mineral-base GR-S compound.

Table 4.2. Example of GR-S mineral-base compound

GR-S	35 parts
Clay	31
Whiting	10
Zinc oxide	16
Wax	1.25
Cumarone resin	3.5
Stearic acid	0.25
Antioxidant	0.62
Accelerators	0.87
Sulfur	0.6

Each of these materials plays a specific part in the development of the properties of the final compound and of its behavior in the many steps of manufacture. Obviously, the number of actual compounds that can be made from even one individual rubber in such a multicomponent system is extremely large.

Hazards to Cable Insulation

Efficient operation of a power cable demands high current loading, resulting in the evolution of heat so that the insulation may be exposed to high temperatures for long periods of time. The tendency is to require higher and higher permissible copper temperatures. This means that insulations must have excellent resistance to aging at high temperatures. Many insulations and jackets must withstand long exposure to sunlight without deterioration. Many cables are exposed to various chemicals and must resist such influences. High-voltage cables generate ozone at the surface of the insulation, hence rubber and plastic compounds for high-voltage service should not be cracked by ozone. Insulation near conductors carrying high voltage may be exposed to the destructive effects of ozone, although itself not under high-voltage stress.

Cables are laid in rivers, harbors, and the ocean for various submarine services, or buried in the ground and exposed to water at some portion of their length. Years ago, it was considered essential to protect such an insulation with a lead sheath to prevent the deterioration of electrical properties through the absorption of water. Today lead sheaths have been largely eliminated, and the insulation is exposed directly to water or buried in the ground. For these applications it must have low water absorption.

The smaller the amount of absorbed water, the less will be its harmful effect on electrical properties. However, in different types of rubber and plastic compounds, the deterioration caused by the absorption of a given amount of water may vary appreciably. This deterioration can be determined in many ways, but most reproducibly by the increase of capacitance with time of exposure. Some of the more advanced specifications now set maximum values, both for actual amount of water absorbed and for the change in capacitance during a water exposure test under accelerated conditions. For example, U.S. Coast Guard Specification 15-C-1 gives the maximum allowable capacitance increases during 70°C soak as 3.35, 7.5, and 12 percent for 1, 7, and 28 days. Table 4.3 shows representative values for various insulations.

Table 4.3. Water absorption (minimum) and stability in water of unoxidized insulating compounds

	Deproteinized Natural Rubber	GR-S	Butyl Rubber	Neoprene	PVC	Polyethylene
Mechanical water absorption, mg/sq in., 7 days, 70°C	7	8	5	25-100	5	1
Capacitance increase at 70°C:						
1-8 days	2.5%	3.5%	3.5%	high	1%	1%
1-9 days	3.5%	5.0%	5.0%	high	1%	1%

Rubber and plastic jackets are used extensively on cables to prevent mechanical damage either during installation or in service. Jackets, of course, should be tough and withstand exposure to sunlight, oils, chemicals, etc., over long periods of time.

Low-temperature flexibility is of importance in many services. Rubber and plastics tend to stiffen and to become brittle at low temperatures. For many services this brittle point should be as low as possible.

Table 4.4 compares the temperature behavior of various insulations. Cracking temperature is determined by slow bending; shattering tests are made by

Table 4.4. Cracking temperature, shattering temperature, and temperature range

	Natural Rubber	GR-S	Butyl Rubber	Neoprene Jacket	PVC	Polyethylene
Cracking temperature, °C	−65	−50	−50	−40	−15 to −55	−70
Shattering temperature, °C	−60	−45	−45	−25	+5 to −35	−40
Temperature range, °C	220	200	220+	200+	115 to 160	175+

impact. Temperature range signifies the interval between brittle and soft state and should be as wide as possible. Figure 4.1 shows the softening with temperature of rubber, polyvinyl chloride, and polyethylene.

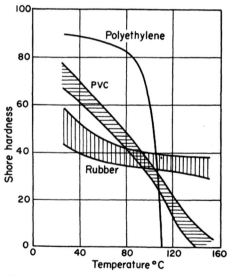

Fig. 4.1. Hardness of some polymers as function of temperature.

For insulations and jackets, the need for superior resistance to oxygen and heat aging cannot be emphasized too strongly. A tire usually wears out from abrasion long before it has been harmed excessively through heat and oxidation. The wire industry needs insulation of long life.

Microbiological Deterioration

Rubber and similar insulations may be destroyed by beetles, termites, and pocket gophers. Fibrous coverings are attacked by fungi, which also cause

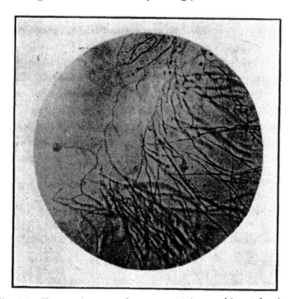

Fig. 4.2. Fungus in agar after removal from rubber tube (vulnerable GR-S insulation buried in soil for 40 days).

excessive surface leakage on hook-up wire, etc., in the tropics. A much less obvious type of attack leading to loss of insulation resistance is made by soil microorganisms. Natural rubber insulation which has failed in soil may show pitting because the base polymer itself can be consumed by certain bacteria, actinomycetes, and fungi.[4,5] Holes are eaten through the wall of the insulation. This does not occur in GR-S, butyl rubber, or neoprene, presumably because the microbes cannot consume them. This more subtle type of failure, which may cause the insulation resistance to drop by six or more decades, is caused by fungus filaments growing through the insulation via microimperfections. Bacteria appear to be unable to penetrate the wall of the synthetics until fungi have paved the way. In our studies, the micropores created by fungi were

[4] J. T. Blake and D. W. Kitchin, *Ind. Eng. Chem.* **41**, 1633 (1949).

[5] J. T. Blake, D. W. Kitchin, and O. S. Pratt, *Trans. Am. Inst. Elec. Engrs.* **69**, Pt. 2, 748 (1950).

detected and marked by electrodeposition of copper.[5] Then tubes of insulation were filled with agar, sealed, sterilized, and buried in active soil.[6] After exposure, the agar was squeezed out and cultured. Figure 4.2 shows fungus filaments in the agar immediately after its removal from a rubber tube and Fig. 4.3 the cultured agar rods. Of the vast number of soil fungi only a few appear able to penetrate rubber insulation, predominantly *Spicaria violacea* Abbott. Species of *Fusarium, Stemphyliopsis, and Metarrhizium* have also been isolated from inside the tubes.

To protect rubber insulations it is necessary to make them fungi-toxic without sacrificing other necessary characteristics. Polyethylene, some PVC, and silicone insulations appear to be inherently resistant to such attack.

Various Cable Dielectrics

Natural rubber and GR-S

GR-S is a generic term covering copolymers of butadiene and styrene, where the latter is the minor constituent. Over 700 varieties have been made on a commercial scale in the government plants, and many thousands experimentally. GR-S was developed as a general-purpose rubber for use in tires, since transportation consumes by far the largest proportion of rubber. The ordinary types of GR-S are not particularly suitable for rubber insulation because the salt-acid method of coagulation (sulfuric acid and sodium chloride) converts the sodium soaps used in polymerization to sodium sulfate. Some sodium chloride and sulfate is retained in the GR-S and give the rubber poor electrical properties. They also produce hygroscopicity so that exposure to water for only a short period of time lowers these inferior values even further. Some of the original types of GR-S had water absorptions as high as 120 mg per sq in.

The electrical properties of GR-S have been improved by manufacturing techniques which reduce the water-soluble salt content to a minimum. This is accomplished by precipitating the polymer as a fine granular material and carefully washing the coagulum. The use of dewatering extruders in the finishing line squeezes out much of the serum and its dissolved salts, so that they are not left in the dried rubber. There is a close relationship between water absorption and inherent electrical quality in GR-S. Synthetic insulating compounds can now be made which are equal to or superior to the best obtained with natural rubber.

[6] J. T. Blake, D. W. Kitchin, and O. S. Pratt, *Trans. Am. Inst. Elec. Engrs.* **72**, Pt. 3, 321 (1953).

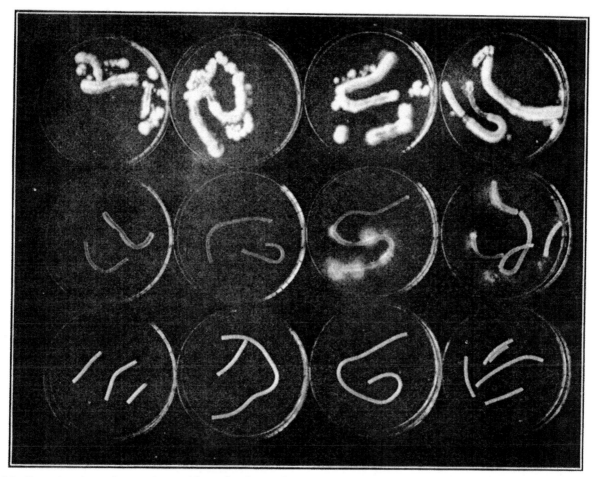

Fig. 4.3. Bacteria cultures in agar from rubber tubes buried for 6 months in soil (top row, natural rubber; middle row, vulnerable GR-S; bottom row, stable GR-S).

Considerable development has made it possible to produce, by polymerization at low temperatures, GR-S "cold rubbers,"[7] which have low water absorption and satisfactory electrical properties. Extensive work has been necessary because of the different polymerization catalyst systems, but excellent electrical grades of cold GR-S are now in common use. Cold rubber, in general, gives better physical properties, including toughness, than can be obtained with GR-S made by the hot polymerization process.

Tensile strength and elongation measurements provide an evaluation of the stability of rubber during aging in accelerated aging tests (Figs. 4.4 and 4.5). The degradation in aging is principally due to oxidation and thermal effects. Many engineers try to relate behavior in an accelerated aging test to a definite service life, but this is impossible since no two rubber products go through the same service history. On the other hand, the wire industry has had over 25 years of experience with aging tests and has learned to interpret them with some degree of confidence.

GR-S insulation, when protected by the appropriate amounts of microcrystalline waxes, exhibits much better resistance to atmospheric aging than natural rubber.

Wire and cable insulation must be free from mechanical impurities which would give spots of low dielectric strength. Natural rubber produced by natives on Far Eastern plantations always has a certain amount of dirt and contamination. GR-S, made in American factories under more carefully controlled conditions, is much more free of contamination. Natural rubber used in insulation, although made from the best grades, needs to be washed extensively and vacuum-dried before compounding; this is unnecessary with GR-S.

Oxidation of rubber during aging increases its water absorption. Natural rubber is susceptible to oxidation,

[7] "Symposium on Low Temperature Rubber," *Ind. Eng. Chem.* 41, 1553 (1949).

and deproteinized rubber which has been used in the past for low water absorption insulation, loses much of its virtue in time if exposed to oxidizing influences. The greater resistance of GR-S to oxidation means that after a moderate period of exposure to moisture and oxidation the best GR-S compounds will show less

bonds in rubber are the points of attack. Butyl rubber thus offers insulation of far greater resistance to heat, oxygen and ozone than GR-S or natural rubber.

The unsaturation of butyl rubber is about one-fiftieth of either natural rubber or GR-S. Since the vulcanization of rubber involves the double bonds, a

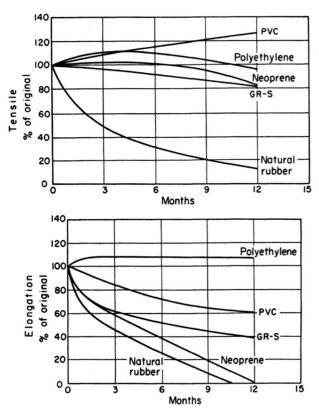

Fig. 4.4. Deterioration in tensile strength and elongation on accelerated aging at 70°C.

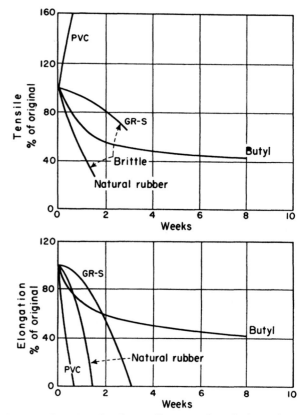

Fig. 4.5. Deterioration in tensile strength and elongation on accelerated aging at 121°C.

deterioration than those made from any variety of natural rubber.

GR-S is better in cable jackets than natural rubber, although neither type has flame and oil resistance. In resistance to heat, light, and oxidation it is superior to natural rubber.

Butyl rubber

Butyl rubber [8] is superior to natural rubber for automobile inner tubes because of its substantially better resistance to the diffusion of gases. This property is of no virtue in wire insulation, but butyl rubber has other inherent properties of value. The butyl rubber molecule has low chemical unsaturation, that is, few double bonds. This makes it very resistant to oxidation, heat and chemical attack, since the double

[8] R. J. Adams and E. J. Buckler, *Trans. Inst. Rubber Ind.* 29, 17 (1953).

more vigorous vulcanizing action is required for a technical cure. Butyl rubber can be vulcanized with sulfur and the conventional organic accelerators, or by quinone dioximes and their derivatives. These latter agents produce greater ozone resistance than vulcanization with sulfur. The relatively large molecular size of the quinone dioxime vulcanizing agents reduces the unsaturation further than sulfur vulcanization is able to do, possibly because the interbond distances are greater in butyl rubber.

Power cables are operated at higher and higher copper temperatures for the sake of increased transmission efficiency; butyl rubber is an ideal insulation for this sort of application. Since butyl rubber is made by a mass polymerization process, in contrast to the emulsion polymerization used for GR-S, the hygroscopic dispersing agents required in GR-S are not present. This means that butyl rubber has inherently low water absorption and is adapted for underwater and underground cables. Naturally, the compounding ingredients have to be selected so that the inherent virtues of butyl rubber can be utilized to their fullest extent.

It might be worth while to mention briefly the resistance to corona. Corona, an electrical discharge at the surface of an insulation, produces ozone. However, some rubber insulations which resist ozone chemically do not resist corona discharge. This implies that a corona discharge has a different effect from ozone and that resistance to both must be provided in a successful high-voltage insulation.

Silicone rubber

Silicone rubber, a fairly recent development, is a material with silicon atoms in the structure (see IIIA, Sec. 3) so that it may be considered to be partially organic and partially inorganic. Like organic rubbers, it may be vulcanized to give an elastic material, although its physical properties are relatively poor. It is susceptible to water absorption, and its electrical properties are only fair. Much progress has been made in the compounding of silicone rubber, its toughness has increased, although there is much room for improvement. Silicone rubber resists deterioration from heat, has an unusually low brittle point, and its usable temperature range exceeds that of any organic rubber. It is expensive, but applied increasingly where high or low temperature or both are encountered.

Polyethylene

Polyethylene, a long-chain saturated hydrocarbon, is flexible and has superb electrical properties. Its dielectric constant is only about 2.3; its power factor at all frequencies only 0.03 percent; both a-c and d-c dielectric strength are high; and the dielectric resistance excellent. As a saturated thermoplastic material it cannot be vulcanized.

Polyethylene is inflammable. Its temperature ceiling for applications lies in the neighborhood of 90°C (see Fig. 4.1). Beyond this point it softens rapidly, and at about 107°C its hardness is zero. Its brittle point lies below −70°C, and it does not shatter at low temperatures. Its hardness is high, it is tough in moderate thicknesses, and its resistance to water is excellent. When compounded with opaque fillers, such as carbon black, it has fairly good sunlight resistance.

Polyethylene has found wide uses in high-frequency applications such as radar cables and also for the insulation of communication and submarine cables. In its use for power cables two weaknesses must be guarded against. Although polyethylene is ozone-proof, it is very susceptible to corona, possibly because the corona raises a portion of the polyethylene above its melting point, and dielectric failure ensues. A polyethylene insulated cable, furthermore, will not withstand overheating by short-circuit currents of more than momentary duration.

Recently polyethylene has been modified by reacting chemically with sulfur dioxide and chlorine. When so treated, polyethylene can be vulcanized since points of chemical attack by vulcanizing agents have been introduced. The vulcanized material has fairly satisfactory physical properties. However, large quantities of both sulfur dioxide and chlorine must be used to obtain good physical properties, and this severely harms the electrical properties. Further developments in this field may produce a vulcanizable polyethylene which has both good electrical and physical properties; thus far this has not been accomplished.

Nylon

Nylon has been used to some extent as a thin-wall electrical insulation. It has relatively poor resistance to water and sunlight. Because of its toughness, nylon serves in thin layers as mechanical protection for tender insulations. For use on wire it usually is plasticized.

Polyvinyl chloride

Polyvinyl chloride has relatively little elasticity (see Fig. 4.1). Being softened by high temperatures, it can be extruded directly onto wire. This simplicity of fabrication makes polyvinyl chloride insulation pre-eminent in the building wire field today.

Plasticizers are needed to make polyvinyl chloride flexible at room temperature. Since most of these substances are volatile to some extent, an insulation of this type loses plasticizer, especially over long periods of heating, and becomes brittle. In addition, plasticizers may diffuse into adjacent materials and change their properties (see IIIB, Sec. 2). Polymeric esters have been developed which are nonvolatile and nondiffusible, but the physical and electrical properties they impart to the finished product are somewhat inferior.

As manufactured commercially, plasticized polyvinyl chloride insulations have good flame and oil resistance, although there is some tendency for certain organic solvents to extract plasticizer. The insulation may have brittle points as low as about −50°C and still be fairly hard at 80°C. Being thermoplastic, it is sensitive to sudden blows. The shattering temperatures are much higher than the bend-brittle points, which means that this type of insulation must be handled rather carefully in cold weather.

Plasticized polyvinyl chloride insulation has power factors varying from 6 to 15 percent, and dielectric constants from 4 to 8. The dielectric strength is excellent, and the dielectric resistance satisfactory. The rather high power factor has restricted these materials to low-voltage applications. Plasticized polyvinyl chloride is used to some extent for cord and cable jackets which must resist flame, sunlight, water, oil, and solvents.

Some newer types of polyvinyl chloride have been extruded recently with little or no plasticizer present. These insulations are moderately flexible in thin walls. Their electrical properties are satisfactory and their resistance to elevated temperatures is much better than that of the plasticized varieties. This is a promising development, and more of this type of material will be used in the future.

Teflon

Polytetrafluoroethylene is structurally similar to polyethylene; all the hydrogen atoms have been replaced by fluorine. If we were to enumerate the electrical and physical properties of the perfect cable dielectric, Teflon would come nearest to being the answer. Its electrical properties are as good as those of polyethylene; in addition, it is flameproof. Its physical properties are excellent, and it probably could be operated satisfactorily at a temperature of 250°C.

As a dielectric it is nearly ideal, but there are two disadvantages. Its cost at the present time is high. Of more importance, its high melting point makes it extremely difficult to process, although some progress is being made in this direction. In order to extrude it on wire, temperatures of about 450°C must be used, and this is close to its decomposition point. Extrusion, even at such high temperature, proceeds only at the rate of a few feet per hour.

Teflon can be made into tapes, and taped insulations with silicone oils between the tapes have been used in some wire applications. This construction has, of course, the disadvantage of not having a solid structure. Progress is being made, however, in the processing of Teflon, and the future will probably see a wider application as wire insulation.

Kel-F

The replacement of one-fourth of the fluorine atoms in Teflon with chlorine results in a much lower melting material. Kel-F can be put on wire in conventional high-temperature machinery such as nylon extruders. It does not have the extreme high-temperature resistance of Teflon or as good electrical properties, but finds some use commercially as compromise material.

Neoprene

Neoprene has rather poor electrical properties but has been used as insulation to some extent for certain low-voltage applications.

Portable cords and cables must be protected mechanically by rubber jackets for heavy-duty service where mechanical abuse and abrasion are present. Before the war these jackets were made of natural rubber and were formulated in the same manner as tire treads in order to get maximum toughness. Such jackets were somewhat sensitive to long exposures to sunlight and weather, and they were inflammable.

The reduction in price of neoprene through quantity production during the war made it available for cable and cord jackets. Neoprene has definite advantages over natural rubber. It is flame-resistant, and this property has made it essential for cables in coal mines. It can be compounded for higher tear and abrasion resistance than natural rubber and resists the action of many chemicals, including oils and solvents. It has excellent resistance to sunlight and weather. In consequence, the wire and cable industry is the largest consumer of neoprene and applies it universally in cable and cord jackets today.

Summary

In the last decade, flexible insulations and jackets on electrical wires and cables have been converted largely from natural rubber to synthetic rubbers and plastics. This conversion has been due to the inherent technical virtues of these new materials in spite of a cost premium at times. Much progress is being made in the development of new and improved materials, and the swing to synthetics will continue.

5 · Problems of the Cable Engineer

By SAMUEL J. ROSCH

As far as the author can recall, this is the first occasion when, in a scientific book, attention is focused on associated factors that must be taken into consideration if a given dielectric is to be used most efficiently. Frequently, the problems of the cable engineer are not so much a result of the electrical shortcomings of a particular dielectric, but rather of the deteriorating influences surrounding a given installation.

Cable Engineering

Cable engineering concerns itself with the theory, design, manufacture, testing, and installation of electrical conductors, both bare and insulated, intended for operation on land, sea, or air. On land, we include installations in underground ducts, direct burial in the soil, in underground or open-pit mines, in industrial and chemical plants, and what is generally considered as low-voltage house wiring. Sea operation includes the vast variety of cables installed under lakes, rivers, and the oceans. Operation in the air includes not only cables installed between poles, towers, or by self-supporting means, but also those insulated conductors used for the wiring of the wide variety of aircraft, ranging from the commercial airliner to the latest high-speed jet-propelled planes.

In addition to these general classes, there is the vast variety of miscellaneous types of cables. They include several hundred used mainly for the wiring of combat vessels by the navies of the world, several hundred types used in the wiring of merchant vessels and ocean liners, and the very large number of types used in the wiring of machine tools, dredgers, electric shovels, switchboards, general electrical appliances, portable and semistationary tools, oil-well drilling and explorations, blasting-cap and shot-fire wiring, apparatus cable, elevator-control cable, railroad signaling and communication, thermostat cable, high-frequency furnace leads, radio and television wiring, coaxial cables, and farm wiring.

Insulated Conductors

The function of an electrical conductor is obviously to transmit or distribute electrical energy. Since this book is concerned with dielectrics, only problems relating to insulated conductors, both single as well as multiple wires and cables, will be considered.

The conductor proper has to carry the current. For higher voltages, it must also be so constructed as to reduce the maximum electrical stress in the insulation next to the conductor. The insulation has to preserve the voltage; it also affects the permissible current-carrying capacity of the conductor. The outer protective covering has to protect the insulation mechanically and guards in addition against deteriorating effects of the environment. All three components have to be discussed because each has a definite bearing on the successful operation of insulated conductors.

The conductor proper

Two basic conductor materials are in use today, copper and aluminum. Although copper has dominated the insulated-conductor field, the last few years have witnessed some inroads by aluminum, chiefly in the lower-voltage classifications.

Conductors are generally furnished solid or stranded in the smaller sizes, and only stranded in the larger sizes. Seven different types of stranding are in use, depending on the degree of flexibility required. Some of these stranded conductors are compressed into sector shapes in order to decrease the over-all diameter of the cable and, with it, the weight of the insulation and of the other components of the completed cable. Figure 5.1 illustrates some of these variations in construction. Whenever a cable is compacted, the individual strands, originally circular, assume shapes far from circular.

Sometimes the individual wires are deliberately drawn in the shape of a predetermined keystone or trapezoid, and then stranded so that the combination forms a circular hollow conductor (Fig. 5.2). This construction, extensively used abroad, minimizes contour irregularities and thereby reduces the maximum stress at the surface of the conductor.

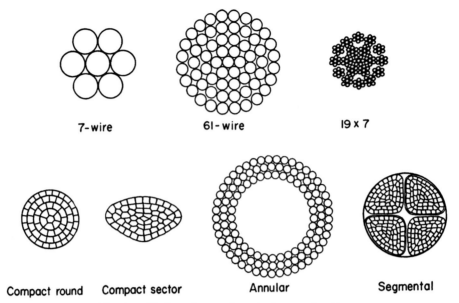

Fig. 5.1. Various types of conductor construction.

Skin effect, proximity, and spirality effects have a tendency to increase the a-c resistance of the larger conductors. This is already a problem at 50 and 60 cycles, but becomes more aggravated with higher power frequencies, as in the 400- and 800-cycle systems recently introduced for certain services. In such systems, the a-c resistance may be lowered through

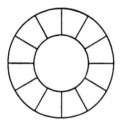

Fig. 5.2. Circular hollow conductor formed from keystone sections.

special types of construction (annular, rope core, or segmental) or by the use of stranded conductors, where each wire has been individually covered with a thin coating of some insulating substance that will survive the stranding operation.

The problem is to choose the most economical conductor construction in achieving the desired technical result.

Sealed conductors

Particularly in the last decade, an urgent need has arisen for cables, in which the voids are sealed throughout in such a manner that air or water applied over a 24-hour period to one end of a 5-ft length at pressures from 25 to 500 lb will not be transmitted through the other end in excess of certain stipulated amounts. This objective must be accomplished on a commercial scale and with materials not adversely affected by the maximum and minimum operating temperatures of the completed cable or by catalytic effects caused by contact with the metal conductor. In addition, the sealing compound must not impair the flexibility of the cable nor the electrical properties of the insulation.

In the past, many millions of feet of such cables have been produced by using one type of sealing material within the strand interstices and a different material in the other parts of the cable cross section. With improved cable designs, higher operating temperatures, and water pressures, the need has arisen for a material that can be used in both places. At the same time, the material should possess a fairly high degree of flame-retardance and be economical in original cost as well as in cost of application. For cables insulated with the ultra-high frequency dielectrics in use today this poses a real problem.

Aluminum conductors

The more extensive use of aluminum conductors in power cables in recent years has created problems not encountered with similar sizes of copper conductors.

Current-carrying capacity

Theoretically, the current-carrying capacity of an insulated aluminum conductor (conductivity 61 percent of that of copper) should be 78 percent of a corresponding copper conductor. This figure is obtained by taking the square root of the ratio of the relative conductivities of the two metals. The National Electrical Code, the universally accepted guide for the wiring of residential and office buildings as well as industrial plants throughout most of the American Continent (it is also the standard of the American Standards Association), has for many years advocated a figure of 84 percent. The uncertainty is further increased in calculating the current-carrying capacity under a given set of operating conditions. The a-c resistance has to be taken into account, a calculation requiring a knowledge of skin effect, proximity effect, and spirality effect. These effects have been fairly well established for copper conductors; for aluminum, where an oxide film forms on the surface of each strand as soon as it is drawn, the a-c, d-c resistance ratio may be less for material of the same size.

The initial trend for using aluminum in the larger conductor sizes was prompted by a temporary copper shortage; it provided insufficient opportunity for determining such basic data by actual measurements. There is obviously a substantial cost differential, not only as far as the insulated cable is concerned, but also in the installation, when we use a current-rating ratio of 78 percent as compared to 84 percent. Engineering committees have been appointed for the purpose of establishing the basic factors, but, until this has been accomplished, this uncertainty in rating will continue to constitute one of the vexing problems for the cable engineer.

Jointing

There is little value in producing a satisfactory aluminum conductor, only to have the consumer encounter difficulties in joining such conductors to each other, to a copper conductor or bus, or within a panel originally built to accommodate the corresponding sizes of smaller copper conductors. Many techniques have been proposed such as soldering, brazing, bolting, or the use of compression fittings. These may be satisfactory initially; but after a number of operating cycles, particularly under temporary overloads involving higher temperatures, the alternating expansion and contraction may cause higher contact resistance, deterioration of the insulation within the region of the joint, and finally a dielectric failure of the cable. These problems require considerable study, and the development of a satisfactory joint for larger conductor sizes does not solve the difficulty automatically for smaller ones.

The electrical insulation

The subject of insulations for conductors is truly a technical paradox. Whereas the technical literature is replete with unrelated articles on a vast variety of insulating materials, only a few books on the subject of insulated cables exist, and nearly all of these are outdated and in no way even begin to reflect current practice. To understand the reason for this apparent neglect, it is necessary to examine some of the problems confronting the cable engineer.

Most of the materials used as insulation for electrical conductors, are basically nonhomogeneous in nature and consist of a blend of materials that have to yield definite and predetermined characteristics. These properties may concern higher operating temperatures, resistance to ozone, moisture or the deteriorating effects occurring with d-c operation in moist locations, lower capacity, lower power factor, higher insulation resistance, higher dielectric strength, ability to be flexed at temperatures as low as $-54°C$, lower cost, ability to be processed at higher speeds without impairment, and so on.

There are other headaches. Occasionally, some essential ingredient of insulation for electrical conductors is taken off the market because the producer finds more lucrative outlets for the use of his plant equipment. This necessitates a lengthy search for a substitute. When found, and certified by laboratory tests, the new material may still yield properties in the final insulation considerably different from the original.

Because of rapid price changes and the possibility that a given ingredient may soon become outmoded, ours is the day of limited inventories. If supplies threaten to be interrupted by strikes, embargoes on common carriers, or work stoppages in supplier plants, it becomes necessary immediately to plan for substitute ingredients so that factory production may be delayed only for a minimum period. In order to anticipate these possibilities and at the same time continue the unending search for new or improved ingredients, it is not unusual to find at the end of the year that the laboratories of a modern cable plant have investigated and often experimented with several thousand new materials. The justification for investigations of many new materials as well as the decision as to their actual use in a given case are some of the difficult responsibilities of the cable engineer.

In Table 5.1, an attempt has been made to assemble a list of the major insulating materials in use throughout the world together with data on maximum voltage ratings, range of operating temperatures in con-

Table 5.1. Survey of present status of cable insulations

Types of Insulation	Maximum a-c Operating Voltage, in kv	Range of Operating Temperatures, in °C		Choice as Compared to Five Years Ago
Rubber compounds—natural	3.0	−40	75 *	Decreasing
Synthetic—buna S and copolymers	3.0	−54	75 *	About the same
Synthetic—latex	0.6	−40	75 †	About the same
Synthetic—butyl	28.0	−40	80	Increasing steadily
Synthetic—neoprene	0.6	−30	90	About the same
Synthetic—silicone	5.0	−40	150	Increase due to military needs
Ozone-resisting—natural + factice	28.0	−15	70 to 75 ‡	About the same
Ozone-resisting—buna S + factice	28.0	−15	70 to 75 ‡	Increasing
Ozone-resisting—butyl	28.0	−40	80	Increasing steadily
Thermoplastics				
Polyvinyl chloride	0.6	−30	105	Rapidly increasing
Blends of polyvinyl chloride and acrylonitrile rubber	0.6	−54	125	Newly developed; trend unpredictable
Polyethylene	15.0 *	−60	80	Increasing steadily
Gutta percha and paragutta	0.6	−10	60	About the same
Teflon	5.0	−54	200	Slight increase due to military needs
Fluorothenes	5.0	−54	150	Slight increase due to military needs
Impregnated paper				
Solid types	95.0	−10	60 to 85 ‡	About the same
Oil-filled	400.0 §	−20	60 to 70 ‡	Slight increase
Gas-filled	275.0	−20	60 to 70 ‡	Slight increase
Internal gas pressure	220.0	−20	60 to 70 ‡	Slight increase
External gas pressure—Pipe types	220.0	−20	60 to 70 ‡	About the same
External oil pressure—Pipe types	220.0	−20	60 to 70 ‡	Steadily increasing
Varnished cloth	28.0	−10	65 to 80 ‡	About the same
Mineral insulation (compressed magnesium oxide)	0.6	0	85 to 150 ‖	Slight increase

* When used in moist locations, the temperature rating is reduced to 60°C for products within the scope of Underwriters Laboratories' jurisdiction.
† Also limited to a maximum conductor size of #6 AWG.
‡ Maximum operating temperature is a function of the working voltage, the higher voltages necessitating the lower operating temperatures.
§ There are two manufacturers who produced cable having a nominal voltage rating of 400 kv a.c. One of these also rates his design as being suitable for 1500 kv d-c service.
‖ Maximum operating temperature depends on the type of termination.

tinuous service and preference as compared with five years ago. A partial review of the data indicates:

(1) At least in some parts of the world, cables are supplied capable of interconnection with the highest voltage overhead transmission lines in use today.

(2) Insulated cables are available for possible future operation up to 1500 kv direct current. Before such cables can be placed in operation, many cable-engineering problems will have to be solved.

(3) A number of insulations of approximately equal voltage ratings compete for the greatest-footage quantities. The choice depends on a number of factors including local interests and ordinances, price fluctuations of ingredients, and types of processing equipment available in a given plant. Frequent price fluctuations, combined with periodic changes in local ordinances, constitute one of the real problems to the cable engineer in advising management as to the types of finished products to be shipped into a given geographic location.

(4) There is a growing trend towards synthetic rubber for voltages up to 15 kv. The use of rubber introduces some vexing problems, the basic cause being shielding considerations.

As is well known, a conductor carrying alternating current will also carry leakage and charging currents. The latter flow not only longitudinally along the conductor but also radially through the insulation towards the outer surface. The magnitude of these radial currents depends upon the nature of the insulating material, the size of the conductor, the voltage to ground, and the nature of the protective outer covering, whether metallic or nonmetallic. If the outer covering is metallic, the radial flow of leakage and charging currents is fairly uniform along the length of the cable, and these currents may be drained to earth by grounding the metallic covering at numerous points. If the outer covering is nonmetallic, the flow of these radial currents will not be uniformly distributed along the length of the cable, but will increase in intensity at portions of the outer surface that provide low-resistance contact to ground.

A typical example is the nonmetallic-sheathed, single conductor cable in underground ducts which are wet in some places and dry in others. Where the nonmetallic sheath is wet, the contact resistance to ground will probably be fairly low and thereby afford effective drainage of the surface charges existing in that portion of the cable. However, the charge built up on the surface of a dry section between two wet points, or between two points of low contact resistance to ground, may assume sufficient magnitude to discharge to ground. If the longitudinal resistance of the surface is fairly high and the condition maintained for a longer period, successive discharges slowly produce carbonized paths and ultimately lead to surface "tracking" which provides the requisite paths to ground. Often the contact resistance of a given radial path to ground dissipates enough energy to cause overheating at that spot. The final release of the surface charge to ground causes "burning" of the nonmetallic sheath and, finally, failure of the dielectric.

The obvious solution is to use a metallic layer underneath the nonmetallic sheath or, if the operating phase-to-ground voltage is not too high, a nonmetallic sheath of high surface resistivity. For cables carrying a phase-to-phase voltage in excess of 5 kv, industry readily assents to metallic shielding. However, in the class of cables operating from 3 to 5 kv, a range which constitutes one of the largest production items, there is a general reluctance to accept the need for such metallic shielding. The claims advanced are that metallic shielding does more harm than good, that the general electrician does not know what to do with the shielding tape and either permits it to "float" or often grounds it only at one end, that the ungrounded end constitutes a hazard due to possible shocks when accidental contact is established, and that an unshielded cable is simpler and less costly to terminate.

As a result, nonmetallic-sheathed cables in the 3 to 5 kv class, are usually furnished with a sheath of high surface resistivity. This is satisfactory where the high surface resistivity is maintained. Often, however, industrial dusts and other contaminating substances attach themselves to the surface and lower the surface resistivity to a point where discharges to ground become prevalent and ultimately lead to failure. The problem of finding a nonmetallic sheathing compound that will be flexible under the low seasonal temperatures encountered during installation in certain geographic locations, will possess high resistance to sunlight, acids, alkali, moisture, oil, and flame propagation, and will retain its high surface resistivity, is one of the challenges confronting the modern cable engineer.

Polyethylene. One of the most interesting cable dielectrics is polyethylene (polythene). Although occasionally blended with other materials, its major use today in communication, ultrahigh-frequency or power work, is in unblended form. This material apparently has everything desired in a dielectric including excellent resistance to a wide variety of chemical media, particularly in an unstressed condition. In Table 5.1 we note that, for power work, cables have been produced for working voltages up to 15 kv. Although a fair number of such cables are in operation both here and abroad, they have in general been limited to power conductors of medium size, the largest (within the author's knowledge) being 4/0 AWG (American Wire Gauge).

It is only natural for the cable engineer to try to extend the use of this dielectric to higher operating voltages. One of the shortcomings of polyethylene is its high coefficient of expansion and contraction especially in the heavier insulation thicknesses. With heavy load cycles, the dielectric tends to shrink permanently away from the conductor, thus leaving ionizable voids at the point of maximum stress. Unless these void formations can be prevented, dielectric failure is bound to occur after a few years of operation. The problem confronting the cable engineer in this case is to develop a simple and economical procedure of accomplishing this result.

High-frequency vulcanization. When we apply any of the high-voltage, rubber-insulating compounds listed in Table 5.1, they are still in the unvulcanized form as they leave the insulating machine before the cable is placed in the vulcanization chamber. Between the insulating and vulcanization process the insulation is soft and possesses little resistance to flattening. Hence decentralization or surface distortion may occur unless prevented by the use of some restraining envelope applied simultaneously with or immediately after the insulating process. The restraining envelope may consist of one or more fabric tapes or an extruded lead sheath, both of which act as a type of "mold" in which vulcanization can take place. Conventional vulcanization of a batch of such cable often consumes two or more hours with the dielectric maintained at fairly elevated temperatures and pressures. During this lengthy period of heating, gases may evolve within the dielectric and, unless precautionary measures are taken, voids result.

These costly precautions could be eliminated if a satisfactory method could be evolved for high-frequency vulcanization of the insulating compound simultaneously with its application. The patent literature is replete with methods trying to accomplish this type of vulcanization; the chemist has supplied us with a wide variety of ultra-accelerators capable of accomplishing vulcanization in a minute or less; but the electronics engineer has failed thus far to find a suitable method for accomplishing the task.

Protective coverings

Not much would be accomplished in providing the best type of dielectric for cable insulation unless safeguards were incorporated against deterioration from environmental influences.

Installations on land by direct burial in soil, in underground ducts, underground or open-pit mines or within chemical and some industrial plants are subjected to distinctive deteriorating influences. For cables buried in the soil, they may be summarized as decomposing organic matter producing soil acids; alkali in the soil; chloride from salt near cattle-feeding areas; fecal matter residue from animals; cinder fill; poor drainage; salt-water leakage; soils known as "muck" or "bottom" land; microorganisms; termites, gophers, and other parasites. For cables in underground ducts, the influences may be alkali or acids in manhole waters; industrial-waste waters filtering into ducts, or containing dissolved oils and gasoline; scoring of cable covering by pebbles or duct irregularities; stray currents, leading to galvanic corrosion; high ambient temperatures caused by adjacent steam mains. For cables in underground or open-pit mines such deteriorating influences are injury due to mechanical impact or abrasion during motion of cable; water; sunlight. For cables in chemical and some industrial plants they are chemical fumes, vapors, and accidental contact with a wide variety of chemicals; unanticipated high ambient loads as a result of process modifications; and attack by rodents. For installations under water (harbors, lakes, rivers, and oceans) the chief hazards include accidental fouling by anchors; sewage and industrial wastes near the shore or docking facilities; water filtering into covering; scoring or abrading caused by excessive tides. Installations in the air are subject to industrial fumes and vapors; alternate exposures to sunlight, rain, snow, and ice; excessive vibration at sub-zero temperatures.

A wide variety of protective coverings are available to the cable engineer:

(1) Sheaths of commercially pure and alloy lead, aluminum, neoprene, other natural and synthetic-rubber compounds, polyethylene, polyvinyl chloride, and thin walls of nylon.

(2) Braids and servings of cotton, jute, hemp, glass, sisal, hawser cord, and seine twine, treated with bituminous or other preservative compounds as protection against different types of deteriorating influences.

(3) Fibrous tapes such as duck, cotton, burlap, glass, and paper, treated similarly with preservatives.

(4) Protective armorings consisting of one or more layers of galvanized iron or nonferrous wires in flat, shaped, or cylindrical form; flat steel tapes, either mild or galvanized; interlocking tape of galvanized or mild steel, or nonferrous material; a braid of galvanized steel wires or nonferrous wires. Such armorings are used in conjunction with some of the other materials previously mentioned, under and/or over the protective armoring material.

(5) A variety of paints and lacquers applied by spray or dip processes, with or without subsequent heat treatment.

The importance of protective coverings in preserving the dielectric and enabling it to render continuous and satisfactory performance cannot be overemphasized. The problem for the cable engineer is which covering or combination of coverings to use under a given set of operating conditions.

A lead sheath and, more recently, an aluminum sheath are the only true impermeable materials against the intrusion of air, moisture, and various liquids. These metals, used by themselves, are subject to various types of corrosion, and both require, under certain installation conditions, additional protective coverings. The coverings used today for this purpose are effective but rather bulky, a factor of great importance in these days of high freight rates. The problem is the development of a simple, light, fairly inexpensive material that can be commercially applied with existing cable equipment. Such material must be capable of protecting the underlying metal, withstanding the abrasion incidental to installation, and the extremes of temperature encountered during installation and operation of the completed cable.

Polyethylene, with its excellent resistance to a wide variety of liquid media, suffers sheath cracking when under physical stress and exposed to metallic soaps and certain oils at elevated temperatures and atmospheric pressure. What is needed is an improved type of polyethylene.

Neoprene is one of the best of the coverings, but whether applied in extruded or laminated form, it has to be vulcanized in the conventional manner. This means that the underlying dielectric is subjected to elevated temperatures during the required time for adequate vulcanization of the neoprene. This may cause some deterioration of its physical properties. The problem could be overcome through the development of a satisfactory method of high-frequency vulcanization of the neoprene jacket.

Calculations

The cable engineer has to calculate many electrical properties of cables in advance of their actual manufacture. Such computations have been made for standard designs and are available in the standards published by the Insulated Power Cable Engineering Association or in the handbooks on cables published by cable manufacturers. However, when the designs are not standard or consideration must be given to special environmental conditions, the properties must be calculated for each special case.

For the purpose of the present discussion, we shall focus attention on only one property, namely, current-carrying capacity. This subject is of considerable importance to the user because the assigned current ratings determine the maximum amount of power that may be transmitted over a given cable continuously or intermittently. If the current ratings are incorrect for the particular conditions, the cable insulation may reach temperatures beyond the safety limit, and premature deterioration results. The formulas required for these calculations are generally not available in any of the textbooks on electrical engineering, nor are they the same for all installation conditions. Typical examples are therefore presented for cables installed in conduit, in air, and in underground ducts.[1-5]

[1] "Current-Carrying Capacity of Impregnated Paper, Rubber and Varnished Cambric Insulated Cables," Insulated Power Cable Engineering Association, Publication No. P-29-226.

[2] "Calculations of Electrical Problems of Underground Cables," D. M. Simmons, *Elec. J.* **29**, 237, 283, 336, 395, 423, 476, 527 (1932).

[3] National Electrical Code for 1951.

[4] Insulated Power Cable Engineering Association, Specifications for Varnished Cambric Insulated Cables.

[5] "Reduction of Sheath Losses in Single Conductor Cables," H. Halperin and K. W. Miller, *Trans. Am. Inst. Elec. Engrs.* **48**, 399 (1929).

Problem I. Current ratings for cables in conduit exposed to air

Special situation. Current rating of three single-conductor 4/0 AWG, 15-kv rubber-insulated, neoprene-jacketed cables installed in conduit in air.

Cable choice (Anaconda General Catalog, Section 6, page 19): Single-conductor 4/0 AWG cables made with 37 strands, each conductor insulated with $19/64$-in. wall of ozone-resisting rubber insulation, shielded, and finished with a $6/64$-in. wall of neoprene jacket or sheath for protective purposes. Cable suitable for operation at 15 kv grounded neutral. Approximate overall diameter = 1.495 in.

Conduit size (from Ref. 3, Tables 11 and 19):

Area of three such cables = $3 \times 1.495^2 \times 0.7854 = 5.266$ sq in.

Conduit fill permitted for three cables = 43%

Conduit area required for three cables = $\dfrac{5.266}{0.43} = 12.25$ sq in.

Nearest size conduit corresponding to 12.25 sq in. is 4-in. diameter; area = 12.72 sq in.

Temperatures:

Copper = $T_c = 80°C$
Ambient air = $T_a = 40°C$

A-c Resistance of conductor:

D-C resistance at 25°C, in ohms per 1000 ft of cable = 0.0509
D-C resistance at 80°C, in ohms per 1000 ft of cable = 0.0617
= $0.0509[1 + 0.00385(80 - 25)]$
A-c/D-c ratio for this 4/0 AWG conductor (Ref. 1, p. 43) = 1.050
∴ A-c resistance at 80°C, in ohms per 1000 ft = $0.0617 \times 1.050 = 0.0648$

Geometric factor G_2 for dielectric loss:

Insulation thickness of $19/64$ in. = 0.297 = T
Diameter of such 4/0 AWG conductor = 0.528 = d

Ratio of $\dfrac{T}{d} = \dfrac{0.297}{0.528} = 0.56$

From Ref. 2, for $\dfrac{T}{d} = 0.56$, $G_2 = 0.76$

Geometric factor G_1 for thermal resistance:

Insulation and jacket thickness = $19/64 + 6/64 = 0.391$ in.
= $T + t$
Diameter of 4/0 AWG conductor = 0.528 in. = d

Ratio of $\dfrac{T + t}{d} = \dfrac{0.391}{0.528} = 0.74$

From Ref. 2 for $\dfrac{T + t}{d} = 0.74$, $G_1 = 0.91$

Dielectric loss calculated from Refs. 1 and 2:

$$W_D = \dfrac{0.000106 \times f \times e^2 \times K \times \cos\phi}{G_2}$$

$$= \dfrac{0.000106 \times 60 \times 8.67^2 \times 5 \times 0.05}{0.76}$$

= 0.157 or 0.16 watts per foot

where $f = 60$; $e = 15/\sqrt{3} = 8.67$; $K = 5$; $\cos\phi = 0.05$

Thermal resistance from the sum of $R_{th} + Q$:

The value of Q for a 4-in. conduit (from Ref. 1, p. 48) = 5.5

$$R_{th} = R_1 + R_s$$

$$R_1 = \frac{0.00522 \rho G_1}{n} = \frac{0.00522 \times 500 \times 0.91}{1} = 2.38$$

$$R_s = \frac{0.00411 B}{D} = \frac{0.00411 \times 1120}{1.495} = 3.08$$

$$R_{th} = R_1 + R_s = 2.38 + 3.08 = 5.46$$

Thermal resistance = $R_{th} + Q = 5.46 + 5.5 = 10.96$

where $\rho = 500$; $G_1 = 0.91$; $B = 1120$ (Ref. 1); $D = 1.495$; n = number of conductors in cable.

Current rating or current-carrying capacity (according to Ref. 1, p. 51):

$$I = \sqrt{\frac{(T_c - T_a) - (T_d - T_a)}{nR_c(R_{th} + Q)}}$$

$$I = \sqrt{\frac{(80 - 40) - (1.75)}{1 \times 0.0000648 \times 10.96}}$$

$$I = 232 \text{ amp}$$

where $T_d - T_a = W_D \times$ thermal resistance
$= 0.16 \times 10.96 = 1.75$
$n = 1$
$T_c = 80°C$; $T_a = 40°C$
$R_{th} + Q = 10.96$
R_c = a-c resistance in ohms/foot

Problem II. Current ratings for cables suspended in air

Special situation. Current rating of a three-conductor 4/0 AWG 15-kv lead-covered paper cable, installed aerially, suspended by suitable messenger cable.

Cable choice (Anaconda General Catalog, Section 8, p. 73): Three-conductor 4/0 AWG compact sector, each conductor insulated with 0.175 in. of impregnated paper followed by a metallic shielding tape, the three shielded conductors being cabled or twisted together so as to form a cylindrical core over which is applied a lead sheath having a thickness of 0.110 in. Cable suitable for operation at 15 kv grounded neutral.

Approximate overall diameter = 1.973 in.

Temperatures:

Copper = $T_c = 81°C$
Ambient air = $T_a = 40°C$

A-c resistance of conductor:

D-c resistance at 25°C, in ohms per 1000 ft = 0.0509
D-c resistance at 81°C, in ohms per 1000 ft = 0.0509[1 + 0.000385(81 − 25)] = 0.0619
A-c/D-c ratio for this type of 4/0 AWG conductor (Ref. 1, p. 9) = 1.050
∴ A-c resistance at 81°C, in ohms per 1000 ft = 1.05 × 0.0619 = 0.0650

Geometric factor G_2 for dielectric loss:

Insulation thickness = 0.175 in.
Conductor diameter for 4/0 AWG round = 0.528 in.

Ratio of $\dfrac{T}{d} = \dfrac{0.175}{0.528} = 0.331$

From Ref. 2, $\dfrac{T}{d} = 0.331$, $G_2 = 0.51$

Geometric factor G_1 for thermal resistance:

Insulation thickness = 0.175 in.
From table in Ref. 2, $G_1 = 0.53$

Dielectric loss (calculated from Refs. 1 and 2):

$$W_D = \frac{0.000106 \times f \times e^2 \times n^2 \times K \times \cos\phi}{nG_2}$$

$$= \frac{0.000106 \times 60 \times 8.67^2 \times 3^2 \times 4 \times 0.023}{3 \times 0.51}$$

$$= 0.259 \text{ or } 0.26 \text{ watts per foot}$$

where $f = 60$; $e = 15/\sqrt{3} = 8.67$; $n = 3$; $K = 4$; $\cos\phi = 0.023$.

Thermal resistance (found by the following procedure and Ref. 1):

$$R_{th} = R_1 + R_s$$

$$R_1 = \frac{0.00522 \rho G_1}{3} = 0.65$$

$$R_s = \frac{0.00411 B}{D} = 2.50$$

$$\therefore R_{th} = 0.65 + 2.50 = 3.15$$

where $\rho = 700$; $G_1 = 0.53$; $B = 1200$; $D = 1.973$

Current rating from Ref. 1:

$$I = \sqrt{\frac{(T_c - T_a) - (T_d - T_a)}{nR_cR_{th}}}$$

$$= \sqrt{\frac{(81 - 40) - (3.15 \times 0.26)}{3 \times 0.0000619 \times 3.15}}$$

$$= 262 \text{ amp}$$

where $T_c = 81$
$T_a = 40$
$T_d - T_a = R_{th}W_D$
$= 3.15 \times 0.26$
R_c = a-c resistance per foot of conductor
$n = 3$
$W_D = 0.26$

Problem III. Current ratings for cables in underground ducts †

Special situation. Current rating of single-conductor 2,000,000 cm, 15-kv lead-covered paper cable, using one cable per duct with three loaded ducts in bank of ducts, the cables operating on a 75 percent load factor.

Cable choice (Anaconda General Catalog, Sec. 8, p. 44): Single-conductor having a cross section of 2,000,000 circular mils and produced in compact segmental construction. Conductor insu-

---

† The current rating for single-conductor cables with one cable per duct is computed on the basis of open-circuited sheath operation with no circulating currents flowing in the cable sheaths. Provision must be made to limit the induced sheath voltage to a maximum of approximately 12 volts to ground. Methods of calculating such sheath currents and voltages, as well as means for limiting sheath voltages, are discussed in Ref. 5.

lated with 0.175 in. of impregnated paper and finished with a 0.120 in. thickness of lead sheath. Cable is suitable for operation at 15-kv grounded neutral.

Approximate conductor diameter = 1.632 inch
Approximate overall diameter = 2.282 inch

Temperatures:

Copper = T_c = 81°C
Ambient earth = T_a = 20°C

A-c resistance of conductor:

D-c resistance at 25°C, in ohms per 1000 ft = 0.00542
D-c resistance at 81°C, in ohms per 1000 ft = 0.00659 = 0.00542[1 + 0.00385(81 − 25)]
A-c/d-c ratio for this type of 2,000,000 CM conductor = 1.052
∴ A-c resistance at 81°C, in ohms per 1000 ft = 1.052 × 0.00659 = 0.00693

Geometric factor G_2:

Insulation thickness = 0.175 in.
Conductor diameter = 1.632 in.

Ratio $\dfrac{T}{d} = \dfrac{0.175}{1.632} = 0.107$

From curve Ref. 2, $G_2 = 0.20$

Dielectric loss calculated from Refs. 1 and 2:

$$W_D = \frac{0.000106 \times f \times e^2 \times K \times \cos\phi}{G_2}$$

$$= \frac{0.000106 \times 60 \times 8.67^2 \times 4 \times 0.022}{0.20}$$

$$= 0.214 \quad \text{or} \quad W_D = 0.21 \text{ watts per foot}$$

where $f = 60$; $e = 15/\sqrt{3} = 8.67$; $K = 4$; $\cos\phi = 0.022$; $G_2 = 0.20$.

Thermal resistance:

$R_{th} = R_1 + R_s + D'$

$R_1 = \dfrac{0.00522\rho G_1}{n}$

$= \dfrac{0.00522 \times 700 \times 0.2}{1} = 0.73$

$R_s = \dfrac{0.00411 \times B}{D}$

$= \dfrac{0.00411 \times 1200}{2.282} = 2.16$

D' = 3 ducts in bank with 75% load factor = 1.90
∴ $R_{th} = 0.73 + 2.16 + 1.90 = 4.79$

where $\rho = 700$; $G_1 = 0.2$; $n = 1$; $B = 1200$; $D = 2.282$; $D' = 1.90$ (Ref. 1).

Thermal resistance for dielectric loss:

$R_{th'} = R_1 + R_s + D'$
R_1 (see above) = 0.73
R_s (see above) = 2.16
D' = 3 ducts in bank with 100% load factor = 3.10
∴ $R_{th'} = 0.73 + 2.16 + 3.10 = 5.99$

Current rating:

$$I = \sqrt{\frac{(T_c - T_a) - (R_{th'} \times W_D)}{n \times R_c \times R_{th}}}$$

$$= \sqrt{\frac{(81 - 20) - (5.99 \times 0.21)}{1 \times 0.00000693 \times 4.79}}$$

$$= 1341 \text{ amp}$$

where $T_c = 81$; $T_a = 20$; $R_{th'} = 5.99$; $R_{th} = 4.79$; $R_c = 0.00693 \times 10^{-3}$ ohms/ft; $n = 1$; $W_D = 0.21$.

Specifications, Testing, and Patents

Conservatively estimated, there are more than thirty specification-writing bodies in the United States and many more in other countries; each of them issues specifications for electrical conductors. Mention is made of foreign specifications because American manufacturers may have to bid to specifications for foreign cable requirements. Occasionally, specifications have been written by incorporating extracts containing the severest requirements from a number of existing specifications, each of the latter originally intended for an entirely different type of product. At other times characteristic properties are extracted from the catalogs of the basic supplier of a given insulating material, properties obtained in the laboratory on pressed or rolled slabs of material and practically impossible to meet in the form of extruded insulation. Yet these properties are incorporated as part of the requirements in finished cable specifications.

Frequently, specifications are quite lengthy due to "window-dressing," in the form of verbatim excerpts from A.I.E.E., I.E.E., V.D.E., or other domestic and foreign engineering association standards, originally developed for an entirely different purpose. These excerpts have no possible bearing on the quality of the product furnished, but their mandatory provisions serve to increase the cost of the product.

Trends in testing are another headache confronting the cable engineer. Especially in the last decade, physicists, chemists, and engineering specialists have attacked the cable and insulation problems, and developed a substantial series of tests, over and above those already in use. No sooner do these tests appear in some technical paper than they are included as criteria in some cable specifications. To protest such proposals is almost always futile so that new tests are pyramided on old ones until today, for some cable products, the cost of testing almost equals the cost of manufacture.

During recent years, large numbers of patents have

been issued dealing with cable constructions, materials, equipment, and manufacturing processes. Frequently, in order to overcome some characteristic deficiency in one of the cable components or the cable as a whole, some slight modification may be made in ingredients, processing, or equipment. The desired results are obtained; the modification is made part of the standard manufacturing procedure. A year or two later it appears that this modification constitutes infringement of some patent dealing with an entirely different problem.

A typical case concerns oil-enclosed electrical switchgear, transformers, and the like, consisting essentially of insulated conductors, the surrounding oil, and a metallic enclosure. An oil-impregnated cable also consists of insulated conductors, surrounding oil, and an enclosure of metal such as lead or aluminum. These are the basic elements in both constructions. The apparatus manufacturer hears of some new oxidation or catalytic inhibitor which would improve the stability of the oil within the enclosure and he ultimately secures a patent for such use. The wording of the patent, however, may contain a claim covering the use of a certain inhibitor for improving properties in an electrical construction consisting of these basic elements. At a later date, the cable manufacturer, totally unaware of what has transpired, incorporates the same inhibitor in his impregnating oil and thus unconsciously infringes on a patent unknown to him. How to keep abreast of the ever-growing list of patents is one of the most vexing problems confronting the cable engineer.

C · Dielectric Materials as Devices

1 · Rectifiers

By CLARENCE W. HEWLETT

Electric power on a commercial scale is simpler and less expensive to produce and distribute as a-c than as d-c. However, there are many applications where direct current is required, as for electrochemical processes, electronic devices, variable-speed motors, and certain types of lighting equipment. In such cases, it becomes necessary to convert alternating into direct current.

An ideal rectifier of 100 percent efficiency would present infinite resistance in one and zero resistance in the opposite direction. Many methods have been proposed and devices developed to accomplish efficient rectification. Each one has advantages and shortcomings. Vibrating contacts and synchronous commutating devices can be made nearly ideal in the sense defined, but they require considerable maintenance and give trouble under transient supply and load variations. Motor generator sets and rotary converters are bulky and expensive, and also inefficient in small sizes. Mercury arc rectifiers, though bulky, are suitable for power applications up to a few thousand volts and currents of 50 to 100 amperes per rectifier element.[1] Recent developments such as the cesium vapor arc have increased the efficiency,[2] but auxiliary apparatus is required to maintain the ionized vapor. Electrolytic rectifiers employing two metal electrodes such as aluminum, lead, or iron in an electrolyte[3] are rarely used now because of bulk, inefficiency, and maintenance considerations.

Thermionic[4] rectifiers (Kenotrons) have served in electronic circuits from the beginning. The electrons are emitted from a hot cathode; hence the reverse current is practically zero. The auxiliary heating of the cathode and the power losses at cathode and anode, however, make them rather inefficient. Kenotrons[4] are mainly used for high-voltage low-current rectification as in X-ray tube operation. The smaller thermionic rectifiers in electronic circuits are being replaced in recent years by semiconducting devices because of lower cost, greater compactness, and higher efficiency. These semiconductor rectifiers, which are dielectric devices in the sense of this book, are the object of our discussion.

[1] F. P. Coffin, *Gen. Elec. Rev.* 16, 691 (1913); I. J. Kaar, *ibid.* 32, 473 (1929); H. D. Brown, *ibid.* 35, 439 (1932).

[2] A. W. Hull, E. E. Burger, and R. E. Turentine, *Gen. Elec. Rev.* 54, 16 (August, 1951).

[3] G. Schulze, *Trans. Faraday Soc.* 9, 266 (1913); H. D. Holler and J. P. Schrodt, *Natl. Bur. Standards (U.S.) Tech. Papers* No. 265 (1924); W. E. Holland, *Trans. Am. Electrochem. Soc.* 53, 195 (1928).

[4] S. Dushman, *Gen. Elec. Rev.* 18, 156 (1915).

Semiconductor Rectifiers

A rectifier of this type has no moving parts and, for many applications, needs no auxiliary apparatus. In general, it consists of a thin layer of a semiconductor sandwiched between and bonded to two metal electrodes. The rectification is due to the unidirectional conduction characteristic at one of the boundaries, called the blocking layer. Examples of such devices are the copper oxide, the copper sulfide, the selenium and the germanium rectifier. The germanium rectifier,[5] a newcomer, is still in a rapid stage of development. In one form it consists of a pellet of germanium soldered to a metal electrode and a fine wire whisker in contact with the exposed polished germanium surface as the other electrode. The polarity for rectification depends on the nature of the impurity in the germanium. This whisker-type rectifier is suitable for currents of a few tens of milliamperes and reverse voltages of about one hundred volts. In another type, a germanium wafer is placed between two metal electrodes; the direction of rectification is determined by the nature and amount of impurity in the germanium and also by the composition of the metal electrodes. The low resistance of the germanium wafer permits current densities in the forward direction of the order of a hundred amperes per square centimeter; the reverse voltages range up to several hundred volts (Fig. 1.1).

In copper sulfide [6] rectifiers, the electrode at the rectifying boundary is magnesium and the rectifying area a small contact spot between sulfide and magnesium, which apparently wanders around. This rectifier operates best above 100°C, its life is relatively short, and the efficiency below 50 percent.

A copper oxide rectifier [7] is made by oxidizing a sheet of copper at about 1000°C. A layer of cuprous oxide a few mils thick is formed on the copper. A thin layer of cupric oxide is removed from the surface of the cuprous oxide, a counter electrode of carbon or metal applied, and the rectifying element thus completed. The current flows when the mother copper is the cathode; the blocking layer lies between the cuprous oxide and the copper electrode. This rectifier has a low forward resistance and will handle satisfactorily about one ampere per square inch. The reverse peak voltage varies from 10 to 30 volts, depending on methods of manufacture. For higher voltages a number of elements may be stacked in series.

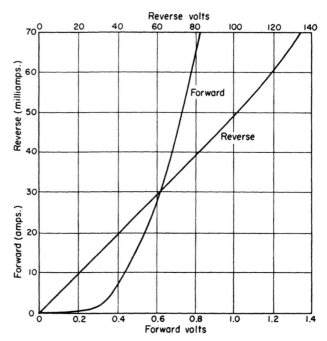

Fig. 1.1. Current-voltage characteristic of germanium rectifier (1.25 cm²).

Selenium rectifiers [8] withstand higher reverse voltages, but the forward resistance is somewhat higher. For a given power output the selenium type weighs several times less than an equivalent copper oxide rectifier, occupies considerably less space, and operates to a higher temperature. Table 1.1 summarizes some characteristics of the various rectifier types.

Table 1.1. **Various semiconductor rectifiers**

Performance	Selenium	Copper Oxide	Germanium Whisker	Germanium Area	Copper Sulfide
Maximum reverse volts per element	30–50	6–20	2–200	50–200	2–3
Maximum direct current output with full wave rectifier circuit, milliamperes/cm²	60	80	2–30	10^5	2×10^3
Maximum total operating temperature, °C	50–80	30–40	30–60	30–60	100–150

[5] E. C. Cornelius, *Electronics* 19, 118 (1946); H. Q. North, *J. Appl. Phys.* 17, 912 (1946); H. C. Torrey and C. A. Whitmer, *Crystal Rectifiers*, McGraw-Hill Book Co., New York, 1948; "Editorial Note regarding Semiconductors," *Bell System Tech. J.* 28, 335 (1949).

[6] M. Bergstein, J. F. Rinke, and C. M. Gutheil, *Phys. Rev.* 36, 587 (1930).

[7] L. O. Grondahl, *Phys. Rev.* 27, 813 (1926), *Revs. Mod. Phys.* 5, 141 (1933); C. E. Hawann and E. A. Harty, *Gen. Elec. Rev.* 36, 691 (1933).

[8] E. Merritt, *Proc. Natl. Acad. Sci.* 11, 572 (1925); H. K. Henisch, *Metal Rectifiers*, Clarendon Press, Oxford, 1949; C. A. Escoffery, *J. Chem. Soc.* 1946, p. 95, *Trans. Electrochem. Soc.* 90, 129 (1946); C. W. Hewlett, U. S. Patents 2,337,329; 2,349,622; 2,354,521; and 2,444,255.

Having been involved in the development of selenium rectifiers, the author will present a more detailed description of the manufacture and performance of these devices.

Selenium Rectifiers

Physical structure

A selenium rectifier element (Fig. 1.2) consists of a base plate of aluminum or iron (0.5 to 1 mm thick) roughened by blasting or chemical etching; a layer of bismuth (2.5×10^{-5} cm thick) or nickel deposited on the roughened surface; a selenium layer (5 to 7×10^{-3} cm thick) containing a small amount of halogen, a blocking layer (10^{-4} to 10^{-5} cm thick) produced by a chemical treatment of the selenium surface with an oxidizing agent (such as hydrogen peroxide, ozone, nitric oxide, etc.) in combination with a subsequent electrical treatment, and an alloy counterelectrode (5

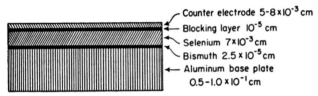

Fig. 1.2. Structure of selenium rectifier.

to 8×10^{-3} cm thick) on top of the blocking layer. Depending upon the use for which the rectifier is intended, the alloy may contain cadmium, bismuth, and tin in various proportions; sometimes a minute amount of thallium is added. When the counterelectrode is the anode, the resistance through the element is high; with the reverse polarity the resistance is low. The corresponding directions of current flow are called the blocking and flow directions respectively.

Methods of manufacture

The blanks of the base metal are punched or sheared from sheet stock and degreased with standard equipment employing a hot grease solvent. A roughening process follows the degreasing. Bismuth is normally applied by evaporation in vacuo, the nickel coating by electroplating.

The selenium is the purest obtainable. A small amount of chlorine, bromine, or iodine (0.1 to 0.01 atomic percent) is added to improve the conductivity and promote a fine-grained, rapidly developing crystal structure during the subsequent heat treatments.

Three main techniques are used in applying the selenium layers: (1) The hot blanks are coated with molten selenium and cooled, or selenium powder is sifted on to the hot blanks. In either case the selenium layer is then subjected to hot-pressing against mica or aluminum foil at a temperature near 140°C. After removal from the press the selenium-coated plates are heated in air at 200° to 216°C for a period of several minutes to an hour. (2) The blanks mounted in vacuo are coated with selenium by evaporation. The rate of deposition and the temperature of the blanks are important factors in determining the final electrical characteristics. After deposition the selenium-coated blanks are heat treated for several minutes to an hour in air at 200° to 216°C. (3) The blanks are spun in molten selenium to coat them uniformly, and cooled rapidly so that the selenium layers are vitreous. The final heat treatment is given after blocking treatment and counterelectrode application, in contrast to methods 1 and 2, where the blocking layer formation is carried out after the final heat treatment.

The counterelectrode is usually sprayed on the treated selenium surface, although in some cases cadmium is evaporated on in vacuo. After counterelectrode application the elements of methods (1) and (2) are subjected to some electrical forming procedure which consists of passing current through the elements for a few minutes to several hours. This forming varies greatly for different manufacturers. Some use direct current in the blocking direction, others alternating current or current pulses. Sometimes, the elements are heated to a definite temperature during forming. There is a tendency to regard the details in these procedures as trade secrets.

Sometimes the final elements are cut from large coated sheets after the heat treatments or even after application of the counterelectrode. Frequently, scored sheets are broken up either before or after forming, or even the individual disks carried through all processing stages. After forming, the elements are tested, assembled into rectifier units, and lacquer-dipped or sprayed to protect them from atmospheric influences.

The rectifying process

For a qualitative understanding of the rectification phenomenon we have to consider briefly some concepts of electrical conduction. A metal contains quasi-free electrons corresponding in number to about the atoms composing it. These electrons are in random motion, but when an electric field is applied a drift velocity is superimposed, and an electric current results. Another way of describing this is to say that the free electrons are in energy levels of a conduction band (Fig. 1.3). An electron can take part in conduction by absorbing energy from the electric field and rising to neighboring unoccupied levels.

Many substances under ordinary circumstances have no free electrons; the conduction band is empty, that is, they are insulators. In other materials electrons can reach the conduction band from a near-by filled band by thermal agitation (intrinsic semiconductors). Holes left in the filled band may also contribute to conduction by moving under the action of the electric field like positive charges. If the gap

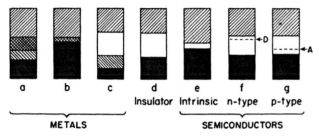

Fig. 1.3. Energy-band picture for metal, insulator, and semiconductor types (black area filled with electrons, white area forbidden regions, shaded areas allowed regions).

between the filled and the conduction band is too large for thermal excitation, certain impurities may cause semiconductivity by inserting acceptor (A) or donor levels (D). Donor atoms transfer electrons to the conduction band and remain fixed as positive charges; thus they produce electron or n-type conduction. Acceptor atoms take electrons from the filled band and become fixed negative charges while the holes in the filled band produce hole or p-type conduction (see Fig. 1.3). p-type and n-type conduction may be distinguished by the direction of the Hall voltage or of the thermo-effect.

Let us consider what happens at the boundary between an n-type and a p-type semiconductor (Fig. 1.4). If the p-type semiconductor is positive, its mobile positive charges (holes) flow toward the n-type semiconductor, and in the n-type material the electrons flow toward the p-type semiconductor. There is a continuous supply of electrons from the metal cathode in contact with the n-type semiconductor, and at the metal anode in contact with the p-type semiconductor there is continuous abstraction of electrons from the semiconductor atoms. At the p-n junction the two types of mobile charges meet and combine as neutral semiconductor atoms. The resistances in the two pieces of semiconductor are determined by the densities and mobilities of these charges. Vice versa, when the p-type semiconductor is negative with respect to the n-type, the mobile positive charges in the p-type semiconductor are retracted toward the cathode, and the mobile electrons in the n-type material toward the anode, and the neighborhood of the p-n junction is cleared of mobile charges. On the p side of the junction remains a region of fixed negative space charge, and on the n side of fixed positive space

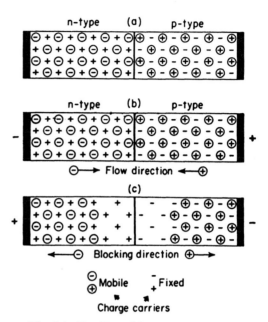

Fig. 1.4. Rectifier with n-p-type junction.

charge. The current flow ceases when the two space charges have been built up to such a thickness that the total driving field lies across them.

This picture of flow and of blocking action is, of course, idealized. Actually, there is still a current flow in the blocking direction due to intrinsic conductivity and diffusion. The physical thickness of the space charge or blocking layer in Fig. 1.4c increases with the applied field, and decreases the greater the density of the fixed charges due to the impurity atoms. The blocking layer extends on both sides of the p-n junction, but the extent is less towards the side of greater charge density.

Correlation of structure and processes of manufacture to the physical picture of rectification

The foregoing simplified outline of the physical phenomena that occur in semiconductors and at the p-n junction is of considerable help in explaining the empirical development of selenium rectifiers. Let us return a moment to consideration of the physical structure and methods of manufacture.

There are two boundaries of the selenium layer: one at the base plate and the other at the counterelectrode. Experiments with metal probes inserted in the selenium layer have established the voltage distribution through the rectifier assembly in the flow and the blocking direction. The total voltage consists of the drops across the base plate boundary, the bulk of the selenium layer, and the counterelectrode boundary.

If sand-blasted uncoated aluminum is used for the base plate, the forward voltage drop at that boundary (curve a) is several times greater than if the base plate carries a thin layer of bismuth (curve b) (Fig. 1.5).

of the direction of current flow and strictly proportional to the magnitude of the current. The resistivity of pure selenium is high (ca. 10^5 ohm-cm). For a selenium layer of 7×10^{-3} cm thick, the resistance for a layer of pure selenium 1 cm² in area would be 700 ohms, which is a few orders of magnitude too high for practical application. For this reason a halogen impurity is added to the selenium which reduces its resistivity to less than 10^3 ohm-cm. The added halogen also reduces the time required for conversion of the selenium from the vitreous to the crystalline form,

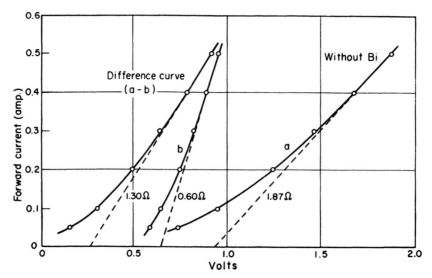

Fig. 1.5. Forward current-voltage curve of selenium rectifier with Al base plate without and with Bi coating.

Moreover, this uncoated boundary exhibits a small rectifying action opposed to that of the counterelectrode. The difference curve $(a - b)$ represents the effect of the aluminum-selenium boundary alone. This boundary volt-ampere characteristic is not linear and may be separated into a linear and a nonlinear section, the linear portion representing a resistance of 1.3 ohms. The resistance of the bulk selenium layer, which dominates in curve b, is not greater than 0.6 ohm.

Selenium is a p-type and aluminum oxide an n-type conductor. When the base plate is positive with respect to the selenium layer, the boundary between the (oxide-coated) aluminum plate and the selenium layer should block the current flow to some extent as observed. The fact that bismuth reduces the voltage drop at the selenium base-plate boundary and also eliminates the feeble rectifying action at this electrode, may indicate that the bismuth atoms penetrate the aluminum oxide layer, converting it from an n-type to a p-type semiconductor.

Probe measurements show that the voltage drop through the bulk of the selenium layer is independent

and influences the fineness of the grain. Unfortunately, the halogen has a deleterious effect on the nature of the blocking layer at the selenium counterelectrode boundary, and on this account a compromise has to be made in the amount of halogen added. The p-type nature of selenium is easily demonstrated by the direction of the thermo-voltage. The mobile carriers, in this case holes, flow from the hot to the cold junction; hence the hot electrode charges up negatively.

The resistivity of a single crystal of selenium depends on the crystallographic direction,[9] and is about one-third as great along the c-axis as perpendicular thereto. In selenium rectifier layers, however, the selenium is multicrystalline, and experiments indicate that a very fine grain structure gives the lowest resistivity, and that the best blocking layers result if the c-axis of the crystals stands parallel to the surface of the layer. In the manufacture of selenium rectifiers, the crystal size and the orientation are controlled by the halogen content, the temperature of the elements

[9] H. W. Henkels, *J. Appl. Phys.* **22**, 916 (1951).

during deposition, the rate of deposition, and the speed, time, and temperature of the heat treating.

The chemical oxidation of the selenium surface, called blocking, results in the production of selenious acid on the selenium surface. The counterelectrode alloy is sprayed on top of this chemically treated layer. In the electrical forming procedure, in which current is passed through the rectifier in the blocking direction, presumably a thin layer of cadmium selenide is built up between the selenium surface and the counterelectrode. This layer is probably an n-type semiconductor, and the boundary between this and the p-type selenium is the p-n junction that gives rise to the rectification observed, blocking when the counterelectrode is positive and allowing flow when the counterelectrode is negative.

If the counterelectrode is applied without preceding chemical oxidation of the selenium surface, the device will still rectify, but the allowable reverse voltage is about one-third of that obtainable with treatment. The forward loss in the oxidized rectifier for equal current output is increased by about 25 percent. Electrical forming of an unoxidized element produces very little effect.

Instead of the chemical oxidation treatment, cadmium selenide or cadmium sulfide may be evaporated on to the selenium surface, and then the counterelectrode applied. After electro-forming, a rectifier of about the same electrical characteristics as the chemically blocked elements will result, but the forward resistance of such an element may be 5 to 10 percent higher.

The blocking layer contributes more to the rectifier losses in the forward direction than the rest of the rectifier element (Fig. 1.6). Curve a represents the forward volt-ampere characteristic of the whole element, b the element exclusive of the blocking layer, and the difference characteristic $(a - b)$ the contribution of the blocking layer to the forward voltage drop of the element. Curve b was obtained as follows: After evaporation of the selenium and before the final heat treatment, two elements were placed in a hot press with the selenium surfaces together and pressed till the selenium surfaces coalesced. This sandwich was then heat treated simultaneously with normal rectifiers. Since the sandwich had twice the selenium thickness of a rectifier element, the measured voltage

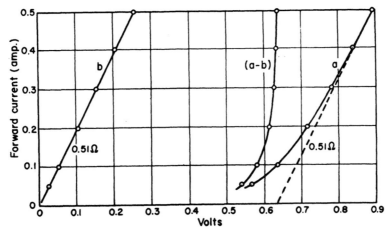

Fig. 1.6. Forward current-voltage characteristic of selenium rectifier split into contributions of selenium proper and of blocking layer.

drops at the various currents were divided by two; these are the values plotted in b. The voltage drop through the sandwich is seen to be strictly linear with current; its half resistance is 0.51 ohm. The dotted tangent on a, drawn with a slope of 0.51 ohm, intersects the abscissa at 0.635 volt. This is the voltage value at which the blocking-layer voltage practically ceases to increase with forward current (see curve $a - b$). If we assume a sine-wave current of average value (0.7 amp) to pass through the rectifier element in the forward half cycle, the average voltage drops for the linear resistance of the selenium and for the blocking layer are 0.36 volt and 0.63 volt, respectively, and corresponding power losses 0.31 and 0.45 watt. The totals are 0.99 volt and 0.76 watt, so that the linear resistance of the rectifier element contributes 35 percent of the total voltage drop and 40 percent of the total power loss, whereas the blocking layer contributes 65 and 60 percent, respectively, even in the forward direction. In general, any procedure that improves the reverse characteristic of a selenium rectifier element will result in an increase in the forward loss of the element.

If a rectifier is placed on a hot plate (70° to 80°C), and current from a 120-v d-c source passed through the element and an incandescent lamp bulb in series (counterelectrode positive), typical forming-speed characteristics for various temperatures are obtained as shown in Fig. 1.7. The characteristic curve NB shows the behavior of an element that does not form a blocking layer.

The average forward voltage drop of an element increases about 25 percent by forming. This forward voltage increases with further use, rapid at first and then leveling off to an almost constant value after

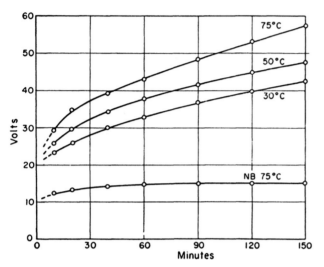

Fig. 1.7. Forming speed characteristics for various temperatures.

several thousand hours. For industrial use, the rating is chosen so that the forward volt drop will not become excessive during 50,000 to 75,000 hours of continuous operation at ambient temperatures below 40°C. The reverse characteristic remains about constant during the life of the rectifier. In some applications long life is not of as great importance as that higher ambient temperatures be withstood. For such applications a higher rating is given the rectifier, with the understanding that the life through aging may be shortened to 1000 or 2000 hours.

One interesting feature of selenium rectifiers is that when they are formed to a certain reverse voltage and subsequently operated at a lower voltage, they will deform in time, and the former voltage can be applied only after reforming. If unused for several months the elements will uniform slightly, but will regain their former rating in a few minutes.

The elements are painted or lacquered to protect the blocking layer from chemical action of gases or vapors. Mercury vapor, for example, is a violent poison for selenium rectifiers. An unpainted element suspended in a closed jar containing a drop of mercury will be almost completely unformed in a few hours.

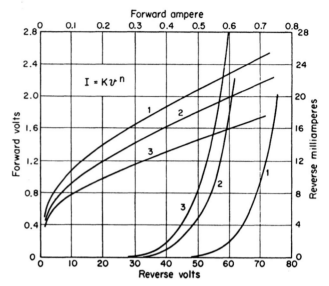

Fig. 1.8. Forward and reverse characteristics of three selenium rectifier elements.

In full-wave rectifier circuits, the selenium rectifiers are subjected to a sine-wave current in the forward half cycle and a sine-wave voltage during the reverse half cycle. Figure 1.8 shows forward and reverse volt-

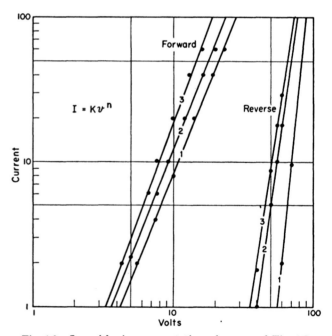

Fig. 1.9. Logarithmic representation of curves of Fig. 1.8.

ampere characteristics for three representative rectifier elements (1-in. square) measured oscillographically.

Rectifier 1 has a high, rectifier 2 a medium, and rectifier 3 a low forward and reverse resistance. In each case, the current-voltage relation can be represented (Fig. 1.9) within the limit of experimental error as

$$I = K \cdot v^n.$$

K and n are constants for a given volt-ampere curve, but vary for different rectifiers and have different values for the forward and reverse characteristics.

For this current-voltage relation, under the circuit conditions mentioned above, the average power dissipation in the rectifier over the forward half cycle is

$$W_F = I_0 \cdot \left(\frac{I_0}{K}\right)^{1/n} \cdot \frac{\Gamma\left(\frac{2n+1}{2n}\right)}{\sqrt{\pi} \cdot \Gamma\left(\frac{3n+1}{2n}\right)},$$

where I_0 represents the peak forward current through the rectifier element, and Γ the gamma function. The average power dissipation during the reverse half cycle is

$$W_R = V_0 \cdot K V_0^n \cdot \frac{\Gamma\left(\frac{n+2}{2}\right)}{\sqrt{\pi} \cdot \Gamma\left(\frac{n+3}{2}\right)},$$

with V_0 the peak reverse voltage across the element.

When the values of K and n are determined from the volt-ampere curves, the power dissipated in the element may be calculated; alternatively, the power dissipation may be specified and the corresponding values of I_0 and V_0 calculated. Table 1.2 gives the values of n and K for the three representative elements and the power dissipated for typical load values:

$I_0 = 0.372$ amp, and $V_0 = 76$ volts; $\overline{W}$ designates the average power loss per half cycle.

If we specify that for the current $I_0 = 0.372$ amp the dissipation in the reverse half cycle be the same as that in the forward half cycle, the peak voltages shown in the last column of Table 1.2 have to be applied. Proper loading is obtained by choosing a total allowable power dissipation, dividing it equally between forward and reverse half cycles, and calculating the corresponding values of I_0 and V_0, after n and K have been determined.

Conclusions

Selenium rectifiers have considerable advantage over copper oxide rectifiers where space and weight considerations are important, since for the same power the weight and volume of a selenium rectifier are only about one-third those of a copper oxide rectifier. The area type germanium rectifier may possibly replace both after mass production methods for germanium ingots of suitable and uniform quality have been developed. The cooling of the small, heavily loaded, germanium rectifier poses special problems in the various applications.

Improved performance of selenium rectifiers may result from an improved blocking layer. Further study is required of the action of semiconducting compounds inserted between selenium and counterelectrode. In all present selenium rectifiers the selenium is polycrystalline, whereas in the recent germanium rectifiers the germanium is a single crystal. Possibly, a single crystal selenium rectifier might be operated at a higher reverse voltage than the present polycrystalline ones.

Table 1.2. Characteristic data for three rectifier elements

Element No.	n Forward	n Reverse	K Forward	K Reverse	W_F	W_R	$\overline{W}$	For $W_R = W_F$ V_0
1	2.41	10.0	0.080	3.37×10^{-21}	0.399	0.389	0.394	76.6
2	2.45	7.17	0.122	3.16×10^{-15}	0.334	2.00	1.167	43.3
3	2.82	6.93	0.193	1.41×10^{-14}	0.270	3.20	1.735	39.3

Only for element No. 1 is the heat dissipation about the same in the forward and reverse half cycles.

2 · Piezoelectric Transducers and Resonators

By KARL S. VAN DYKE

Transducer versus Resonator

The piezoelectric effect (see IB, Sec. 5) provides the essential mechanism for an electromechanical transducer converting between electrical and mechanical energy. If the device as a whole is a converter and the piezoelectric property of the incorporated crystal its principal mechanism, the unit is designated as a piezoelectric *transducer*. In another widely used piezoelectric device the transducer property assumes a secondary role whereas the resonating property of the crystal plays the primary one (Fig. 2.1). Such a

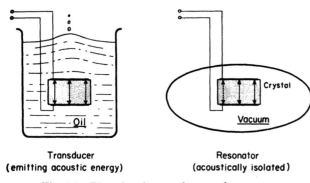

Fig. 2.1. Piezoelectric transducer and resonator.

piezoelectric *resonator* is an acoustic resonating system in which the elastic waves in the crystal set up a standing-wave pattern. The crystal's piezoelectric property in this instance introduces the acoustic energy into the resonator from an eternal circuit and passes it back to such circuit. Hence, the transducer is an energy-passing device whereas the resonator stores as much of the energy as possible in its vibrations. The resonator in consequence has two electric circuit terminals marking the single electric channel of approach and is intentionally isolated on the acoustic side. The transducer has both input and output channels, the one for electric energy transfer to the device with two real electric terminals, the other for acoustic energy transfer.

For purposes of discussion it is convenient to endow the acoustic channel with a fictitious *terminal pair* as signifying a gate for the transfer of energy in this form. The pair of electric terminals marks the place at which potential difference and current must be matched between an external circuit and the internal medium. The acoustic terminal pair similarly represents the boundary between internal and external acoustic media and marks a corresponding matching of stress and velocity across the boundary.

The Transducer

Analogy between transducer and electric circuit coupler

The coupling equations between the electric circuit and the acoustic medium may be written:

$$I = yV + \tau T,$$
$$\dot{S} = \tau V + YT. \quad (2.1)$$

Here the electric variables are electric potential difference and current (V and I) at the input terminals, whereas the acoustic variables are stress and time derivative of strain (T and $\dot{S}$) at the interface with the load. The parameters of the transducer are the electric and acoustic admittances of the device (y and Y) and the conversion parameter (τ).

Equations 2.1 are analogous to the well-known coupling equations between two pairs of terminals: [1]

$$I_1 = y_{11}V_1 + y_{12}V_2,$$
$$I_2 = y_{12}V_1 + y_{22}V_2. \quad (2.2)$$

Here y_{11} and y_{22} are the admittances looking into the device from one or the other pair of terminals with

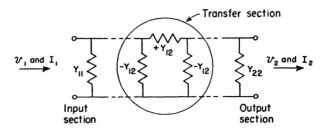

Fig. 2.2. Circuit coupler (admittance approach).

[1] See *Electric Circuits*, by the E. E. Staff, Massachusetts Institute of Technology; published by The Technology Press and John Wiley and Sons, New York, 1940, pp. 449–463.

the second pair short-circuited. y_{12} is the short-circuit transfer admittance (Fig. 2.2).

Special attention is called to the central section of this diagram, which represents the coupling device proper, the ideal coupler, built up solely from elements having positive and negative values of the transfer parameter y_{12}.† For the transducer instead of the circuit coupler this central section consists of positive and negative values of the conversion parameter τ (see Eqs. 2.1). This central section, or ideal transducer, couples the electric input admittance with the acoustic output admittance.

The piezoelectric conversion device as a π-network

The piezoelectric equations of state likewise describe a coupling mechanism:

$$\dot{S} = j\omega s^E T + j\omega\, d\, E,$$
$$\dot{D} = j\omega\, d\, T + j\omega \epsilon^T E. \tag{2.3}$$

As written, they link the time derivatives of strain S and dielectric displacement D to stress T and field strength E and apply to an elementary volume of material of uniform elastic, dielectric, and piezoelectric properties. As a result of the differentiation, the coefficients have the general form of admittances, the elastic compliance s^E, the dielectric permeability ϵ^T, and the piezoelectric conversion parameter d, each being multiplied by $j\omega$.‡ For the present the inertial effects of the medium are neglected.

Figure 2.3 shows in analogy to the equivalent circuit of Fig. 2.2 the relationships described by Eqs. 2.3, under the assumption that the volume element is a unit cube with one face rigidly held in place. This assumption leads directly to the numerical equivalence between stress T and applied force f; time change of strain, S, and the velocity v of the free face; time change of electric displacement, $\dot{D}$, and current I; and electric field strength E and potential difference V.

---

† This combination of elements of equal values but opposite signs has a number of interesting properties. One of these is that it appears from either side as a short circuit when the other terminals are open circuited, and as an open circuit when the others are shorted. The same statement holds for the corresponding T network type of the pure coupler introduced later. The T network is the equivalent of an ideal transformer of unity turns-ratio, with $M = -h/\omega^2$, where M is the coefficient of mutual inductance and h is the piezoelectric coefficient. The separation of an actual shunt element into two parallel branches such as y_{11} and $-y_{12}$ is purely fictitious, for analysis only.

‡ The superscripts specify what variable is to be held constant when the coefficient is measured.

The behavior of the piezoelectric device (Fig. 2.3) is as follows: An elastic compliance of the medium, s^E, diverts some of the mechanical energy input into purely elastic channels; hence only a part of it enters the ideal transducer involving the property d. Again, after conversion to electric form there is a second diversion whereby a part of the energy flows into the purely dielectric mechanism of the medium, denoted by ϵ^T, instead of to an external load. The proportions in the two diversions depend on the admittance of the external load in relation to the internal elastic and

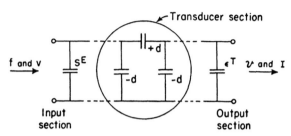

Fig. 2.3. Equivalent circuit of transducer (admittance approach).

dielectric loads. Flow of energy in the direction opposite to that traced here, namely, from the electric terminals to the acoustic, is subject to the same branch-point divisions.

Although we have considered diversionary paths, it will be noted that both are, in general, purely reactive and do not represent a dissipation of energy within the transducer. The transfer of energy through a piezoelectric transducer when impedances are matched is a highly efficient process.

The piezoelectric conversion device as a T-network

Equations 2.2 and 2.3 have emphasized the admittance aspect of the device. A corresponding treatment from the impedance approach is often equally appropriate in the handling of a piezoelectric problem. Equations 2.4 and 2.5 and Figs. 2.4 and 2.5 describe

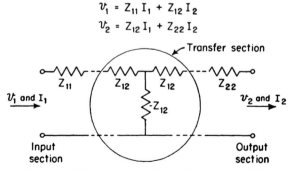

Fig. 2.4. Circuit coupler (impedance approach).

this situation. Here c^D and β^S are the elastic stiffness and dielectric impermeability of the crystal, and h is a second piezoelectric parameter describing, in place of d, the piezoelectric conversion property. In Fig. 2.5

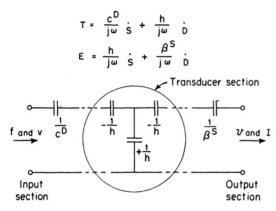

Fig. 2.5. Equivalent circuit of transducer (impedance approach).

the elements are again pictured as capacitances. The elastic stiffness and dielectric impermeability appear as obstacles which stand in the way of full application of the applied force to the converting mechanism h and to the full external use of the generated emf.

Which of the two views, the π- (d-type) or the T- (h-type) network and which of four alternative formulations † of the piezoelectric equations of state is favored for the solution of a problem depends upon the particular situation.[2]

The admittances and impedances describing the material are thus far treated as purely reactive. Departure from this may be expected, especially in ferroelectrics where hysteresis losses appear.

† The remaining two pairs of equations of state are here written in the conventional form without the differentiation with respect to time which has been employed in Eqs. 2.3 and Fig. 2.5:

g-Type
$$S = s^D T + g_t D$$
$$-E = gT - \beta^T D$$

e-Type
$$-T = -c^E S + e_t E$$
$$D = eS + \epsilon^S E$$

The full set of equations of state is due to H. G. Baerwald (OSRD Report No. 287, Contr. No. OEMsr-120, 1941) and was adopted in "Standards on Piezoelectric Crystals, 1949" (*Proc. IRE* 37, #12, December, 1949). Reference should be made to the Standards for discussion of units, symbols, conventions as to crystallography and axes, and typical drawings of crystals of each class of symmetry.

[2] The development of equivalent networks involving a perfect transformer instead of capacitances is carried through by W. P. Mason, *Electromechanical Transducers and Wave Filters*, D. Van Nostrand and Co., New York, 1942, Chapter VI.

Some comments on the piezoelectric equations

No consideration has been given thus far to the tensor nature of stress and strain or to the vector nature of electric field and dielectric displacement (see IB, Sec. 5). These greatly expand the equations of state beyond the simple forms here used. The foregoing analysis is usually appropriate in practical problems as a first approximation, provided the principal stress and the corresponding strain are used together with the electric field and displacement components coupled to them.

Piezoelectricity exists in crystals of certain symmetry, and the directions of the coupled elastic and electric components are prescribed by this symmetry.[3] In the more complete analysis of a piezoelectric problem it becomes therefore necessary to consider a complex array of components of stress and strain and the resultant electrical responses arising. The handling of this situation becomes greatly simplified through the systematized listing of these properties in matrix form [4] and through the rules of matrix multiplication. One set of these equations of state, the d-type (not differentiated as in Eqs. 2.3), is shown here as an example both in the schematic form of Fig. 2.6 and in full (Eqs. 2.4). In the actual matrices for quartz and

TENSOR FORM	BLOCK FORM (schematic)	
The d-type equations	The d-type matrix	
	T	E
$S = s^E T + d_t E$	S: $s^E \times 10^{11}$ m²/newton	$d_t \times 10^{11}$ coulomb/newton
$D = dT + \epsilon^T E$	D: $d \times 10^{11}$ coulomb/newton	$\epsilon^T \times 10^{11}$ farad/m

Fig. 2.6. Equations of state (extensional stress and strain counted positive).

ammonium dihydrogen phosphate (Table 2.1), a number of the positions in the 9 by 9 matrices are left blank. The numerical values of these parameters are

[3] See W. G. Cady, *Piezoelectricity*, McGraw-Hill Book Co., New York, 1946; W. P. Mason, *Electromechanical Transducers and Wave Filters*, D. Van Nostrand Co., 1942; R. A. Heising, *Quartz Crystals for Electrical Circuits*, D. Van Nostrand Co., 1946; P. Vigoureux and C. F. Booth, *Quartz Vibrators and Their Applications*, His Majesty's Stationery Office, London, 1950; Symposium on Quartz Oscillator-Plates, *Am. Mineralogist* 30, 205 (1945).

[4] Tables of such matrices of elastic, dielectric, and piezoelectric constants for all materials whose complete properties are known have been prepared by Van Dyke and Gordon, Piezoelectric Data, Suppl. Vol. 1, Tenth Report to U. S. Army Signal Corps, Contract DA36-039-sc-73 (1953). The mathematical manipulation of matrix equations for piezoelectricity is treated by W. L. Bond, *Bell System Tech. J.*, 22, 1 (1943).

$$S_1 = s_{11}{}^E T_1 + s_{12}{}^E T_2 + s_{13}{}^E T_3 + s_{14}{}^E T_4 + s_{15}{}^E T_5 + s_{16}{}^E T_6 + d_{t11} E_1 + d_{t12} E_2 + d_{t13} E_3$$

$$S_2 = s_{21}{}^E T_1 + s_{22}{}^E T_2 + s_{23}{}^E T_3 + s_{24}{}^E T_4 + s_{25}{}^E T_5 + s_{26}{}^E T_6 + d_{t21} E_1 + d_{t22} E_2 + d_{t23} E_3$$

$$S_3 = s_{31}{}^E T_1 + s_{32}{}^E T_2 + s_{33}{}^E T_3 + s_{34}{}^E T_4 + s_{35}{}^E T_5 + s_{36}{}^E T_6 + d_{t31} E_1 + d_{t32} E_2 + d_{t33} E_3$$

$$S_4 = s_{41}{}^E T_1 + s_{42}{}^E T_2 + s_{43}{}^E T_3 + s_{44}{}^E T_4 + s_{45}{}^E T_5 + s_{46}{}^E T_6 + d_{t41} E_1 + d_{t42} E_2 + d_{t43} E_3$$

$$S_5 = s_{51}{}^E T_1 + s_{52}{}^E T_2 + s_{53}{}^E T_3 + s_{54}{}^E T_4 + s_{55}{}^E T_5 + s_{56}{}^E T_6 + d_{t51} E_1 + d_{t52} E_2 + d_{t53} E_3 \quad (2.4)$$

$$S_6 = s_{61}{}^E T_1 + s_{62}{}^E T_2 + s_{63}{}^E T_3 + s_{64}{}^E T_4 + s_{65}{}^E T_5 + s_{66}{}^E T_6 + d_{t61} E_1 + d_{t62} E_2 + d_{t63} E_3$$

$$D_1 = d_{11} T_1 + d_{12} T_2 + d_{13} T_3 + d_{14} T_4 + d_{15} T_5 + d_{16} T_6 + \epsilon_{11}{}^T E_1 + \epsilon_{12}{}^T E_2 + \epsilon_{13}{}^T E_3$$

$$D_2 = d_{21} T_1 + d_{22} T_2 + d_{23} T_3 + d_{24} T_4 + d_{25} T_5 + d_{26} T_6 + \epsilon_{21}{}^T E_1 + \epsilon_{22}{}^T E_2 + \epsilon_{23}{}^T E_3$$

$$D_3 = d_{31} T_1 + d_{32} T_2 + d_{33} T_3 + d_{34} T_4 + d_{35} T_5 + d_{36} T_6 + \epsilon_{31}{}^T E_1 + \epsilon_{32}{}^T E_2 + \epsilon_{33}{}^T E_3$$

Rationalized units assumed.

Table 2.1. d-Type matrices of the piezoelectric, elastic and dielectric constants of right α quartz and of ammonium dihydrogen phosphate (ADP) (unrotated axes) in mks units (rationalized)

The entries in the matrices are

$$s^E \times 10^{11},\ \epsilon^T \times 10^{11},\ (d \text{ and } d_t) \times 10^{11}$$

Right α quartz (specific gravity, 2.65)

	T				E		
S	1.269	−0.169	−0.154	0.431		−0.23	
	−0.169	1.269	−0.154	−0.431		0.23	
	−0.154	−0.154	0.971				
	0.431	−0.431		2.005		−0.067	
					2.005	0.862	0.067
					0.862	2.88	0.46
D	−0.23	0.23		−0.067		4.0	
					0.067	0.46	4.0
							4.1

ADP (specific gravity, 1.80)

	T				E		
S	1.8	0.7	−1.1				
	0.7	1.8	−1.1				
	−1.1	−1.1	4.3				
				11.3		0.167	
					11.3		0.167
						16.2	5.17
D				0.167		54	
					0.167		54
						5.17	13.8

zero as required by the symmetry of these crystals. The crystallographic structure of quartz and ADP and the orientations of several cuts for resonators or transducers are shown in Figs. 2.7 and 2.8.

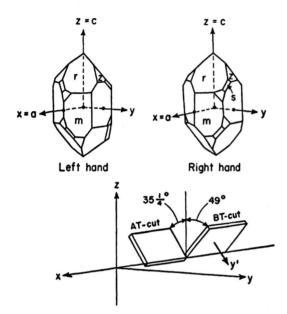

Fig. 2.7. Left and right quartz and orientation of *AT*- and *BT*-cut.

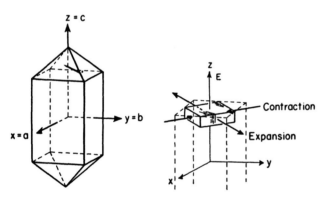

Fig. 2.8. ADP (ammonium dihydrogen phosphate) crystal and transducer plate.

In the representations of Figs. 2.3 and 2.5 an elemental unit-cube transducer was assumed, inertialess and with uniform stress, strain, field, and displacement throughout the volume. In a practical unit there are variations of vibrational amplitude and of electric field from point to point of the standing-wave pattern, and integrations of the basic equations of state over the whole volume are required. In general, the network analogues of the complete transducer will be of the same form, though with parameters of different magnitudes. The network parameters will then involve, in addition, the dimensions of the piece of piezoelectric material, its density, and any constraints or loading introduced by the mounting. These last incidentally introduce dissipative and inductive elements into the networks.[5]

The Piezoelectric Resonator

In many uses of the transducer property of piezoelectrics, as in microphones, phonograph pick-ups, and ultrasonic generators, the primary function is energy transfer, and impedance matching is an important consideration for effective design. In the most extensive application of piezoelectrics, however, as resonators [6] for frequency control or filter use, the crystal serves in sequence as transducer, acoustic medium, and again as transducer. The central feature of the device is the excellence of its resonator quality, a Q unmatched by any other circuit element.†

The two-terminal resonator represents in its overall electric characteristics the resonance properties of its internal elasto-inertial system. Its external response is that of an extraordinarily selective electric resonating network. The fact that the energy conversion occurs twice in the device makes the parameters of the equivalent electric network of the resonator depend on the square of the piezoelectric conversion constant. The circuit element becomes a high- or a low-impedance unit at resonance dependent on whether the piezoelectric coupling is small or large.

The acoustic resonator is inherently a standing-wave device with many resonances, each corresponding to some relationship between wavelength and the dimensions of the unit. For the fundamental mode the determining dimension equals one half wave length of the elastic wave. In solid materials different types of waves exist. Any deformation that is coupled piezoelectrically to a component of internal electric dis-

---

[5] For actual transducer design see W. J. Fry, J. M. Taylor, and B. W. Henvis, *Design of Crystal Vibrating Systems for Projectors and Other Applications*, Dover Publications, New York, 1948; A. C. Keller, *Trans. Am. Inst. Elec. Engrs.* 66, 840 (1947); W. Roth, *Proc. IRE*, 37, 750 (1949); F. E. Borgnis, *A General Theory of the Acoustic Interferometer for Plane Waves*, Tech. Rep. No. 3, Contract Nonr-220(02), Task No. 384-404, submitted by W. G. Cady, Project Director, California Institute of Technology, January, 25, 1952; W. P. Mason, *Piezoelectric Crystals and Their Application to Ultrasonics*, D. Van Nostrand Co., New York, 1950.

[6] W. G. Cady, "The Piezoelectric Resonator" (abstr.), *Phys. Rev. 17*, 531 (1921); *Proc. IRE* 10, 83 (1922).

† Q values of 50,000 to 2,000,000 are common for quartz, and values as high as 6,000,000 have been attained, while the Q of ordinary resonance circuits rarely exceeds a few hundred.

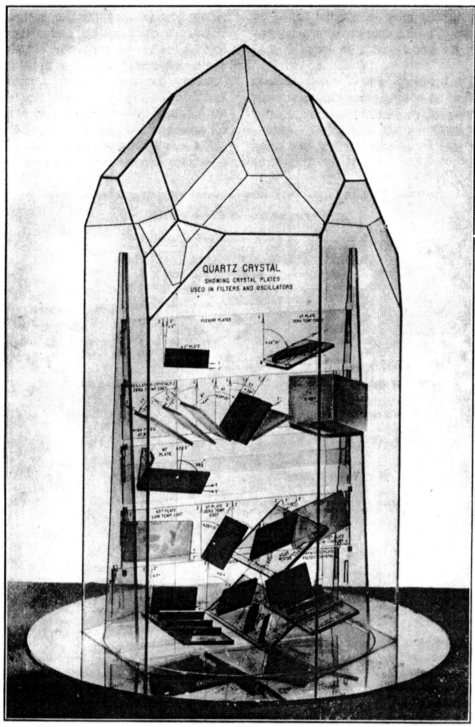

;. 2.9. Quartz crystal model and various cuts for filter and oscillator plates. (Courtesy Bell Telephone Laboratories.) The angle of cut has an important effect on the operating conditions.

placement offers a connection between internal elastic waves and circuit voltages. Piezoelectric resonators in common use depend upon extensional, shear, or flexural resonance. The standing waves in some resonators lie along the field direction, in others at right angles.

Resonators for frequency control are commonly made of quartz because of its extremely sharp acoustic resonance and excellent mechanical properties, notably its permanence, hardness, and insolubility. Quartz also offers compensation between positive and negative temperature coefficients of elastic stiffness in different directions so that frequency standards with zero temperature coefficients can be realized. The choices of resonating direction in the quartz, of types of strain in the vibration, and of the direction of the field lead to various geometries of the resonating piece and to prescribed orientations in the crystal structure, depending on the application. A wide range of cuts—specific orientation of plates or bars within the crystal lattice—have been developed, each of them most suited for some specific application (Fig. 2.9).

For each quartz resonator at any one of its resonances the response to external circuit voltage and cur-

Table 2.2. The equivalent parameters of some quartz resonators

(Electrodes in contact with crystal)

1. *X-Cut Bar in Longitudinal Extension along Y*

$l = 4 \times 10^{-2}$ m along Y Electroded faces
$b = 5 \times 10^{-3}$ m along Z are $l \times b$
$T = 5 \times 10^{-3}$ m along X $Q_1 = 10^5$

Constants of Quartz

$s_{22}^E = 1.269 \times 10^{-11}$
$\epsilon_{11}^S = 3.9 \times 10^{-11}$
$e_{12} = 0.17$

Formulas for Thin Rod   *Computed Values*

$L_1 = \rho l t / 8 e^2 b$ 460 h
$C_1 = 8 e^2 b l s^E / \pi^2 t$ 0.015 $\mu\mu$f
$R_1 = \omega_0 L_1 / Q_1$ 1600 ohms
$C_0 = l b \epsilon^S / t$ 1.6 $\mu\mu$f
$F_1 = \dfrac{1}{2\pi\sqrt{L_1 C_1}} = \dfrac{1}{2l\sqrt{\rho s^E}}$ 67.2 kc/sec

2. *Rectangular Plate in Thickness-Shear in Plane Including the Field (Y')*

$l = 10^{-2}$ m along X Electroded faces
$b = 10^{-2}$ m along Z' are $l \times b$
$t = 10^{-3}$ m along Y' $Q_1 = 10^5$

Constants of Quartz

	AT-Cut	BT-Cut
c_{66}^D	0.298×10^{11}	0.674×10^{11}
ϵ_{22}^S	4.0×10^{-11}	4.0×10^{-11}
e_{26}	0.097	0.093

Formulas for Thin Plate as Section of Infinite Sheet   *Computed Values*

	AT-Cut	BT-Cut
$L_1 = \rho t^3 / 8 e^2 l b$	0.35 h	0.40 h
$C_1 = 8 e^2 l b / \pi^2 t c^D$	0.026 $\mu\mu$f	0.010 $\mu\mu$f
$R_1 = \omega_0 L_1 / Q_1$	37 ohms	62 ohms
$C_0 = l b \epsilon^S / t$	4.0 $\mu\mu$f	4.0 $\mu\mu$f
$F_1 = \dfrac{1}{2\pi\sqrt{L_1 C_1}} = \dfrac{1}{2t\sqrt{\rho/c^D}}$	1.68 Mc/sec	2.5 Mc/sec

All constants are given rationalized mks units. Density of quartz is 2.65×10^3 kg/m³. Falling off of vibration toward edges in finite plate may nearly double L_1 (halve C_1), depending on precise contours and boundary conditions.

Formulas adapted from K. S. Van Dyke, "The Piezoelectric Resonator and Its Equivalent Network," *Proc. Inst. Radio Engrs.* **16**, 757 (1928).

rent is given by the equivalent network of Fig. 2.10.[7] The L_1, C_1, R_1 series elements represent the motional properties of the resonator (inertial, elastic, frictional); the capacitance C_0 represents the purely dielectric property of the quartz at constant strain. This by-pass capacitance becomes the limiting parameter in designing resonators for very high frequencies.

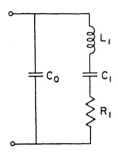

Fig. 2.10. Equivalent network of quartz resonator.

Table 2.2 gives the circuit parameters of the network of Fig. 2.10 and the numerical constants for a quartz bar in longitudinal extensional resonance and for a quartz plate operating as thickness-shear resonator of the AT- and BT-cut types (see Fig. 2.9).

the admittance circle.† Since the resonator is a very high-Q device, the admittance of the dielectric capacitance C_0 in the network may be regarded as constant over the narrow frequency range of a resonance curve. With this approximation the locus of the admittance is a circle of diameter $1/R_1$, tangent to the axis of imaginaries and with its center displaced from the axis of reals in the positive direction by the fixed amount $j\omega C_0$ (Fig. 2.11). Increasing frequency leads around the circle in a clockwise sense. The two half-power points A and E of the motional admittance branch $L_1 C_1 R_1$ correspond to frequencies $(1 - 1/2Q_1)f_1$ and $(1 + 1/2Q_1)f_1$, where f_1 is the resonance frequency of this branch (point C). Resonance and antiresonance of the two branches together, defined by the conditions of zero reactance, are indicated by points D and G on the circle. The points of maximum and of minimum admittance of the combined branches are found as B and I. The striking feature of sharp crystal resonance is the rapid rate of circle traversal in the vicinity of resonance. For a Q of 6×10^6 (a quartz resonator of exceptional quality)[8] the frequency difference between the two quad-

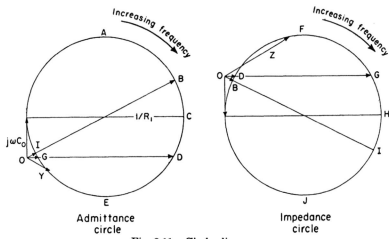

Fig. 2.11. Circle diagrams.

The bar is X-cut with its length along y. The plates are considered as rotated Y-cuts, that is, rotated about x, away from the Y-cut position. For the bar the electric field is along x; for the plates along y'. Electrodes are assumed to be in contact with the crystals, covering the full area of the faces.

Admittance locus

The frequency dependence of the admittance of the equivalent network (Fig. 2.10) can be traced by use of

[7] K. S. Van Dyke, *Phys. Rev.* **25**, 895 (1925); *Proc. Inst. Radio Engrs.* **16**, 742 (1928); D. W. Dye, *Proc. Phys. Soc.* (London) **38**, 399 (1926).

rantal points A and E is but one six-millionth of the resonance frequency.

The impedance locus of the same network is, of course, also a circle obtained by inverting the first circle upon a unit circle centered at the origin. The direction of the maximum and minimum impedance vectors (B and I) makes an angle of the same magnitude with the axis of reals as that for maximum and

† This representation was adopted for crystal resonator work by Cady at the beginning of the piezoelectric resonator art, following its earlier use in telephone-receiver analysis.

[8] M. Waltz and K. S. Van Dyke, *J. Am. Acoust. Soc.* **19**, 732 (1947).

minimum admittance, but of opposite sign. The direction of increasing frequency is again clockwise.

Measurements on piezoelectric resonators

The resonator is a device of high electric impedance except within a narrow frequency range near series resonance. Parallel resonance occurs close to series resonance and has a very high impedance. The result is a wide swing in impedance values, many hundreds of thousands fold, within a frequency range deviating only a few parts in a million from the resonance frequency. In measuring the electrical properties of the resonator as well as in planning its use, this extraordinary rate of variation must be considered.

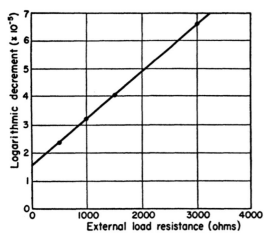

Fig. 2.12. Decrement of quartz resonator in vacuum with resistance load (X-cut bar, 40 mm long, cross section 5×5 mm, 67.5 kc/sec. $Q = 196{,}000$; $R_1 = 950$ ohms, $L_1 = 395$ henries).

The impedance at series resonance for some quartz resonators falls to but a few ohms. On the other hand, many resonators are used close to their parallel resonance when the impedance may be in the hundreds of thousands of ohms.

Various devices are used for evaluating the properties of quartz resonators for frequency-control application. Some of them are little more than test oscillators duplicating the circuit in which the crystal shall be used, and the test is one of satisfactory crystal and circuit performance. Other test oscillators [9] operate with substitution of calibrated electrical components for the crystal until identity of frequency and amplitude response is achieved. Thus, the parameters of the complete equivalent network may be determined.

Let us describe two direct methods of measuring the resonator properties of crystals. In the first, a transient method, the resonator is set in vibration and the amplitude of the generated voltage is observed by use of a linear amplifier and scope as the vibrations die away. The procedure is repeated for several values of the external resistance load. The plot of logarithmic decrement against resistance is a straight line for small resistances; its slope depends upon L_1, and its intercept is R_1 (Fig. 2.12). The method [10] presents some technical difficulties but is particularly suited for resonators of extremely high Q.

A standard method [11] for determining circuit parameters consists in using an impedance bridge. The chief difficulty here is the need for a driver of sufficient constancy in frequency to permit a steady point-by-point balance over the very sharp resonance curve. For measurements on extremely high-Q resonators, the driver itself must be crystal-controlled, with the attendant difficulties of varying the frequency. Recourse is made to heterodyne methods for determining small deviations in frequency with respect to a reference crystal oscillator. A high-frequency counter in which the time gate for the counting interval is crystal-controlled provides a ready reading device for the frequency of the driver. Some commercial signal generators have a sufficient short-time stability for balancing the bridge and registering the frequency on a high-frequency impulse counter.

Because of the extreme range of impedances near crystal resonance, a single commercial bridge covers only a portion of the resonance range. The General Radio 916A Bridge, in which the balancing network is a condenser in parallel with a resistor, is best suited for small values of impedance, and thus may be balanced in the immediate vicinity of resonance. The General Radio Twin-T Impedance Measuring Circuit (Type 821-A), a null transmission network, on the other hand is adapted to high impedances. Thus, the first bridge covers the region at the extreme right on the admittance circle, the second bridge regions well around to the left on the upper and lower portions of the circle.

The reduction of resonance data to a straight-line form [12] has the merit of showing by its linearity the existence of but a single resonance in the immediate frequency region. Secondary resonances, of which a multiplicity exist, commonly introduce serious sources of error into the determination of the properties of the resonator as well as interfere with its use in its many applications. Relative shift of the frequencies of secondary and fundamental resonances sometimes brings in interference between modes, leads to false determination of temperature coefficients, and introduces such complexity into the resonance curves that a network interpretation is hazard-

[9] L. F. Koerner, *Proc. IRE* **39**, 16 (1951); C. W. Harrison, *Bell System Tech. J.* **24**, 217 (1945).

[10] K. S. Van Dyke, *Proc. Inst. Radio Engrs.* **23**, 386 (1935).

[11] E. A. Gerber, *Proc. Inst. Radio Engrs.* **41**, 1103 (1953).

[12] G. D. Gordon, *Thickness-Shear Resonators*, Thesis, Wesleyan University, 1950. (Suppl. Vol. Second Quarterly Report of Wesleyan University—Contr. DA36-039-sc-73 (1950) to U. S. Army Signal Corps.)

ous. Such shifts in the resonance spectrum may occur from temperature changes (even those caused by the crystal currents), from changes in the mounting or pressure, and with extremely slight modification of the crystal dimensions.

A third method of measuring network parameters places the crystal as the series transmission element between two low resistances as legs of a π-section. Voltage and current readings at series resonance and the difference between the frequencies of series and parallel resonance suffice to yield the full network for a single isolated resonance. The method, adapted by Mason from transmission measurement practice, is widely used as more rapid than those described above.

Frequency standards

The sharpness of the resonance peaks and their independence from external influences make quartz resonances ideal frequency standards. Extremely clean surfaces, permanence of surface finish, proper mounting, evacuation of the container, temperature control, and the use of only small currents and small vibrational amplitudes are essentials for ultimate stability. Frequency stabilities of the order of parts in a billion have been achieved.

For the highest precision, obtained with the standard crystal clocks [13] in various laboratories, it is necessary to extend the control of external conditions to circuit and crystal and to keep the crystal vibrating continuously. The clock rate suffers a disturbance for an extended period after any interruption. Not all crystal resonators for frequency control, however, are intended to achieve such extreme accuracy. Very high in its frequency constancy was the generator of the Loran [14] navigation system developed during the Second World War. In this unit a GT-cut quartz plate was used. Adjustment of a trimming condenser could set the crystal on frequency and, when the crystal had been adequately aged, a short-time (10 minutes) frequency stability of 3 parts in 10^9 was attained.

At the lower end of the constancy scale were the crystals used for combat communication during the same war. Here quartz crystals were needed by the million † to establish communication channels instantaneously and with certainty. The equipment offered very limited space, no temperature control, and had to take the hard knocks of service conditions. A common overall frequency tolerance here was ± 2 parts in 10,000 for temperature variations from $-50°C$ to $+90°C$. This would be a very low order of accomplishment for a hand-tailored crystal in a constant-temperature oven, but was in that day a ready solution to the problem of instantaneous netting of integrated military groups.

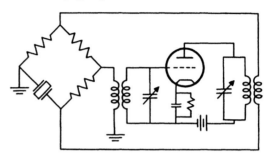

Fig. 2.13. Bridge-stabilized Meacham oscillator.

Frequency control by crystal resonators is accomplished by many different circuit arrangements. An illustration of a circuit in which the oscillation frequency is close to the series resonance of the resonator is the Meacham bridge oscillator of Fig. 2.13. In the so-called Pierce and Miller types (Fig. 2.14) the oscillation occurs near parallel resonance.

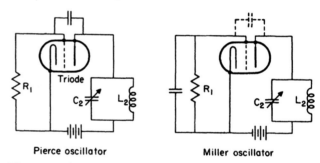

Fig. 2.14. Oscillator types operating near parallel resonance.

Crystal resonators

Crystal resonators are widely incorporated into the electric wave filter circuits of carrier telephony and telegraphy. Here again the crystal offers a sharply selective circuit component. One such filter is shown in Fig. 2.15.[15]

Crystal resonators have been made for resonances scattered from 1 kc per sec to several hundred megacycles per sec. For the lowest frequencies a ring [16] of quartz containing a small gap is cut and used as a flexing unit (Fig. 2.16). Up the frequency scale we shift

---

[13] C. F. Booth, *Proc. Inst. Elec. Engrs.* (London) **98**, Pt. 3, 1 (1951); W. A. Marrison, *Bell System Tech. J.* **27**, 510 (1948).

[14] J. A. Pierce, A. A. McKenzie, and R. H. Woodward, *Loran*, M.I.T. Radiation Laboratory Series, McGraw-Hill Book Co., New York, 1948.

† Seventy-five million units were made during the Second World War.

[15] W. P. Mason, *Bell System Tech. J.* **16**, 423, Fig. 1 (1937).

[16] Post Office Engineering Department, "Further Development of Quartz Flexual Vibrators," *Radio Rep.* No. 1956, Feb. 27, 1950.

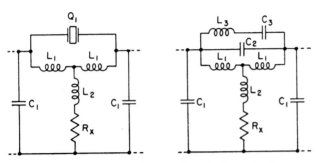

Fig. 2.15. Crystal filter and equivalent circuit.

to flexing bars, then extensional bars, then face shear, followed by thickness shear of plates and finally thickness extension of thin plates. Overtone modes in which several half waves of the acoustic vibration occur to

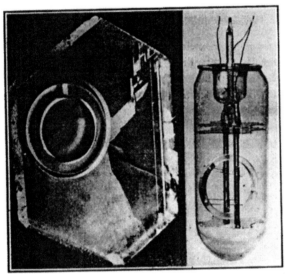

Fig. 2.16. Z-cut split ring of quartz as low-frequency resonator.

the thickness of the plate are commonly employed for frequencies above 30 Mc per sec (Table 2.3).

Table 2.3. Types of vibration in quartz resonators and approximate frequency range to which each is adaptable

Split ring in flexure	0.5–5 kc/sec	Diameters	1–5 cm
Bar in flexure	1–70	Lengths	2–8 cm
Rod in torsion	30–300	Lengths	2–5 cm
Ring in compression	100–500	Diameters	1–5 cm
Plate in face shear	100–1000	Faces	0.5–4 cm
Plate in thickness shear	1000–20,000	Thickness	0.025–0.5 mm
Overtone modes of same	3000–100,000	Thickness	0.025–0.5 mm

The Ultrasonic Generator

When the piezoelectric device is to be used for transferring considerable power from an electric circuit to an acoustic medium, a multiple array of crystals may be used instead of a single crystal transducer. Applications in the ultrasonic field include the underwater soundheads of submarine echo-ranging systems and the industrial and laboratory generators of ultrasonic energy for the agitation of solutions, for the preparation of emulsions, the control of polymerization, and for the destruction of bacteria, to name a few uses.

In principle all the individual crystals in the multiple array behave alike. The extent of the array is designed with the acoustic problem in mind of securing the desired sharpness of the beam. The amplitude of the acoustic beam at the transducer surface is limited to what the individual crystal generator can stand

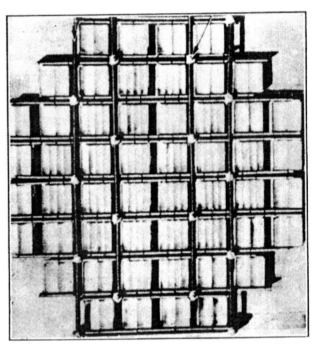

Fig. 2.17. Sonar transducer made from ADP crystal blocks.

without fracture. Whether electrical input shall be at high or at low voltage in providing power depends on whether the individual crystal blocks are connected electrically in series or in parallel, and upon the dimensions of the individual block. The 10½-in. array of ADP crystal blocks of Fig. 2.17 is that of the QJA transducer of sonar which delivers at least ½ watt per cm² of crystal area to the liquid medium on 100 to 150 watt electrical input in the frequency range 10 to 30 kc per sec.[17]

The recently developed ceramic piezoelectric elements seem more likely to become of use in ultrasonic generators than as frequency-control devices. They provide a greater degree of piezoelectric coupling than

[17] A. C. Keller, *Trans. Am. Inst. Elec. Engrs.* **66**, 1217 (1947).

ADP, and afford the nearest approach to rochelle salt in this respect. They are free, however, from the troublesome anomalies of rochelle salt which during the Second World War led to its general discard in favor of ADP because of the destructive effects of overload and high temperature. A characteristic of the ferroelectric ceramic transducer elements to date is the dependence of the parameters of their equivalent networks upon amplitude and load because of the temperature rise which occurs during operation.

Single- or double-element transducer elements, in contrast to multiple, each element using either single crystals or prepolarized aggregates, are in wide use as phonograph pick-ups, microphones, and earphones. A recent example of the development of very large mechanical amplitudes is Mason's [18] use of a solid tapered horn that produces amplitudes up to 1.5 mils at 18 kc per sec from a ceramic transducer (Fig. 2.18). The

[18] W. P. Mason and R. F. Wick, *J. Am. Acoust. Soc.* 23, 209 (1951); W. P. Mason and S. D. White, *Bell System Tech. J.* 31, 469 (1952).

device gives promise as a fast-acting chipping or abrasive tool.

Ultrasonic amplitudes can be increased far beyond those at the surface of the generator by focusing the

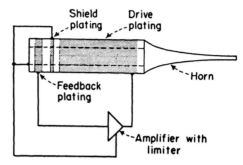

Fig. 2.18. Ceramic, tapered horn transducer of Mason.[18]

beam [19] through the use of a concave spherical radiating face. One advantage which ceramics offer over the single crystal is that they may be molded to shape for such focusing action.

[19] G. W. Willard, *J. Am. Acoust. Soc.* 21, 360 (1949).

3 · Magnetic and Dielectric Amplifiers

By DUDLEY A. BUCK

Magnetic amplifiers are rapidly coming into widespread use in military, scientific, and industrial control applications. Developments of the last decade have been stimulated by the advent of commercially available "rectangular-loop" magnetic materials and motivated by the quest for amplifiers which are more reliable and more rugged than vacuum-tube amplifiers.

Magnetic amplifiers have been under development for many years. In fact, in 1916 they looked more promising for some applications than did vacuum tubes. This is illustrated by the following excerpt from a 1916 discussion between Lee de Forest, vacuum-tube pioneer, and E. F. W. Alexanderson, magnetic-amplifier pioneer: [1]

LEE DE FOREST: I would like to direct comparison between this high power radio telephone method [which uses a large magnetic amplifier and was developed by the General Electric Co.] and the method recently developed by the Western Electric Company and tried out at Arlington, where the audion amplifier principle was worked to the limit. In that case they

[1] Excerpt from discussion following "A Magnetic Amplifier for Radio Telephony," presented by E. F. W. Alexanderson before the Institute of Radio Engineers, February 2, 1916; *Proc. Inst. Radio Engrs.* 4, No. 2, p. 101 (1916).

started, as Mr. Alexanderson did, with a microphone and amplified the voice currents through an audion amplifier, and kept on amplifying them. In the first oscillating audion circuit the radio frequency currents were modulated by the microphone, then these modulated radio frequency currents were amplified in a bank of some twenty bulbs, and finally the current from this bank was amplified in a bank of 500 bulbs.

This ensemble of amplifying audion tubes put a total energy of 11 kilowatts in the antenna; beautifully controlled by the voice, it is true, but having the modest upkeep renewal cost of something like ten thousand dollars per month. So as a practical engineering proposition, there is absolutely no comparison between that method and the method worked out by the General Electric Company. I believe that I am entitled to express that sort of opinion of the audion if anyone is! This is the situation as it stands today. No one can say, however, that the situation will not be altered very materially in one, two, or three years, after we learn how to build oscillions for large power outputs, say for 5 or 10 kilowatts each. That will create a very different situation. It is difficult to say, therefore, which of the two methods discussed possesses the greatest practical promise.

As we now know, de Forest's audion was destined to replace magnetic amplifiers for this and many other applications. Since 1916 the use of vacuum tubes has become the accepted technique for producing amplification, and researchers now study the magnetic amplifier primarily as a substitute for the vacuum-tube amplifier. Magnetic amplifiers, reliable and rugged, are proving to be good substitutes in many applications.

Dielectric amplifiers, under development for less than a decade, are being studied and thought of primarily as the electrical dual of magnetic amplifiers, once again, however, as vacuum-tube substitutes. Unfortunately, dielectric amplifiers lack the stimulus of commercially available "rectangular-loop" ferroelectric materials. As a result, dielectric amplifiers are still in the laboratory stage, and current research in the field is largely devoted to development of materials.

Amplifiers in General

Before describing how a piece of iron alloy or a thin sheet of barium titanate can act as an amplifier, let us consider briefly amplifiers in general. (The term *amplifier* here refers to a power amplifier.)

Power amplifiers are devices with three terminal-pairs (Fig. 3.1). One terminal-pair is associated with

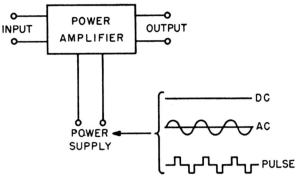

Fig. 3.1. Power amplifier, a three-terminal-pair device.

the input, or signal to be amplified, one with the output, or load, and the third with the power supply. Power amplification, or power gain, may be defined as the average signal power in the output divided by the average signal power in the input, without reference to the average power delivered by the power supply.

The power supply contains no information, that is, it contains no meaningful fluctuating component that is to be amplified. The power supply can be a direct voltage, a sinusoidal voltage, a sequence of pulses, or any other wave form for which the amplifier is designed to operate. Vacuum-tube and transistor amplifiers normally use d-c power supplies, whereas magnetic and dielectric amplifiers, by their very nature, must use a-c or pulsed power supplies.

Nonlinearity in the properties of ferromagnetic and ferroelectric materials makes possible their use in devices with three terminal-pairs which are capable of supplying more average signal power at the output terminals than is supplied to them at the input terminals. These devices are called magnetic and dielectric amplifiers.

Their output is generally a sequence of pulses, sine waves, or portions of sine waves, whose frequency is the power-supply frequency and whose amplitude, when averaged, is the amplified signal. In the following discussion it will be assumed that whatever is connected to the output terminals is capable of accepting the output wave form in the form of pulses and then of averaging, or filtering, to meet its own requirements. Stated in a different way, it might be said that such amplifiers should be called modulators instead of amplifiers, the amplified signal being contained in the envelope of a modulated carrier, and that the load demodulates the output wave form. For some of the amplifiers to be described, the output wave form will be rectified into unidirectional pulses; demodulation will then consist of merely filtering.

In a like manner, the input signal can be a direct voltage, the envelope of a modulated carrier (of the same power-supply frequency), or a series of unidirectional pulses.

Subsequent sections will show that there are two types each of magnetic and of dielectric amplifiers. Without explaining their circuit operation in detail, let us first show in simple schematic drawings how a nonlinear element can be connected with a power supply and a load to form a basic amplifier circuit. The nonlinear circuit elements with which we shall deal are nonlinear inductance, represented by a winding on a magnetic core; nonlinear capacitance, represented by a condenser made with a ferroelectric dielectric; and, for completion, nonlinear resistance, represented by a vacuum tube.

For each of these three nonlinear elements, there are two basic amplifier circuits: one with the nonlinear element, the load, and the power supply in series, and the other with all three in parallel. The two basic circuits for the vacuum tube are the series-fed and the shunt-fed vacuum-tube amplifier (Fig. 3.2). The tube plays the role of a variable resistor controlled by the grid voltage. This resistance, as it changes, affects the output power in the series circuit by controlling the current, in the parallel circuit by controlling the voltage.

The series-fed amplifier requires a power supply with as low an internal impedance as possible, so that changes in the vacuum-tube resistance will cause large changes in the series current and therefore in the output power. If a power supply with a very high internal impedance is substituted, the amplifier would not work, the vacuum tube having little or no control over the series current. The power supply itself would determine the current by the ratio of its voltage to its internal impedance. Such a high-impedance power supply represents a current source. For the series-fed vacuum-tube amplifier, the other extreme is wanted, a voltage source.

The shunt-fed amplifier, on the other hand, requires a high-impedance power supply so that changes in the tube resistance will cause large changes in the load voltage and therefore large changes in the output power. Clearly, here the power supply should approach a current source. The impedance of the current-source power supply is often made exceedingly high over some frequency range by inserting an inductor (radio-frequency choke) in series with the power supply. Similarly, the impedance of a voltage-source power supply is often lowered by shunting it with a condenser (plate-supply by-pass condenser).

When comparing these two circuits, we might say that in the series-fed amplifier, voltage across the tube subtracts from the load voltage; in the shunt-fed amplifier, voltage across the tube adds to the load voltage. The dual can also be stated: in the first circuit, current in the tube is the load current; in the second circuit, current in the tube subtracts from (shunts) the load current.

Magnetic amplifiers, too, can be divided into two categories: saturable inductors (Fig. 3.2C) in which changes in flux in a magnetic core produce a voltage which subtracts from the load voltage, and saturable transformers (Fig. 3.2D) in which changes in flux in a magnetic core produce a voltage which adds to the load voltage.

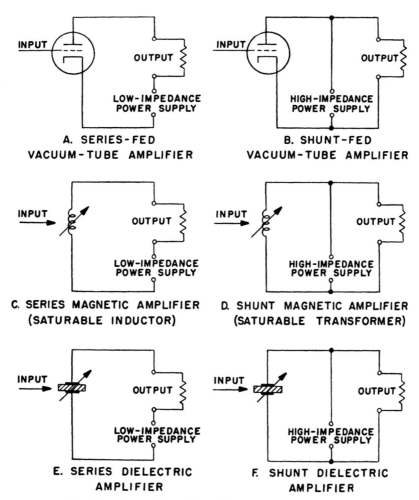

Fig. 3.2. Comparison of amplifiers (equivalent circuits).

The equivalent circuit of the saturable inductor follows directly from the series-fed vacuum-tube amplifier circuit if the tube is replaced by a nonlinear inductor. Similarly, the equivalent circuit of the saturable transformer follows from the shunt-fed vacuum-tube amplifier. Series and shunt dielectric amplifiers are obtained if the tube is replaced by a nonlinear capacitor.

Let us now look more closely at the operation of these circuits.

Magnetic Amplifiers

Saturable inductors

A simple saturable inductor employs a winding on a magnetic core in series with the output and the a-c power supply (Fig. 3.3). Voltage across this winding, due to flux changes in the core, subtracts from the voltage available to the output. As long as flux in the core is changing, the output voltage is very low. If during the cycle the core should reach saturation so that flux can no longer change, the voltage across this winding drops to zero, and the full power-supply voltage is applied to the output. The input voltage is

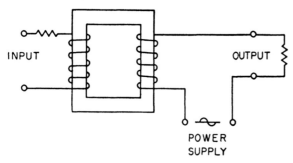

Fig. 3.3. Single-core saturable inductor.

connected to a second winding on the magnetic core which controls the point in the cycle at which saturation occurs.

Minimum output will result with the input circuit de-energized. As the power supply voltage goes through its cycle, a small magnetizing current flows. Assume that the design is such that flux in the core does not quite reach saturation in either direction; the core is traversing its largest possible hysteresis loop (Fig. 3.4, the dotted loop) without going into the flat region at the top and the bottom of the loop.† Under

† A hysteresis loop can be defined as the plot of magnetic induction (B) versus magnetic field (H) for a ferromagnetic material under sinusoidal excitation. Figure 3.4 does not show a B-H loop, but rather a Φ-NI loop, a plot of magnetic flux in a given core versus ampere-turns in the driving winding. The B-H loop is useful for discussing the properties of a magnetic

this condition, flux will be changing continuously throughout the cycle, inducing a voltage opposite and nearly equal to the power-supply voltage at all times. Therefore the output power will be low.

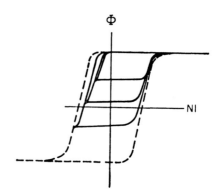

Fig. 3.4. Asymmetrical hysteresis loops of a rectangular-loop core material (ferrite MF-1118-259).

Maximum output will occur when the input circuit is energized with enough current to keep the core saturated at all times. Under this condition, changes in flux are negligibly small and the output receives the full power-supply voltage throughout the cycle, as if the core had been removed. Current in the output circuit is limited only by the ratio of power-supply voltage to load impedance.

This amplifier can operate in the range between minimum and maximum output. Starting from minimum output, where flux in the core does not quite reach saturation in either direction, we may add to the magnetic field in one direction with current in the input winding so that the core saturates in this direction at some time in the cycle. When saturation occurs, the current will jump suddenly to a value limited only by the voltage and impedance in the output circuit. If the load is resistive, this pulse of current will be shaped somewhat like a portion of a sine wave, falling to zero as the power-supply voltage falls to zero. Such a pulse will occur once each cycle.

Within the operating range the magnetic material traverses an asymmetrical hysteresis loop (Fig. 3.4, the solid loops), reaching into saturation in one direction only. As the input voltage is increased, the magnetic material is driven further into saturation in this direction; the hysteresis loop becomes smaller, and saturation occurs earlier in the cycle. The current

material, whereas the Φ-NI loop is useful for discussing the properties of a given magnetic core. The two loops will differ because of geometry effects such as the finite outside to inside diameter ratio of a ring-shaped core which causes material along the outer periphery to operate in a lower field than material along the inner periphery.

pulse therefore becomes a larger fraction of a sine wave, lasting longer and rising to a higher peak value. The increase in input voltage thus increases the average output power. Figure 3.5 shows this action for three input voltages. These wave forms (which also

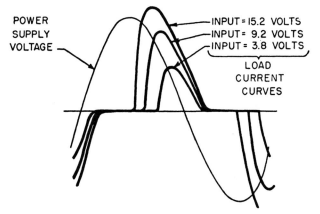

Fig. 3.5. Observed wave forms of resistance-loaded saturable inductor using "Deltamax" cores. (After Wilson.[2])

have negative-going pulses) were traced on a slightly more complicated amplifier employing two cores instead of one.[2]

Before describing the two-core saturable inductor circuit, it might be well to examine the single-core saturable inductor with respect to power gain. Power gain can be defined as the average signal power in the output divided by the average signal power in the input. We might ask what losses might be expected in the input circuit. At minimum output, the input is de-energized; the control current is zero, and therefore the input power appears to be zero. At maximum output, the core is always saturated, so that there is no induced voltage across the input winding. The input current is high, but, since the voltage is zero, the input power is again zero. It would appear as if these two operating points gave infinite power gain. The difficulty becomes clear when we consider the series resistance in the input circuit. The series resistance should be very high in the minimum-output condition (input circuit de-energized) so that induced current will not flow and cause large power losses. At maximum output (high control current) the input resistance should be very low to prevent losses. Clearly, these two requirements are not compatible.

The two-core saturable inductor circuit (Fig. 3.6) overcomes this difficulty. It consists of two single-core saturable inductors connected with their react-

[2] T. G. Wilson, Series-Connected Magnetic Amplifier with Inductive Loading, NRL Report 3923, Naval Research Laboratory, Washington, D. C., Jan. 9, 1953.

ance windings (the windings in series with the load) in series, and their input windings in series opposition. Operation is much like that of the single-core saturable inductor. At minimum output both cores traverse a hysteresis loop which does not quite reach saturation. Flux changes continuously over the cycle, causing a voltage nearly as large as the power-supply voltage, leaving near-zero voltage for the load. A current in the input winding will cause one core to saturate on one half of the power-supply cycle, and the other core to saturate on the other half (since the input windings are in series opposition). The higher the input current, the earlier in the cycle saturation occurs, and therefore the higher the average output power. The wave forms of Fig. 3.5, as mentioned, show this action for three input voltages.

The main difficulty with this circuit is that load current must always flow through a winding on an unsaturated core, since one core saturates at a time. Current is thereby induced in the control circuit. This induced current in the control circuit imposes most of the practical limitations on this magnetic amplifier. If resistance in the control circuit is lowered so as to increase the power gain, the amplifier becomes sluggish in its response to changes in the input voltage. If the resistance is raised so as to make the amplifier respond rapidly to changes, the power gain becomes low.

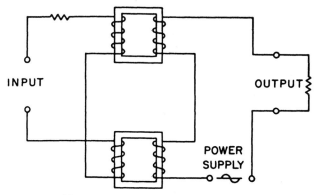

Fig. 3.6. Two-core saturable inductor.

The relationship between response time and power gain can be improved by positive feedback as will be discussed later.

Saturable transformers

Figure 3.7 illustrates a simple one-core saturable transformer. The operation is similar, but inverse in control logic, to the operation of the saturable inductor. Flux changes in the core produce the output voltage rather than a voltage which subtracts from the output voltage. The asymmetrical hysteresis loops of Fig. 3.4 can also be used to describe the operation of the saturable transformer. However, traversal of a small loop now means a small output voltage whereas traversal of a large loop means a large output voltage. This circuit gives maximum output with the input de-energized and minimum output when the flux in the core is saturated—just the opposite of the saturable inductor. When flux in the core is changing, the saturable transformer behaves very much like

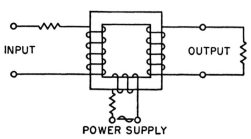

Fig. 3.7. Single-core saturable transformer.

an ordinary transformer, but when the magnetic core saturates, the power supply and the load are effectively uncoupled.

Power-supply considerations

A search of magnetic-amplifier literature would show that saturable inductors are most commonly used in power-control applications, whereas saturable transformers are most commonly used in low-power computer applications. Power-supply considerations are primarily responsible for this difference.

A saturable inductor delivers more power to its load by lowering the circuit impedance. Lowering of the impedance will exact more power only if the power supply's own internal impedance is still lower. It is because commercial power is distributed at a very low impedance that saturable inductors are most commonly used in power-control applications.

A saturable transformer delivers more power to its load by raising the circuit impedance. As the core comes out of saturation and starts down the steep part of its hysteresis loop, the power supply delivers more power because it sees a higher impedance. Because high-impedance power supplies (pentode driver tubes) are common in computer applications, saturable transformers are used in computers.

Frequently, several magnetic amplifiers operate from a common power supply. Because minimum interaction is desired, saturable inductors should be connected in parallel, saturable transformers in series.

Feedback

A portion of the output voltage or current is often added to or subtracted from the input so as to produce positive or negative feedback. Figure 3.8 illustrates the effect of feedback on the operating curve of a magnetic amplifier. Negative feedback lowers the gain but improves the linearity of the amplifier. Positive feedback, on the other hand, increases the gain,

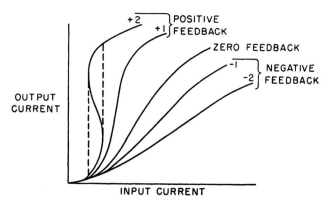

Fig. 3.8. Effect of feedback on magnetic amplifier operation.

but reduces linearity. In many circuits, positive feedback can be increased until instability results, causing a "snap-action" (dotted lines). The output current jumps between two values as the input current passes through a certain range. This phenomenon is sometimes called "triggering."

Positive feedback increases not merely the gain of an amplifier but the ratio of gain to response time. Higher gains result for a given response time or faster response for a given gain. Positive feedback thus increases the "performance" of a magnetic amplifier, but at the expense of linearity. With positive feedback the amplifier becomes more nonlinear, dependent on the exact shape of the hysteresis loop. For this reason modern "high-performance" magnetic amplifiers depend on close control of the magnetic properties of the core material.

Figure 3.9 illustrates one of the many possible circuit arrangements with feedback. The output current

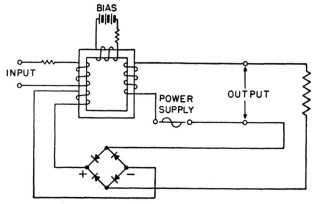

Fig. 3.9. Saturable inductor with external feedback.

is rectified and the direct current passed through a second input winding, the feedback winding. The load can be placed on the a-c side of the rectifier (as shown) or on the d-c side of the rectifier. The same circuit arrangement can provide feedback for a saturable transformer (Fig. 3.10).

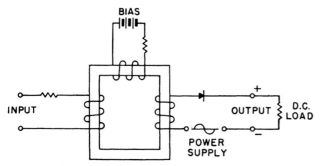

Fig. 3.10. Saturable transformer with external feedback.

This scheme, where a portion of the output is fed back to the input via an external loop, is known as *external feedback*. In contrast, *internal feedback* can be introduced by a rectifier in series with the load; a unidirectional current flows in the output circuit

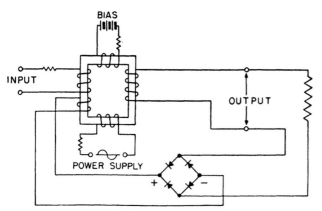

Fig. 3.11. Saturable inductor with internal feedback.

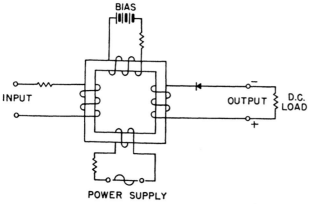

Fig. 3.12. Saturable transformer with internal feedback.

(Figs. 3.11 and 3.12). Internal feedback gives the same effect as external feedback, with an equal number of turns on the input and feedback windings. Magnetic amplifiers with internal feedback are sometimes called "self-saturating" amplifiers. Figures 3.13 and 3.14 illustrate two common saturable inductor cir-

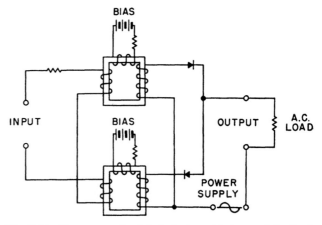

Fig. 3.13. Two-core saturable inductor with internal feedback for a-c load.

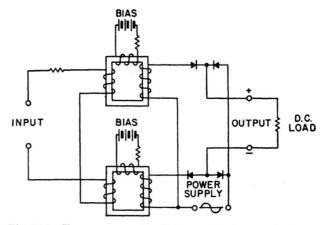

Fig. 3.14. Two-core saturable inductor with internal feedback for d-c load.

cuits employing two cores with internal feedback, the last one with full-wave rectification.

Core and circuit variations

The operating point of a magnetic amplifier can be shifted by the addition of a bias voltage. The bias voltage can be connected to a separate winding on the magnetic core as shown in Figs. 3.9 through 3.14, or in series with one of the existing windings. Sometimes it is desirable to use the bias voltage to turn an amplifier "half-way-on" so that negative input voltages produce a drop in the average output power whereas positive input voltages produce a rise. Operation of this type is analogous to "Class A" operation of a vacuum-tube amplifier. The input to a magnetic amplifier can be thought of as the net effect of input voltage and bias.

More complicated magnetic cores are often encountered in practice. The two-core combinations may be combined into a single three-legged core to facilitate fabrication. In addition, three-legged and four-legged cores can be built wherein magnetic fluxes are added and subtracted in the various legs just as voltages are added and subtracted in windings.

Many circuit variations have been made, some introducing other types of feedback and feedback control, others employing saturable inductors in bridges or adding condensers which allow certain windings to be operated near resonance.[3]

Magnetic amplifiers with half-cycle response time

Interaction between output and input circuits limits the performance of the magnetic amplifiers described in the preceding sections; compromises must be made between gain and response time. A magnetic amplifier is currently under development which responds in one-half cycle of the power-supply frequency. It achieves a high degree of isolation between output and input circuits which makes gain and response time relatively independent.

To explain its operation, let us assume that a toggle switch is placed in the input and the output circuit. The toggle switch on the input side is closed during one-half cycle while the input current changes the flux in one direction. The toggle switch in the output circuit is closed during the next half cycle while the load current drives the flux in the opposite direction. If both switches are never closed simultaneously, the isolation between output and input is complete.

Operation of such a circuit would be as follows. Assume that the load current always drives the core to saturation in one direction. Starting from saturation in this direction, let a pulse of input voltage drive the core away from saturation during one-half cycle. The amount of flux change in this direction will depend on the size of the input voltage pulse—the larger the pulse, the greater the flux change. The core will now hold, or "remember," this flux change while the switches are changed over, and on the subsequent half cycle, load current will drive the core back to saturation. The flux change during the output half cycle can be only

[3] See J. G. Miles, *Types of Magnetic Amplifiers—A Survey*, Engineering Research Associates, St. Paul, Minn., Jan. 24, 1951; *Trans. Am. Inst. Elec. Engrs.* 70, Pt. 2, 2104 (1951).

as large as the flux change during the input half cycle. The input voltage thus controls the output voltage, just as with the other types of magnetic amplifiers. This amplifier, with its toggle switches, gives an exceedingly high gain-to-response-time ratio. The response time, however, is always one-half cycle.[†]

Toggle switches are, of course, much too slow for this type of operation, but they can be approximated by rectifiers; a rectifier conducting current in the forward direction corresponds to a closed switch, whereas a rectifier with a back voltage across it corresponds to an open switch. Saturable inductors using this half-cycle operation have been studied by Ramey,[4] and saturable transformers by the Harvard Computation Laboratory.[5]

Figure 3.15 illustrates a saturable inductor with half-cycle response time. Load current flows in the

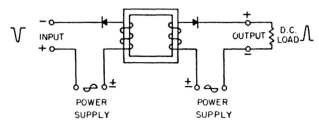

Fig. 3.15. Half-cycle response time, saturable inductor. (After Ramey.[6])

output circuit during one half cycle, passing through the rectifier in the forward direction. On the next half cycle, the power supply voltage swings to the opposite polarity, cutting off the rectifier with a back voltage and thereby disconnecting the load from the core. While the load is disconnected, the input circuit "sets" the flux in the core. The less the flux changes in a saturable inductor, the higher the output power. Maximum output power will result during the output half cycle if no current flows in the input circuit during its half cycle. On the input half cycle the flux changes by an amount determined by the power-supply voltage in the input circuit *minus the input voltage*. On the other half cycle, the power supply volt-

[†] A response time of one-half cycle assumes that changes in the input voltage are synchronized with the start of the input half cycle. Actually, an extra half cycle would be wasted if a change of input voltage occurred at the start of the output half cycle. The input voltage would then have to wait a half cycle for the toggle switch in the input circuit to close.

[4] R. A. Ramey, *Trans. Am. Inst. Elec. Engrs.* 70, Pt. 2, 1214, 2124 (1951); NRL Reports 3799 and 3869, Naval Research Laboratory, Washington, D. C.

[5] H. H. Aiken, "Investigation for Design for Digital Calculating Machinery," Harvard Computation Laboratory Reports. See especially Reps. 3 and 4, 1949.

age in the input circuit, as it reverses polarity, supplies a back voltage across the rectifier. This effectively removes the input circuit from the core.

Figure 3.16 illustrates a saturable transformer with half-cycle response time. Here the control logic is

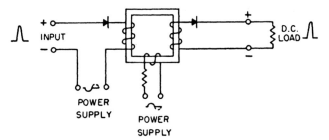

Fig. 3.16. Half-cycle response time, saturable inductor (pulse-operated).

inverted. As the core is "set" on the input half cycle, a voltage is induced in the output circuit which cuts off the rectifier. The half cycle of power-supply voltage which formerly cut off the rectifier is no longer necessary. Both power supplies can thus operate with unidirectional pulses; this type of amplifier is therefore called a "pulse-operated" magnetic amplifier.

An amplifier of this sort with a modification in the input circuit (Fig. 3.17), that eliminates the input

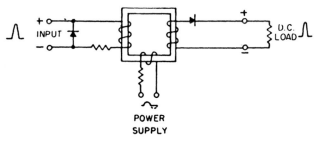

Fig. 3.17. Pulse-operated saturable transformer in computer service.

power supply at the expense of somewhat higher losses, is in wide use in information-handling machines. One large digital computer is being built almost entirely with magnetic amplifiers of this type.[6]

Materials and performance of magnetic amplifiers

Magnetic amplifiers have been used most extensively with 60-cycle and 400-cycle a-c power supplies. They are performing such tasks as the control of d-c motor fields and armatures, control of a-c motors, control of rectifier power supplies, etc.

Recently, in connection with computer service, the trend is to step up the power-supply frequencies.

[6] See Progress Reports on the Harvard Mark IV Computer.

Some magnetic amplifiers are in operation with power-supply frequencies in the region between 10^4 and 10^5 cps.

In the low-frequency applications, magnetic amplifiers are commercially available which deliver anywhere from a few watts to over a million watts[7] of output power and with amplification sometimes exceeding 10,000. They have proved themselves rugged and reliable in mobile service. In this field, they are being used in increasing numbers both in conjunction with and as substitutes for vacuum-tube amplifiers. As d-c amplifiers, they exhibit low drift as compared to their vacuum-tube counterparts.

Chiefly responsible for the recent widespread use of magnetic amplifiers are new core materials with rectangular hysteresis loops. Once a laboratory curiosity, rectangular hysteresis loops are now built into commercially available core materials by processes of grain orientation and domain orientation. The most widely used of these alloys are listed in Table 3.1 with certain of their trade names and approximate compositions.[8]

Table 3.1. Magnetic metals of current interest for magnetic amplifiers

Hipersil, Trancor XXX, Silectron Type C
(grain-oriented silicon steels–3% Si, 97% Fe)

Mu-Metal
(75% Ni, 2% Cr, 5% Cu, 18% Fe)

Supermalloy
(79% Ni, 5% Mo, 15% Fe)

4-79 Mo Permalloy
(79% Ni, 4% Mo, 16% Fe)

Deltamax, Hipernik V, Orthonik, Permeron, Orthonol
(50% Ni, 50% Fe)

These core materials are generally available in the form of thin ribbons (0.005 in. or less) of any desired width. Rectangular hysteresis loops are exhibited if the magnetic field is applied in the direction for which the grains or domains have been oriented. This is usually in the plane of the ribbon parallel to the edges, so that fabrication of a core requires simply winding the ribbon on a spool. Thin insulating layers separate the wraps of tape so as to minimize eddy currents.

The rectangular hysteresis loop with its steep sides causes eddy-current effects to show up at relatively low frequencies. Eddy currents produce a "skin ef-

[7] E. L. Harder, *Electronics 25*, No. 10, 115 (1952).
[8] F. Benjamin, "Improvements Extend Magnetic Amplifier Applications," *Electronics 25*, No. 6, 119 (1952).

fect" which limits the depth of flux penetration, thereby effectively lowering the cross-sectional area of the magnetic flux path. The remedy is to use thinner and thinner ribbon as the frequency is raised. Ribbons as thin as 1/8 mil are in use.

An alternate solution to the eddy-current problem is to use non-metals with a much higher volume resistivity. These "ferrites," composed of oxides of iron and other metals, are manufactured by ceramics processes. Figure 3.18 is a hysteresis-loop photograph made with one of these rectangular-loop ferrites.

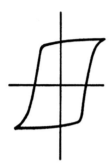

Fig. 3.18. Hysteresis loop of a rectangular-loop ferrite (MF-1326B).

These ferrites have not yet loops as rectangular as those of metallic alloys, and they have the disadvantage of lower saturation induction and higher coercive force. Ferrites, however, promise to extend the frequency range of magnetic amplifiers.

Considering the computer applications of magnetic amplifiers, pulse-operated, we find it more natural to think of the eddy-current phenomenon in the time domain than in the frequency domain. *Switching time* is then defined as the time required for the material to traverse one-half of its hysteresis loop. Switching time is a function of the applied magnetic field. With relatively high fields (twenty-five times the coercive force) switching times as short as 10^{-7} second have been observed for 1/4-mil 4-79 Mo Permalloy ribbon-wound cores (see also IIIC, Sec. 4).

Heating of the core material due to hysteresis loss is a second factor which limits the frequency at which magnetic amplifiers can be operated. Each time the material traverses a hysteresis loop, an amount of energy is dissipated which is equal to the area of the loop. This energy is converted into heat, and a temperature rise results. Even before the temperature rises to a damaging value, changes in the magnetic properties of the core, such as a drop in the saturation flux density, may occur.

Dielectric Amplifiers

A power amplifier, as we have seen, can be built which utilizes for its operation the nonlinear magnetic properties of a ferromagnetic material. In a similar manner, the nonlinear electrical properties of a ferroelectric material can be used to build a power amplifier. This latter type, called a dielectric amplifier, is the electrical dual of the magnetic amplifier.

Before describing dielectric amplifier circuits and their operation, let us consider briefly the electrical properties of ferroelectric materials (see also IB, Sec. 5).

Ferroelectric materials

Ferromagnetic materials owe their nonlinearity primarily to the existence within the material of domains of permanent magnetic dipoles. Recently, materials have been discovered which have domains of permanent electric dipoles [9-11] (see IB, Sec. 5). These materials, named ferroelectrics, exhibit hysteresis loops in the D-E plane similar to those of the ferromagnetic materials in the B-H plane (Fig. 3.19).

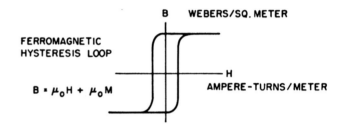

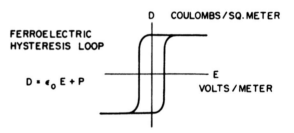

Fig. 3.19. Comparison of ferromagnetic and ferroelectric hysteresis loops.

Ferroelectricity is exhibited by three groups of materials whose representatives are the tetrahydrate of potassium-sodium tartrate (rochelle salt), dihydrogen

[9] A. von Hippel and co-workers, NDRC Contract OEMsr-191, Reports 7 (August, 1944) and 11 (October, 1945), Laboratory for Insulation Research, Massachusetts Institute of Technology.
[10] A. von Hippel, *Revs. Mod. Phys.* **22**, 221 (1950).
[11] D. A. Buck, "Ferroelectrics for Digital Information Storage and Switching," Digital Computer Laboratory Report R-212, Massachusetts Institute of Technology, June 5, 1952.

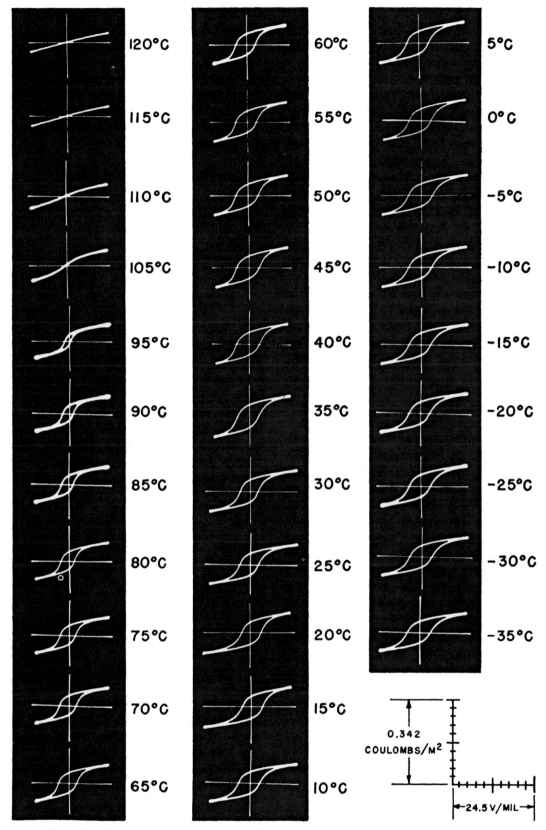

Fig. 3.20. Hysteresis loops of barium-titanate ceramic as a function of temperature.

potassium phosphate and arsenate, and barium titanate. Materials of the last group have been considered for use in dielectric amplifiers. Unlike the others, barium titanate can be prepared in the form of a rugged ceramic which exhibits ferroelectricity over a wide temperature range, and which, when compounded with other titanates, can be tailored to a wide variety of electrical properties. This third group has recently been augmented by the addition of the ferroelectric tantalates, niobates, and tungstates.[12]

The hysteresis loops of Fig. 3.20 are those of ceramic barium titanate (Glenco body X-18). They are not nearly so rectangular as the hysteresis loops of the magnetic cores used for magnetic amplifiers. If we desire hysteresis loops of greater rectangularity, the only available method at present is to abandon ceramics and grow single crystals of barium titanate.

As with the ferromagnetic materials, the hysteresis loop disappears as the material is heated through the Curie temperature, and the electric dipoles are no longer able to align themselves spontaneously against the randomizing action of thermal vibrations (Fig. 3.20).

Dielectric amplifier circuits

Considering dielectric amplifiers as the dual of magnetic amplifiers, interchanging everywhere the words *voltage* and *current*, we immediately encounter the rather troublesome fact that a single ferroelectric condenser is the dual of a magnetic core with only one winding. The dual of a magnetic core with several windings cannot be represented by a single ferroelectric condenser. The engineer recognizes this as a distinct handicap, for he must often incorporate additional circuit elements, linear condensers and linear transformers, to obtain multiple connections to the ferroelectric, d-c isolation between terminal pairs, and impedance matching.

There are two basic dielectric amplifier circuits just as there are two basic magnetic amplifier circuits, one in which the current through the dielectric adds to the output current, and one in which current through the dielectric subtracts from (shunts) the output current.

A simple dielectric amplifier of the first type (Fig. 3.21) can be made by replacing the reactance winding of the saturable inductor of Fig. 3.3 with a saturable ferroelectric condenser. The operation can then be described similarly to that of the saturable transformer of Fig. 3.7. The saturable condenser is in series with the power supply and the load, as was the saturable inductor. It passes a current which contributes to the load current whenever its charge changes. Therefore, the output power is greatest when the ferroelectric condenser traverses the largest possible hysteresis loop and smallest when the ferroelectric condenser is biased into a saturated region.[13]

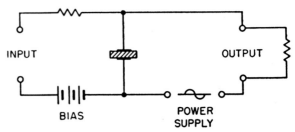

Fig. 3.21. Simple one-element saturable condenser.

An improvement can be made by using more than one ferroelectric condenser. In one such circuit (Fig. 3.22), the ferroelectric condensers are located in two legs of a bridge and linear condensers in the other two legs.

Quite commonly, dielectric amplifiers incorporate linear inductances which allow the circuit to operate near resonance (Fig. 3.23). One such mode of operation is obtained when the resonant circuit has too large a capacitance to resonate at the power-supply

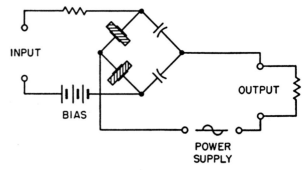

Fig. 3.22. Bridge-connected saturable condensers.

frequency with zero input voltage. An increase of the input voltage drives the ferroelectric condensers towards saturation, effectively lowering their capacity and bringing the circuit into tune.

Dielectric amplifiers have been built which operate with power gains greater than 50, using power-supply

---

[12] B. Matthias, "Phase Transitions in Ferroelectrics," in *Phase Transformations in Solids*, R. Smoluchowski (Editor), John Wiley and Sons, New York, 1951.

[13] S. Roberts, "Barium Titanate and Barium-Strontium Titanate as Nonlinear Dielectrics," Thesis, Massachusetts Institute of Technology, 1946.

frequencies in the 2-Mc region.[14] Power output levels, however, are generally in the milliwatts.

A dearth of suitable materials seems to be the one thing which is holding back the full-scale development of dielectric amplifiers. The ceramics, inexpensive and rugged, do not exhibit saturation regions as sharply defined as is desired. Grown single crystals, although they have most of the desirable characteristics, are not yet readily available. Hysteresis loss in both single-crystal and polycrystalline barium titanate is higher by far than in most magnetic materials on a per-unit-volume basis. For this reason, barium titanate for dielectric amplifiers is generally

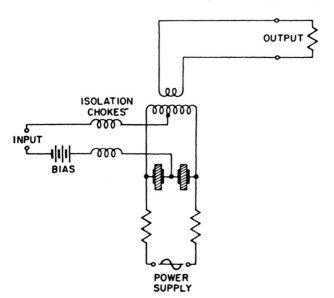

Fig. 3.23. Dielectric amplifier operated near resonance.

made with its Curie temperature slightly below room temperature, so that it can always operate slightly above the Curie temperature where the hysteresis loss is very low. It has been found that in this region the dielectric constant of a barium titanate condenser can be changed by changing the d-c voltage, and therefore the material behaves like a nonlinear capacitor.

In spite of present disadvantages, the future of dielectric amplifiers is indeed exciting. The packaging possibilities alone offer something that is not possible with magnetic circuits, for nature made magnetic fields divergenceless, so that closed paths—rings or shells—are needed for the lines of flux, whereas electric fields can start and stop on any material within which we can mobilize charge carriers. We can therefore think of many small ferroelectric condensers being fabricated side by side on a single thin sheet of barium

[14] H. Urkowitz, "A Ferroelectric Amplifier," *Philco Report* No. 199-M, Philco Research Division, Philadelphia, Pa.

titanate with electrodes produced by silk-screening, evaporation, or photoengraving processes.

Summary

Table 3.2 summarizes the various types of amplifiers. The three basic circuit elements—inductance,

Table 3.2. Power amplifiers

		Controlled Circuit Element		
		Inductance	Capacitance	Resistance
Input	Magnetic field	Magnetic amplifier		Magneto-resistive amplifier
	Electric field		Dielectric amplifier	Electro-resistive amplifier (vacuum tube)

capacitance, and resistance—are controlled by a magnetic field or an electric field. In the magnetic amplifier the inductance of a circuit is controlled by a changing magnetic field. In the dielectric amplifier the capacitance is controlled by a changing electric

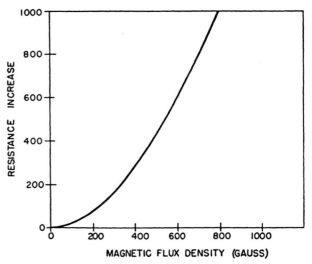

Fig. 3.24. Magneto-resistance of bismuth at 4.2°K. (After Alers, personal communication.)

field. In the vacuum-tube amplifier the resistance is controlled by a changing electric field. A magnetoresistive amplifier might be built utilizing the anomalously high magnetoresistance † of bismuth at low temperatures (Fig. 3.24). There are still two blank squares in the table.

† Magnetoresistance is the tendency of materials to change their resistance when placed in a magnetic field. Most materials exhibit this property to a small extent, but in bismuth it is high and becomes higher as the temperature is lowered. A 35-kilogauss field will produce a 3-to-1 increase in resistance at room temperature, but this factor increases to over 10^5 to 1 at the temperature of liquid helium.

In conclusion, we might take the liberty of rephrasing Lee de Forest's eloquent sentences:

This is the situation as it stands today. No one can say, however, that the situation will not be altered very materially in one, two, or three years, after we learn to build [rectangular-loop ferroelectric ceramics and other improved materials]. That will create a very different situation. It is difficult to say, therefore, which of the methods discussed possesses the greatest practical promise.

4 · Memory Devices

By WILLIAM N. PAPIAN

In most communications and control systems it is necessary, at times, to "remember" or to "store" some of the information which is being processed. This need is particularly obvious in digital computer systems; many pieces of information, such as instructions and numbers, have to be stored away for later use in the computation or control problem at hand.

The modern, high-speed, electronic digital computer does its computation in binary arithmetic. (For example, the binary equivalent of the decimal number 25 is 11001.) These binary numbers are usually stored in a grouping of memory cells called a memory "register." Each cell of the register need store only one binary digit; it may be likened to a toggle switch which can store one of two positions, "on" or "off," or, in binary notation, "ONE" or "ZERO."

One of the most important problems in digital computer design today is to develop a very large memory within which arbitrary selection of one out of the many registers is accomplished rapidly and efficiently.

Storage in a Stationary Field

We shall discuss here the storage of binary information in a stationary magnetic or electric field only; moving-part devices (such as rotating magnetic drums) and electromagnetic delay lines will not be treated.

As introduction to the elementary concepts of information storage in a stationary field, let us consider the possibility and disadvantages of using linear capacitors or inductors for such storage. The polarity of the charge on a linear capacitor may be defined in terms of binary states, say positive for a ONE, and negative for a ZERO. "Reading," or sensing, the information state of the capacitor can then be accomplished either by allowing the voltage across the capacitor to open and close an electronic "gate," or by applying a "write ZERO" voltage pulse to the capacitor and observing whether it reverses its polarity or not (such a reversal would result in an observable pulse of current in the connecting wires). This type of memory has two important practical limitations. One is that the charge leaks off, partly through the dielectric, mostly through the outside circuit. Second, the linear capacitor offers no help in the problem of selecting one out of a multitude of memory cells; some type of selection matrix becomes necessary [1] containing at least as many nonlinear elements (such as crystal diodes) as there are memory cells. It will be seen later that a good part of the selection mechanism may be "built into" the memory cells in the form of nonlinear characteristics, and the loss of information by charge leakage may be avoided by using a material having a high residual, or remanent, charge.

The dual of linear-capacitor storage is storing each binary digit of information in the magnetic field of a linear inductor. Again selection is difficult and loss of information by the gradual collapse of the magnetic field even more troublesome.

Three general types of memories now exist which use some of the hysteretic properties of ferromagnetic materials such as remanence and nonlinearity to solve the information-retention and (or) the selection problem. These are (1) a parallel-access array with an external selection mechanism; (2) the Harvard stepping register which uses time (or sequence) as its selection coordinate; and (3) multi-coordinate arrays

[1] A. W. Holt, "A Very Rapid Access Memory Using Diodes and Capacitors," Paper 180, delivered at the 1952 National Convention of the Institute of Radio Engineers in New York.

with built-in partial selection. These will be discussed later after a short review of the basic memory property of a hysteretic device.

Basic memory property of a hysteretic device

Figure 4.1a is a photograph of the 60-cycle flux-current characteristic of a ring-shaped ferrite core which has a "rectangular" hysteresis loop. (The loop may be considered as quasi-static, and, to a rough approximation, it may be taken as the B-H, or flux-density versus magnetic-field-intensity, characteristic.) The loop intersects the vertical axis at the remanent flux points,

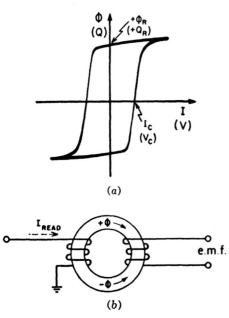

Fig. 4.1. (a) Hysteresis loop (Ferramic 1118-259); (b) Core (schematic).

$+\Phi_R$ and $-\Phi_R$. At these points, the current I is zero, and the core acts like a small permanent magnet. For all practical purposes the flux remains at Φ_R as long as the current remains zero or very small. The two remanent points may be assigned the binary numbers ONE and ZERO. A number may be "written" by applying a large current pulse of the proper polarity and duration; the number will be "stored" for an indefinite period of time and can be "read" by observing $d\Phi/dt$, the emf produced across a winding on the core during the application of the "read" current pulse, as indicated in Fig. 4.1b. The core, when used in this manner, may be classed as a biremanent memory device. Note that the remanent flux property of the core makes it possible to maintain a flux magnitude and polarity without current or magnetizing force. Loss of information by the collapse of the magnetic field, as for the linear inductor, is thus avoided.

The remanent charge property of a ferroelectric capacitor may be used to avoid the analogous loss of information due to the collapse of the electric field in a capacitor memory device. The charge-voltage characteristic of such a capacitor should resemble Fig. 4.1a, with the substitutions of charge Q for flux Φ and voltage V for current I. The stored charge remains at Q_R even though the capacitor is short-circuited and the voltage reduced to zero.

The extreme nonlinearity of the hysteresis loop is also an aid in the selection problem. Any memory element which is insensitive to an applied field of a given magnitude, but responds to twice or three times that magnitude, can be operated with many others in a large matrix or array where element selection is accomplished at the junction of two or three physical co-ordinates. This is the third type of memory in our list. It solves the onerous selection problem elegantly.

Some Memory Systems

Single-co-ordinate register selection, parallel access array

Where the digits of a stored word are wanted in parallel, that is, simultaneously, and the memory volume is not too large, a linear, or single-co-ordinate, register selection is practical. The principle of such a system is illustrated schematically in Fig. 4.2. The rings represent small ferromagnetic cores whose hysteresis loops exhibit "rectangularity" and high remanent flux. Register selection in this small (two-register, two-digit) memory is accomplished by the single-pole, two-position switch. In the example shown it makes circuit connections to the upper register. During the "read" operation, the circuit looks effectively like Fig. 4.2a. "Read" current is supplied through the selection switch to cores a and b, and the resultant induced voltages are fed out to the digit busses. The register is left full of ZERO'S, or "cleared," and ready for a "write" operation. The effective schematic during the "write" operation is illustrated in Fig. 4.2b. Here, "write" currents are supplied only by those digit busses which hold a ONE. In the figure the left-hand digit bus holds a ONE; it supplies "write" current, as shown, which passes through core a and through the selection switch to ground. The upper register is left with the binary number 10 in it. Nonlinear elements in the form of rectifiers act as isolators to prevent "sneak" currents from taking their complex paths through the array; were they absent from Fig. 4.2b, current could flow from the left-hand digit input through cores c, d, and

b, in that order, and then to ground through the selection switch.

The remanent flux aspect of the hysteresis loops of the cores allows information retention at zero driving currents, but no attempt is made to use core characteristics to aid in the selection problem. This type of memory uses at least one diode per stored digit plus

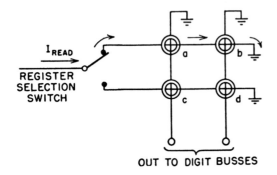

(a) READ

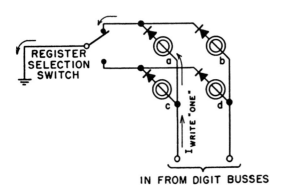

(b) WRITE

Fig. 4.2. Single-co-ordinate register selection.

a single-pole, n-position, electronic selection switch, where n is equal to the desired number of registers. The number of diodes and the size and current-capacity of the switch become prohibitive as the size of the memory is increased above a few hundred registers.

Time or sequence selection

The Computation Laboratory of Harvard University has done a large amount of pioneering work in the application of magnetic cores to digital computer problems. Its principal contribution was the development of a device known as the Static Magnetic Delay Line.[2] It might also be called a shifting register

[2] A. Wang and W. Woo, *J. Appl. Phys.* **21**, 1 (1950).

or stepping register, and consists of a line of cascaded memory cores so arranged that information can be shifted or stepped down the line sequentially, at a rate determined by a primary pulse source, until the information comes out at the end. As it is now being used at Harvard, the line requires two cores and four rectifiers per stored binary digit (see IIIC, Sec. 3).

The line is basically a serial memory device; parallel access to a word may be accomplished by using a group of lines, one line for the corresponding digits of the stored words. Words can then be stored and read only in time sequence, so that the co-ordinate for register selection may be said to be time or sequence. Cores with high remanent flux are used, but no attempt is made to utilize the core's nonlinearity for carrying a part of the selection burden.

The Static Magnetic Delay Line serves well as a memory of the serial type or as buffer memory between computer elements which run at different speeds. It is not, however, a strong contender for high-speed memory devices, where arbitrary-access and large scale storage are required.

Multi-co-ordinate register selection

Register selection may be accomplished at the junction of two or more physical co-ordinates. Ferromagnetic and ferroelectric memories of this type are being developed at the Massachusetts Institute of Technology,[3] at R.C.A. Laboratories,[4] at Bell Telephone Laboratories,[5] and possibly elsewhere. This application puts more stringent requirements on the cores than most applications. The memory system will be described in some detail, and the material requirements derived.

Figure 4.3 shows one of the internal hysteresis (Φ versus I) loops of the magnetic-ferrite cores used in a successful memory array. The selection scheme in this array depends on the fact that a current of magnitude I_m can change the flux in a core from $-\Phi_R$ to $+\Phi_R$, whereas half that current ($I_m/2$) has a negligible effect. This is a consequence of the high degree of nonlinearity shown by the sharpness of the "knees" of the loop.

If, for example, nine such cores are arranged in a planar array, as in Fig. 4.4, and currents $I_m/2$ are caused to flow coincidentally in the selected lines x_3

[3] J. W. Forrester, *J. Appl. Phys.* **22**, 44 (1951).
[4] J. A. Rajchman, *RCA Rev.* **XIII**, 2 (June, 1952).
[5] J. R. Anderson, "Ferroelectric Materials as Storage Elements for Digital Computers and Switching Systems," Conference paper delivered at the Winter General Meeting of the American Institute of Electrical Engineers, New York, Jan. 24, 1952.

and y_2, core F is the only core in the array subjected to the full magnetizing force (I_m). The others have

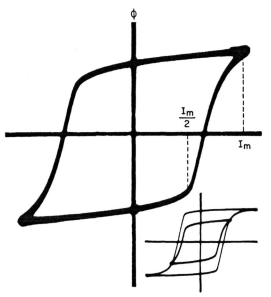

Fig. 4.3. Φ-I loop (Ferramic 1118-259).

either half or zero magnetizing force impressed. Each core operates as a coincidence device, and the only core whose flux is significantly affected is the one at the junction of the selected lines. Output voltages are taken from the sensing windings, S, after suitable mixing.

Extension to three dimensions may be accomplished by stacking such two-dimensional arrays along a z-axis, with corresponding x lines connected in common and corresponding y lines connected in common. In this arrangement the application of $I_m/2$ to an x line and a y line results in the selection of a *line* of cores (a register) parallel to the z-axis. To "write ZERO" in some of the cores of the selected register requires only to "inhibit" them by an opposing polarity current of half magnitude.

Variations on this selection scheme have been devised which impose slightly less stringent requirements on the core. It is possible to have the core discriminate only between currents which bear 3:1, 5:1, or greater ratios to each other;[6] advantage may be taken of these greater current ratios to reduce switching times, increase signal-to-noise ratios, or relax some of the hysteresis-loop requirements. Reduced discrimination implies an increase in the electronic equipment of the system.

Figure 4.5 is a photograph of a planar ferrite memory array operated at the Massachusetts Institute of Technology. The 256 small memory cores occupy the square in the center of the picture. The x and y driving windings consist merely of the wire grid upon which the cores are mounted (Fig. 4.6). Size

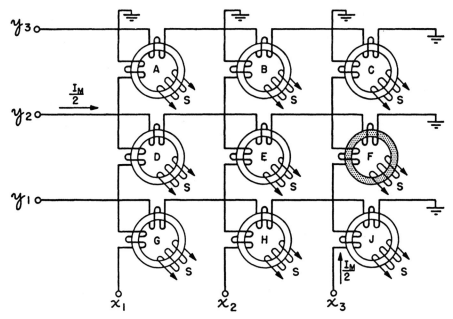

Fig. 4.4. Two-dimensional array of cores.

may be judged from the heavy co-ordinate wires, which are AWG No. 20 enameled magnet wire.

[6] R. R. Everett, "Selection Systems for Magnetic Core Storage," Engineering Note E-413, August 7, 1951, an internal document of the Digital Computer Laboratory, Massachusetts Institute of Technology.

Fig. 4.5. Ferrite memory plane.

Memory Devices

Fig. 4.6. Close-up of ferrite memory plane.

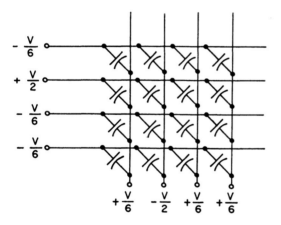

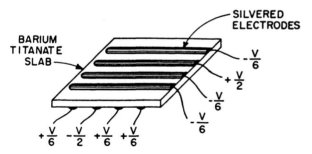

Fig. 4.7. Ferroelectric memory.

The dual of this type of memory is not difficult to visualize. One possible arrangement is illustrated in Fig. 4.7. The voltages shown provide selection with a 3:1 voltage ratio, since the barium titanate ceramics available have not yet sufficiently rectangular Q-V loops to operate as coincident-voltage memory cells at 2:1 ratios. The barium titanate slab containing the experimental 8-by-8 memory may be seen at the top of Fig. 4.8. In the upper compartment of the

Fig. 4.8. Ferroelectric memory.

"cage" a ferritic core "current transformer" serves in the sensing circuitry.

Material Evaluation and Testing

In order to judge the performance of ferromagnetic and ferroelectric materials, relevant criteria must be established describing quantitatively the ability of the parts of the memory system to process information accurately and rapidly. The ONE-to-ZERO output-signal ratio and the switching (or flux-reversing) time tell a great deal about a simple magnetic-core memory unit of the non-coincident-current type. Criteria for the coincident-current unit are somewhat more involved, and will be discussed in some detail.

Information retention and loop shapes

An "undisturbed ONE" is defined as the $-\Phi_R$ flux state of the core and an "undisturbed ZERO" as the $+\Phi_R$ state (Fig. 4.9). Let a "read" pulse of current

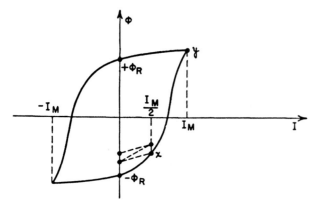

Fig. 4.9. Paths of operation of a magnetic memory unit.

be arbitrarily fixed at $+I_m$ so that reading a ONE results in a large flux change (large output pulse) and reading a ZERO gives a small output pulse. A ZERO is then "written" by a $+I_m$ pulse also, and a ONE is written by a $-I_m$ pulse.

Recall that selecting one core in a two- or three-dimensional array results in the application of $I_m/2$, called a "half-selecting" pulse, to cores elsewhere in the array. The application of repeated half-selecting read pulses to a core containing a ONE tends to run the remanent state of that core up along the Φ-axis (dashed lines in Fig. 4.9) disturbing or destroying its information. When the hysteresis loop is properly rectangular, the operating point moves up the axis only a short distance above the point $-\Phi_R$, and the core operates satisfactorily for the coincident-current scheme.

A core which contained an undisturbed ONE and has been subjected to a large number of half-selecting read pulses is considered to hold a "disturbed ONE." By the foregoing reasoning, the disturbed-ONE output is smaller than the undisturbed-ONE output. In an analogous manner, repeated half-selecting write-ONE pulses will run the core's operating point from $+\Phi_R$, downward, increasing the size of a ZERO output pulse. A disturbed-ZERO output is larger than an undisturbed-ZERO output.

Since half-selecting disturbances reduce the output signal from a core containing a ONE and increase it from a core containing a ZERO, the ratio of the disturbed-ONE output to the disturbed-ZERO output, the "disturbed-signal ratio," is a critical measure of a core's performance as a coincident-current memory unit. This ratio approaches infinity in the ideal case of a square loop; it should be much greater than one for a reasonable discrimination between the binary digits.

The application of a half-selecting pulse to a core results in a voltage output or noise, called a "half-selected output." The ratio of a disturbed-ONE output to a half-selected output, called the "half-selected signal ratio," is another important criterion of operation. Like the disturbed-signal ratio, it approaches infinity in the ideal case and should be much greater than one for satisfactory operation.

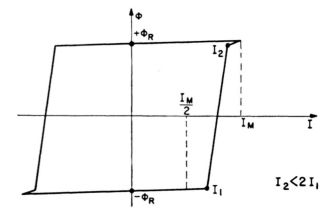

Fig. 4.10. Idealization of a Φ-I loop.

According to the idealized hysteresis loop of Fig. 4.10 necessary conditions for coincident-current operation are:

$$\left.\begin{aligned} I_M &> I_2, \\ \frac{I_M}{2} &< I_1, \end{aligned}\right\} \text{hence } I_2 < 2I_1.$$

(I_1 and I_2 are the points at which the Φ-I curve changes direction abruptly.) Experimental results give qualitative support to this general shape criterion for the hysteresis loop.

Pulse testing

The previous considerations lead directly to the magnetizing pulse patterns desired for signal testing. Two of these patterns are illustrated in Fig. 4.11; mode a consists of alternate-polarity, full-amplitude pulses for checking an undisturbed-ONE; mode b checks a disturbed-ONE by interspersing a large number of half-amplitude pulses between the negative (write-ONE) pulse and the positive (read) pulse. The response to the half-selecting pulses may also be observed during mode b operation. Other modes check for undisturbed and disturbed ZEROS. (Pulse amplitudes, lengths, spacing, and the number of intervening half-selecting pulses can be varied independently in these tests.)

The hysteresis loops are generally observed at low frequencies by usual methods. Response times are taken as equal to the lengths of the disturbed-ONE output pulses; the length of the test pulses is kept somewhat longer than this.

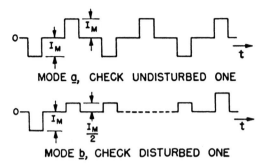

Fig. 4.11. Core-testing pulse patterns of I.

Many cores were tested and the choice narrowed to two types of cores: the one metallic, the other a magnetic ferrite. The metallic core (Allegheny Ludlum one-mil Selectron tape) had excellent signal ratios; its response time, during the reading of a ONE, was about 25 microseconds. Scope traces of the disturbed-ONE and ZERO modes are shown in Fig. 4.12 with arrows pointing to the pertinent output pulses. The

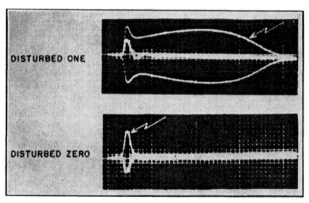

Fig. 4.12. Coincident-current test results, metallic core.

disturbed-signal ratio and the half-selected signal ratio are both obtainable from these traces. (The time scale is 5 microseconds per large division.) The negative trace in the upper photograph is the core output when the ONE is written; the heavy center trace combines the large number of half-selected outputs which follow the writing; and the final disturbed-ONE output shows as the large positive pulse. The positive trace in the lower photograph is the output from the read or write-ZERO operation; the outputs due to the interspersed half-selecting write-ONE pulses merge in the heavy negative trace.

A very promising material for high-speed work is a magnetic ferrite obtained from the General Ceramics and Steatite Company. It has fair signal ratios and a response time near one-half microsecond. Figure 4.13 shows scope photographs for this core under two con-

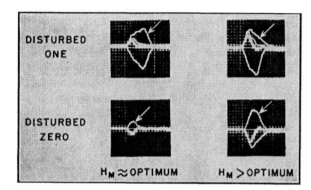

Fig. 4.13. Coincident-current test results, ferritic core (horizontal scale ca. ½ μsec per large division).

ditions: I_M adjusted to optimum amplitude (first column) and $I_m >$ optimum (second column). In the second case an unsatisfactory disturbed-signal ratio close to one results.

Table 4.1 summarizes the important test results for

Table 4.1. Core comparisons

Core Criteria	Best Metallic		Best Ferritic		
	1950	1952 *	1950	1952 †	Ideal
Disturbed-signal ratio	13	18	6	10	$\to \infty$
Nonselecting-signal ratio	16	20	3	10	$\to \infty$
Response time (μsec)	25	10	½	¾	$\to 0$

* 79-4 Molybdenum Permalloy core; ⅛-mil ribbon of Magnetics, Inc.

† Ferramic 1118 of General Ceramic and Steatite Company.

the two types of cores. (The signal ratios are taken from the voltage-time areas of the output pulses; this gives a rather pessimistic, but fundamental, measure of a core's characteristics.) Note that the major improvements have come to the metallic cores as a reduction in response time and to the ferrites as increases in signal ratios.

One additional important core characteristic is the amount of total single-turn driving current required (I_M) for operating a coincident-current unit. It depends on the size of the core and its coercivity. The current values are roughly 0.2 and 1.0 amp, respectively, for metallic and ceramic cores now operating at Massachusetts Institute of Technology.

Development program

The improvement of ferromagnetic and ferroelectric materials for computer circuits is being carried on at two levels. A large amount of empirical development work is being carried on by commercial manufacturers. Fundamental investigations of ferrites are undertaken at the Massachusetts Institute of Technology by the Laboratory for Insulation Research and the Digital Computer Laboratory. The goal is reached when we are able to specify explicit synthesis procedures for materials of prescribed characteristics.

Conclusion

Storage of binary information in the stationary field of a ferromagnetic ring or a ferroelectric slab promises a satisfactory solution to one of the most vexing problems in the digital computer field, the problem of "remembering" and making available a very large number of binary "words" rapidly and efficiently. The remanent flux and nonlinear characteristics of the "rectangular" hysteresis loop simplify the information-retention and the word-selection aspects of the problem tremendously.

Highly rectangular hysteresis loops, high speeds, low losses, low driving requirements, excellent quality and uniformity, and reasonable cost are among the characteristics desired in these materials for the applications in view. Freedom from aging and deterioration are also extremely important.

Improvements in core materials during the last two years have been encouraging, and much future progress is expected.†

---

† Much progress has occurred since the above article was prepared. Full-scale memories of the multicoordinate type are now in computer operation at Massachusetts Institute of Technology (W. N. Papian, "The MIT Magnetic-Core Memory," *Proceedings of the Joint Eastern Computer Conference*, December, 1953, published by the Institute of Radio Engineers); improved ferrite cores are available in quantity from various sources, notably the General Ceramics Corporation; and a powerful supporting effort under the direction of D. R. Brown at Massachusetts Institute of Technology has succeeded in demonstrating that hundreds of thousands of cores can be individually tested on a production basis and that cores with desired characteristics can be synthesized in a scientific manner and reproduced in quantity with a high degree of uniformity [D. R. Brown and E. Albers-Schoenberg, *Electronics* 26, No. 4, 146 (1953)].

Two banks of computer memory are operating in the Whirlwind computer (M. F. Mann, R. R. Rathbone, and J. B. Bennett, "Whirlwind I Operation Logic," *Report R-221*, May 1, 1954, an internal document of the Digital Computer Laboratory, M.I.T.). Each bank contains 1024 registers with 17 binary digits per register, making a total of 17,408 cores in each bank. The cores, which are made of General Ceramics material MF-1326B, are very small (80 mils o.d.). Single-turn switching currents are approximately 850 ma, and single-turn output voltages are about 0.1 v. Switching time, under these conditions, is about 1.2 μsec. Core selection was made on the basis of a series of pulse tests, approximately four per core, and resulted in a yield of approximately 40 percent of those shipped to us by the producer. These core memories replace the electrostatic storage tubes previously used, and they have resulted in greatly increased reliability and speed, and reduced maintenance and floor space requirements.

A third bank of core memory, four times as large as each of the two in Whirlwind, is also operational. Any one of its 4096 binary words (17 digits) may be read and rewritten in 6 μsec. Its reliability has been demonstrated during the last four months in a full-scale test computer.

The confidence now placed in the ability of ferrite manufacturers to produce large quantities of satisfactory cores at reasonable prices is such that new memories requiring over a million cores per bank are being seriously contemplated.

The research described in this and the preceding contribution by D. A. Buck is supported jointly by the Army, Navy, and Air Force under contract with the Lincoln Laboratory, Massachusetts Institute of Technology.

IV · DIELECTRIC REQUIREMENTS OF THE ARMED SERVICES

In closing this book, *Dielectric Materials and Applications*, it seems appropriate to present brief statements by the Armed Services on their special interest in this subject and an explanation of some of the more outstanding motivations behind this interest. Many of us have grumbled over specifications and "red tape." By hearing about the problems and worries behind the scene, we will be better able to interpret these distress signals and give help with a new kind of understanding.

A. v. H.

A · The Air Force

By FRED H. BEHRENS

In opening the discussion for the United States Air Force, the presentation of needs and applications of dielectrics will primarily be restricted to airborne electronic equipment, where size and weight are of nearly equal importance to other requirements. Ground-equipment applications are largely common with those of the Naval and Ground Forces and will be discussed by these Services.

Air Force's Interest in Dielectrics

Miniaturization of airborne electronic equipment

Military requirements are steadily emphasizing operation at higher speeds, higher altitudes, and under all weather conditions. This operation, in turn, requires electronic aids to carry out the many complicated flight functions, such as navigation, bombing, gun-fire control, and missile guidance. In fact, the precision required and the extremely short time spans available for executing essential functions at such speeds, require the Air Force to rely more and more upon automatic equipment which will either direct the air crews' actions or even carry out the entire operation without human aid. In a typical run against a target at near sonic speeds, the time span from target detection and identification, through necessary sighting and preparatory operations, and on to final action may be only a few seconds or, at most, about a minute. Under adverse weather conditions, electronic aids may be the only means of navigating the aircraft through its mission and back to a safe landing at home base. Hence the emphasis on the development of more and better electronic equipment.

Naturally, such equipment is complicated and poses difficult design problems in view of the limited space and weight-carrying capacity of airplanes. The only practicable answer is miniaturization of the airborne electronic gear.

The normal operating temperature range of miniaturized equipment will necessarily have to be substantially broader than that for conventional equipment, even if cooling means are used. Studies at Wright Air Development Center have established a range from $-65°$ to $+200°C$ as representative. Over this temperature range the bulk of electronic components will have to operate satisfactorily. As a result, the range $-65°$ to $+200°C$ has been adopted as the current goal for all airborne electronic component development for the United States Air Force. Higher temperatures are required in isolated instances where internal hot spots, aerodynamic heating, and other influences demand still more stable materials. Thus this temperature requirement alone would justify a great interest of the Air Force in dielectrics, since the dielectric materials set, in general, the temperature range at which equipment may operate. Even a cursory survey of existing materials will show that comparatively few can qualify for the coming applications.

Reliability in airborne electronic equipment

A further major consideration in the Air Force's interest in dielectrics is equipment reliability. Data sensing, guidance control, navigation, and communications through electronic means make an airplane and its crew dependent on the proper functioning of literally thousands of tiny circuit elements. Malfunctioning of almost any of these elements can result in an aborted mission, if not more serious consequences. It is clear that with so many elements the chance of one failure in the time span of even one flight or mission is great, when all components are of normal reliability and average life span. There appear only two ways out of this dilemma, simplification of equipment or increase of the reliability of equipment and components. Normal development may be expected to simplify the present equipment to some degree, but the chances are that the increased need for still more electronic equipment will neutralize much of this developmental gain. The only alternative is to increase the reliability of all components. Some of this increased reliability can undoubtedly be realized through improved design, more conservative rating, and better quality control in manufacture. But a great proportion of this increased reliability must necessarily come from the development and more intelligent use of better dielectric materials.

The word "better" may need further definition as applied to this case. Regardless of the form of the dielectric material (solid, liquid, or gas), to be "better" for these applications, it must be uniformly stable, both in physical and in dielectric properties over the full range of operating temperatures specified above. The materials must retain that stability under varying atmospheric pressure and moisture conditions, in operational aging, and under exposure, in some cases, to radiations of various kinds. Furthermore, the dielectrics must retain stable and dependable properties when subjected to the various processing and fabrication activities which convert them into dependable finished components.

Some Fields of Special Interest

Radome dielectrics

Probably the most extensive single dielectric development effort sponsored directly by the Air Force is in radome materials. This is true, not because it is necessarily of the greatest relative importance, but rather because the Air Force is still a young organization. Except for caring for particularly pressing special interest developments such as radomes, it still depends on others for a large portion of its dielectric development.

The radome, a radar window in the streamlined exterior surface of the aircraft, makes heavy demands upon the constituent dielectric materials. It must be transparent to microwave energy, of low dielectric constant and loss, high strength, and stable in electrical, physical, and serviceability properties over a wide range of temperatures, since as an exterior part of the aircraft it is subject to aerodynamic heating and heavy structural loading. It must be moldable to any shape dictated by aerodynamic and microwave optical design considerations, and light in weight. The usual radome is a composite sandwich structure, made up of thin, high-strength inner and outer skins enclosing a low-density bonded core which provides proper spacing and stabilization of these skins.

The principal Air Force radome materials program is aimed at caring for the high-operating temperatures of streamlined supersonic aircraft and guided-missile applications. One group of materials, principally organic, will serve where aerodynamic heating forces the temperature up to 260°C. A second group of materials, principally inorganic, is being developed to care for applications well above 260°C.

The materials in the first group include a new low-dielectric constant, low-loss glass fiber for plastic laminate reinforcement. The dielectric constant is slightly under 4.0 with a loss tangent of approximately 0.002 at a frequency near 10,000 Mc. This glass reinforcement is intended to be used with either the common polyester laminating resins, or with the newer high-temperature resins such as triallyl cyanurates. It may also serve for many other general-purpose applications. A high-temperature organic foam of approximately 10 lb per cu ft density is being developed for the new foam-in-place sandwich radome fabrication process. The ultimate goal, good working strength and stability at 260°C, has not yet been reached, but the new foams will operate at temperatures slightly over 200°C ($\kappa' \sim 1.2$; tan $\delta \sim 0.002$ at ca. 10,000 Mc and 25°C). For the higher temperature range, some low-density ceramics are promising ($\kappa' \sim 1.45$; tan $\delta \sim 0.001$; compressive strength ~ 200 psi).

High-temperature dielectrics

A number of development projects for high-temperature dielectrics are being co-sponsored by the Air Force and the Army Signal Corps. They include extrudable dielectrics, insulation for hook-up wire, liquid dielectrics, molding materials, and dielectric films.

Conductive elastomer for precipitation of static

A special-purpose material needed for radomes and other antenna housings is a surface coating for the dual function of protecting from rain erosion and draining off static charge, which, collecting on the insulating surfaces, causes severe radio interference.

The desired erosion protection can be achieved with a layer of neoprene cement 10 mils thick. This resilient material absorbs the impact energy of the raindrops quite effectively, and, if the neoprene is loaded with a low-loss filler instead of lossy carbon, the transmission of even 3 cm waves is not much impaired. A conductive surface for static charge drainage can be added by a very thin overlay of conductive neoprene (2 mils). This conductive surface layer can be tolerated if its resistance is greater than 0.25 megohm per square (it will drain static adequately if its surface resistance is less than approximately 10 megohms per square). Thus, if the conductive material is thin and the resilient coating deep enough, the objective can be realized; but the material must remain stable in service. Thus far we have only marginally satisfactory materials from the stability point of view. The surface resistance of the conductive overlay tends to rise above desired limits on heat aging, weather exposure, etc.

Ice-phobic surface treatment

An important problem for the Air Force is surface anti-icing by the development of ice-phobic surfaces or surface treatments. Ice accretions on the radome obviously degrade the transmission effectiveness of the radome, but, even worse, introduce intolerable directional errors of guidance equipment.

There are other approaches to the anti-icing and de-icing problem, such as the pulsating de-icer boot, the thermal de-icing system, the spraying of freezing-point depressant liquids, and de-icing agents. Each of these approaches, unfortunately, involves some rather serious compromises which an ice-phobic surface treatment approach could circumvent. It is recognized that unsuccessful work has been done along this line in the past by various investigators. There is, however, experimental evidence that silicone oils and greases on radome surfaces reduce ice adhesion to low values. These treatments rapidly wear off in flight, but they may point a way to a successful solution.

Conclusion

Considerations of miniaturization of electronic equipment and of providing the essential increases in component reliability of airborne electronic equipment are the prime motivations behind Air Force's interests in dielectrics development. The immediate range of operating temperatures for most dielectric material types is $-65°C$ to $+200°C$. Extension of this range is required for internal insulations subject to local hot spots and for external materials, especially radomes, liable to aerodynamic heating.

B · The Army

By CHARLES P. LASCARO

In the course of modern warfare, Signal Corps equipment may be airborne or shipborne, in addition to the usual means of ground transportation. Equipment may be dropped or stacked on a beach for a considerable period of time, exposed to the effects of salt spray, sand, and sea water. In guided-missile and high-altitude meteorological use, it may be carried to outer space. Obviously, the Army has many requirements in common with the Navy and the Air Force and certain others which must receive specific consideration.

Basis for Army Requirements

Before describing actual requirements, it might be helpful to nonservice members to explain briefly the basis for Army Service requirements. The Army Field Forces, strategically prepared to fight under worldwide environmental conditions, prescribe the military characteristics of our communication equipments. The responsibility for the development and design of such equipment rests with the Signal Corps Engineering Laboratories at Fort Monmouth, New Jersey. Here the design engineers devise systems and component equipment. In order to meet desired military characteristics, the equipment designer may need improved component parts and materials which are not yet available. Since equipment can be produced only with developed components and materials, the lack of such parts and materials represents a loss of military characteristics and a sacrifice of certain objectives in military strategy. The component-parts engineer and the materials engineer must therefore keep abreast of equipment-design trends and anticipate by an appreciable margin the future needs of the services in terms of material requirements. This is especially important since new parts and materials require usually a long development time before production. Frequently, dielectrics of improved properties mean great strides in the improvement of equipment performance and reliability. If certain requirements appear unrealistic to the design engineer, it must be realized that they represent ultimate objectives which will buy improved performance needed by the Army Field Forces.

The basis for many of the present-day dielectric material requirements are new systems and devices, such as guided missile control and instrumentation; new wire, radio, and television equipments; high-altitude meteorological devices; electronic computers; high-power radar and broad-band countermeasures equipment; and radiological detection.

Requirements and Development Programs

Reliability and miniaturization are at present the most important requirements, as already stressed by the spokesman of the Air Force.

Reliability

To meet the extremes of worldwide field service, the performance of dielectrics under the following conditions and certain combinations thereof must be known and controlled as closely as possible.

Temperature. Operation from $-55°$ to $200°C$ and storage to $-65°C$. Predominantly, however, Signal Corps requirements lie within the range of $-55°$ to $85°C$, and dielectrics that withstand these temperatures may be fully utilized in many applications. A few critical situations call for performance up to $125°$ or $150°C$, and the current practice of sealing and miniaturizing leads the trend towards higher temperatures between $85°$ and $200°C$.

Relative humidity. 0 to 100 percent R.H., including condensation. In general, a 100-megohm minimum impedance level should be maintained under the wettest conditions and for indefinite periods; for certain critical applications, higher insulation resistance is required. Insulator bodies for high-voltage and special circuit applications must have nonwetting surfaces and be arc resistant. For low-voltage circuits as in wire communication, lower-impedance levels down to 1 megohm are acceptable.

Shock and vibration. Fabricated and assembled dielectric bodies must be resistant to the effects of shock, bounce, and vibration, especially at low temperatures. The prime requirements for dielectric materials, in general, are not those pertaining to mechanical and shock-resisting strength since proper design and fabrication techniques can usually protect mechanically weak but electrically attractive materials. A more rugged material, however, if obtainable, will eliminate the need for extra precautions. Studies of complete equipments have indicated that the worst conditions arise during transportation; here equipment elements may be subjected to accelerations up to 10 g's. A four-foot drop test is usually applied to assembled equipments; vibration frequencies do not exceed 33 cps. For equipments subject to ballistic impact and other critical applications, this average shock, bounce, and vibration may be greatly exceeded. Here, extra protection is achieved through equipment design. Summarizing, it is desirable to have ruggedness an inherent feature in all elements and materials, but dielectrically attractive materials can be used if no other alternatives exist.

Resistance to atomic radiation. Dielectric materials should be resistant to the effects of atomic radiation. Especially radiological measuring instruments such as dosimeters should not change their high-insulation values significantly.

Surface contamination. Material surfaces should be resistant to fungus growth and other forms of contamination, such as industrial dusts, gases and vapors, lubricants, and salt spray. The full effects of such contamination are normally not anticipated, since materials are usually tested and evaluated in a cleaned condition.

Service life. Past experience indicates that equipment, after storage for 5 to 8 years in shelters and warehouses in all parts of the world, may suddenly be put to intensive field use for periods up to 3 or 4 years.

Critical materials. Often dielectrics are utilized that may be in critical supply during an emergency period. The use of such materials should be minimized, and a suitable substitute found.

Miniaturization

The necessity to miniaturize is secondary to reliability, but current circuitry and equipment-design practices include both factors. The close proximity of parts in miniature assemblies makes the following characteristics highly desirable in plastic or other dielectric parts used to mount, hold, and insulate components.

Low dielectric constant ($\kappa' \leq 3$) to reduce distributed capacitance effects in all interconnecting, mounting, or packaging media.

High dielectric constant ($\kappa' \geq 15$) in certain miniature-type capacitors, coupled with thermal stability and low-temperature coefficient of κ'. Such materials should be capable of fabrication into thin flexible films or slabs (down to 0.25 mil).

Higher dielectric strength because of the closer spacing of conducting elements. Terminal boards are smaller, the miniature inserts are closely spaced, dielectrics are used as insulating barriers (between components and chassis). In addition, high dielectric strength is of particular importance in achieving miniaturization of components.

Heat resistance because the high concentration of dissipative components in miniature assemblies means considerably higher temperatures than in conventional-size assemblies. A goal of 200°C is therefore the requirement in many new dielectric developments for the miniaturization program.

Lower power factors at higher frequencies since the dielectric losses are a problem in many high-frequency applications in miniature assemblies.

Special problems

(1) Present ASTM-method test data on the dielectric strength of suitable materials are not applicable to high frequencies. Breakdown data are needed for such materials over a frequency range from 1000 Mc to the top limit of the equipment. Specifically, the interrelation between breakdown characteristics and material parameters of insulating materials at various pulse patterns at radar frequencies must be known so that design data can be made available to development engineers.

(2) Increased weight, low thermal conductivity, special circuit and mechanical assembly design, and other factors prevent the electronic industry from taking advantage of casting resins for complete imbedding of miniaturized equipment. Casting resins should be developed which incorporate as many of the following characteristics as possible: low weight, high thermal conductivity, high dielectric strength, high insulation resistance, low moisture absorption, low shrinkage, long pot life, low corrosivity, high thermal aging resistance, and the ability to absorb vibration effects.

(3) Many ceramic-type capacitors made with high-dielectric-constant bodies must be severely derated because of poor load-life characteristics. Capacitors operating at least 10,000 hours under working stress of 30 to 60 volts per mil are required. It is needed to determine the chemical and physical factors which control the degradation of the dielectric under prolonged exposure to high temperature and voltage stress and to determine the mechanism by which the degradation takes place.

(4) For capacitors in critical circuit applications, a high-constant body of low temperature coefficient is required. The following list of specifications is indicative:

κ'	150
Power factor	0.0005 at 1 Mc, $-40°$ to 100°C
Temperature coefficient of κ'	± 30 ppm, $-40°$ to 100°C
Volume resistivity	10^{12}, $-40°$ to 200°C
Dielectric strength	250 volts per mil, -40 to 200°C
Life test	1000 hours, 200°C
	Voltage increased in steps 5 to 8 weeks
	10, 15, 20, 25, 30 volts per mil, 1 week each
Thickness	0.001

(5) For insulator use in critical high-frequency circuits, an ultra low-loss ceramic-type body is required (tan $\delta \sim 0.0004$) which will meet Grade L-6 of JAN-I-10 requirements. Wollastonite shows promise, but has not yet proved itself.

(6) The majority of dielectric failures occur on the surface of insulators. Much has been done to improve surfaces by glazing, surface treatments, and increasing length of surface path, but little has been done to establish the effects of various contaminants on insulation failures.

C · The Navy

By THOMAS D. CALLINAN

The Navy's needs for improved dielectrics arise from its mission of guarding the seas by maintenance of combat efficiency above, below, and at the water's surface. This mission requires the application of electricity in the confined quarters of ships, submarines, lighter-than-air craft, and carrier-based planes. The dielectric qualities of materials are paramount in the mind of the developer of novel instruments; space, weight, and logistic factors are pre-eminent in the architect's mind; and the reliability and effectiveness of the equipment are decisive considerations in the commander's mind. Emphasis here is placed on dielectric problems specific to the Navy in terms of uses, physical and chemical parameters, environment, safety, and combat efficiency.

Power generation

The large surface craft of the fleet use turbogenerators as the means of transforming fuel into controllable power. Such a system operates at high efficiencies, permits split-second control, long life, minimum lubrication problems, and high-space utilization. An a-c system is used generally except in standard submersibles, where a combination of diesel-generator and battery-supplied direct current is employed.

In the generation of electric power on shipboard, three factors endanger the electric system and thus the combat efficiency and safety of the craft: corrosion, electric breakdown, and fracturing of the insulation due to mechanical vibration. Metallurgists and protective-coatings experts are trying to solve the first problem; the second, dealing with the impulse breakdown of dielectrics, is being investigated sporadically; and the relationship of mechanical stability to electrical life is still in the discussion stage.

The development of long-lived, *compact* generators rendering the utmost in reliability and safety has lagged to the detriment of weight, space, and economy. Overdesign and safety factor ratings (based on inadequate fundamental knowledge), often exceeding 250 percent, are an encroachment on the fighting power of the craft. Development of laws relating to the impulse breakdown of solid insulation and a more precise knowledge of the nature of mechanical deterioration of dielectrics are most needed at this time if engineers are to be relieved of the present long-drawn trial-and-error technique for testing every new varnish, coating compound, and insulant for service life.

Power distribution

The power generated must be distributed efficiently as energy for ship propulsion, gun control, pumps, lighting, winches, and stabilizing equipment; for use in damage-control operations such as air-cleaning apparatus, auxiliary lighting, and air-pumping equipment; and for the transmission of intelligence through various intercommunication networks, as well as radio, radar, and special devices such as sonar. Problems of ship-board distribution arise from the reaction of the components to their environment and from the diversity of the system; the former is of special interest to the development engineer concerned with materials, whereas the latter is a problem in systems co-ordination.

D-c distribution

Because submarines are employed in individual operations, they should be constructed in such a way that their crews can repair them with a minimum of equipment. Thus bus-bars are used where cables might be more efficient. Crowded quarters necessitate insulating the buses, but arguments against cables still hold. To solve the insulation problem, the Navy requires a coating material which can be applied to bus-bars already installed. It must bond to copper, be applicable with a brush or knife, dry without the evolution of much solvent in thickness of 20 mils. The material should be tough, adhere to the conductor when subjected to shock, and not leak electrically under the moist conditions of submarine existence.

A-c distribution

Between conductors in confined quarters, high resistance is required and high dielectric strength at rated and impulse voltages. That high values of breakdown may be obtained by using impregnated laminated structures is part of classical engineering; the low heat dissipation that results necessitates the development of thermally more stable materials. We should have cables capable of withstanding temperatures up to 250°C and maintaining their flexibility at temperatures as low as −55°C.

System co-ordination

The separation in power and communication requirements has led to the development of individual equipments without regard for the overall situation. Improvements of importance have been made in the quality of components for power purposes. However, the growing burden placed on the power installation by ever-increasing communications facilities has resulted in demands on power distribution facilities and space for which the power specialists were unprepared. Miniaturization of components, the re-evaluation of safety factors in design, and the development of such special devices as the transistor may offer some remedies.

Communications equipment

Communications equipment has as its purpose the rapid, effective, and accurate transmission of combat command decisions, and the detection and identification of hostile craft. Telephonic radio, radar, and acoustic techniques are employed in equipment built with the emphatically requested qualities of ruggedness and reliability under battle conditions. Needless to say, the effect of a shock wave under battle conditions on a liquid-filled component, on the vibration of a glass tube or a cold solid cable has been the source of justified worry to many a communications officer.

The problems arising with the use of communications equipment frequently concern the qualities of components. The criticism that military requirements are excessive is basically a disagreement over safety factors in design. Requests as to thermal stability, exceptional dielectric strength, low losses, stupendous resistances to environmental conditions are justified in part by the history of failure in service. The remedy is thought to lie in a further raising of the specification standards.

However, this tendency for overdesign does not give the commander of a craft operational certainty; once a piece of equipment has been put into service, it begins to deteriorate. Time alone does not measure this decreasing quality, for the accelerated deterioration caused by combat operations is impressive. The saving of lives, materials, and vast sums of money spent for replacements, spare parts stocks, and rescue operations could be prodigious, if we could judge the quality of insulation during service. A nondestructive test for evaluating the state of insulation and its present expected service life is urgently needed and of the utmost importance for the continued and improved safety and combat efficiency of the fleet.

Logistics

The logistic problems associated with equipment include the employment of dielectrics indigenous to the United States. The necessity of importing critical materials from overseas weakens our position in time of war. The development of a thermally stable, coated wire, requiring, for example, cadmium as an ingredient, would vitiate the use of such wiring in fields of high neutron flux. The development of a substitute, if its later production proves difficult and requires a large labor force, is not a logistically sound replacement of a critically needed item.

Special devices

Special devices like sonar depend on the transformation of electrical energy into mechanical energy, and vice versa; the mass of electronic gear increases prodigiously with each request for improved ranging, and simultaneously decreases in reliability by the same factor. Since the heart of the device is the piezoelectric crystal or ceramic, the limitations of the instrument are those of these materials. Much improvement, especially in output pattern and Curie temperature, is desired. Furthermore, better protective domes for housing these transducers are much needed. They should be of improved transmissivity, and capable of withstanding the action of sea water and the vibration of the craft.

Possibly the greatest interest in the development of dielectrics per se centers at present on dielectric amplifiers. A high-impedance device, in contrast to the low-impedance magnetic amplifier, it will have applications where only manual control now serves. The systematic preparation of ferroelectrics with enhanced square-loop hysteresis characteristics in place of the moderately useful s-loops now obtainable is a prerequisite for a successful development of this device.

Deterioration of dielectrics at sea

The deterioration of dielectrics at sea includes the embrittlement of the insulation, the corrosion of leads and terminals, and the deposition of salt spray on insulating surfaces. Embrittlement results in mechanical failure of the dielectric followed by electric rupture through the crack; corrosion results in broken contacts, spurious electrolytic cell formation, and embrittlement; salt deposition leads to arc-over and leakage to ground. The embrittlement of dielectric substances like plasticized resins and varnishes is due to the volatilization of the plasticizer; moisture and saline atmospheres accelerate this process. The embrittlement of nonplasticized resins is due to chemical transformation abetted by the concomitant action of heat and moisture. The action of sea water on copper produces copper salts. When fine wire is attacked in this way, it becomes less ductile, very fragile, and easily broken by a slight mechanical shock. The deposition of salt spray on normally nonconducting surfaces, such as panel board, leads at high humidities to arc-over and to attenuation in radio circuitry.

The development of insulating resins, flexible per se, or made flexible by nonmigrating plasticizers; the development of noncorroding conductors, possibly by the judicious incorporation of inhibitor elements; and, finally, the prevention of salt deposits on surfaces are challenging problems.

Conclusions

The desire for reliability of materials at sea is a primary consideration. Requirements for higher thermal stability, ruggedness, and improved electrical breakdown properties are often enhanced by the empirical nature of safety factors. These factors can be reduced appreciably when and if a nondestructive test has been developed for evaluating insulation after it has been placed in service. The object of such test would be to indicate the expected life of the insulant under the foreseeable conditions of combat. In the field of material synthesis, the dielectric amplifier and ferroelectric transducer of higher service temperature and improved performance are two of the most needed items.

V · TABLES OF DIELECTRIC MATERIALS

LABORATORY FOR INSULATION RESEARCH
MASSACHUSETTS INSTITUTE OF TECHNOLOGY

The Laboratory for Insulation Research assumed in the Second World War the practical task of acting as a clearing house for information about dielectric materials and their uses. It issued *Tables of Dielectric Materials* for aiding government agencies, engineers, and manufacturers in the proper application of dielectrics and in the development of better products. These tables, of which four volumes have appeared thus far, were issued in limited circulation as "Technical Reports" of the Laboratory. They summarize our measurements on important dielectrics made in this country. The selection of these materials is undertaken with the full cooperation of the manufacturers concerned.

We are fully aware that these data should be expanded, especially towards higher temperatures and frequencies; that additional measurements on d-c conductivity, breakdown strength, and other electric parameters would be of great value; that a graphical presentation of the characteristics, accompanied by essential information on composition, properties, methods of handling, and recommended uses, should be given as started in Volume II of the "Tables"; and finally, that a real "dielectric analysis" of the materials should be undertaken, linking the dielectric response to composition and structure.

At this juncture, it seems appropriate to make Volume IV of the "Tables," issued in January, 1953, more freely available by including it in this book as Part V. Its quantitative information on more than 600 important materials will prove helpful in translating the theoretical concepts and practical experience presented in this treatise on *Dielectric Materials and Applications* into further progress on a widening front.

A. von Hippel

Contents

Dielectric Parameters	294
Measurements and Accuracy	300
Tabulated Dielectric Data	301
I. *Solids*	301
A. *Inorganic*	301
1. Crystals	301
2. Ceramics	303
a. Steatite Bodies	303
b. Titania and Titanate Bodies	304
c. Porcelains	306
d. Miscellaneous Ceramics	306
e. Ferrites	307
3. Glasses	309
4. Mica and Glass	313
5. Miscellaneous Inorganics	313
B. *Organic* (with or without Inorganic Components)	315
1. Crystals	315
2. Simple Noncrystals	315
3. Plastics	315
a. Phenol-formaldehyde	315
b. Phenol-aniline-formaldehyde	320
c. Aniline-formaldehyde	321
d. Melamine-formaldehyde	321
e. Urea-formaldehyde	323
f. Benzoguanamine-formaldehyde	323
g. Polyamide Resins	323
h. Cellulose Derivatives	323
(1) Acetates	323
(2) Propionate	325
(3) Nitrate	325
(4) Methyl Cellulose	325
(5) Ethyl Cellulose	325
i. Silicone Resins	325
j. Polyvinyl Resins	327
(1) Polyethylene	327
(2) Polyisobutylene	328
(3) Polyvinyl chloride-acetate	328
(4) Polyvinylidene and vinyl chloride	331
(5) Polychlorotrifluoroethylene	331
(6) Polytetrafluoroethylene	331
(7) Polyvinyl alcohol-acetate	333
(8) Polyvinyl acetals	334
(9) Polyacrylates	334
(10) Polystyrene	335
(11) Miscellaneous polystyrenes	337
(12) Styrene copolymers, linear	338
(13) Styrene copolymers, cross-linked	339
(14) Polystyrene plus fillers	341
(15) Polychlorostyrenes	341
(16) Poly-2,5-dichlorostyrene plus fillers	342
(17) Polyvinylcyclohexane	346
(18) Poly-α-vinylnaphthalene	346
(19) Poly-2-vinylpyridine	346
(20) Poly-N-vinylcarbazole	346
k. Polyesters	346
m. Alkyd Resins	348
n. Epoxy Resins	349
o. Miscellaneous Plastics	350
4. Elastomers	351
a. Natural Rubber	351
b. Gutta-percha	351
c. Balata	351
d. Cyclized Rubbers	351
e. Buna Rubbers	352
f. Butyl Rubbers	352
g. Nitrile Rubbers	353
h. Neoprene	353
i. Thiokol	354
j. Silicone Rubbers	354
5. Natural Resins	355
6. Asphalts and Cements	356
7. Waxes	356
8. Woods	359
9. Miscellaneous Organics	360
II. *Liquids*	361
A. Inorganic	361
B. Organic	362
1. Aliphatic	362
2. Aromatic	363
3. Petroleum Oils	365
4. Silicones	366
Supplementary High-Temperature Data	368
Supplementary High-Humidity Data	369
Graphs of Dielectric Constant and Loss Tangent versus Temperature:	371
Inorganic Crystals	371
Steatites	377
Titania and Titanate Bodies	387
Porcelains	393
Miscellaneous Ceramics	400
Glasses	400
Mica and Glass	405
Plastics	408
Phenol-formaldehyde Resins	408
Melamine-formaldehyde Resins	411

Graphs, Plastics (*continued*)		Polyesters	419
Polyamide Resin	412	Alkyd Resins	422
Cellulose Propionate	413	Epoxy Resin	424
Ethyl Cellulose	414	Elastomers	425
Polychlorotrifluoroethylene	415	Silicone Rubbers	425
Miscellaneous Vinyls	416	Company Index	426
Styrene Copolymers, Linear	417		
Styrene Copolymers, Cross-linked	418	Materials Index	429

Dielectric Parameters

1. ϵ'/ϵ_0, the dielectric constant or permittivity relative to vacuum, also designated in the literature as K, κ, ϵ, D.C., etc.
2. $\tan \delta$ or $\tan \delta_d$, the dielectric-loss tangent or dissipation factor, also designated in the literature as D.F., $1/Q$, and, when losses are low, as power factor or $\cos \theta$.
3. μ'/μ_0, the magnetic permeability relative to vacuum, also given in the literature as μ' or μ_0.
4. $\tan \delta_m$, the magnetic-loss tangent.
5. ρ, the a-c volume resistivity in ohm-cm. This parameter is used in these tables only for very high-loss materials.

Transformation to Other Parameters. The dielectric-loss factor relative to vacuum, ϵ''/ϵ_0, is the product of the dielectric loss tangent and ϵ'/ϵ_0. Alternating-current volume conductivity, σ, is given by

$$\sigma = \frac{1}{\rho} = \frac{f(\epsilon'/\epsilon_0) \tan \delta}{1.8 \times 10^{12}} \text{ [mho-cm]} \quad (f \text{ in cps}).$$

A chart is given (page 295) for approximate calculations of σ or ρ from the data given in the tables.

The magnetic-loss factor relative to vacuum, μ''/μ_0, is the product of the magnetic-loss tangent and μ'/μ_0 (in analogy to the dielectric-loss factor). In the literature, the loss factor is sometimes given as $1/\mu_0' Q$ or, in our notation, $\tan \delta_m/(\mu'/\mu_0)$.

The attenuation constant, α, for propagation in free space is

$$\frac{2\pi}{\lambda_0} \left[\frac{\mu' \epsilon'}{\mu_0 \epsilon_0} \frac{1 - \tan \delta_d \cdot \tan \delta_m}{2} ([1 + \tan^2 (\delta_d + \delta_m)]^{1/2} - 1) \right]^{1/2},$$

which for nonmagnetic dielectrics reduces to

$$\frac{2\pi}{\lambda_0} \left(\frac{\epsilon'}{\epsilon_0}\right)^{1/2} \left[\frac{(1 + \tan^2 \delta_d)^{1/2} - 1}{2} \right]^{1/2}.$$

Charts for finding α in terms of ϵ'/ϵ_0 and $\tan \delta_d$ are included in this section (pages 296–298). They apply also to magnetic dielectrics when the product $(\mu'/\mu)(\epsilon'/\epsilon_0)$ is substituted for ϵ'/ϵ_0 and an equivalent combined loss tangent $\tan \delta_e$ is used instead of $\tan \delta_d$. A graph on page 299 gives $\tan \delta_e$ for values of $\tan \delta_d$ and $\tan \delta_m$ in the range 0.1 to 10. For smaller values, $\tan \delta_e$ is the sum of $\tan \delta_d$ and $\tan \delta_m$.

The phase constant β for propagation in free space is

$$\frac{2\pi}{\lambda_0} \left[\frac{\mu' \epsilon'}{\mu_0 \epsilon_0} \frac{1 - \tan \delta_d \cdot \tan \delta_m}{2} ([1 + \tan^2 (\delta_d + \delta_m)]^{1/2} + 1) \right]^{1/2},$$

which reduces for nonmagnetic materials to

$$\frac{2\pi}{\lambda_0} \left(\frac{\epsilon'}{\epsilon_0}\right)^{1/2} \left[\frac{(1 + \tan^2 \delta_d)^{1/2} + 1}{2} \right]^{1/2}.$$

The intrinsic impedance Z of the material is

$$377 \left(\frac{\mu^* \epsilon_0}{\mu_0 \epsilon^*}\right)^{1/2}.$$

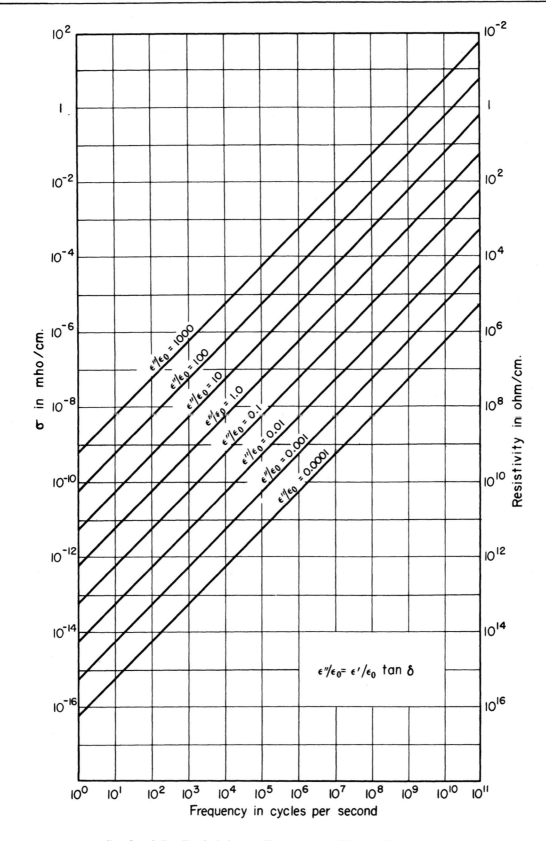

Conductivity-Resistivity as Function of ϵ''/ϵ_0 and Frequency

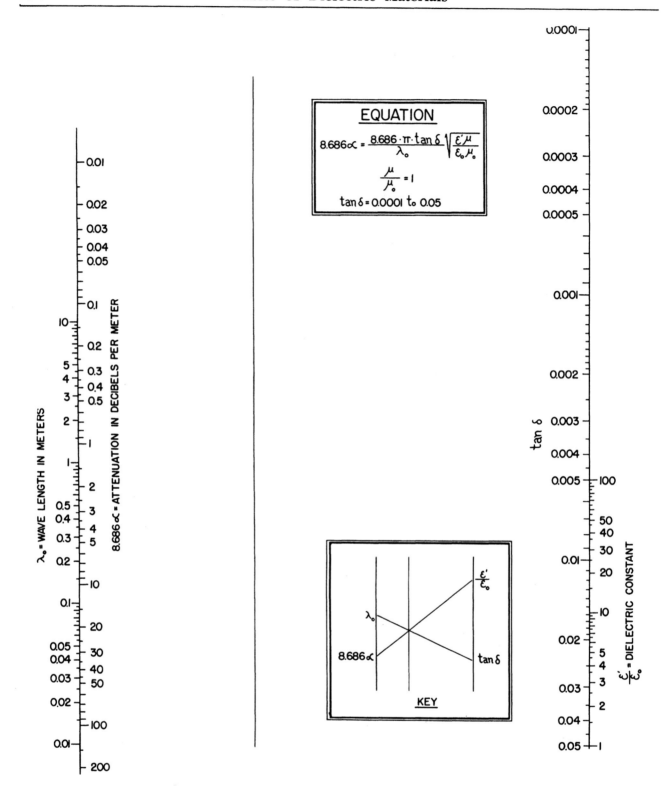

Decibel Loss per Meter for Low-Loss Dielectrics (tan $\delta \ll 1$)

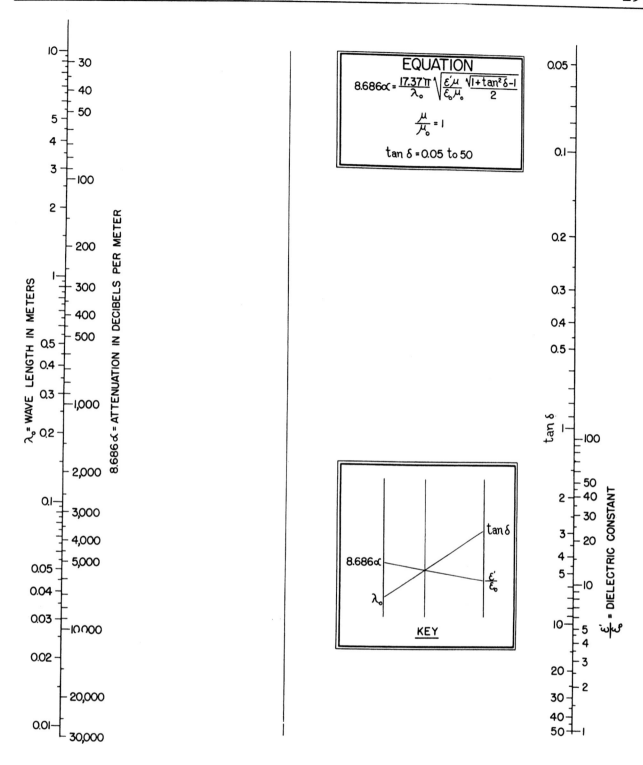

Decibel Loss per Meter for Medium-Loss Dielectrics (tan δ comparable to 1)

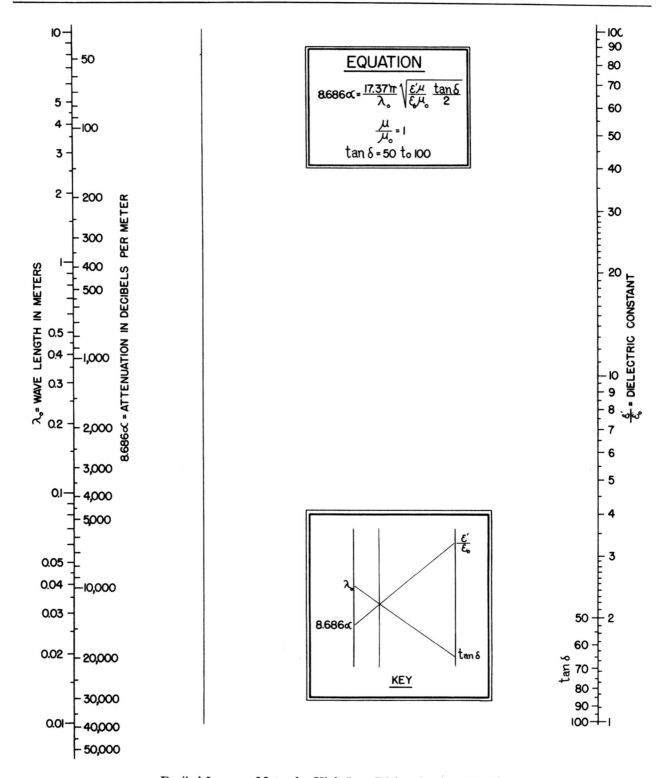

Decibel Loss per Meter for High-Loss Dielectrics (tan δ ≫ 1)

Dielectric Parameters

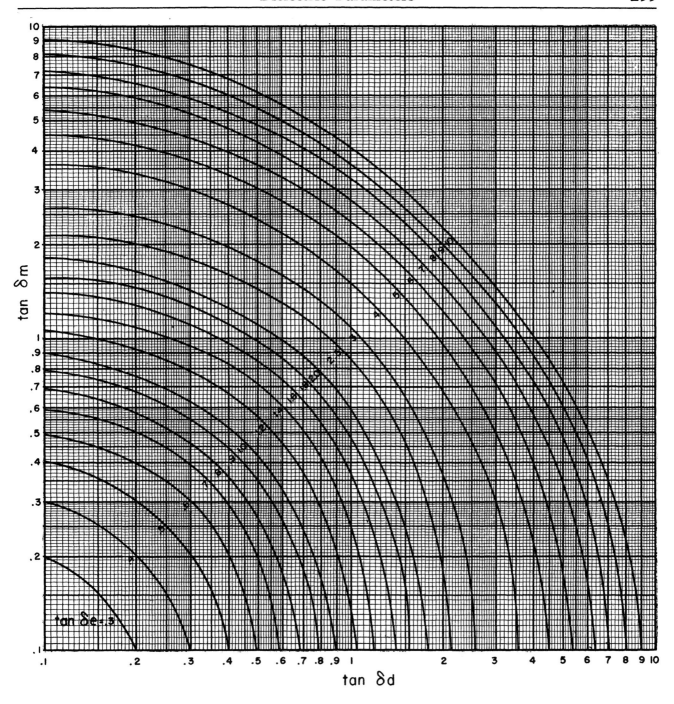

Equivalent Combined Loss Tangent for Magnetic Materials

Measurements and Accuracy†

The measurements have been made, in general, on only one batch of the material and refer, unless otherwise specified, to samples dried over phosphorous pentoxide. Owing to the wide variety of materials and improvements in techniques, no figures of general validity can be given concerning the accuracy of these measurements. For ϵ'/ϵ_0 the nominal accuracy is ±2 percent; the accuracy trends are toward ±1 percent for rigid, low-loss materials (tan δ < 0.005) and ±5 percent for high-loss materials (tan δ > 1). For tan δ_d the nominal accuracy is ±5 percent; for high-loss materials the error may be ±10 percent. For very low-loss materials (tan δ < 0.002), the accuracy is ±0.0001 when the losses are given as multiples of 0.0001. When the loss is expressed in multiples of 0.00001, the error may be ±0.00003. For μ'/μ_0, the nominal accuracy is ±5 percent; for tan δ_m, ±10 percent.

Field Strengths. Linear dielectrics not field-strength sensitive were measured at field strengths of approximately 50 volts per centimeter in the frequency range 10^2 to 10^8 cps and at lower field strengths at higher frequencies. At high temperatures (near 500°C) many of these materials show field-strength sensitivity, particularly at low frequencies. The effects may be wholly or partly due to space charge polarization.

The ferroelectric and ferromagnetic materials, unless otherwise noted, were measured at field strengths in their linear region. The values thus measured are the *initial* permittivity and permeability. Typical field strengths are 40 volts per centimeter for titanate ceramics, 0.01 volt per centimeter for KDP crystals, 0.01 oersted for the ferrites.

† The measurements were made by W. B. Westphal, group leader of the dielectric measurements group, ably supported by three staff operators: Helen M. Dunn, Patricia A. Fergus, and Elizabeth McCarty. Invaluable help in editing and correcting was given by Dr. L. G. Wesson and Hariot B. Armstrong.

This work was sponsored under O.N.R. Contracts jointly by the Navy Department (Office of Naval Research), the Army Signal Corps, and the Air Force (Air Matériel Command).

Tabulated Dielectric Data

I. SOLIDS

Values for tan δ are multiplied by 10^4; frequency given in c/s.

A. Inorganic

1. Crystals

	T°C		1×10^2	1×10^3	1×10^4	1×10^5	1×10^6	1×10^7	1×10^8	3×10^8	3×10^9	1×10^{10}	2.5×10^{10}
Ice[a]	-12	ϵ'/ϵ_0	----	----	----	4.8	4.15	3.7	----	----	3.20	3.17	
		tan δ	----	----	----	8000	1200	180	----	----	9	7	
Snow[b]	-20	ϵ'/ϵ_0	----	3.33	1.82	1.24	1.20	1.20	----	1.20	1.20	1.26*	
		tan δ	----	4920	3420	1400	215	40	----	12	2.9	4.2*	
Snow[c]	-6	ϵ'/ϵ_0	----	----	----	1.9	1.55	----	----	----	1.5		
		tan δ	----	----	----	15300	2900	----	----	----	9		
Aluminum oxide, sapphire[d] (field ⊥ optical axis)	25	ϵ'/ϵ_0	8.6	8.6	8.6	8.6	8.6	8.6	8.6	8.6	----	----	8.6
		tan δ	<10	<2	<10	<10	<10	<10	<10	<1	----	----	14
(field ∥ optical axis)		ϵ'/ϵ_0	10.55	10.55	10.55	10.55	10.55	10.55	10.55	10.55			
		tan δ	<10	<2	<10	<10	<10	<10	<10	<1			
Ammonium dihydrogen phosphate (field ⊥ optical axis)[e]	25	ϵ'/ϵ_0	56.4	56.0	55.9	55.9	55.9	55.9	55.9	55.9	----	13.7	
		tan δ	400	46	4.6	<5	<5	<5	<5	<10	----	50	
(field ∥ optical axis)		ϵ'/ϵ_0	16.4	16.0	15.4	14.7	14.3	14.3	----	----	----		
		tan δ	2400	240	70	70	60	10	----	----	----		
Lithium fluoride[f]	25	ϵ'/ϵ_0	9.00	9.00	9.00	9.00	9.00	9.00	----	9.00	----	9.00	
		tan δ	15	<3	<2	<2	<2	<2	----	.7	----	1.8	
	80	ϵ'/ϵ_0	9.11	9.11	9.11	9.11	9.11	9.11	----	----	----	9.11	
		tan δ	120	20	11	4	<2	<2	----	----	----	3.3	
Magnesium oxide[g]	25	ϵ'/ϵ_0	9.65	9.65	9.65	9.65	9.65	9.65	9.65	----	----		
		tan δ	<3	<3	<3	<3	<3	<3	<3	----	----		
Potassium bromide[f]	25	ϵ'/ϵ_0	4.90	4.90	4.90	4.90	4.90	4.90	----	4.90	----	4.90	
		tan δ	7	7	8	4.5	<2	<2	----	<1	----	2.3	
	87	ϵ'/ϵ_0	4.97	4.97	4.97	4.97	4.97	4.97	----	4.97	----	4.97	
		tan δ	16	7	11	9	5	3	----	2.4	----	3.5	
Potassium dihydrogen phosphate (field ⊥ optical axis)[e]	25	ϵ'/ϵ_0	44.5	44.3	44.3	44.3	44.3	44.3	4.43	----	----	----	
		tan δ	98	15	<5	<5	<5	<5	<5	----	----	----	
(field ∥ optical axis)		ϵ'/ϵ_0	21.4	20.7	20.5	20.3	20.2	20.2	20.2	----	----	----	
		tan δ	170	24	<20	<5	<5	<5	<5	----	----	----	
Selenium, multi-crystal-line[h]	25	ϵ'/ϵ_0	----	----	----	----	----	----	----	11.0	10.4	----	7.5
		tan δ	----	----	----	----	----	----	----	2500	1540	----	1100

a. From conductivity water. b. Freshly fallen snow. c. Hard-packed snow followed by light rain. d. Linde Air. e. Brush.
f. Fresh crystals (Harshaw). g. Norton. h. Lab. Ins. Res.
* -6°C.

I. Solids, A. Inorganic,
1. Crystals (cont.)

Values for tan δ are multiplied by 10^4; frequency given in c/s.

	T°C		1×10^2	1×10^3	1×10^4	1×10^5	1×10^6	1×10^7	1×10^8	3×10^8	3×10^9	1×10^{10}	2.5×10^{10}
Sodium chloride[a]	25	ϵ'/ϵ_0	5.90	5.90	5.90	5.90	5.90	5.90					5.90
		tan δ	<1	<1	<1	<2	<2	<2					<5
	85	ϵ'/ϵ_0	6.35	6.11	6.00	5.98	5.98	5.98					5.97
		tan δ	170	240	70	6	<2	<2					<3.9
Sulfur[b] (100)	25	ϵ'/ϵ_0	3.75	3.75									
		tan δ	<5	<5									
(010)	25	ϵ'/ϵ_0	3.95	3.95									
		tan δ	<5	<5									
(001)	25	ϵ'/ϵ_0	4.45	4.45									
		tan δ	<5	<5									
Sulfur, sublimed[c]	25	ϵ'/ϵ_0	3.69	3.69	3.69	3.69	3.69	3.69	----	----	3.62	3.58	
		tan δ	3	2	<2	<2	<2	<2	----	----	.4	1.5	
Thallium bromide[d]	25	ϵ'/ϵ_0	31.1	30.3	30.3	30.3	30.3	30.3					
		tan δ	1300	128	13.3	2	1	.4					
Thallium iodide[d]	25	ϵ'/ϵ_0	22.3	21.8	21.8	21.8	21.8	21.8					
		tan δ	950	120	12	1.2	.5	.5					
	193	ϵ'/ϵ_0	----	----	----	----	----	37.3					
		tan δ	----	----	----	----	----	820					
Thallium bromide - chloride[d,e]	25	ϵ'/ϵ_0	32.9	32.0	32.0	31.8							
		tan δ	4500	441	46	5.9							
Thallium bromide - iodide[d,f]	25	ϵ'/ϵ_0	32.9	32.8	32.5	32.5	32.5	32.5					
		tan δ	29800	2800	277	37	4	.4					
Titanium dioxide, rutile[g]	25	ϵ'/ϵ_0	87.3	86.7	86.4	86	85.8	85.8					
(field ⊥ optical axis)		tan δ	110	32	9	4	2	2					
(field ∥ optical axis)	25	ϵ'/ϵ_0	----	----	200	170	170	160					
		tan δ	----	----	3500	600	80	16					

a. Fresh crystals (Harshaw). b. Measured with field ⊥ to cuts indicated. Grown at Lab. Ins. Res. c. U.S.P. d. Grown at Eng. Res. and Dev. Lab., Fort Belvoir, Va. e. KRS-6, 60% TlBr, 40% TlCl. f. KRS-5, 42% TlBr, 58% TlI. g. Linde Air.

I. Solids, A. Inorganic

2. Ceramics
a. Steatite Bodies

Values for tan δ are multiplied by 10^4; frequency given in c/s.

	T°C		1×10^2	1×10^3	1×10^4	1×10^5	1×10^6	1×10^7	1×10^8	3×10^8	3×10^9	1×10^{10}	2.5×10^{10}
AlSiMag A-35[a]	23	ϵ'/ϵ_o	6.10	5.96	5.89	5.86	5.84	5.80	5.75	----	5.60	----	5.36
		tan δ	150	100	70	50	38	35	37	----	41	----	58
	85	ϵ'/ϵ_o	6.84	6.37	6.11	5.96	5.86	5.80	5.75	----	5.50	----	
		tan δ	890	370	175	103	77	50	50	----	47	----	
AlSiMag A-196[a]	25	ϵ'/ϵ_o	5.90	5.88	5.84	5.80	5.70	5.65	5.60	----	5.42	5.24	5.18
		tan δ	30	59	79.5	55	30.5	19	16	----	18	26	38
	81	ϵ'/ϵ_o	5.90	5.88	5.84	5.80	5.70	5.65	5.60	----	5.42		
		tan δ	58	40	46.5	70.5	66	40.5	24	----	18		
AlSiMag 211[a]	25	ϵ'/ϵ_o	6.00	5.98	5.98	5.97	5.97	5.96	5.96	----	----	5.90	
		tan δ	92	34	12	6	5	4	4	----	----	14	
AlSiMag 228[a]	25	ϵ'/ϵ_o	6.40	6.40	6.40	6.40	6.36	6.30	6.20	----	5.97	5.93	5.83
		tan δ	15.6	20	20	15.6	12.4	11.2	10	----	13	19.5	42
	81	ϵ'/ϵ_o	6.52	6.46	6.40	6.40	6.36	6.30	----	----	5.95		
		tan δ	35.6	22	18	21.5	18.4	11.8	----	----	11		
AlSiMag 243[a]	22	ϵ'/ϵ_o	6.30	6.30	6.28	6.25	6.22	6.17	6.10	----	5.78	5.76	5.75
		tan δ	12.5	4.5	4.0	<9	3.7	3.5	3	----	6	8.5	12
	85	ϵ'/ϵ_o	6.37	6.37	6.37	6.36	6.32	6.28	----	----	5.88		
		tan δ	21	13.7	8.0	<9	3.7	3.5	----	----	6		
AlSiMag 393[b]	24	ϵ'/ϵ_o	4.95	4.95	4.95	4.95	4.95	4.95	4.95	----	----	4.95	4.91
		tan δ	38	23	16	12	10	10	10	----	----	9.7	14
Ceramic F-66[c]	25	ϵ'/ϵ_o	6.22	6.22	6.22	6.22	6.22	6.22	6.22	----	6.22	----	6.2
		tan δ	14.5	9	5	2	1	1.5	3	----	5.5	----	11
Steatite Type 302[d]	25	ϵ'/ϵ_o	5.80	5.80	5.80	5.80	5.80	5.80	5.80	----	5.8	5.8	
		tan δ	32	20	16	13	12	12	12	----	19	36	
Steatite Type 400[d]	25	ϵ'/ϵ_o	5.54	5.54	5.54	5.54	5.54	5.54	5.54	----	5.5	5.5	
		tan δ	160	100	72	60	50	45	39	----	39	53	
Steatite Type 410[d]	25	5.77	5.77	5.77	5.77	5.77	5.77	5.77	5.77	----	5.7	5.7	
		tan δ	55	30	16	9	7	6	6	----	8.9	22	
Steatite Type 452[d]	25	ϵ'/ϵ_o	8.15	8.15	8.15	8.15	8.15	8.15	8.15	----	8.15	8.15	
		tan δ	65	28	17	12	10	10	10	----	20	31	

a. Magnesium silicate (Am. Lava). b. 93% Al_2O_3, 6% SiO_2, 1% MgO (Am. Lava). c. 60% talc, 15% kaolin, 17.5% $BaCO_3$, 7.5% $MgCO_3$ (Bell). d. Centralab.

304   Tables of Dielectric Materials

I. Solids, A. Inorganic (cont.)

2. Ceramics (cont)

Values for tan δ are multiplied by 10^4; frequency given in c/s.

	T°C		1×10^2	1×10^3	1×10^4	1×10^5	1×10^6	1×10^7	1×10^8	3×10^8	3×10^9	1×10^{10}	2.5×10^{10}
a. Steatite Bodies (cont.)													
Steatite Body 7292[a]	25	ϵ'/ϵ_0	6.55	6.55	6.54	6.53	6.53	6.53	6.53	6.53	6.52	6.51	
		tan δ	14	7	4.8	3.9	4.9	5.2	6.2	6.8	9	10.9	
Crolite #29[b]	24	ϵ'/ϵ_0	6.04	6.04	6.04	6.04	6.04	6.04	----	----	5.90	5.71	
		tan δ	25	19	15	13	11	10	----	----	24	30	
b. Titania and Titanate Bodies													
Ceramic NPOT96[c]	25	ϵ'/ϵ_0	29.5	29.5	29.5	29.5	29.5	29.5	29.5	----	----	28.9	
		tan δ	12	4.9	3.3	2.5	1.6	1.7	2	----	----	20	
Ceramic N750T96[c]	25	ϵ'/ϵ_0	83.4	83.4	83.4	83.4	83.4	83.4	83.4	----	----	83.4	
		tan δ	5.7	4.5	3.5	2.5	2.2	2.3	4.6	----	----	14.6	
Ceramic N1400T110[c]	25	ϵ'/ϵ_0	131	130.8	130.7	130.5	130.2	130.2	130.0				
		tan δ	6.7	5.5	3.3	1.4	3.0	5.5	7.0				
Body T106[c]	25*	ϵ'/ϵ_0	1518	1508	1480								
		tan δ	31	87	99								
	25**	ϵ'/ϵ_0	1308	1280	1260	1245	1232	1220	1210				
		tan δ	98	120	140	100	85	150	420				
TI Pure R-200[d]	26	ϵ'/ϵ_0	100	100	100	100	100	100	100			90	
		tan δ	23	15	6.2	4	3	2.5	2.5			20	
	78	ϵ'/ϵ_0	97.5	97	97	97	97	97					
		tan δ	130	60	33	13	4.5	2.3					
Tam Ticon T-J, T-L and T-M[e]	24	ϵ'/ϵ_0	96	96	96	96	96	96	----	----	96	----	91
		tan δ	8	4.5	2	1	2	3	----	----	3.4	----	33
	78	ϵ'/ϵ_0	90	89	88.5	88	87.5	87					
		tan δ	55	16	4	4	4	4					
Tam Ticon MC[f]	25	ϵ'/ϵ_0	13.9	13.9	13.9	13.9	13.9	13.9	13.9	13.9	13.8	13.8	13.7
		tan δ	15	11	9	7	4	4	5	9	17	28	65

a. Gen. Ceramics and Steatite. b. Al_2O_3, SiO_2, MgO, CaO, BaO (Crowley). c. Amer. Lava. d. Rutile (Dupont, fired at Lab. Ins. Res.). e. Rutile (Titanium Alloy, fired at Lab. Ins. Res.). f. Magnesium titanate (Titanium Alloy, fired at Lab. Ins. Res.).

* - 3/4" Disk.   ** 0.2" disk cut from 3/4" disk.

Dielectric Data

I. Solids, A. Inorganic (cont.)

2. Ceramics (cont.)
b. Titania and titanate bodies (cont.)

Values for tan δ are multiplied by 10^4; frequency given in c/s.

	T°C		1×10^2	1×10^3	1×10^4	1×10^5	1×10^6	1×10^7	1×10^8	3×10^8	3×10^9	1×10^{10}	2.5×10^{10}
Titania Ceramic[a]	25	ϵ'/ϵ_0	----	----	----	----	33	----	----	60			
		tan δ	----	----	----	----	5.2	60.5	----	11			
Titania Ceramic[b]	25	ϵ'/ϵ_0	----	----	----	----	----	8.3					
		tan δ	----	----	----	----	----						
Tam Ticon C[c]	25	ϵ'/ϵ_0	167.8	167.7	167.7	167.7	167.7	167.7	----	166.8	165	165	
		tan δ	13.6	4.4	2.5	2	2	2	----	5	23	85	
	82	ϵ'/ϵ_0	157	156	156	156	156	156					
		tan δ	400	49	11	6	6	6					
Tam Ticon S[d]	25	ϵ'/ϵ_0	234	233	232	232	232	232	232	----	----	230	
		tan δ	21	11	8	5	2	1	1	----	----	28	
Tam Ticon B[e]	25	ϵ'/ϵ_0	1240	1200	1170	1153	1143	1140	----	1100	600	150	100
		tan δ	360	130	150	120	105	75	----	500	3000	5000	6000
Tam Ticon BS[f] (discontinued)	25	ϵ'/ϵ_0	8700	8500	8200	8100	8000	8000	8000	7000	2000	200	
		tan δ	230	200	230	90	48	98	400	900	5000	4000	
Mix 71 BaTiO$_3$[g]	25	ϵ'/ϵ_0	1740	1720	1680	1650							
		tan δ	63	107	170	150							
Mix 72 BaTiO$_3$ 90%[g] SrTiO$_3$ 10%	25	ϵ'/ϵ_0	1310	1307	1300	1280	1275						
		tan δ	31	43	70	77	46						
Mix 73 BaTiO$_3$ 80%[g] SrTiO$_3$ 20%	25	ϵ'/ϵ_0	2163	2160	2120	2020	1960						
		tan δ	94	155	280	200	75						
Mix 74 BaTiO$_3$ 70.9%[g] SrTiO$_3$ 29.1%	25	ϵ'/ϵ_0	2990	2970	2910	2840	2820						
		tan δ	47	61	88	67	40						
Mix 75 BaTiO$_3$ 60%[g] SrTiO$_3$ 40%	25	ϵ'/ϵ_0	1125	1122	1120	1110	1090						
		tan δ	110	72	45	30	19						

a. 84.3% TiO_2, 11.1% $MgCO_3$ =4.6% $MgOSiO_2$ (Lab. Ins. Res.). b. 84.6% TiO_2, 15.4% kaolin (Lab. Ins. Res.). c. Calcium titanate (Titanium Alloy; fired at Lab. Ins. Res.). d. Strontium titanate (Titanium Alloy; fired at Lab.Ins.Res.). e. Barium titanate (Titanium Alloy; fired at Lab. Ins. Res.). f. 79% barium, 21% strontium titanate (Titanium Alloy; fired at Lab. Ins. Res.). g. Thin sheet (Glenco).

I. Solids, A. Inorganic (cont.)

2. Ceramics (cont.)

Values for tan δ are multiplied by 10^4; frequency given in c/s.

b. Titania and titanate bodies (cont.)

	T°C		1×10^2	1×10^3	1×10^4	1×10^5	1×10^6	1×10^7	1×10^8	3×10^8	3×10^9	1×10^{10}
Mix 76 BaTiO₃ 50%[a] SrTiO₃ 50%	25	ϵ'/ϵ_o	740	738	735	732	730	----	----	----	----	----
		tan δ	10.7	6	7	21	28	----	----	----	----	----
Mix 77 BaTiO₃ 68.1%[a] SrTiO₃ 28.1% MgZrO₃ 2.8%	25	ϵ'/ϵ_o	892	890	890	890	890	----	----	----	----	----
		tan δ	16	10	12	25	8	----	----	----	----	----

c. Porcelains

	T°C		1×10^2	1×10^3	1×10^4	1×10^5	1×10^6	1×10^7	1×10^8	3×10^8	3×10^9	1×10^{10}
Zirconium porcelain Zi-4[b]	25	ϵ'/ϵ_o	6.44	6.40	6.35	6.32	6.32	6.30	6.30	6.30	6.23	6.18
		tan δ	59	40	31	27	23	21	25	27	45	57
Porcelain, wet process[c]	25	ϵ'/ϵ_o	6.47	6.24	6.08	5.98	5.87	5.82	5.80	5.75	----	5.51
		tan δ	280	180	130	105	90	115	135	140	----	155
Porcelain, dry process[c]	25	ϵ'/ϵ_o	5.50	5.36	5.23	5.14	5.08	5.04	5.04	5.02	----	4.74
		tan δ	220	140	105	85	75	70	78	98	----	156
Coors AI-200[d]	25	ϵ'/ϵ_o	8.83	8.83	8.82	8.80	8.80	8.80	8.80	----	8.79	8.79
		tan δ	14	5.7	4.8	3.8	3.3	3.2	3.0	----	10	18
Porcelain #4462[e]	25	ϵ'/ϵ_o	8.99	8.95	8.95	8.95	8.95	8.95	8.95	8.93*	8.90	8.80
		tan δ	22	9.1	6.0	3.0	2.0	2.0	4.0	9*	11	14
Coors AB-2[d]	25	ϵ'/ϵ_o	8.22	8.18	8.17	8.17	8.16	8.16	8.16	----	8.14	8.08
		tan δ	20	13.4	11.4	10.5	9.0	7.5	9.0	----	16	27
AlSiMag 491[f]	25	ϵ'/ϵ_o	----	----	----	----	8.74	----	----	----	8.60	8.50
		tan δ	----	----	----	----	22	----	----	----	17	23

d. Miscellaneous Ceramics

	T°C		1×10^2	1×10^3	1×10^4	1×10^5	1×10^6	1×10^7	1×10^8	3×10^8	3×10^9	1×10^{10}
Beryllium oxide[g]	25	ϵ'/ϵ_o	4.61	4.47	4.41	4.34	4.28	4.24	4.23	----	----	4.20
		tan δ	170	84	74	72	38	19	12.5	----	----	5
Porous Ceramic AF-497[h]	25	ϵ'/ϵ_o	----	----	----	----	----	----	----	----	----	1.472
		tan δ	----	----	----	----	----	----	----	----	----	17

a. Thin sheet (Glenco). b. Coors. c. Knox. d. Aluminum oxide (Coors). e. Aluminum oxide (Frenchtown Porcelain). f. Aluminum oxide (Amer. Lava). g. Norton. h. 63% diatomaceous-earth, 32% anthracite coal, 5.5% whiting (Stupakoff).
*Frequency = 1×10^9.

Dielectric Data

I. Solids, A. Inorganic, 2. Ceramics (cont.)

e. Ferrites

Frequency given in c/s; temperature = 25°C.

Freq.	Crowloy 20[a] ϵ'/ϵ_o	tan δ_d	μ'/μ_o	tan δ_m	Crowloy 70[a] ϵ'/ϵ_o	tan δ_d	μ'/μ_o	tan δ_m	Crowloy BX113[a]* ϵ'/ϵ_o	tan δ_d	μ'/μ_o	tan δ_m	Crowloy BX114[a] ϵ'/ϵ_o	tan δ_d	μ'/μ_o	tan δ_m
10^2	35400	2.41	----	----	440000	26.7	400	----	19.4	.59	----	----	29	.58	21	----
4×10^2	9900	3.94	----	----	----	----	----	----	----	----	----	----	19.2	----	----	----
10^3	5430	2.65	120	----	350000	3.35	400	----	14.7	.20	14.5	----	19.2	.30	17.5	----
10^4	1590	1.47	120	----	250000	.78	400	----	13.0	.066	14.5	----	14.9	.15	15.6	----
10^5	480	1.6	120	----	123000	.64	400	----	12.8	.024	14.5	----	12.9	.06	14.5	----
10^6	----	----	----	----	----	----	440	.28	12.6	.010	14	----	12.6	.024	----	----
3×10^6	----	----	108	.08	----	----	147	1.53	----	----	13	<.1	----	----	16	<.1
10^7	18	1.1	84	.48	16400	1.4	92	.24	12.5	.005	12,47	<.1,.02	12.4	.007	18.5	<.1
2×10^7	----	----	----	----	----	----	----	----	----	----	12	<.2	----	----	----	----
3×10^7	----	----	47	.92	----	----	----	----	----	----	12.8,46	.05,.03	----	----	20.7	<.1
6.5×10^7	----	----	35	1.16	----	----	41	1.92	----	----	51	.25	----	----	23	<.1
8×10^7	----	----	24.6	1.58	----	----	29	2.36	----	----	49.4	.46	----	----	24	<.1
10^8	----	----	17.9	2.15	----	----	18	3.35	12.5	.0026	37	.74	12.4	.0037	26	<.1
1.2×10^8	----	----	15.9	1.98	----	----	21	2.58	----	----	34	.9	----	----	30	.13
1.4×10^8	----	----	24.6	1.43	----	----	----	----	----	----	28.5	1.15	----	----	31	.32
3×10^8	----	----	----	----	----	----	----	----	----	----	17.9	1.5	----	----	17.6	1.61
9×10^8	12	.07	----	----	----	----	----	----	----	----	5.34	3.4	----	----	3.25	4.46
10^9	12	.07	.98	5.8	----	----	----	----	----	----	1.74	1.9	----	----	1.34	2.48
3×10^9	12	.04	.28	7.2	42	----	.19	10.6	12	<.012	----	----	12	<.01	----	----

a. H. L. Crowley.

*Two samples were measured; they differed as shown by the two sets of values at 10^7 and 3×10^7. Lower μ'/μ_o values were obtained on sample cut from 2" disk, higher values on 3/4" disk. ϵ^*/ϵ_o values did not differ.

Tables of Dielectric Materials

I. Solids, A. Inorganic, 2. Ceramics (cont.)

e. Ferrites (cont.)

Frequency given in c/s; temperature = 25°C.

Freq.	Ferramic A[a,b] ϵ'/ϵ_o	tan δ_d	μ'/μ_o	tan δ_m	Ferramic B[a] ϵ'/ϵ_o	tan δ_d	μ'/μ_o	tan δ_m	Ferramic C[a] ϵ'/ϵ_o	tan δ_d	μ'/μ_o	tan δ_m	Ferramic D[a] ϵ'/ϵ_o	tan δ_d	μ'/μ_o	tan δ_m
10^3	9.82	.112	19.6	----	20800	4.5	----	----	110000	2.4	240	----	35000	84	380	----
10^4	9.30	.026	19.6	----	19700	4.8	94	----	32000	1.44	240	----	35000	8.1	380	----
10^5	9.13	.0076	19.5	----	15500	.78	----	----	22000	1	----	----	----	----	----	----
10^6	8.99	.0032	19.4	----	8600	.65	----	----	13500	1.32	242	.015	17600	.77	510	<.05
2×10^6	----	----	20.0	----	----	----	108	.027	----	----	259	----	----	----	550	.18
3×10^6	----	----	22	----	----	----	----	----	----	----	267	.068	11000	.80	565	.47
4×10^6	----	----	23	----	----	----	----	----	----	----	----	----	----	----	411	.55
7×10^6	----	----	----	----	----	----	----	----	----	----	----	----	----	----	112	2.76
10^7	8.87	.0016	24.6	.08	1060	2.9	107	.36	2850	2.0	195	.85	6000	.90	52	4.0
2×10^7	----	----	25.4	.177	----	----	----	----	----	----	78	1.6	----	----	17.7	5.8
3×10^7	----	----	----	----	----	----	62	.67	----	----	33.5	2.7	----	----	12.9	6.2
5×10^7	8.5	<.1	12.5	1.00	----	----	----	----	----	----	----	----	----	----	----	----
10^8	----	----	7.2	1.29	51	1.56	14.4	1.12	64	.70	16.8	.61	485	2.9	----	----
3×10^8	----	----	2.8	1.6	----	----	----	----	----	----	----	----	106	3.1	1.09	18.4
5×10^8	----	----	2.0	1.2	----	----	----	----	----	----	----	----	----	----	----	----
10^9	8.5	<.1	3.0	.67	----	----	----	----	----	----	----	----	----	----	----	----
2×10^9	----	----	1.0	2.0	----	----	----	----	----	----	----	----	----	----	----	----
3×10^9	----	----	.6	2.0	20	.68	.29	10	24.6	1.14	.18	22	35	3.3	.15	20
10^{10}	8.5	<.1	1.0	----	15	.52	.42	.37	18.7	.58	.51	1.26	30	1.35	1.9	1.0

Freq.	Ferramic G[a] ϵ'/ϵ_o	tan δ_d	μ'/μ_o	tan δ_m	Ferramic H[a] ϵ'/ϵ_o	tan δ_d	μ'/μ_o	tan δ_m	Ferramic I[a] ϵ'/ϵ_o	tan δ_d	μ'/μ_o	tan δ_m	Ferramic J[a] ϵ'/ϵ_o	tan δ_d	μ'/μ_o	tan δ_m
10^3	----	----	450	----	670000	7.2	720	----	12000	2.9	890	----	----	----	256	----
10^4	23.5	.60	400	----	152500	4.5	720	----	7800	.81	890	----	17.5	1.95	256	----
10^5	15.5	.216	360	----	22400	4.2	720	----	930	3.1	890	<.1	13.8	.37	256	----
10^6	13.6	.069	340	.05	1600	8.2	690	.18	90	4.7	1050	.29	12.6	.062	256	.015
10^7	13.0	.026	245	.66	----	----	300	1.33	26	2.1	215	1.6	12.4	.015	196	.66
3×10^8	----	----	----	----	----	----	----	----	13.6	.114	10.7	1.03				
3×10^9	----	----	----	----	----	----	----	----	12	.012	.90	.04				

a. Gen. Ceramics and Steatite. b. Data partly from Rado, Wright and Emerson, Phys. Rev. 80, 273 (1950).

Dielectric Data

I. Solids, A. Inorganic (cont.)

3. Glasses

Values for tan δ are multiplied by 10^4; frequency given in c/s.

Glasses	T°C		1×10^2	1×10^3	1×10^4	1×10^5	1×10^6	1×10^7	1×10^8	3×10^8	3×10^9	1×10^{10}	2.5×10^{10}
Phosphate Glass #2043x[a]	25	ϵ'/ϵ_o	5.25	5.25	5.25	5.25	5.25	5.25	5.24	5.23	5.17	5.00	4.93
		tan δ	22	18	16	15	14	16	20	25	46	42	34
Phosphate Glass #2279x[b]	25	ϵ'/ϵ_o	4.92	4.92	4.92	4.92	4.92	4.92	4.92	----	4.9	4.9	4.9
		tan δ	26	19	16	14	14	13	14	----	18	19	21
Borosilicate Glass[c]	25	ϵ'/ϵ_o	4.05	4.05	4.05	4.05	4.05	4.05	4.05	----	4.05	4.05	
		tan δ	13.6	8.6	4.4	4.4	5.8	7.0	8.0	----	10.6	15.5	
Corning 0010[d]	24	ϵ'/ϵ_o	6.68	6.63	6.57	6.50	6.43	6.39	6.33	----	6.1	5.96	5.87
		tan δ	77.5	53.5	35	23	16.5	15	23	----	60	90	110
Corning 0014[e]	25	ϵ'/ϵ_o	6.78	6.77	6.76	6.75	6.73	6.72	6.70	6.69	----	6.64	
		tan δ	23.1	17.2	14.4	12.2	12.4	13.8	17.0	19.5	----	70	
Corning 0080[f]	23	ϵ'/ϵ_o	8.30	7.70	7.35	7.08	6.90	6.82	6.75	----	6.71	6.71	6.62
		tan δ	780	400	220	140	100	85	90	----	126	170	180
Corning 0090[g]	20	ϵ'/ϵ_o	9.15	9.15	9.15	9.14	9.12	9.10	9.02	----	8.67	8.45	8.25
		tan δ	12	8	7	7	8	12	18	----	54	103	122
Corning 0100[h]	25	ϵ'/ϵ_o	7.18	7.17	7.16	7.14	7.10	7.10	7.07	----	7.00	6.95	6.87
		tan δ	24	16	13.5	13	14	17	24	----	44	63	106
Corning 0120[i]	23	ϵ'/ϵ_o	6.75	6.70	6.66	6.65	6.65	6.65	6.65	----	6.64	6.60	6.51
		tan δ	46	30	20	14	12	13	18	----	41	63	127
Corning 1770[j]	25	ϵ'/ϵ_o	6.25	6.16	6.10	6.03	6.00	6.00	6.00	----	5.95	5.83	5.44
		tan δ	49.5	42	33	26	27	34	38	----	56	84	140
Corning 1990[k]	24	ϵ'/ϵ_o	8.40	8.38	8.35	8.32	8.30	8.25	8.20	----	7.99	7.94	7.84
		tan δ	4	4	3	4	5	7	9	----	19.9	42	112
Corning 1991[m]	24	ϵ'/ϵ_o	8.10	8.10	8.08	8.08	8.08	8.06	8.00	----	7.92	7.83	
		tan δ	12	9	6	5	5	7	12	----	38	51	
Corning 3320[n]	24	ϵ'/ϵ_o	5.00	4.93	4.88	4.82	4.79	4.78	4.77	----	4.74	4.72	4.7
		tan δ	80	58	43	34	30	30	32	----	55	73	120
Corning 7040[o]	25	ϵ'/ϵ_o	4.84	4.82	4.79	4.77	4.73	4.70	4.68	----	4.67	4.64	4.52
		tan δ	50	34	25.5	20.5	19	22	27	----	44	57	73
Corning 7050[p]	25	ϵ'/ϵ_o	4.88	4.84	4.82	4.80	4.78	4.76	4.75	----	4.74	4.71	4.64
		tan δ	81	56	43	33	27	28	35	----	52	61	83

a. Contains 2% iron oxide (Am. Optical). b. Aluminum-zinc phosphate. (ca. 70% P_2O_5)(Am. Optical). c. 73.2% SiO_2, 24.8% B_2O_3 (Components and Systems Lab., Air Matériel Command). d. Soda-potash-lead silicate ca. 20% PbO. e. Lead-barium glass. f. Potash-lead-silicate. h. Potash-soda-barium silicate. i. Soda-potash-lead-silicate. j. Lime-alumina-silicate. k. Iron-sealing glass. m. 45% SiO_2, 14% K_2O, 6% Na_2O, 3% PbO, 5% CaO (iron-sealing glass). n. Soda-potash-borosilicate. o. Soda-potash-borosilicate. p. Soda-borosilicate (ca. 70% SiO_2).

I. Solids, A. Inorganic (cont.)

3. Glasses (cont.)

Values for tan δ are multiplied by 10^4; frequency given in c/s.

Material		T°C	1×10^2	1×10^3	1×10^4	1×10^5	1×10^6	1×10^7	1×10^8	3×10^8	3×10^9	1×10^{10}	2.5×10^{10}
Corning 7052[a]	$\varepsilon'/\varepsilon_0$	23	5.20	5.18	5.14	5.12	5.10	5.10	5.09	----	5.04	4.93	4.85
	tan δ		68	49	34	26	24	28	34	----	58	81	114
Corning 7055	$\varepsilon'/\varepsilon_0$	25	5.45	5.41	5.38	5.33	5.31	5.30	5.27	5.25	----	5.08	----
	tan δ		45	36	30	28	28	29	38	49	----	130	----
Corning 7060[b]	$\varepsilon'/\varepsilon_0$	25	5.02	4.97	4.92	4.86	4.84	4.84	4.84	----	4.82	4.80	4.65
	tan δ		89	55	42	40	36	30	30	----	54	98	90
Corning 7070[c]	$\varepsilon'/\varepsilon_0$	23	4.00	4.00	4.00	4.00	4.00	4.00	4.00	4.00	4.00	4.00	3.9
	tan δ		6	5	5	6	8	11	12	12	12	21	31
	$\varepsilon'/\varepsilon_0$	100	4.17	4.16	4.15	4.14	4.13	4.10	----	----	4.00	4.00	----
	tan δ		50	22	13	10	9	11	----	----	19	21	----
Corning 7230[d]	$\varepsilon'/\varepsilon_0$	25	3.88	3.86	3.85	3.85	3.85	3.85	----	----	3.76	----	----
	tan δ		33	23	16	13	11	12	----	----	22	----	----
Corning 7570	$\varepsilon'/\varepsilon_0$	25	14.58	14.56	14.54	14.53	14.52	14.50	14.42	14.4	----	14.2	----
	tan δ		11.5	13.5	15.9	16.5	19.0	23.5	33	44	----	98	----
Corning 7720[e]	$\varepsilon'/\varepsilon_0$	24	4.74	4.70	4.67	4.64	4.62	4.61	----	----	----	4.59	----
	tan δ		78	42	29	22	20	23	----	----	----	43	----
Corning 7740[f]	$\varepsilon'/\varepsilon_0$	25	4.80	4.73	4.70	4.60	4.55	4.52	4.52	----	----	4.52	4.50
	tan δ		128	86	65	54	49	45	45	----	----	85	96
Corning 7750[f]	$\varepsilon'/\varepsilon_0$	25	4.45	4.42	4.39	4.38	4.38	4.38	----	----	4.38	4.38	----
	tan δ		45	33	24	20	18	19	----	----	43	54	----
Corning 7900[g]	$\varepsilon'/\varepsilon_0$	20	3.85	3.85	3.85	3.85	3.85	3.85	----	3.85	3.84	3.82	3.82
	tan δ		6	6	6	6	6	6	----	6	6.8	9.4	13
	$\varepsilon'/\varepsilon_0$	100	3.85	3.85	3.85	3.85	3.85	3.85	----	3.85	3.84	3.82	----
	tan δ		37	17	12	10	8.5	7.5	----	7.5	10	13	----
Corning 7911	$\varepsilon'/\varepsilon_0$	25	----	----	----	----	----	----	----	----	----	3.82	----
	tan δ		----	----	----	----	----	----	----	----	----	6.5	----
Corning 8460[h]	$\varepsilon'/\varepsilon_0$	25	8.35	8.30	8.30	8.30	8.30	8.30	8.30	----	8.10	8.06	8.05
	tan δ		11	9	7.5	7	8	10	16	----	40	57	60
Corning 8830	$\varepsilon'/\varepsilon_0$	25	5.38	5.28	5.20	5.11	5.05	5.01	5.00	4.97	----	4.83	----
	tan δ		204	130	91	73	60	54	57	63	----	99	----

a. Soda-potash-lithia-borosilicate.   b. Soda-borosilicate (Pyrex).   c. Low alkali potash-lithia-borosilicate.   d. Aluminum borosilicate.
e. Soda-lead borosilicate.   f. Soda-borosilicate (ca. 80% SiO_2).   g. 96% SiO_2.   h. Barium borosilicate.

Dielectric Data

I. Solids, A. Inorganic (cont.)
3. Glasses (cont.)

Values for tan δ are multiplied by 10^4; frequency given in c/s.

	T°C		1×10^2	1×10^3	1×10^4	1×10^5	1×10^6	1×10^7	1×10^8	3×10^8	3×10^9	1×10^{10}	2.5×10^{10}
Corning 8871[a]	25	ϵ'/ϵ_0	8.45	8.45	8.45	8.45	8.45	8.43	----	8.40	8.34	8.05	7.82
		tan δ	18	13	9	7	6	7	----	14	26	49	70
Corning 9010	25	ϵ'/ϵ_0	6.51	6.49	6.48	6.45	6.44	6.43	6.42	6.40	----	6.27	
		tan δ	50.5	36.2	26.7	22.7	21.5	22.6	30	41	----	91	
Corning Lab. No. 189CS	25	ϵ'/ϵ_0	19.2	19.2	19.2	19.1	19.0	19.0	19.0	----	17.8		
		tan δ	12.5	13	16.5	21	27	37	57	----	124		
"E" Glass[b]	23	ϵ'/ϵ_0	6.43	6.40	6.39	6.37	6.32	6.25	6.22	----	----	6.11	
		tan δ	42	34	27	18	15	17	23	----	----	60	
Foamglas[c]	23	ϵ'/ϵ_0	90.0	82.5	68.0	44.0	17.5	9.0	----	----	----	5.49	
		tan δ	1500	1600	2380	3200	3180	1960	----	----	----	455	
Fused Silica 915c[d]	25	ϵ'/ϵ_0	3.78	3.78	3.78	3.78	3.78	3.78	3.78	3.78	----	3.78	
		tan δ	6.6	2.6	1.1	0.4	0.1	0.1	0.3	0.5	----	1.7	
Fused Silica 915c[d,e]	25	ϵ'/ϵ_0	3.78	3.78	3.78								
		tan δ	0.08	0.13	<0.05								
Fused quartz[f]	25	ϵ'/ϵ_0	3.78	3.78	3.78	3.78	3.78	3.78	3.78	----	3.78	3.78	3.78
		tan δ	8.5	7.5	6	4	2	1	1	----	0.6	1	2.5
Soda-Silica Glasses[g]													
9% Na_2O, 91% SiO_2	25	ϵ'/ϵ_0	6.4	6.2	5.7	----	5.4	----	----	5.1	----	5.05	4.9
		tan δ	2500	820	400	----	130	----	----	100	----	130	160
12% Na_2O, 88% SiO_2	25	ϵ'/ϵ_0	8.2	6.7	6.1	----	5.6	----	----	5.2	----	5.15	5.04
		tan δ	1900	600	300	----	150	----	----	100	----	155	170
16% Na_2O, 84% SiO_2	25	ϵ'/ϵ_0	9.4	7.4	6.6	6.2	5.9	5.7	----	5.5	----	5.27	
		tan δ	3000	960	500	250	165	130	----	110	----	180	
20% Na_2O, 80% SiO_2	25	ϵ'/ϵ_0	10.8	8.3	7.3	6.8	6.6	6.3	----	5.9	----	5.6	6.1
		tan δ	4000	1500	670	360	220	180	----	140	----	200	280
25% Na_2O, 75% SiO_2	25	ϵ'/ϵ_0	13	9.7	8.4	----	7.6	----	----	6.7	----	6.3	
		tan δ	6700	2400	1000	----	310	----	----	170	----	220	
30% Na_2O, 70% SiO_2	25	ϵ'/ϵ_0	18	12	10.4	----	8.5	----	----	7.5	----	7.2	7.0
		tan δ	11000	3900	1300	----	400	----	----	190	----	240	350

a. Alkaline lead silicate. b. Owens-Corning. c. Soda-lime (Pittsburgh-Corning). d. SiO_2 (Corning). e. Sample B-135.
f. SiO_2 (General Electric). g. Composition in mole % of oxides as mixed (Lab. Ins. Res.).

312 Tables of Dielectric Materials

I. Solids, A. Inorganic (Cont.) Values for tan δ are multiplied by 10^4; frequency given in c/s.

3. Glasses (cont.)
Alkali-silica glasses[a]

	T°C		1×10^2	1×10^3	1×10^4	1×10^5	1×10^6	1×10^7	1×10^8	3×10^8	3×10^9	1×10^{10}
12.8% Li_2O, 87.2% SiO_2	25	ϵ'/ϵ_0	9.94	6.54	5.45	5.1	4.95	4.92	----	4.9	----	4.80
		tan δ	9700	3600	1000	310	174	124	----	79	----	102
12.8% Na_2O, 87.2% SiO_2	25	ϵ'/ϵ_0	8.09	6.61	6.00	5.8	5.66	5.57	----	5.4	----	5.33
		tan δ	3050	1370	450	240	159	126	----	118	----	182
12.8% K_2O, 87.2% SiO_2	25	ϵ'/ϵ_0	7.53	6.49	6.25	6.17	6.09	6.02	----	5.8	----	5.8
		tan δ	502	360	200	121	90	80	----	99	----	220
12.8% K_2O, 87.2% SiO_2 Quenched	25	ϵ'/ϵ_0	7.10	6.63	6.30	6.12	6.10	6.08	----	5.95	----	5.75
		tan δ	730	470	270	160	119	103	----	106	----	240
12.8% Rb_2O, 87.2% SiO_2	25	ϵ'/ϵ_0	5.39	5.32	5.23	5.22	5.21	5.20	----	5.15	----	5.05
		tan δ	98	89	58	46	41	38	----	59	----	120
6.4% Li_2O, 6.4% Na_2O, 87.2% SiO_2	25	ϵ'/ϵ_0	5.15	5.08	5.05	5.05	5.04	5.03	----	5.00	----	4.95
		tan δ	145	87	47	28	19	17	----	26	----	52
6.4% Li_2O, 6.4% Na_2O, 87.2% SiO_2 Quenched	25	ϵ'/ϵ_0	5.15	5.09	5.05	5.05	5.04	5.02	----	4.98	----	4.94
		tan δ	150	87	47	27	19	21	----	31	----	57
3.3% Li_2O, 6.6% K_2O, 71% SiO_2	25	ϵ'/ϵ_0	5.23	5.19	5.17	5.15	5.14	5.10	----	5.07	----	5.04
		tan δ	53	47	37	28	24	24	----	40	----	83
3.3% Li_2O, 6.6% K_2O, 91% SiO_2 Quenched	25	ϵ'/ϵ_0	5.38	5.35	5.28	5.26	5.23	5.20	----	5.15	----	5.08
		tan δ	102	74	44	30	30	33	----	49	----	102
6.4% Na_2O, 6.4% K_2O, 87.2% SiO_2	25	ϵ'/ϵ_0	5.68	5.62	5.58	5.56	5.56	5.54	----	5.51	----	5.50
		tan δ	102	75	42	31	25	23	----	40	----	115
6.4% Na_2O, 6.4% K_2O, 87.2% SiO_2 Quenched	25	ϵ'/ϵ_0	5.70	5.62	5.58	5.56	5.56	5.55	----	5.54	----	5.53
		tan δ	129	89	47	36	33	34	----	44	----	160

a. Composition in mole % of oxides as mixed (Lab. Ins. Res.).

Dielectric Data

I. Solids A. Inorganic (cont.)

Values for tan δ are multiplied by 10^4; frequency given in c/s.

4. Mica and Glass

	T°C		1×10^2	1×10^3	1×10^4	1×10^5	1×10^6	1×10^7	1×10^8	3×10^8	3×10^9	1×10^{10}	2.5×10^{10}
Mycalex 2821[a] (discontinued)	25	$\varepsilon'/\varepsilon_0$	7.50	7.50	7.50	7.50	7.50	7.45	7.45	----	----	7.09	
		tan δ	60	28	17	13	10	9	9	----	----	27	
Mycalex 400[b]	25	$\varepsilon'/\varepsilon_0$	7.47	7.45	7.42	7.40	7.39	7.38	----	----	----	7.12	
		tan δ	29	19	16	14	13	13	----	----	----	33	
	80	$\varepsilon'/\varepsilon_0$	7.64	7.59	7.54	7.52	7.50	7.47	----	----	----	7.32	
		tan δ	150	85	50	25	16	14	----	----	----	57	
Mycalex K10[c]	24	$\varepsilon'/\varepsilon_0$	9.5	9.3	9.2	9.1	9.0	9.0	----	----	11.3*	11.3*	
		tan δ	170	125	76	42	26	21	----	----	40	40	
Mykroy Grade 8[d]	25	$\varepsilon'/\varepsilon_0$	6.87	6.81	6.76	6.74	6.73	6.73	6.72	----	6.68**	6.96**	6.66
		tan δ	95	66	43	31	26	24	25	----	38	48	81
Mykroy Grade 38[d]	25	$\varepsilon'/\varepsilon_0$	7.71	7.69	7.64	7.61	7.61	7.61	----	----	7.68**	8.35**	
		tan δ	43	33	27	24	21	14	----	----	35	40	

5. Miscellaneous Inorganics

	T°C		1×10^2	1×10^3	1×10^4	1×10^5	1×10^6	1×10^7	1×10^8	3×10^8	3×10^9	1×10^{10}	2.5×10^{10}
Ruby mica[e]	26	$\varepsilon'/\varepsilon_0$	5.4	5.4	5.4	5.4	5.4	5.4	5.4	----	5.4		
		tan δ	25	6	3.5	3	3	3	2	----	3		
Canadian mica (field ⊥ sheet)	25	$\varepsilon'/\varepsilon_0$	6.90	6.90	6.90								
(field ∥ sheet)	25	$\varepsilon'/\varepsilon_0$	11.5	8.7	7.3								
		tan δ	2300	980	400								
Marble S-3030[f] (after drying over P_2O_5)	25	$\varepsilon'/\varepsilon_0$	15.6	12.8	11.4	10.6	10.0	9.5	9.1	8.8	----	8.6	
		tan δ	2000	1100	630	390	360	370	290	250	----	120	
(after baking)	25	$\varepsilon'/\varepsilon_0$	9.45	9.42	9.29	9.25	9.07	8.98	8.85	----	----	8.53	
		tan δ	180	130	81	75	110	120	120	----	----	110	
(after 60% humidity for 3 weeks)	25	$\varepsilon'/\varepsilon_0$	----	----	----	----	----	9.35	8.95				
		tan δ	----	----	----	----	----	500	250				
Selenium, amorphous[g]	25	$\varepsilon'/\varepsilon_0$	6.00	6.00	6.00	6.00	6.00	6.00	6.00	6.0	6.00	6.00	6.00
		tan δ	18	4	<3	<5	<3	<2	<2	<5	1.8	6.7	13
Quinterra[h] (Measured at 50% R.H.)	25	$\varepsilon'/\varepsilon_0$	5.75	4.80	4.1	3.3	3.1						
		tan δ	1770	1500	1240	590	250						
Quinorgo #3000[i] (Measured at 50% R.H.)	25	$\varepsilon'/\varepsilon_0$	8.75	6.4	4.9	4.0	3.3						
		tan δ	2540	2310	2150	1350	870						

a. Mica, glass (Gen. Elec.). b. Mica, glass (Mycalex). c. Mica, glass, TiO_2 (Mycalex). d. Mica, glass (Electronic Mech.). e. Muscovite. f. Class A (Tenn. Marble). g. Amer. Smelt. and Refining. h. Asbestos fiber, chrysotile, Type I (Johns-Manville). i. 85% chrysotile asbestos, 15% organic material (Johns-Manville).

*Not corrected for variations of density. **Samples nonhomogeneous.

Tables of Dielectric Materials

I. Solids A. Inorganic (cont.)
5. Misc. Inorganics (cont.)

	T°C		1×10^2	1×10^3	1×10^4	1×10^5	1×10^6	1×10^7	1×10^8	3×10^8	3×10^9	1×10^{10}
						Frequency given in c/s.						
Sandy soil, dry	25	ϵ'/ϵ_o	3.42	2.91	2.75	2.65	2.59	2.55	----	2.55	2.55	2.53
		tan δ	.196	.08	.034	.020	.017	.016	----	.0100	.0062	.0036
2-18% moisture	25	ϵ'/ϵ_o	3.23	2.72	2.50	2.50	2.50	2.50	----	2.50	2.50	2.50
		tan δ	.64	.13	.056	.030	.025	.025	----	.026	.03	.065
3.88% moisture	25	ϵ'/ϵ_o	----	----	----	5.0	4.70	4.50	----	4.50	4.40	3.60
		tan δ	----	1500	----	19	1.75	.3	----	.03	.046	.12
16.8% moisture	25	ϵ'/ϵ_o	----	----	----	----	20	20	----	20	20	13
		tan δ	----	----	367	----	4	.35	----	.03	.13	.29
Loamy soil, dry	25	ϵ'/ϵ_o	3.06	2.83	2.69	2.60	2.53	2.48	----	2.47	2.44	2.44
		tan δ	.07	.05	.035	.03	.018	.014	----	.0065	.0011	.0014
2.2% moisture	25	ϵ'/ϵ_o	----	----	18	----	6.9	4	----	3.5	3.5	3.50
		tan δ	----	2.1	1.6	----	.65	.45	----	.06	.04	.03
13.77% moisture	25	ϵ'/ϵ_o	----	----	970	----	----	14.5	----	20	20	13.8
		tan δ	----	8490	----	----	----	1.3	----	.16	.12	.19
Clay soil, dry	25	ϵ'/ϵ_o	4.73	3.94	3.27	2.79	2.57	2.44	----	2.38	2.27	2.16
		tan δ	.12	.12	.12	.10	.065	.04	----	.020	.015	.013
20.09% moisture	25	μ'/μ_o	----	----	1000	----	----	21.6	----	20	11.3	
		tan δ_m	----	7800	----	----	----	1.7	----	.52	.25	
Magnetite soil, dry	25	ϵ'/ϵ_o	----	----	----	----	1.09	1.09	----	1.09	1.005	
		tan δ	----	----	----	----	----	----	----	.025	.099	
		ϵ'/ϵ_o	3.95	3.75	3.62	3.52	3.50	3.60	----	3.50	3.50	
		tan δ_d	.041	.029	.022	.017	.015	.012	----	.018	.022	
4.9% moisture	25	ϵ'/ϵ_o	----	----	$\rho = 2\times10^5$ ohm cm	----	12	10	----	9.0	8.3	
		tan δ_d	----	----		----	9	1.0	----	.12	.22	
11.2% moisture	25	ϵ'/ϵ_o	----	----	$\rho = 2\times10^4$ ohm cm	----	----	30	----	30	30	
		tan δ_d	----	----		----	----	4	----	.2	.32	
Amplifilm[a]	25	ϵ'/ϵ_o	4.34	4.32	4.30	4.30	4.30	4.29				
		tan δ	.00314	.0022	.0019	.0017	.0017	.0018				

a. Bentonite with organic binder (Aircraft Marine).

Dielectric Data

I. Solids, B. Organic, with or without inorganic components – Values for tan δ are multiplied by 10^4; frequency given in c/s.

1. **Crystals**

	T°C		1×10^2	1×10^3	1×10^4	1×10^5	1×10^6	1×10^7	1×10^8	3×10^8	3×10^9	1×10^{10}	2.5×10^{10}
α-Trinitrotoluene[a]	26	ϵ'/ϵ_o	2.64	2.64	2.64	2.64	2.64	2.64					
		tan δ	36	20	11	6	4.5	4					
α-Trinitrotoluene at ca. 20% humidity[b]	25	ϵ'/ϵ_o	3.80	3.09	2.93	2.90	2.89	2.89	2.89	2.89	2.86	2.82	
		tan δ	2600	900	200	40	20	17	28	39	18	14	
Santicizer 9[c]	25	ϵ'/ϵ_o	4.19	3.76	3.56	3.51	3.47	3.43	3.35				3.21
		tan δ	1140	560	350	185	120	110	71				28
Naphthalene[d]	25	ϵ'/ϵ_o	-----	2.85	2.85	2.85	2.85	2.85	-----	-----	2.8	-----	
		tan δ	-----	19	9	5	3	2	-----	-----	2.1	-----	
Orthoterphenyl[e]	25	ϵ'/ϵ_o	2.8	2.8	2.8	2.8	2.8	2.8	-----	-----	-----	2.74	
		tan δ	3.7	4.7	6.0	6.3	6	5	-----	-----	-----	2.7	
Metaterphenyl[e]	25	ϵ'/ϵ_o	-----	2.86	2.86	-----	2.86	2.86	2.86	-----	2.83	-----	
		tan δ	-----	4.5	4.5	-----	9.5	10.1	5	-----	0.7	-----	
Paraterphenyl[e]	25	ϵ'/ϵ_o	-----	(2.95–3.20 depending on degree of crystallization)					-----	-----	2.95	-----	
		tan δ	-----	2	1	<2	<2	<2	<2	-----	2.1	-----	
Aroclor 1268[f]	25	ϵ'/ϵ_o	2.64	2.64	2.64	2.64	2.64	2.64					
		tan δ	14	3.5	2	<3	<6	<6					

2. **Simple Noncrystals**

	T°C		1×10^2	1×10^3	1×10^4	1×10^5	1×10^6	1×10^7	1×10^8
Aroclor 5460[g]	25	ϵ'/ϵ_o	2.43	2.43	2.42	2.42	2.41	2.41	2.41
		tan δ	<3	4	16	20	18	13	8
Aroclor 4465[h]	25	ϵ'/ϵ_o	2.81	2.80	2.79	2.79	2.79	2.79	2.79
		tan δ	45	17	6.6	4	2.4	1.5	1.0

3. **Plastics**

  a. Phenol-formaldehyde

	T°C		1×10^2	1×10^3	1×10^4	1×10^5	1×10^6	1×10^7	1×10^8	3×10^9	1×10^{10}	2.5×10^{10}
Bakelite BM-120[i] (preformed and preheated)	25	ϵ'/ϵ_o	4.87	4.74	4.62	4.50	4.36	4.16	3.95	3.70	3.68	3.55
		tan δ	300	220	200	210	280	350	380	438	410	390

a. Chem. Lab. M.I.T. b. War Dept., Picatinny Arsenal. c. o- and p-toluene sulfonamides (Monsanto). d. Eastman Kodak: recryst. and resubl. Lab. Ins. Res. e. Monsanto, recryst. Lab. Ins. Res. f. Nonachlorobiphenyl (Monsanto). g. Nonachloroterphenyls (Monsanto). h. Chlorinated mixture of biphenyl and terphenyl (Monsanto). i. 46% wood flour, 8% misc. (Bakelite).

Tables of Dielectric Materials

I. Solids, B. Organic (cont.)
3. Plastics (cont.)
a. Phenol-formaldehyde (cont.)

Values for tan δ are multiplied by 10^4; frequency given in c/s.

Material	T°C		1×10^2	1×10^3	1×10^4	1×10^5	1×10^6	1×10^7	1×10^8	3×10^8	3×10^9	1×10^{10}	2.5×10^{10}
Bakelite BM-120[a] (not preformed or preheated)	27	ϵ'/ϵ_0	5.50	5.15	4.90	4.65	4.45	4.30	----	----	3.70	3.55	
		tan δ	740	460	345	320	350	415	----	----	400	500	
	57	ϵ'/ϵ_0	7.80	6.35	5.70	5.30	4.90	4.65	4.5	----	4.15		
		tan δ	2950	1150	530	380	430	470	480	----	530		
	88	ϵ'/ϵ_0	18.2	8.5	6.5	5.7	5.2	5.0	4.7	----	4.40		
		tan δ	7600	3700	1400	600	400	420	470	----	700		
Bakelite BM-250[b] (preformed and preheated)	25	ϵ'/ϵ_0	37	22	12	7.2	5.3	5.0	----	----	----	----	5.0
		tan δ	3000	3700	3900	2900	1250	220	----	----	----	----	320
Bakelite BP-48-306[c]	24	ϵ'/ϵ_0	8.2	7.15	6.5	5.9	5.4	4.9	4.4	----	3.64	3.52	
		tan δ	1350	820	630	560	600	730	770	----	519	366	
Bakelite BM-16981[d]	25	ϵ'/ϵ_0	7.6	6.1	5.4	5.1	4.9	4.8	4.7	----	4.6	4.5	
(not preformed or preheated)		tan δ	2300	1000	500	300	200	130	100	----	100	120	
Bakelite BM-16981[d]	25	ϵ'/ϵ_0	4.82	4.73	4.66	4.62	4.60	4.59	4.58	----	4.57	4.57	
(preformed and preheated)		tan δ	140	120	100	70	52	50	60	----	90	85	
Bakelite BM-16981[d]	25	ϵ'/ϵ_0	5.05	4.87	4.80	4.79	4.72	4.67	4.62	----	----	4.52	
(powder preheated)		tan δ	190	160	130	90	72	60	56	----	----	82	
Laminated Fiberglas BK-174[e]	24	ϵ'/ϵ_0	14.2	9.8	7.2	5.9	5.3	5.0	4.8	4.54	4.40	4.37	
		tan δ	2500	2600	1600	880	460	340	260	240	290	360	
Catalin 200 base, white[f]	22*	ϵ'/ϵ_0	8.6	8.2	7.9	7.5	7.0	6.5	----	----	4.89		
		tan δ	400	290	330	400	500	650	----	----	1080		
	85*	ϵ'/ϵ_0	53.5	34	24	17	12	9.5	----	----	5.81		
		tan δ	98000	20000	5200	1800	1000	1100	----	----	1910		
	25**	ϵ'/ϵ_0	6.9	6.7	6.6	6.5	6.3	5.8	----	5.1	4.61		
		tan δ	130	170	260	330	380	480	----	878	926		

a. 46% wood flour, 8% misc. (Bakelite). b. 66% asbestos fiber (Bakelite). c. 100% (Bakelite). d. Mica-filled (Bakelite).
e. 31.6% Bakelite's BV-17085, 68.4% Fiberglas (Owens Corning). f. Catalin.
* Measured December, 1943. **Same material measured after two years at room temperature and humidity.

I. Solids, B. Organic 3. Plastics (cont.)

a. Phenol-formaldehyde

Values for tan δ are multiplied by 10^4; frequency given in c/s.

(cont.)	T°C		1×10^2	1×10^3	1×10^4	1×10^5	1×10^6	1×10^7	1×10^8	3×10^8	3×10^9	1×10^{10}	2.5×10^{10}
Catalin 500 base, yellow[a]	22*	ϵ'/ϵ_0	9.7	9.4	8.8	8.1	7.3	6.3	----	4.90	4.72		
		tan δ	280	320	400	550	760	1200	----	996	870		
	25**	ϵ'/ϵ_0	4.92	4.86	4.78	4.69	4.57	4.40	4.2	4.02	3.7	3.41	
		tan δ	66	100	150	230	330	460	580	620	520	360	
Catalin 500 base, standard[a]	25	ϵ'/ϵ_0	20.0	15.2	12.6	10.8	9.6	8.4	----	5.79	4.77	4.43	
		tan δ	2800	1500	1000	860	940	1200	----	1540	1250	1300	
Catalin 700 base (Prystal)[a]	25*	ϵ'/ϵ_0		44	16	11	8	6			4.74		
		tan δ	16500	8000	4100	2300	1600	1400	----	----	1530		
	80*	ϵ'/ϵ_0	----	----	51	18.5	10.0	7.4	----	----	4.77		
		tan δ	----	----	13200	6600	3700	2400	----	----	1640		
	25**	ϵ'/ϵ_0	58	24	14	9.6	8.0	6.6	5.0	4.1	3.7		
		tan δ	14000	6400	3100	1900	1500	1400	950	740	500		
Durez 1601, natural[b]	26	ϵ'/ϵ_0	5.09	4.94	4.80	4.68	4.60	4.55	4.51	----	4.48	4.47	
		tan δ	270	210	132	100	80	70	64	----	62	62	
Durite 500[c]	24	ϵ'/ϵ_0	5.10	5.03	4.95	4.86	4.78	4.72	4.72	----	4.71	4.70	
		tan δ	130	104	86	78	82	102	115	----	126	128	
Formica XX[d]	26	ϵ'/ϵ_0	5.23	5.15	4.96	4.78	4.60	4.32	4.04	----	3.57***	3.55+	
(field ⊥ to laminate)		tan δ	230	165	170	230	340	490	570	----	600***	700+	
Formica LE[e]	26	ϵ'/ϵ_0	6.50	5.70	5.30	5.00	4.75	4.35	3.95	----	3.35***	3.25+	
(field ⊥ to laminate)		tan δ	1350	600	430	400	410	480	500	----	400***	460+	
Grade YN-25[f]	25	ϵ'/ϵ_0	3.73	3.65	3.61	3.61	3.59	3.47	----	3.24++	3.20	3.12	
(field ∥ to laminate)		tan δ	121	132	200	207	187	187	----	183++	178	173	
(field ⊥ to laminate)	25	ϵ'/ϵ_0	3.87	3.77	3.68	3.61	3.55	3.47					
		tan δ	166	150	182	204	182	191					
Panelyte Grade 776[g]	25	ϵ'/ϵ_0	4.25	4.18	4.10	3.95	3.87	3.73	3.52	3.40	3.22	3.12	3.05
		tan δ	190	150	160	200	280	370	400	390	370	340	310

a. Catalin. b. 67% mica (Durez). c. 65% mica, 4% lubricants (Durite). d. 50% paper laminate (Formica). e. 40% cotton fabric (Formica). f. 50% Nylon fabric (Formica). g. Paper base (St. Regis).

*Measured December, 1943. **Same material measured after two years at room temperature and humidity. ***Rod stock in coaxial line. +Rod stock in H_{11} (TE_{11}) mode of circular guide. ++Freq. = 1×10^9.

Tables of Dielectric Materials

I. Solids, B. Organic (Cont.).
3. Plastics (cont.)
a. Phenol-formaldehyde (cont.)

Values for tan δ are multiplied by 10^4; frequency given in c/s.

Material	T°C		1×10^2	1×10^3	1×10^4	1×10^5	1×10^6	1×10^7	1×10^8	3×10^8	3×10^9	1×10^{10}	2.5×10^{10}
Micarta #254[a]	-13	ϵ'/ϵ_o	4.9	4.7	4.5	4.4	4.2	4.0	3.4	----	3.34	3.13	
		tan δ	250	220	240	280	370	390	370	----	286	215	
	25	ϵ'/ϵ_o	5.30	4.95	4.81	4.66	4.51	4.20	3.85	3.78	3.43	3.25	3.21
		tan δ	760	330	260	270	360	470	550	470	505	410	380
	82	ϵ'/ϵ_o	9.2	6.3	5.5	5.3	5.1	4.9	----	----	4.02	3.21	
		tan δ	4100	1820	690	360	300	400	----	----	984	773	
Micarta #496[b] (field ⊥ to laminate)	25	ϵ'/ϵ_o	8.6	7.0	6.3	5.6	5.2	4.8	4.38				
		tan δ	2200	1100	630	500	480	630	710				
Micarta #496[b] (field ∥ to laminate)	25	ϵ'/ϵ_o	----	----	----	----	----	----	----	3.95*	3.78*	3.62*	
		tan δ	----	----	----	----	----	----	----	510*	550*	570*	
Phenolic paper laminate JH-1410[c]	25	ϵ'/ϵ_o	4.21	4.17	4.12	4.06	3.94	3.81	3.62				
		tan δ	146	100	115	162	222	295	339				
Resinox 10231[d]	25	ϵ'/ϵ_o	11.4	9.3	7.6	6.2	5.4	5.0	4.8	4.8**	4.7	4.6	
		tan δ	1390	1410	1440	1180	700	400	290	310**	330	340	
Resinox 10900[e]	25	ϵ'/ϵ_o	4.80	4.64	4.45	4.37	4.24	4.23	4.08	4.05**	4.03	4.02	
		tan δ	275	270	230	180	135	110	93	93**	100	96	
Taylor Grade GGG[f] (field ∥ laminate)	25	ϵ'/ϵ_o	7.25	6.08	4.67	4.14	4.00	3.9	3.9	3.9**	3.9	3.9	
		tan δ	1300	1600	1500	500	150	110	140	200**	250	290	
(field ⊥ laminate)	25	ϵ'/ϵ_o	3.95	3.91	3.85	3.82	3.75	3.68	3.64				
		tan δ	119	88	91	104	110	114	110				
Dilecto (Mecoboard)[g]	25	ϵ'/ϵ_o	4.20	3.98	3.79	3.62	3.46	3.37	3.23	3.18	3.11	3.08	
		tan δ	400	344	318	304	263	223	216	215	220	229	
	90	ϵ'/ϵ_o	11.90	8.40	6.23	5.50	5.32	4.34	----	3.35	----	3.30	
		tan δ	2310	1640	1020	584	570	970	----	660	----	470	

a. Cresylic acid-formaldehyde, 50% α-cellulose (Westinghouse). b. Cresylic acid-formaldehyde, 50% cotton drilling. (Westinghouse).
c. 35% paper (Catalin). d. 53% filler (Monsanto). e. 35% mica, 18% filler (Monsanto). f. 40% random glass mat (Taylor).
g. 45% cresol-phenol formaldehyde, 15% tung oil, 15% nylon (Continental Diamond).
*Samples turned from sheet stock. **Freq = 1×10^9.

Dielectric Data

I. Solids, B. Organic (Cont.)
3. Plastics (cont.)
a. Phenol-formaldehyde (cont.)

Values for tan δ are multiplied by 10^4; frequency given in c/s.

	T°C		1×10^2	1×10^3	1×10^4	1×10^5	1×10^6	1×10^7	1×10^8	3×10^8	3×10^9	1×10^{10}
Dilecto (hot punching) XXX-P-26[a]	25	ϵ'/ϵ_0	14.7	8.61	6.68	5.76	5.05	4.60	4.10	3.78	3.45	3.35
		tan δ	6420	2970	1380	840	720	705	690	680	500	480
(Field ∥ sheet)	90	ϵ'/ϵ_0	----	----	----	----	----	----	----	4.36	----	3.72
		tan δ	----	----	----	----	----	----	----	950	----	770
	120	ϵ'/ϵ_0	----	----	----	----	----	----	----	4.76	----	4.27
		tan δ	----	----	----	----	----	----	----	990	----	820
(Field ⊥ sheet)	25	ϵ'/ϵ_0	4.29	4.21	4.14	4.01	3.89	3.73	3.56	3.35		
		tan δ	137	119	137	180	250	340	390	440		
	90	ϵ'/ϵ_0	5.74	5.09	4.77	4.57	4.32	4.15				
		tan δ	1990	620	360	284	273	308				
	120	ϵ'/ϵ_0	7.42	5.38	4.80	4.52	4.34	4.21				
		tan δ	6700	1910	630	380	290	322				
Micarta #299[b] (field ⊥ to laminate)	24	ϵ'/ϵ_0	5.36	5.29	5.20	5.10	5.01	4.92	4.80			
		tan δ	270	130	115	122	138	150	170			
Micarta #299[b] (field ∥ to laminate)	24	ϵ'/ϵ_0	----	----	----	----	----	----	----	4.54**	4.34*	4.59**
		tan δ	----	----	----	----	----	----	----	149*	188*	230**
Durite #221X[c]	24	ϵ'/ϵ_0	6.70	5.70	5.30	4.90	4.55	4.30	4.15	----	3.65	
		tan δ	2000	820	520	470	430	400	390	----	350	
	84	ϵ'/ϵ_0	13.8	7.0	5.8	5.2	4.9	4.5	4.3	----	3.72	
		tan δ	9000	3300	1200	550	460	470	420	----	800	
Corfoam 114[d]	25	ϵ'/ϵ_0	----	----	----	----	----	----	----	----	----	1.36
		tan δ	----	----	----	----	----	----	----	----	----	750
Expanded plastic board CF-7B[e]	25	ϵ'/ϵ_0	----	----	----	----	----	----	----	----	1.21	
		tan δ	----	----	----	----	----	----	----	----	80	
Expanded phenolic board (dark)[e]	25	ϵ'/ϵ_0	----	----	----	----	----	----	----	----	1.185	
		tan δ	----	----	----	----	----	----	----	----	58	

a. 45% cresol-phenol formaldehyde, 15% tung oil, 40% α paper (Conti. Diamond). b. Cresylic acid-formaldehyde, 60-65% glass fabric (Westinghouse). c. Phenol-furfuraldehyde (Durite). d. Expanded phenolic, den. 0.202 (Rezolin). e. Sponge Rubber Prod.
*Stacked sheets in coaxial line. **Stacked sheets in H_{11} (TE_{11}) mode of circular guide.

I. Solids, B. Organic (Cont.)
3. Plastics (cont.)
b. Phenol-aniline-formaldehyde

Values for tan δ are multiplied by 10^4; frequency given in c/s.

	T°C		1×10^2	1×10^3	1×10^4	1×10^5	1×10^6	1×10^7	1×10^8	3×10^8	3×10^9	1×10^{10}	2.5×10^{10}
Bakelite BM-262[a]	25	ϵ'/ϵ_0	4.85	4.80	4.74	4.72	4.67	4.66	4.65	----	----	4.55	4.5
(preformed and preheated)		tan δ	98	82	75	60	55	55	57	----	----	105	89
Bakelite BM-262[a]	26	ϵ'/ϵ_0	4.9	4.8	4.75	4.7	4.7	4.65	4.65	----	4.58	4.45	
(not preformed or preheated)		tan δ	235	165	113	83	70	65	80	----	90	104	
	84	ϵ'/ϵ_0	5.8	5.4	5.1	5.0	4.9	4.8	4.7	----	4.58	----	
		tan δ	680	440	280	210	160	135	120	----	120	----	
Bakelite BM-1895[b]	25	ϵ'/ϵ_0	4.80	4.72	4.70	4.67	4.64	4.61	4.58	4.55	4.53	4.43	4.35
(preformed and preheated)		tan δ	87	77	80	65	52	52	80	98	102	100	130
Bakelite BM-1895[b]	28	ϵ'/ϵ_0	5.0	4.9	4.75	4.65	4.55	4.5	4.5	----	4.44	4.44	
(not preformed or preheated)		tan δ	138	122	106	90	72	57	60	----	91	96	
	84	ϵ'/ϵ_0	5.8	5.3	5.0	4.8	4.65	4.6	4.6	----	4.6	----	
		tan δ	740	470	310	225	175	140	125	----	132	----	
Durez 11863[c]	-12	ϵ'/ϵ_0	4.68	4.59	4.52	4.48	4.42	4.39	----	----	4.32	4.31	
(preformed and preheated)		tan δ	97	84	66	52	43	38	----	----	44	47	
	25	ϵ'/ϵ_0	4.76	4.70	4.64	4.60	4.55	4.50	4.48	----	4.45	4.42	
		tan δ	108	100	82	62	52	48	52	----	69	76	
	80	ϵ'/ϵ_0	5.26	5.03	4.87	4.74	4.63	4.55	----	----	4.40	4.38	
		tan δ	504	290	195	145	125	113	----	----	100	106	
Durez 11863[c]	26	ϵ'/ϵ_0	5.44	5.22	5.04	4.90	4.78	4.69	----	----	4.52	4.49	
(not preformed or preheated)		tan δ	760	390	230	160	115	88	----	----	63	63	
Formica Grade MF-66 Fiberglas[d]	25	ϵ'/ϵ_0	4.53	4.50	4.43	4.38	4.31	4.24	4.11	4.09	3.90	3.88	3.85
		tan δ	106	95	107	102	95	109	160	195	260	290	300
	79	ϵ'/ϵ_0	4.94	4.75	4.66	4.59	4.51	4.44	4.35	----	4.20	4.10	
		tan δ	350	192	135	110	105	110	130	----	346	702	
Resinox 7934[e]	25	ϵ'/ϵ_0	4.62	4.46	4.34	4.28	4.20	4.13	4.10	4.05	4.04	4.04	
(preformed and preheated)		tan δ	220	210	178	130	99	88	85	84	84	83	

a. 62% mica (Bakelite). b. 59.5% mica, 8.5% misc. (Bakelite). c. 43% mica, 5% misc.; discontinued, substitute 12810 (Durez). d. 40% glass mat (Formica). e. 60% mica (Monsanto).

Dielectric Data   321

I. Solids, B. Organic (Cont.)
3. Plastics (cont.)

Values for tan δ are multiplied by 10^4; frequency given in c/s.

	T °C		1×10^2	1×10^3	1×10^4	1×10^5	1×10^6	1×10^7	1×10^8	3×10^8	3×10^9	1×10^{10}	2.5×10^{10}
c. Aniline-formaldehyde													
Cibanite E [a]	25	ϵ'/ϵ_0	3.58	3.57	3.56	3.48	3.41	3.41	3.41	----	3.40	3.40	
		tan δ	29	41	69	85	79	49	39	----	29	49	
Dilectene 100 [b]	22	ϵ'/ϵ_0	3.6	3.6	3.55	3.55	3.5	3.5	3.5	----	3.44	----	3.42
(1943 product)		tan δ	33	45	76	90	72	41	32	----	39	----	56
	84	ϵ'/ϵ_0	3.6	3.6	3.6	3.6	3.55	3.5	3.5	----	3.44	----	
		tan δ	36.5	34	37	50	90	83	61	----	47	----	
Dilectene 100 [b]	25	ϵ'/ϵ_0	3.68	3.68	3.66	3.60	3.58	3.50	3.50	----	3.44	3.44	
(1944 product)		tan δ	33	32	57	68	61	47	33	----	26	45	
d. Melamine-formaldehyde													
Formica Grade FF-41 [c]	26	ϵ'/ϵ_0	6.15	6.00	5.95	5.85	5.75	5.65	5.5	----			
(sheet stock)		tan δ	400	119	86	93	115	165	200	----			
Formica Grade FF-41 [c]	26	ϵ'/ϵ_0	----	----	----	3.7	3.7	3.65	3.6	----		3.3	
(rod stock)		tan δ	----	----	----	50	48	80	111	----		115	
Formica Grade FF-55 [d]	24	ϵ'/ϵ_0	6.55	6.35	6.20	6.05	6.00	5.90	5.80	5.63	----	----	4.73
		tan δ	400	142	85	85	108	155	220	250	----	----	440
Formica FF-55 [d]	24	ϵ'/ϵ_0	----	----	----	----	----	----	----	5.55*	5.31*	5.4**	4.8
(field ⊥ to laminate)		tan δ	----	----	----	----	----	----	----	275*	328*	399**	345
Formica FF-55 [d]													
(field ∥ to laminate)													
GMG Melamine [e]	25	ϵ'/ϵ_0	8.2	7.0	6.7	6.6	6.4	6.3	6.1	----			
(field ⊥ to laminate)		tan δ	1900	690	190	100	110	180	230	----			
GMG Melamine [e]	25	ϵ'/ϵ_0	----	----	----	----	----	----	----	5.4			
(field ∥ to laminate)		tan δ	----	----	----	----	----	----	----	230			
Melmac Resin 592 [f]	27	ϵ'/ϵ_0	6.70	6.25	5.85	5.50	5.20	4.90	4.70	----	4.67	4.59	
		tan δ	590	470	410	375	347	326	360	----	410	434	
	57	ϵ'/ϵ_0	8.15	6.95	6.35	5.85	5.40	5.10	4.90	----	4.75		
		tan δ	1250	750	490	400	350	320	350	----	480		
	88	ϵ'/ϵ_0	21.8	11.8	8.0	6.5	6.0	5.8	5.5	----	4.90		
		tan δ	7400	3400	1650	750	520	470	380	----	520		
Melmac Type 1077 [g]	28	ϵ'/ϵ_0	7.00	6.90	6.75	6.50	6.20	5.90	----	----	4.8	4.7	
(Ivory WB 48)		tan δ	240	130	140	190	280	440	----	----	900	1000	

a. 100% (Ciba). b. 100% (Continental-Diamond). c. 55% filler (Formica). d. Glass fiber mat (Formica). e. Glass Lamicoid #6038 (Mica Insulator). f. Mineral filler (Am. Cyanamid). g. 25% alpha pulp, Zn stearate (Am. Cyanamid).
*Stacked sheets in coaxial line. ** Stacked sheets in H_{11} (TE_{11}) mode of circular guide.

I. Solids. B. Organic 3. Plastics (Cont.)

Values for tan δ are multiplied by 10^4; frequency given in c/s.

d. Melamine-formaldehyde (cont.)	T°C		1×10^2	1×10^3	1×10^4	1×10^5	1×10^6	1×10^7	1×10^8	3×10^8	3×10^9	1×10^{10}	2.5×10^{10}
Melmac Molding Compound 1500[a]	25	ϵ'/ϵ_0	6.50	6.31	6.18	6.01	5.85	5.53	5.10	4.37*	4.20	4.10	
		tan δ	308	173	162	221	320	415	500	523*	520	505	
Melmac Molding Compound 1502[b]	25	ϵ'/ϵ_0	8.43	6.95	6.37	6.04	5.77	5.36	4.90	4.40*	4.20	4.00	
		tan δ	2020	960	460	350	410	510	560	620*	620	520	
Polyglas M[c]	24	ϵ'/ϵ_0	5.58	5.53	5.49	5.40	5.32	5.22	----	----	4.86**	5.22**	
		tan δ	140	71	64	69	93	160	----	----	339**	660**	
	80	ϵ'/ϵ_0	8.4	6.5	5.9	5.8	5.8	5.8	----	----	5.68**	5.32**	
		tan δ	4000	1400	280	95	65	110	----	----	498**	721**	
Micarta #259[d] (field ⊥ to laminate)	24	ϵ'/ϵ_0	6.18	6.07	5.97	5.90	5.81	5.72	5.55	----	----	4.80++	
		tan δ	190	125	105	85	108	145	200	----	----	322++	
Micarta #259[d] (field ∥ to laminate)	24	ϵ'/ϵ_0	----	----	----	----	----	----	----	4.83***	4.75***	5.12+	4.70++
		tan δ	----	----	----	----	----	----	----	150***	260***	380+	400++
Panelyte 140[e] (field ⊥ to laminate)	24	ϵ'/ϵ_0	6.15	6.05	6.00	5.93	5.82	5.70	5.53	5.5	4.70	4.70	4.62
		tan δ	135	93	120	155	155	170	215	260	360	360	260
Panelyte 140[e] (field ∥ to laminate)	25	ϵ'/ϵ_0	----	----	----	----	----	----	----	4.95	4.73	----	4.60
		tan δ	----	----	----	----	----	----	----	210	250	----	260
Plaskon Melamine[f]	24	ϵ'/ϵ_0	7.73	7.57	7.40	7.26	7.00	6.45	6.0	5.73	4.93	4.60	
		tan δ	190	122	150	240	410	640	850	985	1028	1100	
	80	ϵ'/ϵ_0	9.86	8.7	8.4	8.4	7.9	7.5	7.0	----	5.5	5.04	
		tan δ	1800	560	210	140	220	410	700	----	1250	1490	
Resimene 803-A[g] (not preformed or preheated)	24	ϵ'/ϵ_0	7.05	6.90	6.75	6.50	6.20	5.70	5.20	5.04	4.53	4.23	
		tan δ	410	210	190	270	440	660	820	859	820	760	
Resimene 803-A[g] (powder preheated)	25	ϵ'/ϵ_0	----	----	----	----	----	----	----	----	----	3.75	3.74
		tan δ	----	----	----	----	----	----	----	----	----	320	300

a. 40% wood flour, 18% plasticizer (Am. Cyanamid). b. 60% melamine, formaldehyde and aniline polymer with wood flour filler (Am. Cyanamid). c. 56.5% Am. Cyanamid's Melmac 7278, 43.5% Owens-Corning's Glass "E" (Hood Rubber). d. 65-70% Fiberglas (Westinghouse). e. Fiberglas (St. Regis). f. α-cellulose (Libbey-Owens-Ford). g. 40% α-cellulose (Monsanto).

* Freq. = 1 x 10^9. ** Sample nonhomogenous; stacked layers. *** In coaxial lines. +In $H_{11}(TE_{11})$ mode circular guide. ++In $H_{01}(TE_{01})$ mode rectangular guide.

Dielectric Data

I. Solids. B. Organic 3. Plastics (cont.) Values for tan δ are multiplied by 10^4; frequency given in c/s.

		T°C		1×10^2	1×10^3	1×10^4	1×10^5	1×10^6	1×10^7	1×10^8	3×10^8	3×10^9	1×10^{10}	2.5×10^{10}
e. Urea-formaldehyde														
Beetle Resin[a]		27	ϵ'/ϵ_0	6.5	6.2	6.05	5.9	5.65	5.4	5.1	----	4.57	4.47	
			tan δ	300	240	200	195	270	420	500	----	555	570	
Plaskon Urea, natural[b]		24	ϵ'/ϵ_0	7.1	6.7	6.4	6.2	6.0	5.7	5.2	5.1	4.79	4.65	
			tan δ	380	280	220	220	310	410	500	530	694	782	
		80	ϵ'/ϵ_0	8.8	7.8	7.4	7.1	6.8	6.6	----	----	5.54	4.27	
			tan δ	940	600	420	320	300	350	----	----	819	1630	
Plaskon Urea, brown[b]		24	ϵ'/ϵ_0	6.65	6.35	6.15	6.00	5.75	5.42	5.0	4.82	4.60	4.55	
			tan δ	320	260	240	250	280	350	420	438	527	742	
f. Benzoguanamine-formaldehyde														
Benzoguanamine Resin[c]		25	ϵ'/ϵ_0	4.72	4.58	4.50	4.40	4.26	4.15	3.96	3.62*	3.52	3.38	
			tan δ	348	189	127	140	196	273	320	342*	336	318	
g. Polyamide Resins														
Nylon 66[d]		25	ϵ'/ϵ_0	3.88	3.75	3.60	3.45	3.33	3.24	3.16	----	3.03		
			tan δ	144	193	233	254	257	244	210	----	128		
Nylon 610[d]		25	ϵ'/ϵ_0	3.60	3.50	3.35	3.24	3.14	3.05	3.0	----	2.84	----	2.73
			tan δ	155	186	208	221	218	205	200	----	117	----	105
		84	ϵ'/ϵ_0	13.5	11.2	9.0	6.3	4.4	3.7	3.4	----	2.94		
			tan δ	2350	1400	1580	2030	1720	1150	670	----	356		
Nylon 610[d]-90% humidity		25	ϵ'/ϵ_0	4.5	4.2	4.0	3.7	3.2	3.05	3.0	----	2.85		
			tan δ	650	640	600	500	380	280	220	----	125		
Nylon FM 10001[d]		25	ϵ'/ϵ_0	3.84	3.76	3.64	3.49	3.36	3.24	3.17	3.06*	3.02	3.02	
			tan δ	115	180	262	275	232	214	175	140*	120	107	
Resin #90S[e]		25	ϵ'/ϵ_0	3.25	2.94	2.80	2.72	2.64	2.61	2.58		2.54	2.53	
			tan δ	1070	500	274	165	114	92	83		62	57	
h. Cellulose Derivatives														
1) Acetates														
LL-1[f]		25	ϵ'/ϵ_0	3.82	3.77	3.67	3.53	3.42	3.30	3.29	3.28	3.24	3.24	3.1
			tan δ	95	150	200	230	230	210	210	220	290	400	310
		85	ϵ'/ϵ_0	3.98	3.96	3.90	3.77	3.58	3.44	----	----	3.30		
			tan δ	34	75	135	210	270	280	----	----	290		

a. Cellulose (Am. Cyanamid). b. α-cellulose (Libbey-Owens-Ford). c. 28% α paper (Am. Cyanamid). d. Hexamethylene-adipamide (Dupont).
e. Code RC-2072 (General Mills). f. 55.4% acetyl (Hercules).
* Freq. = 1×10^9.

I. Solids, B. Organic, 3. Plastics (cont.)
h. Cellulose Derivatives (cont.)
1) Acetates (cont.)

Values for tan δ are multiplied by 10^4; frequency given in c/s.

	T°C		1×10^2	1×10^3	1×10^4	1×10^5	1×10^6	1×10^7	1×10^8	3×10^8	3×10^9	1×10^{10}	2.5×10^{10}
Lumarith XFA-H4[a]	25	ϵ'/ϵ_0	4.61	4.51	4.36	4.13	3.79	3.57	3.37	3.20*	3.13	3.08	
		tan δ	152	240	340	428	480	423	380	365*	340	325	
Fibestos 2050TYA C-1686[b]	26	ϵ'/ϵ_0	4.75	4.53	4.30	3.82	3.60	3.52	3.35				
		tan δ	180	305	380	425	430	400	350				
Tenite I 008A H$_2$[c]	26	ϵ'/ϵ_0	4.65	4.55	4.40	4.20	3.96	3.72	3.47		3.16		3.08
		tan δ	102	173	286	370	430	465	410		310		350
Tenite I 008A H$_4$[c]	26	ϵ'/ϵ_0	4.55	4.48	4.33	4.14	3.90	3.63	3.40		3.25	3.16	3.11
		tan δ	80	175	270	345	393	405	380		310	300	300
Tenite I 008A M[c]	26	ϵ'/ϵ_0	4.97	4.86	4.70	4.52	4.25	3.88	3.57		3.24		
		tan δ	123	175	285	400	495	560	550		380		
Tenite I 008A MH[c]	26	ϵ'/ϵ_0	4.87	4.80	4.60	4.40	4.18	3.80	3.5		3.26	3.09	
		tan δ	117	170	290	380	465	525	550		370	330	
Tenite I 008A S[c]	26	ϵ'/ϵ_0	5.14	5.06	4.90	4.64	4.30	3.96	3.65		3.23		
		tan δ	100	160	280	405	530	620	580		420		
Tenite I 008A S$_4$[c]	26	ϵ'/ϵ_0	5.34	5.28	5.15	4.90	4.57	4.20	3.75		3.20		3.1
		tan δ	85	135	230	370	540	690	700		480		380
Tenite II 205A H$_2$[d]	26	ϵ'/ϵ_0	3.54	3.50	3.44	3.38	3.28	3.18	3.05		2.80		
		tan δ	78	107	158	174	178	180	190		267		
Tenite II 205A H$_4$[d]	26	ϵ'/ϵ_0	3.54	3.48	3.42	3.37	3.30	3.18	3.08	3.00	2.91	2.82	
		tan δ	50	97	156	175	178	175	170	179	280	298	
Tenite II 205A MH[d]	27	ϵ'/ϵ_0	3.60	3.56	3.50	3.40	3.30	3.22	3.10		2.87		
		tan δ	80	117	170	192	200	200	200		310		
Tenite II 205A MS[d]	27	ϵ'/ϵ_0	3.67	3.64	3.55	3.45	3.40	3.28	3.15		2.92	2.83	
		tan δ	23	115	178	213	220	227	250		370	360	
Tenite II 205A S$_2$[d]	27	ϵ'/ϵ_0	3.80	3.75	3.66	3.58	3.48	3.38	3.20		2.98		
		tan δ	82	95	155	204	225	247	300		460		
Tenite II 205A S$_4$[d]	27	ϵ'/ϵ_0	3.83	3.80	3.74	3.66	3.58	3.44	3.30		3.08		2.8
		tan δ	80	90	150	200	235	260	320		520		360

a. 28% plast. (Celanese). b. 26% plasticizer (Monsanto). c. 23-31% plasticizer, pigments, dyes (Tenn. Eastman). d. 5-15% plasticizer, pigments, dyes (Tenn. Eastman).

* Freq. = 1×10^9.

Dielectric Data

I. Solids, B. Organic 3. Plastics (cont.)

Values for tan δ are multiplied by 10^4; frequency given in c/s.

h. Cellulose Derivatives (cont.)	T°C		1×10^2	1×10^3	1×10^4	1×10^5	1×10^6	1×10^7	1×10^8	1×10^9	3×10^9	1×10^{10}	2.5×10^{10}
2) Propionate Forticel[a]	25	ϵ'/ϵ_0	3.52	3.48	3.41	3.33	3.23	3.18	3.08	2.95	2.88	2.87	
		tan δ	60	101	157	188	175	164	192	247	289	308	
3) Nitrate Pyralin[b]	27	ϵ'/ϵ_0	10.8	8.4	7.5	7.0	6.6	6.1	5.2		3.74	3.32	
		tan δ	6400	1000	450	400	640	930	1030		1650	1310	
	78	ϵ'/ϵ_0	----	7.5	6.7	6.3	6.2	6.1	5.2		4.0		
		tan δ	----	7000	1500	600	640	930	1030		1620		
4) Methyl Cellulose Methocel[c]	22	ϵ'/ϵ_0	7.6	6.8	6.4	6.1	5.7	4.9	4.3		3.35		
		tan δ	1280	570	330	400	650	1020	1000		550		
5) Ethyl Cellulose Ethocel LT5[c]	25	ϵ'/ϵ_0	3.11	3.09	3.05	3.02	3.01	2.96	2.90	2.77	2.74	2.70	
		tan δ	75	65	63	76	113	150	160	170	210	250	
Lumarith #2236l[d]	24	ϵ'/ϵ_0	3.10	3.06	3.02	2.99	2.92	2.87	2.80		2.74	2.67	2.65
		tan δ	48	48	52	67	115	158	160		196	256	300
	84	ϵ'/ϵ_0	3.00	3.00	2.90	2.85	2.80	2.80	2.80		2.79	2.66	
		tan δ	78	70	66	66	90	140	200		214	402	
i. Silicone Resins Formica G7[e] (field ⊥ laminate)	25	ϵ'/ϵ_0	3.73	3.73	3.73	3.73	3.73	3.73	3.73				
		tan δ	17	16	15	14	12	15	19				
(field ∥ laminate)	25	ϵ'/ϵ_0	5.18	4.72	4.37	4.09	3.99	3.95	3.94	3.92	3.90	3.90	
		tan δ	790	575	570	530	97	25	22	40	55	45	
Formica G6[e] (field ∥ laminate)	25	ϵ'/ϵ_0	3.99	3.91	3.87	3.85	3.82	3.82	3.82	3.81	3.79	3.74	
		tan δ	210	110	96	130	46	22	27	37	51	61	
(field ⊥ laminate)	25	ϵ'/ϵ_0	3.79	3.79	3.79	3.79	3.79	3.79					
		tan δ	13	11.5	10.5	10	10.5	13.3					

a. 8% plast. (Celanese). b. 25% camphor (DuPont). c. Dow. d. 13% plast. (Celanese). e. Fiberglas laminate (Formica).

I. Solids, B. Organic, 3. Plastics (cont.)

1. Silicone Resins (cont.)

Values for tan δ are multiplied by 10^4; frequency given in c/s.

	T°C		1×10^2	1×10^3	1×10^4	1×10^5	1×10^6	1×10^7	1×10^8	1×10^9	3×10^9	1×10^{10}
DC 996 cured 16 hrs. at 150°C[a]	25	ϵ'/ϵ_0	3.08	3.04	3.02	3.01	2.99	2.98	2.96			
		tan δ	83	74	65	52	42	38	31			
DC 996 cured 16 hrs. at 250°C[a]	25	ϵ'/ϵ_0	2.90	2.90	2.90	2.90	2.90	2.90	2.90			
		tan δ	14	15	16.5	17.5	18	17	16.5			
DC 2101[b] (discontinued)	24	ϵ'/ϵ_0	2.90	2.90	2.90	2.90	2.90	2.90	2.90			
		tan δ	70	56	49	46	45	45	45			
Polyglas S[c]	24	ϵ'/ϵ_0	3.60	3.59	3.58	3.57	3.57	3.56	----	----	3.55	3.53
		tan δ	11	11	12	13	15	19	31	----	40	46
	80	ϵ'/ϵ_0	3.60	3.59	3.58	3.57	3.57	3.56	----	----	3.55	3.51
		tan δ	27	17	11	10	10	14	16.5	----	46	49
Molding Compound XM-3[d]	25	ϵ'/ϵ_0	4.03	4.00	3.99	3.98	3.95	3.93	3.92	----	3.85	3.85
		tan δ	51	52	50	47	38	42	50	----	67	94
Dilecto (Silicone glass laminate) GB-261S (field ⊥ sheet)[e]	25	ϵ'/ϵ_0	3.83	3.82	3.81	3.80	3.80	3.79	3.79			
		tan δ	12.8	13	13	12.3	12.9	14.8	23.5			
	90	ϵ'/ϵ_0	3.74	3.73	3.71	3.69	3.63	3.56				
		tan δ	21	20	8.5	21	39	26				
	200	ϵ'/ϵ_0	3.50	3.62	3.44	3.42	3.40	3.35				
		tan δ	339	74.7	32.6	22.7	31.4	30				
(field ∥ sheet)	25	ϵ'/ϵ_0	4.97	4.48	4.43	3.95	3.92	3.87	3.87	3.82	3.79	3.76
		tan δ	580	560	440	185	55	29	23	42	57	78
	220	ϵ'/ϵ_0	4.68	4.26	4.18	4.18	----	----	----	----	----	3.87**
		tan δ	1620	526	300	195	----	----	----	----	----	75**
	25*	ϵ'/ϵ_0	4.05	3.93	3.90	3.85						
		tan δ	104	60	42	31						

a. Methyl, phenyl, and methyl phenyl polysiloxane resin (Dow Corning). b. Cross-linked organo-siloxane polymer (Dow Corning). c. 16% DC 2101, 84% Corning's 790 glass powder (Lab. Ins. Res.). d. 35% methyl and phenyl polysilicone resin, 45% glass fibers, 19% silica filler (Dow Corning). e. 50% DC-2103, 50% staple fibre glass base (Cont. Diamond).
*After temperature run. ** 200°C.

Dielectric Data

I. Solids B. Organic 3. Plastics (cont.) Values for tan δ are multiplied by 10^4; frequency given in c/s.

	T°C		1×10^2	1×10^3	1×10^4	1×10^5	1×10^6	1×10^7	1×10^8	1×10^9	3×10^9	1×10^{10}	2.5×10^{10}
1. Silicone Resins (cont.)													
Dilecto (silicone glass	25	ϵ'/ϵ_0	3.56	3.56	3.56	3.55	3.54	3.54	3.54				
laminate) GB 112S[a]		tan δ	13.5	12.9	12.9	11.1	8.5	11.4	18.6				
(field ⊥ sheet)	90	ϵ'/ϵ_0	3.41	3.39	3.38	3.38	3.38	3.38					
		tan δ	26.4	16.2	14.9	16.1	11	9.8					
	200	ϵ'/ϵ_0	3.18	3.18	3.18	3.17	3.15	3.10					
		tan δ	236	43	23	195	21	15.4					
(field ∥ sheet)	25	ϵ'/ϵ_0	6.55	5.72	4.94	3.96	3.84	3.82	3.80	3.78	3.76	3.70	
		tan δ	1250	910	1080	516	128	37	25	39	52	87	
	215	ϵ'/ϵ_0	10.93	8.60	6.65	4.92	----	----	----	----	----	3.78	
		tan δ	2000	2080	1800	1410	----	----	----	----	----	80	
	25*	ϵ'/ϵ_0	5.71	5.16	4.54	3.98							
		tan δ	624	880	940	640							
DC 2103 Laminate (XL-48)[b]	25	ϵ'/ϵ_0	3.94	3.92	3.90	3.90	3.90	3.90	3.88	----	3.88**	3.89**	
		tan δ	18	16	15	14	16	26	37	----	67**	99**	
Taylor Grade GSC[c]	25	ϵ'/ϵ_0	5.07	4.14	4.11	4.04	4.03	4.01	----	3.91	3.90	3.85	
(field ∥ laminate)		tan δ	99	72	83	44	20	15	----	30	39	40	
(field ⊥ laminate)	25	ϵ'/ϵ_0	3.78	3.77	3.77	3.77	3.77	3.74	3.74				
		tan δ	13.4	12.5	11	9.5	9.3	11.9	18				
Taylor Grade GSS[d]	25	ϵ'/ϵ_0	4.10	4.00	3.94	3.94	3.94	3.94	3.94	3.92	3.92	3.75	
(field ∥ laminate)		tan δ	160	160	100	34	21	20	30	44	55	63	
(field ⊥ laminate)	25	ϵ'/ϵ_0	3.74	3.74	3.74	3.74	3.74	3.74	3.74				
		tan δ	15	14	13	12	13	15	22				
DC 2104 Laminate (XL-269)[e]	25	ϵ'/ϵ_0	4.14	4.14	4.13	4.13	4.13	4.11	4.10	----	4.07**	4.06**	
		tan δ	32	29	26	22	22	29	34	----	71**	83**	
J. Polyvinyl Resins													
1) Polyethylene													
Polyethylene DE-3401[f]	25	ϵ'/ϵ_0	2.26	2.26	2.26	2.26	2.26	2.26			2.26	2.26	
		tan δ	<2	<2	<2	<2	<2	<2			3.1	3.6	
Polythene A-3305[g]	24	ϵ'/ϵ_0	2.25	2.25	2.25	2.25	2.25	2.25			2.25	2.25	2.24
(Now replaced by		tan δ	5	<3	<3	<5	<4	<3			3	4	6.7
Alathon)	80	ϵ'/ϵ_0	2.25	2.25	2.25	2.25	2.25	2.25			2.25	2.20	
		tan δ	<5	<3	<3	<5	<4	<3			7.2	8.5	

a. 50% DC-2103, 50% continuous-filament glass base (Cont. Diamond). b. 45% methyl and phenyl polysiloxane resin, 55% ECC-261 Fiberglas (Dow-Corning). c. 45-50% DC-2103, 50-55% cont.-filament glass cloth (Taylor). d. 45-50% DC-2103, 50-55% staple fibre glass cloth (Taylor). e. 35% methyl and phenyl polysiloxane resin, 65% ECC-181 fiberglas (Dow-Corning). f. 0.1% antioxidant (Bakelite). g. 100% polyethylene (DuPont).

*Measured after temperature run. **field ∥ laminate, at other frequencies field ⊥ sheet.

I. Solids, B. Organic 3. Plastics J. Polyvinyl Resins (cont.)

Values for tan δ are multiplied by 10^4; frequency given in c/s.

1) Polyethylene (cont.)

Polyethylene[a]	T°C		1×10^2	1×10^3	1×10^4	1×10^5	1×10^6	1×10^7	1×10^8	3×10^8	3×10^9	1×10^{10}	2.5×10^{10}
milled 3', 125°C	23	ϵ'/ϵ_0	2.25	2.25	2.25	2.25	2.25	2.25	2.25	----	2.25	2.24	
		tan δ	<2	<2	<2	<2	<2	<2	<2	----	5.8	6.6	
milled 30', 125°C	23	ϵ'/ϵ_0	2.26	2.26	2.26	2.26	2.26	2.26	2.26	2.26	2.25	2.25	
		tan δ	<5	5	6	7	8	9	10	10.6	11.7	11.9	
milled 30', 190°C	-12	ϵ'/ϵ_0	2.38	2.37	2.36	2.35	2.35	2.34	2.33	----	2.32	2.30	
		tan δ	24	21	19	18	21	28	39	----	36	22	
	23	ϵ'/ϵ_0	2.38	2.37	2.36	2.36	2.35	2.34	2.33	2.33	2.32	2.31	
		tan δ	28	28	27	27	28	30	42	51	50	44	

2) Polyisobutylene

Polyisobutylene[b]	25	ϵ'/ϵ_0	2.23	2.23	2.23	2.23	2.23	2.23	2.23	----	2.23		
		tan δ	4	1	1	<2	1	1	3	----	4.7		
Run 5047-2 Copolene B[c]	25	ϵ'/ϵ_0	2.3	2.3	2.3	2.3	2.3	2.3	2.3	----	2.3		
		tan δ	16	8	4	1	1	7	14	----	11		

3) Polyvinyl chloride-acetate

Vinylite QYNA[d]	20	ϵ'/ϵ_0	3.18	3.10	3.02	2.96	2.88	2.87	2.85	----	2.84		
		tan δ	130	185	225	210	160	115	81	----	55		
	47	ϵ'/ϵ_0	3.60	3.52	3.41	3.28	3.14	3.02	2.92	----	2.81		
		tan δ	100	166	240	261	228	162	110	----	77		
	76	ϵ'/ϵ_0	3.92	3.83	3.68	3.3	3.0	2.87	2.8	2.8			
		tan δ	180	220	320	400	350	270	190	175			
	96	ϵ'/ϵ_0	6.60	5.30	4.40	3.7	3.3	2.8	2.7	2.7	----	2.6	
		tan δ	1500	1400	1200	980	740	500	320	280	----	180	
	110	ϵ'/ϵ_0	9.9	8.6	6.8	5.6							
		tan δ	1030	1330	1780	1900							
Vinylite VG-5544[e]	25	ϵ'/ϵ_0	7.72	7.20	6.40	5.25	4.13	3.45	3.05	2.99	2.94	2.82	2.80
		tan δ	570	640	1060	1500	1550	1200	650	460	185	159	150
Vinylite VG-5901, black[f]	25	ϵ'/ϵ_0	6.5	5.5	4.6	3.9	3.4	3.1	3.0	2.94	2.88	2.83	
		tan δ	1020	1180	1190	1000	740	500	280	200	106	105	

a. Bakelite. b. 100% Esso Lab.). c. 61% polyisobutylene B-100, 39% Marbon B (Am. Phenolic). d. 100% polyvinyl chloride (Bakelite).
e. 40% polyvinyl chloride-acetate, 40% plast., 14% misc. (Bakelite). f. 62.5% polyvinyl chloride-acetate, 29% plast., 8.5% misc. (Bakelite).

I. Solids, B. Organic 3. Plastics J. Polyvinyl Resins (cont.) Values for tan δ are multiplied by 10^4; frequency given in c/s.

3) Polyvinyl chloride (cont.)

	T°C		1×10^2	1×10^3	1×10^4	1×10^5	1×10^6	1×10^7	1×10^8	3×10^8	3×10^9	1×10^{10}	2.5×10^{10}
Vinylite VG-5904, black[a]	25	ϵ'/ϵ_0	8.1	7.5	6.6	5.4	4.3	3.7	3.3	3.1	2.94	2.83	
		tan δ	550	710	1060	1390	1400	1100	670	500	340	320	
Vinylite VU-1900, clear[b]	24	ϵ'/ϵ_0	6.55	5.65	4.70	3.90	3.30	2.95	2.80	----	2.65	2.62	2.62
		tan δ	1000	1150	1300	1180	880	560	310	----	131	104	110
	79	ϵ'/ϵ_0	10.3	8.15	7.5	6.5	5.5	4.3	3.4	----	2.84	2.60	
		tan δ	7300	1250	720	800	1550	2700	1550	----	498	351	
Vinylite VYHH[c]	22	ϵ'/ϵ_0	3.20	3.12	3.06	3.00	2.91	2.88	2.83	----	2.79		
		tan δ	100	130	155	150	140	110	90	----	76		
	47	ϵ'/ϵ_0	3.56	3.48	3.38	3.27	3.16	3.02	2.9	----	2.79		
		tan δ	110	142	190	227	206	152	114	----	92		
Vinylite VYNS[d]	20	ϵ'/ϵ_0	3.10	3.08	3.02	2.95	2.90	2.85	2.8	----	2.74		
		tan δ	115	140	170	170	140	105	80	----	63		
Vinylite VYNW[e]	20	ϵ'/ϵ_0	3.20	3.15	3.05	2.96	2.90	2.84	2.8	----	2.74		
		tan δ	135	165	197	190	150	110	80	----	59		
Geon 2046[f]	23	ϵ'/ϵ_0	6.95	6.10	5.05	4.13	3.55	3.15	3.00	2.97	2.89	2.83	
		tan δ	820	1100	1320	1200	890	570	300	211	116	116	
	80	ϵ'/ϵ_0	9.1	8.8	8.3	7.6	6.5	5.0	4.0	----	3.06	2.90	
		tan δ	250	300	410	680	1540	2800	1500	----	484	328	
Geon 80365[g]	25	ϵ'/ϵ_0	3.67	3.65	3.58	3.52	3.42	3.39	3.34				
		tan δ	68	95	133	145	130	115	120				
Geon 80384[h]	25	ϵ'/ϵ_0	3.43	3.34	3.23	3.14	3.05	3.96	3.95				
		tan δ	120	184	224	217	158	114	101				

a. 54% polyvinyl chloride-acetate, 41% plast., 5% misc. (Bakelite). b. 64.5% polymer of 95% vinyl chloride and 5% vinyl acetate, 32% Flexol D.O.P., 3.5% misc. (Bakelite). c. Polymer of 87% vinyl chloride and 13% vinyl acetate (Bakelite). d. Polymer of 91% vinyl chloride and 9% vinyl acetate (Bakelite). e. Polymer of 95% vinyl chloride and 5% vinyl acetate (Bakelite). f. 59% polyvinyl chloride, 30% dioctyl phosphate, 6% stabilizer, 5% filler (Goodrich). g. 71% polyvinyl chloride, 10.5% filler, 5% plasticizer, 8.5% stabilizer (Goodrich). h. 87.8% polyvinyl chloride, 10.5% stabilizer (Goodrich).

I. Solids B. Organic 3. Plastics j. Polyvinyl Resins (cont.) Values for tan δ are multiplied by 10^4; frequency given in c/s.

3) Polyvinyl Chloride (cont.)

	T °C		1×10^2	1×10^3	1×10^4	1×10^5	1×10^6	1×10^7	1×10^8	3×10^8	3×10^9	1×10^{10}
Koroseel 5CS-243[a]	27	ϵ'/ϵ_0	6.1	5.65	5.00	4.15	3.60	3.20	2.9	----	2.73	2.62
		tan δ	790	1000	1300	1250	930	560	300	----	112	120
Lucoflex[b]	25	ϵ'/ϵ_0	----	----	----	----	----	----	2.75			
		tan δ	----	----	----	----	----	----	170			
Polyvinyl chloride 1006[c]	25	ϵ'/ϵ_0	6.1	----	4.55	----	3.3					
		tan δ	760	----	1100	----	760					
Polyvinyl chloride 1018[d]	25	ϵ'/ϵ_0	6.2	----	4.95	----	3.15					
		tan δ	630	----	950	----	920					
Polyvinyl chloride 1216[e]	25	ϵ'/ϵ_0	6.1	----	4.4	----	3.2					
		tan δ	1170	----	1110	----	450					
Polyvinyl chloride 1406[f]	25	ϵ'/ϵ_0	6.05	----	4.5	----	3.6					
		tan δ	870	----	940	----	480					
Ultron Wire Compound UL300[g]	25	ϵ'/ϵ_0	3.31	3.22	3.10	2.98	2.85					
		tan δ	150	210	250	220	155					
Ultron Wire Compound UL1004[h]	25	ϵ'/ϵ_0	6.4	----	4.65	----	3.3					
		tan δ	810	----	1210	----	740					
Ultron Wire Compound UL2 4001[k]	25	ϵ'/ϵ_0	6.7	----	4.7	----	3.5					
		tan δ	1100	----	1210	----	700					
Polyvinyl chloride W-174[m]	25	ϵ'/ϵ_0	5.45	4.77	4.17	3.75	3.52	3.25	3.00			
		tan δ	815	930	880	740	550	425	415			
Polyvinyl chloride W-175[n]	25	ϵ'/ϵ_0	5.94	5.20	4.51	3.90	3.44	3.37	3.04			
		tan δ	840	960	960	865	580	430	300			
Polyvinyl chloride W-176[p]	25	ϵ'/ϵ_0	6.21	5.52	4.70	3.96	3.53	3.28	3.00			
		tan δ	730	940	1070	960	720	520	500			

a. 63.7% polyvinyl chloride, 33.1% di-2-ethylhexylphthalate, lead silicate (Goodrich). b. Unplasticized polyvinyl chloride (Lucoflex Plastic). c. 57.5% polymer, 12.6% fillers, 28.7% plasticizers (Monsanto). d. 52.4% polymer, 15.1% fillers, 31.4% plasticizers (Monsanto). e. 57.5% polymer, 10.4% fillers, 31.6% plasticizers (Monsanto). f. 59.4% polymer, 10.7% fillers, 29.7% plasticizers (Monsanto). g. 100% polymer (Monsanto). h. 64.7% polymer, 2% filler, 32.5% plasticizers (Monsanto). k. 60.1% polymer, 7.8% fillers, 31.2% plasticizers (Monsanto). m. 65% Geon 101, 35% Paraplex G-25 (Rohm and Haas). n. 65% Geon 101, 35% Paraplex G-50 (Rohm and Haas). p. 65% Geon, 35% Paraplex G-60 (Rohm and Haas).

Dielectric Data

I. Solids B. Organic 3. Plastics j. Polyvinyl Resins (cont.) Values for tan δ are multiplied by 10^4; frequency given in c/s.

3) Polyvinyl Chloride (cont.)

	T°C		1×10^2	1×10^3	1×10^4	1×10^5	1×10^6	1×10^7	1×10^8	3×10^8	1×10^9	3×10^9	1×10^{10}	2.5×10^{10}
Plasticell	25	ϵ'/ϵ_0	1.04	1.04	1.04	1.04	1.04	---	1.04	1.04*		1.04	1.04	
		tan δ	21	11	<15	<15	10	---	10	18*		55	50	
Ensolite M2240[b]	25	ϵ'/ϵ_0	1.51	1.36	1.29	1.24	1.16							
		tan δ	3200	610	330	340	420							
Ensolite M2239[b]	25	ϵ'/ϵ_0	1.48	1.40	1.31	1.25	1.20							
		tan δ	2850	770	280	250	340							
Ensolite 3036[b]	25	ϵ'/ϵ_0	1.30	1.24	1.19	1.17	1.16							
		tan δ	368	263	187	137	111							
(field ⊥ plane of sample)	25	ϵ'/ϵ_0								1.14*		1.12	1.10	
		tan δ								87*		86	30	
(field ‖ plane of sample)	25	ϵ'/ϵ_0												
		tan δ												

4) Polyvinylidene and Vinyl chloride

	T°C		1×10^2	1×10^3	1×10^4	1×10^5	1×10^6	1×10^7	1×10^8	3×10^8	1×10^9	3×10^9	1×10^{10}	2.5×10^{10}
Saran B-115[c]	23	ϵ'/ϵ_0	4.88	4.65	4.17	3.60	3.18	2.97	2.82	---	2.71	2.70		
		tan δ	450	630	885	845	570	310	180	---	72	51		
	84	ϵ'/ϵ_0	5.13	4.94	4.85	4.71	4.40	3.75	3.2	---	2.76			
		tan δ	800	210	130	320	780	1300	900	---	242			

5) Polychlorotrifluoroethylene

	T°C		1×10^2	1×10^3	1×10^4	1×10^5	1×10^6	1×10^7	1×10^8	3×10^8	1×10^9	3×10^9	1×10^{10}	2.5×10^{10}
Kel-F[d]	26	ϵ'/ϵ_0	2.72	2.63	2.53	2.46	2.43	2.35	---	2.30	2.30	2.29		
		tan δ	210	270	230	135	82	60	---	30	28	39		
Kel-F Grade 300[d]	25	ϵ'/ϵ_0	2.82	2.76	2.65	2.50	2.46	2.42	2.36	2.35	2.34	2.33		
		tan δ	148	225	212	140	96	75	54	51	66	59		
Kel-F Grade 300-P25[e]	25	ϵ'/ϵ_0	2.84	2.75	2.68	2.58	2.51	2.45	2.37	2.35	2.31	2.26		
		tan δ	126	207	234	204	175	214	186	150	93	93		

6) Polytetrafluoroethylene

	T°C		1×10^2	1×10^3	1×10^4	1×10^5	1×10^6	1×10^7	1×10^8	3×10^8	1×10^9	3×10^9	1×10^{10}	2.5×10^{10}
Dilecto (Teflon Laminate GB-112T)[f]	25	ϵ'/ϵ_0	2.76	2.74	2.74	2.74	2.73	2.73	2.73					
		tan δ	8.9	6.1	6	6	5.8	6.2	11.8					
(field ⊥ laminate)	25	ϵ'/ϵ_0	2.48	2.46	2.46	2.46	---	2.44						
		tan δ	80	36	18	14	---	2.5						
	250	ϵ'/ϵ_0	2.70	2.68	2.69	2.69								
	25**	tan δ	4.7	3.9	4.7	4.8								

a. Expanded polyvinyl chloride (Sponge Rubber). b. Modified polyvinyl chloride (U. S. Rubber). c. Polyvinylidene and vinyl chlorides (Dow). d. Polychlorotrifluoroethylene (Kellogg). e. Plasticized polychlorotrifluoroethylene (Kellogg). f. 65-68% Teflon, 32-35% continuous-filament glass base (Cont. Diamond).

*Freq. = 1×10^9. **After heating.

Tables of Dielectric Materials

I. Solids, B. Organic 3. Plastics J. Polyvinyl Resins (cont.) Values for tan δ are multiplied by 10^4; frequency given in c/s.

6) Polytetrafluoroethylene (cont.)

Material	T°C		1×10^2	1×10^3	1×10^4	1×10^5	1×10^6	1×10^7	1×10^8	3×10^8	3×10^9	1×10^{10}	2.5×10^{10}
Dilecto (Teflon Laminate GB-112T)[a]	25	ϵ'/ϵ_o	3.35	3.24	3.18	3.17	3.16	3.15	3.15	3.15**	3.15	3.10	
		tan δ	361	245	169	108	42	23	24	31**	38	47	
(field ‖ laminate)	250	ϵ'/ϵ_o	----	----	----	----	----	----	----	----	----	3.10	
		tan δ	----	----	----	----	----	----	----	----	----	70	
	25*	ϵ'/ϵ_o	3.26	3.25	3.24	3.21							
		tan δ	66	14.3	17.3	27							
Teflon[b]	22	ϵ'/ϵ_o	2.1	2.1	2.1	2.1	2.1	2.1	2.1	2.1	2.1	2.08	2.08
		tan δ	<5	<3	<3	<3	<2	<2	<2	1.5	1.5	3.7	6
	100	ϵ'/ϵ_o	2.04	2.04	2.04	2.04	2.04	2.04	----	----	----	2.04	
		tan δ	10	4	2	<3	<2	<2	----	----	----	5.1	
Chemelec M1405[c]	25	ϵ'/ϵ_o	2.50	2.50	2.50	2.50	2.50	2.50	2.50	2.50	2.50		
		tan δ	18	5.1	2.7	3	5	7.2	9	8	6.8		
Chemelec M1406[d]	25	ϵ'/ϵ_o	----	----	----	----	72	----	----	----	12.7		
		tan δ***	----	----	>10000	170	62	----	----	----	.33		
		ρ	----	520	520	2400	410				145		
Chemelec M1407[e]	25	ϵ'/ϵ_o	3.42	3.02	2.85	2.74	2.71	2.65	2.63				
		tan δ	917	700	350	180	150	147	158				
Chemelec M1411[f] (field ⊥ sheet)	25	ϵ'/ϵ_o	2.14	2.14	2.14	2.14	2.14	2.14	2.14				
		tan δ	18.5	9.6	7	6.8	7	9.2	10				
(field ‖ sheet)	25	ϵ'/ϵ_o	----	----	----	----	----	----	----	----	2.52	2.50	
		tan δ	----	----	----	----	----	----	----	----	25	28	
Chemelec M1412[g]	25	ϵ'/ϵ_o	----	2.35	----	----	----	----	----	----	2.35		
		tan δ	----	12	----	----	----	150	43	----	15		
Chemelec M1414[h]	25	ϵ'/ϵ_o	----	----	----	----	>100	11.8	4.8	----	33		
		tan δ***	----	----	>10000	>1000					3.9		
		ρ	----	----	95	95	95	95	80		4.8		

a. 65-68% Teflon, 32-35% continuous-filament glass base (Cont. Diamond). b. Polytetrafluoroethylene (DuPont). c. 75% Teflon, 25% calcium fluoride (U.S.Gasket). d. 80% Teflon, 20% carbon (U.S. Gasket). e. 88% Teflon, 12% ceramic (U.S.Gasket). f. 75% Teflon, 25% Fiberglas (U.S.Gasket). g. 75% Teflon, 25% glass (U.S.Gasket). h. 75% Teflon, 25% graphite (U.S.Gasket).
*After temperature run. **Freq. = 1 x10^9. ***tan δ not multiplied by 10^4.

Dielectric Data

I. Solids, B. Organic 3. Plastics j. Polyvinyl resins (cont.) Values for tan δ are multiplied by 10^4; frequency given in c/s.

	$T^\circ C$		1×10^2	1×10^3	1×10^4	1×10^5	1×10^6	1×10^7	1×10^8	3×10^8	3×10^9	1×10^{10}	2.5×10^{10}
6) Polytetrafluoro-ethylene (cont.)													
Chemelec M1418-2[a] (desiccated 48 hrs. P_2O_5)	25	ϵ'/ϵ_0	2.30	2.20	2.20	2.15	2.15	2.14	2.14				
		tan δ	600	200	45	4	2	2.5	5				
(dried in oven)	25	ϵ'/ϵ_0	2.15	2.15	2.15	2.15	2.15						
		tan δ	11	6	2.7	1.5	1.8						
Chemelec M1418-5[b]	25	ϵ'/ϵ_0	2.27	2.23	2.18	2.18	2.16	2.16	2.16				
		tan δ	450	118	32	8.8	5.7	5.9	7.1				
Chemelec M1422[c]	25	ϵ'/ϵ_0	2.72	2.72	2.72	2.72	2.72	2.72	2.72				
		tan δ	11.9	7.7	4.5	2.8	2.0	1.8	2.4				
Chemelec M1423[d]	25	ϵ'/ϵ_0		1.308								1.30	
		tan δ		1.81								62	
7) Polyvinyl alcohol-acetate													
Elvanol 51A-05[e]	25	ϵ'/ϵ_0	8.2	7.8	7.2	6.2	5.2	4.5			3.74	3.50	3.46
		tan δ	430	440	580	720	900	1000			550	502	620
	85	ϵ'/ϵ_0	400	100	33	16	10	7.3			4.67		
		tan δ	15000	13000	9000	3600	2300	2000			1770		
Elvanol 50A-42[f]	23	ϵ'/ϵ_0	14.0	10.4	8.0	6.6	5.7	5.0			3.75		
		tan δ	4050	1850	1150	1000	950	900			715		
	68	ϵ'/ϵ_0	85	26	14	9.5	7.5	5.8					
		tan δ	9600	8100	4000	1600	1400	1500					
	100	ϵ'/ϵ_0	3000	400	50	18	13	8.7			5.6	4.1	
		tan δ	15000	22000	20000	4800	2700	2600			2300	3000	
Elvanol 70A-05[f]	26	ϵ'/ϵ_0			58	27	14	7			4.12		
		tan δ			8200	5900	4100	2700			840		
Elvanol 72A-51[f]	26	ϵ'/ϵ_0			45.5	22.0	12.5	7.5			3.89	3.8	3.8
		tan δ			7600	5100	3500	2500			640	636	730
Elvacet 42A-900[g]	25	ϵ'/ϵ_0	3.09	3.07	3.05	3.02	2.98	2.94	2.90		2.88		
		tan δ	49	50	52	56	65	49	37		28		
	85	ϵ'/ϵ_0	7.3	7.15	5.9	3.75	3.25	2.95	2.90		2.87		
		tan δ	180	590	2400	1830	830	560	200		85		

a. 90% Teflon, 10% quartz (U.S.Gasket). b. 75% Teflon, 25% quartz (U.S.Gasket). c. 80% Teflon, 20% titanium dioxide (U.S.Gasket). d. Air-filled Teflon (U.S.Gasket). e. Polyvinyl alcohol-acetate, 11-14% acetoxyl (DuPont). f. Polyvinyl alcohol-acetate, 0-1.5% acetoxyl (DuPont). g. Polyvinyl acetate (DuPont).

I. Solids, B. Organic 3. Plastics j. Polyvinyl Resins (cont.) Values for tan δ are multiplied by 10^4; frequency given in c/s.

8) Polyvinyl acetals	T°C		1×10^2	1×10^3	1×10^4	1×10^5	1×10^6	1×10^7	1×10^8	3×10^8	3×10^9	1×10^{10}	2.5×10^{10}
Formvar, Type E[a]	26	ϵ'/ϵ_0	3.16	3.12	3.08	3.00	2.92	2.85	----	----	2.76	----	2.7
		tan δ	54	100	154	190	190	165	----	----	113	----	115
	88	ϵ'/ϵ_0	3.55	3.5	3.4	3.25	3.1	2.95	2.85	----	2.80	----	----
		tan δ	60	83	102	145	213	310	300	----	227	----	----
Alvar 11/90[b]	25	ϵ'/ϵ_0	3.17	3.14	3.09	3.03	2.96	2.90	2.82	2.80	2.73	2.70	2.65
		tan δ	65	70	100	140	180	160	125	119	136	175	175
	84	ϵ'/ϵ_0	3.84	3.54	3.40	3.25	3.05	2.90	----	----	2.74	----	----
		tan δ	575	390	270	195	180	225	----	----	196	----	----
Butvar, Low OH[c]	27	ϵ'/ϵ_0	2.69	2.67	2.65	2.63	2.61	2.59	----	----	2.51	2.48	2.46
		tan δ	38	40	56	88	124	138	----	----	111	107	99
Butvar 55/98[c]	27	ϵ'/ϵ_0	3.04	3.02	2.98	2.94	2.86	2.75	2.67	----	2.62	----	----
		tan δ	41	59	108	161	215	216	177	----	172	----	----
9) Polyacrylates													
Lucite HM-119[d] (now replaced by HM-140)	-12	ϵ'/ϵ_0	3.0	2.9	2.8	2.7	2.63	2.60	2.59	----	2.58	2.57	----
		tan δ	330	250	190	140	102	70	55	----	35.4	34	----
	23	ϵ'/ϵ_0	3.20	2.84	2.75	2.68	2.63	2.60	2.58	----	2.58	2.57	2.57
		tan δ	620	440	315	220	145	100	67	----	51.3	49	32
	81	ϵ'/ϵ_0	3.97	3.45	3.08	2.86	2.72	2.62	2.59	----	2.58	2.57	----
		tan δ	600	820	720	540	380	220	130	----	77	95	----
Lucite, sintered[e] (obsolete)	27	ϵ'/ϵ_0	2.29	2.12	2.01	2.00	2.00	2.00	2.00	----	1.82	----	----
		tan δ	408	290	195	116	66	49	40	----	52	----	----
Gafite cast polymer[f] (sheet sample)	25	ϵ'/ϵ_0	3.25	3.12	3.02	2.97	2.90	2.88	2.83	----			
		tan δ	310	240	183	155	115	77	64	----			
(rod sample)	25	ϵ'/ϵ_0	----	----	----	----	----	----	----	2.75		2.75	
		tan δ	----	----	----	----	----	120	----	99		83	
Plexiglas[g]	27	ϵ'/ϵ_0	3.40	3.12	2.95	2.84	2.76	2.71	----	2.66	2.60	2.59	
		tan δ	605	465	300	200	140	100	----	62	57	67	
	80	ϵ'/ϵ_0	4.30	3.80	3.34	3.00	2.80	2.70	----	----	2.56	----	
		tan δ	700	895	800	520	320	210	----	----	79	----	

a. Polyvinyl formal (Shawinigan). b. Polyvinyl acetal (Shawinigan). c. Polyvinyl butyral (Shawinigan). d. Polymethyl methacrylate (DuPont). e. DuPont. f. Methyl and alpha-chloroacrylate (Gen. Aniline). g. Polymethyl methacrylate (Rohm and Haas).

I. Solids, B. Organic 3. Plastics j. Polyvinyl Resins (cont.) Values for tan δ are multiplied by 10^4; frequency given in c/s.

9) Polyacrylates (cont.)	T°C		1×10^2	1×10^3	1×10^4	1×10^5	1×10^6	1×10^7	1×10^8	3×10^8	3×10^9	1×10^{10}	2.5×10^{10}
Polyethyl methacrylate[a]	22	ϵ'/ϵ_0	2.90	2.75	2.65	2.60	2.55	2.53	-----	-----	2.51	2.50	2.5
		tan δ	420	294	185	118	90	75	-----	-----	75	97	83
	80	ϵ'/ϵ_0	3.87	3.36	2.86	2.70	2.61	2.57	2.55	-----	2.49	2.48	
		tan δ	810	1060	960	710	400	260	140	-----	91	135	
Polybutyl methacrylate[a]	-12	ϵ'/ϵ_0	2.50	2.46	2.44	2.42	2.41	2.40	-----	-----	2.38		
		tan δ	190	120	68	42	33	25	-----	-----	48		
	24	ϵ'/ϵ_0	2.82	2.62	2.52	2.47	2.43	2.42	-----	-----	2.38	2.36	
		tan δ	605	360	200	125	80	55	-----	-----	44	46	
	81	ϵ'/ϵ_0	3.70	3.52	3.14	2.82	2.60	2.48	-----	-----	2.39		
		tan δ	150	650	1200	980	470	220	-----	-----	185		
Polyisobutyl methacrylate[a]	25	ϵ'/ϵ_0	2.70	2.68	2.63	2.55	2.45	2.45	2.42	2.40	2.39	2.38	2.37
		tan δ	111	70	50	37	35	46	52	47	31	39	52
	80	ϵ'/ϵ_0	2.9	2.7	2.5	2.5	2.44	2.42	2.42	2.40	2.39	2.38	
		tan δ	830	600	360	210	100	80	70	65	54	59	
Polycyclohexyl methacrylate[b]	25	ϵ'/ϵ_0	2.58	2.52	2.52	2.52	2.52	2.48	2.47	-----	2.46	-----	2.46
		tan δ	46	47.5	60	50	28	23	25	-----	34.9	-----	41
	84	ϵ'/ϵ_0	2.63	2.60	2.55	2.47	2.42	2.39	2.37	-----	2.37		
		tan δ	150	124	98	81	68	38	32	-----	49		
10) Polystyrene Polystyrene[c] (commercially molded) Sheet Stock	25	ϵ'/ϵ_0	2.56	2.56	2.56	2.56	2.56	2.56	2.55	2.55	2.55	2.54	2.54
		tan δ	<0.5	<0.5	<0.5	0.5	0.7	<2	<1	3.5	3.3	4.3	12
	80	ϵ'/ϵ_0	2.54	2.54	2.54	2.54	2.54	2.54	2.54	2.54	2.54	2.53	
		tan δ	9	2	<1	<2	<2	<2	<3	2.7	4.5	5.3	
Rod Stock													
Sample A	25	tan δ	-----	-----	-----	-----	-----	-----	-----	-----	2.5	3.6	4.8
Sample B	25	tan δ	-----	-----	-----	-----	-----	-----	-----	-----	12	------	22
Sample B	80	tan δ	-----	-----	-----	-----	-----	-----	-----	-----	16		

a. DuPont. b. Polaroid. c. For sheet stock, various samples used for different frequencies; for rod stock, ϵ'/ϵ_0 is the same as for sheet stock (Plax).

336 Tables of Dielectric Materials

I. Solids. B. Organic 3. Plastics J. Polyvinyl Resins (cont.) Values for tan δ are multiplied by 10^4; frequency given in c/s.

10) Polystyrene (cont.)	T°C		1×10^2	1×10^3	1×10^4	1×10^5	1×10^6	1×10^7	1×10^8	3×10^8	3×10^9	1×10^{10}	2.5×10^{10}
Styron C-176[a]	25	ϵ'/ϵ_0	2.55	2.55	2.55	2.55	2.55	2.55	2.55	2.55	2.54	2.54	2.54
		tan δ	<3	<2	<2	<2	<2	<2	<3	1.7	2.3	3.0	5.3
	80	ϵ'/ϵ_0	2.55								2.5	2.5	
		tan δ	12								5.0	6.5	
Styron C-176, 30% humidity[b]	25	ϵ'/ϵ_0									2.54		
		tan δ									3.2		
Styron C-176, 60% humidity[b]	25	ϵ'/ϵ_0									2.54		
		tan δ									4.9		
Styron C-176, 90% humidity[b]	25	ϵ'/ϵ_0									2.54	2.54	
		tan δ									7.7	7.3	
Styron C-176 + 0.5% paraffin, 90% humidity[b]	25	ϵ'/ϵ_0									2.54		
		tan δ									2.5		
Styron 411-A[c] (formerly Exp. Plastic Q-247)	22	ϵ'/ϵ_0	2.55	2.55	2.55	2.55	2.55	2.55	2.55	2.55	2.54	2.54	2.54
		tan δ	<3	<2	<2	<2	<2	<2	<3	2.1	3.1	3.7	5.3
	79	ϵ'/ϵ_0	2.55	2.55	2.55	2.55	2.55	2.55	2.55		2.54	2.54	
		tan δ	12	2.5	<2	<2	<2	2	3		5.2	6	
Styron 475[b]	25	ϵ'/ϵ_0	2.62	2.61	2.59	2.56	2.56	2.55	2.55	2.53*	2.53	2.53	
		tan δ	3.6	3.0	2.4	2.6	4.2	7.6	11	19*	36	17	
Styron 666[b]	25	ϵ'/ϵ_0	2.54	2.54	2.54	2.54	2.54	2.54	2.54	2.54**	2.53	2.52	
		tan δ	1.75	1.1	<1	<1	.7	1.2	2	2.7*	3.1	3.4	
Styron 671[b]	25	ϵ'/ϵ_0	2.55	2.55	2.55	2.55	2.54	2.54	2.54	2.54**	2.54	2.54	
		tan δ	.97	.64	<.5	<.5	.2	.38	.57	1.47*	1.76	2.1	

a. Polystyrene (Dow). Similar values given by Dow's C-244 and Q-247.1; Catalin's Loalin #1 Color 4000; Bakelite's IMS-4A-621, XRS-23-B10012-44q, XRS-23-B10012-52Q, XRS-23-B10016-D63, XRS-23-B10012-D107; Monsanto's Samples D-277, D-279 (extra purity), D-334.
b. Polystyrene (Dow). c. Polystyrene (Dow). Similar values given by Monsanto's Lustron Res. Sample D-276.
*Freq. = 1×10^9.

I. Solids, B. Organic 3. Plastics j. Polyvinyl Resins (cont.)   Values for tan δ are multiplied by 10^4; frequency given in c/s.

10) Polystyrene (cont.)

	T°C		1×10^2	1×10^3	1×10^4	1×10^5	1×10^6	1×10^7	1×10^8	3×10^8	3×10^9	1×10^{10}	2.5×10^{10}
Fibers Q-107[a]	26	ϵ'/ϵ_0	2.14	2.14	2.14	2.14	2.14	2.14	2.14	----	2.11		
		tan δ	7.6	6.3	4.6	3	3	3	4	----	6.3		
Foam Q-103, 90% humidity[a]	27	ϵ'/ϵ_0	----	----	----	----	----	----	----	----	1.03		
		tan δ	----	----	----	----	----	----	----	----	<0.3		
Styrofoam 103.7[b]	25	ϵ'/ϵ_0	1.03	1.03	1.03	1.03	1.03	1.03	----	----	1.03	1.03	
		tan δ	<2	<1	<1	<1	<2	<2	----	----	1	1.5	
Polystyrene cast in vacuo[c]	25	ϵ'/ϵ_0	2.55	2.55	2.55	2.55	2.55	2.55	----	----	2.55		
		tan δ	1	1	0.5	<2	<2	<3	----	----	2.4		
Polystyrene cast in air[c]	25	ϵ'/ϵ_0	----	2.50	2.50	2.50	2.50	2.50	----	----	2.48		
		tan δ	----	9	5.5	<5	2	8	----	----	12.8		
Experimental Plastic Q764.6[a]	25	ϵ'/ϵ_0	2.63	2.58	2.58	2.58	2.58	2.58	2.58	2.55*	2.51	2.51	
		tan δ	2.5	1.3	1.8	2.5	2.3	1.4	2.8	7.3*	8.2	6.5	
Experimental Plastic Q767.2[a]	25	ϵ'/ϵ_0	2.98	2.95	2.92	2.90	2.86	2.83	2.80	2.77*	2.75	2.75	
		tan δ	62	73	80	74	65	55	43	36*	38	43	
Styramic #18[d]	22	ϵ'/ϵ_0	2.68	2.66	2.65	2.65	2.65	2.65	2.65	2.65	2.65	2.63	2.63
		tan δ	29	18.4	5	<2	<2	<2	1.6	1.9	2.2	2.3	4
	78	ϵ'/ϵ_0	2.7	2.7	2.67	2.65	2.65	2.65	2.65	----	2.63		
		tan δ	91	67	25	6.5	4	4	3	----	2.0		

11) Misc. Polystyrenes

	T°C		1×10^2	1×10^3	1×10^4	1×10^5	1×10^6	1×10^7	1×10^8	3×10^8	3×10^9	1×10^{10}	2.5×10^{10}
Exp. Plastic Q817.1[e]	25	ϵ'/ϵ_0	2.60	2.60	2.60	----	2.59	2.58	2.57	2.57*	2.57	2.56	
		tan δ	.7	.6	.61	----	.7	1	2.1	4.8*	4.5	3	
Exp. Plastic Q406[f]	25	ϵ'/ϵ_0	2.60	2.59	2.58	2.57	2.56	2.55	2.52	2.50*	2.49	2.49	
		tan δ	7.3	6.2	4.7	3.6	2.2	1.5	1.2	1.9*	2.0	2.6	
Poly-p-xylylene[g]	25	ϵ'/ϵ_0	2.45	2.45	2.45	2.45	2.45	2.45	----	----	2.45		
		tan δ	2	3.4	3.9	<2	0.7	3.7	----	----	9		

a. Dow. b. 99.75% polystyrene, 0.25% filler (Dow). c. From Dow's N-100 styrene (Lab. Ins. Res.). d. 50% polystyrene, 50% chlorinated diphenyl (Monsanto). e. Poly alpha-methylstyrene (Dow). f. Polyvinyltoluene (Dow). g. Pressed fibers (Polaroid).

*Freq. = 1×10^9.

338 Tables of Dielectric Materials

I. Solids, B. Organic 3. Plastics j. Polyvinyl Resins (cont.) Values for tan δ are multiplied by 10^4; frequency given in c/s.

12) Styrene copolymers

linear	T°C		1×10^2	1×10^3	1×10^4	1×10^5	1×10^6	1×10^7	1×10^8	3×10^8	3×10^9	1×10^{10}	2.5×10^{10}
Styrene-2,4-dimethyl- styrene copolymer[a]	25	ϵ'/ϵ_0	2.53	2.53	2.53	2.53	2.52	2.52	2.52	2.51*	2.50	2.49	
		tan δ	1.6	1.3	0.8	0.7	0.9	1.2	1.5	4.0*	5.4	2.8	
Styrene-acrylonitrile copolymer[a]	25	ϵ'/ϵ_0	2.96	2.95	2.92	2.87	2.80	2.78	2.77	2.77*	2.77	2.76	
		tan δ	59	63	67	67	64	50	41	40*	41	45	
Darex copolymer 3[b]	25	ϵ'/ϵ_0	2.53	2.53	2.52	2.51	2.50	2.49	2.49	----	2.48	2.45	
		tan δ	23.7	21.5	30.9	38.9	34.8	26.8	23.8	----	21.3	20.9	
Darex copolymer 43R[b]	25	ϵ'/ϵ_0	2.54	2.54	2.54	2.54	2.54	2.54	2.54	----	2.54	2.53	
		tan δ	5.5	2.8	2.2	2.8	3.9	6.2	8.2	----	9.8	13	
Darex copolymer I-34[b]	25	ϵ'/ϵ_0	2.55	2.54	2.54	2.53	2.52	2.52	2.51	----	2.50	2.48	
		tan δ	20.4	12.7	17.5	19.5	12	15	14	----	11.6	12	
Darex copolymer I-43[b]	25	ϵ'/ϵ_0	2.56	2.56	2.56	2.56	2.55	2.55	2.55	----	2.55	2.54	
		tan δ	7.3	2.5	2.3	2.2	3	4	5.5	----	8.9	11.6	
Styralloy 22[c]	-12	ϵ'/ϵ_0	2.4	2.4	2.4	2.4	2.4	2.4	----	----	2.4	2.4	2.4
		tan δ	8	8	4	15	36	52	----	----	28	18	18
	23	ϵ'/ϵ_0	2.4	2.4	2.4	2.4	2.4	2.4	2.40	2.40	2.4	2.4	2.4
		tan δ	9	6	5	7	12	30	52	46	32	24	
	81	ϵ'/ϵ_0	2.4	2.4	2.4	2.4	2.4	2.4	----	----	2.4	2.4	
		tan δ	16	10	8	6	6	13	----	----	51	55	
Marbon S[d]	25	ϵ'/ϵ_0	2.62	2.62	2.62	2.61	2.60	2.60	2.57	2.55*	2.55	2.54	
Code 7206		tan δ	17.4	14.1	13.8	13.8	18.3	22.1	22.2	20.6*	20	19	
Marbon S-1	25	ϵ'/ϵ_0	2.55	2.55	2.54	2.54	2.54	2.54	2.53	2.52*	2.52	2.52	
Code 7254[d]		tan δ	10	6	3.5	3.9	6.1	7.9	8.3	10*	11	12.5	
Marbon 8000[e]	25	ϵ'/ϵ_0	2.56	2.56	2.56	2.56	2.56	2.56	2.56	2.52*	2.51	2.51	
		tan δ	5.4	4	3.2	3.0	4.0	5.0	7.2	9.0*	9.7	10	
Marbon 9200[f]	25	ϵ'/ϵ_0	2.60	2.57	2.56	2.56	2.56	2.56	2.55	2.52*	2.52	2.52	
		tan δ	5.4	3.1	3	4.9	6.7	9	10	11.8*	11.9	11.5	

a. Am. Cyanamid. b. Dewey and Almy. c. Copolymer of butadiene and styrene (Dow). d. Butadiene-styrene copolymer, ca. 10% butadiene (Marbon). e. Butadiene-styrene copolymer ca. 15% butadiene (Marbon). f. Butadiene-styrene copolymer ca. 14% butadiene (Marbon).

*Freq. = 1×10^9.

Dielectric Data

I. Solids, B. Organic 3. Plastics j. Polyvinyl Resins (cont.) Values for tan δ are multiplied by 10^4; frequency given in c/s.

12) Styrene copolymers, linear (cont.)	T°C		1×10^2	1×10^3	1×10^4	1×10^5	1×10^6	1×10^7	1×10^8	1×10^9	3×10^9	1×10^{10}	2.5×10^{10}
Piccolastic D-125[a]	25	ϵ'/ϵ_0	2.58	2.58	2.58	2.58	2.58	2.58	2.58	----	2.55	2.52	
		tan δ	2	1.5	1	1	1	1.5	3	----	5	5	
Styrene (50%) and 1,3,5-trivinyl-2,4,6-trichlorobenzene (50%) copolymer	25	ϵ'/ϵ_0	2.70	2.70	2.70	2.70	2.70	2.70	2.70	2.66	2.65	2.65	
		tan δ	8.5	8.9	9.7	10	11	16.2	26	20	21	19	
Styrene (50%) and 1,4-divinyl-2,3,5,6-tetrachlorobenzene (50%) copolymer[b] S-40[c]	26	ϵ'/ϵ_0	2.70	2.70	2.70	2.70	2.70	2.70	2.70	2.70	2.70	2.70	
		tan δ	6	7.4	8.2	7.8	8.2	12.0	22	13.2	13.8	13.6	
S-40[c]	25	ϵ'/ϵ_0	2.40	2.40	2.40	2.39	2.39	2.39	2.38	----	2.37	2.35	
		tan δ	21	14	13	12	8	6	6	----	6.4	7.1	
S-60[c]	25	ϵ'/ϵ_0	2.50	2.49	2.48	2.46	2.46	2.45	2.45	----	2.44	2.44	
		tan δ	27	24	20	15	10	6	4	----	6.5	7.4	
13) Styrene copolymers cross-linked													
Copolymer 8012[d]	25	ϵ'/ϵ_0	2.58	2.58	2.58	2.58	2.58	2.57	2.55	2.54	2.53	2.52	
		tan δ	2.7	1.3	.87	.7	1.1	2	3.8	6.8	8.8	8.9	
Plastic EK2784[d]	25	ϵ'/ϵ_0	2.59	2.59	2.59	2.56	2.56	2.56	2.56	2.55	2.54	2.52	
		tan δ	2.6	1.5	1.2	1.4	1.8	2.3	4.2	8.8	6.7	4.7	
Exp. Plastic Q-166[e]	23	ϵ'/ϵ_0	3.44	3.40	3.38	3.36	3.32	3.22	3.05	----	2.71	2.62	
		tan δ	45	55	70	110	180	350	420	----	315	250	
	80	ϵ'/ϵ_0	3.25	3.23	3.21	3.19	3.16	3.12	3.05	----	2.85	2.63	
		tan δ	550	95	41	38	55	110	250	----	570	550	
Exp. Plastic Q-166[e] plus Fiberglas	23	ϵ'/ϵ_0	4.46	4.42	4.37	4.30	4.21	4.09	3.95	----	3.78	3.70	3.6
		tan δ	160	110	96	120	180	250	320	----	241	231	200
	79	ϵ'/ϵ_0	4.5	4.4	4.4	4.3	4.2	4.1	4.0	----	3.8	3.7	
		tan δ	880	320	140	85	80	130	220	----	415	470	
Exp. Plastic Q-200.5[f]	26	ϵ'/ϵ_0	2.55	2.55	2.55	2.55	2.55	2.55	2.55	----	2.55	----	2.54
		tan δ	4	< 2	< 2	< 2	2	3.4	4	----	5.2	----	24
	100	ϵ'/ϵ_0	2.56	2.56	2.56	2.56	2.55	2.55	----	----	2.55*	----	
		tan δ	20	7	2	< 2	< 2	< 3	----	----	7.6*	----	

a. Methylstyrene-styrene copolymer (Penn. Ind. Chem.). b. Sprague (two different castings measured in the ranges $10^2 - 10^8$ and $10^9 - 10^{10}$ c/s.) c. Standard Oil Dev. Co. N.J. d. Catalin. e. Dow. f. Cross-linked with o-, m-, p-divinylbenzenes (Dow).

*Measurements made at 78°C.

340 Tables of Dielectric Materials

I. Solids, B. Organic 3. Plastics j. Polyvinyl Resins (cont.)   Values for tan δ are multiplied by 10^4; frequency given in c/s.

13) Styrene copolymers cross-linked (cont.)

Material	T°C		1×10^2	1×10^3	1×10^4	1×10^5	1×10^6	1×10^7	1×10^8	3×10^8	3×10^9	1×10^{10}	2.5×10^{10}
Exp. Plastic Q-344[a]	24	ϵ'/ϵ_0	2.40	2.40	2.39	2.38	2.38	2.38	2.38	----	2.34	2.31	2.31
		tan δ	11	20	50	20	9	7	7	----	9	8.6	10
	80	ϵ'/ϵ_0	2.4	2.4	2.39	2.38	2.38	2.38	2.35	----	2.33	2.31	
		tan δ	21	14	12	16	45	42	19	----	19	15	
Exp. Plastic Q-475.5[a]	22	ϵ'/ϵ_0	2.51	2.51	2.51	2.51	2.51	2.51	2.51	----	2.50	2.49	
		tan δ	4	2	2	4	7	9	9	----	6.3	8	
	80	ϵ'/ϵ_0	2.48	2.48	2.48	2.48	2.48	2.48	2.48	----	2.47	2.47	
		tan δ	6	4	2	2	4	9	14	----	16	14	
Pliolite S5[b]	25	ϵ'/ϵ_0	2.58	2.58	2.58	2.55	2.55	2.55	2.55	2.53*	2.51	2.51	
		tan δ	6.8	3.3	2.1	2.5	4.7	7.3	7.8	7.6*	7.2	6.7	
Pliolite S3[b]	25	ϵ'/ϵ_0	2.58	2.58	2.58	2.58	2.58	2.58	2.58	2.55*	2.52	2.52	
		tan δ	7.7	3.5	2.1	1.8	2.6	3.6	6.2	15*	8.8	7.5	
Pliolite S6B[b]	25	ϵ'/ϵ_0	2.58	2.58	2.58	2.58	2.58	2.58	2.58	2.53*	2.52	2.51	
		tan δ	6	2.9	2.4	1.9	2	3.7	6.8	10.6*	11	8	
Pliolite S6[b]	25	ϵ'/ϵ_0	2.58	2.58	2.58	2.58	2.58	2.58	2.58	2.57	2.52	2.52	
		tan δ	6.6	2.8	2.1	2.1	2.1	3	5.5	16	7.6	6.3	
Resin C[c]	25	ϵ'/ϵ_0	2.51	2.51	2.51	2.51	2.50	2.50	2.47	----	2.47	2.46	
		tan δ	8	10	13	27	34	31	29	----	30	39	
Rexolite 1422[d]	25	ϵ'/ϵ_0	2.55	2.55	2.55	2.55	2.55	2.55	2.55	2.55*	2.54	2.54	
		tan δ	2.1	1.1	1	1.1	1.3	2	3.8	4.6*	4.8	4.7	
Bureau of Standards Casting Resin[e]	25	ϵ'/ϵ_0	2.64	2.62	2.62	2.62	2.62	2.62	2.62	2.60	2.59		
		tan δ	20	15.6	11	7	4.7	9	11	11	5		
Vibron 140[f]	25	ϵ'/ϵ_0	2.59	2.59	2.59	2.58	2.58	2.58	2.58	----	2.58	2.57	
		tan δ	4	5	8	12	16	19	20	----	19	17.5	
Vibron 141[f]	23	ϵ'/ϵ_0	2.64	2.64	2.64	2.63	2.62	2.62	2.62	----	2.62	2.62	
		tan δ	6	7	1'	19	33	39	34	----	28	28	

a. Cross-linked polystyrene (Dow). b. Goodyear. c. Esso C oil in styrene-divinylbenzene solution (Polaroid). d. Rex Corp.
e. 32.5% polystyrene, 53.5% poly-2,5-dichlorostyrene, 13% hydrogenated terphenyl, 0.5% divinylbenzene (U.S. Bur. Stand.).
f. U. S. Rubber.
*Freq. = 1×10^9.

Dielectric Data

I. Solids B. Organic 3. Plastics J. Polyvinyl resins (cont.) Values for tan δ are multiplied by 10^4; frequency given in c/s.

14) Polystyrene plus fillers

fillers	T°C		1×10^2	1×10^3	1×10^4	1×10^5	1×10^6	1×10^7	1×10^8	3×10^8	3×10^9	1×10^{10}	2.5×10^{10}
Polystyrene 91%,	26	ϵ'/ϵ_0	3.85	3.85	3.85	3.85	3.85	3.84	----	3.81	3.60	3.46	
carbon 9%[a]		tan δ	17	24	30	33	32	37	----	310	386	344	
Polystyrene 70%,	25	ϵ'/ϵ_0	----	----	----	----	----	----	----	11*	9.1	8.8	
carbon 30%[a]		tan δ	----	----	----	----	----	----	----	2300*	2500	1100	
Polystyrene 50%,	25	ϵ'/ϵ_0	----	----	----	----	----	----	----	25.9*	20.8	19.4	
carbon 50%[a]		tan δ	----	----	----	----	----	----	----	7300*	5600	2800	
(molded at 1000 p.s.i.)	25	ϵ'/ϵ_0	----	----	----	----	----	----	----	36*	25	24	
(molded at 10000 p.s.i.)	25	tan δ	----	----	----	----	----	----	----	4600*	6200	3000	
Lustrex Loaded Glass Mat[b]	25	ϵ'/ϵ_0	3.04	3.04	3.02	3.00	2.97	2.97	2.96				
(field ⊥ plane of laminate)		tan δ	13.8	8.8	7.4	7	8.3	10.6	13.8				
(field in plane of laminate)	25	ϵ'/ϵ_0	----	----	----	----	----	----	----	----	3.07	3.07	
		tan δ	----	----	----	----	----	----	----	----	31	36	
Polyglas P⁺(experimental)[c]	24	ϵ'/ϵ_0	3.36	3.36	3.36	3.36	3.36	3.36	----	----	3.35	3.33	3.32
		tan δ	8	7	7	7	7	7	----	----	7.8	8.4	14.0
	79	ϵ'/ϵ_0	3.36	3.36	3.36	3.36	3.36	3.36	----	----	3.35	3.33	
		tan δ	30	17	12	9	8	7	----	----	14	19	
Polyglas P⁺(technical)[d]	25	ϵ'/ϵ_0	3.38	3.38	3.38	3.38	3.38	3.38	----	3.36	3.35	3.34	
		tan δ	9	9	9	9	9	9	----	9.6	11.7	19	

15) Polychlorostyrenes

	T°C		1×10^2	1×10^3	1×10^4	1×10^5	1×10^6	1×10^7	1×10^8	3×10^8	3×10^9	1×10^{10}	2.5×10^{10}
Plastic CT-8[e]	24	ϵ'/ϵ_0	2.61	2.61	2.60	2.60	2.60	2.60	2.6	----	2.60	----	2.59
		tan δ	<2	<2	<2	<2	<2	2	2.5	----	3.1	----	29
	80	ϵ'/ϵ_0	2.61	2.61	2.60	2.60	2.60	2.60	2.6	----	2.60	----	
		tan δ	6	2	<3	<3	<3	3	4	----	7.3	----	
Plastic CQ-10DM[f]	25	ϵ'/ϵ_0	2.70	2.70	2.70	2.70	2.70	2.70	2.69	2.67	2.66	----	
		tan δ	5	5	5	6	8	10	11	11	10.8	----	
	100	ϵ'/ϵ_0	2.66	2.65	2.65	2.65	2.65	2.65	2.65	2.65	2.65	----	
		tan δ	33	17	8	6	7	11	16	17	17	----	

a. Cabot's #9 (Lab. Ins. Res.). b. 30% fiberglas (Monsanto). c. 18.6% Dow's C-244 polystyrene, 81.1% Corning's 790 glass powder, 0.25% paraffin, 0.1% Dow Corning's Ignition Sealing Compound No. 4 (Lab. Ins. Res.). d. Id. except that Monsanto's polystyrene was used (Monsanto). e. 97% poly-2,5-dichlorostyrene (Mathieson). f. Copolymer of 50% 2,4,-25% 2,5-, 25% 2,3-, 2,6- and 3,4-dichlorostyrenes (Mathieson).

*Freq. = 1×10^9.

I. Solids B. Organic 3. Plastics 1. Polyvinyl Resins (cont.) Values for tan δ are multiplied by 10^4; frequency given in c/s.

	T°C		1×10^2	1×10^3	1×10^4	1×10^5	1×10^6	1×10^7	1×10^8	3×10^8	3×10^9	1×10^{10}	2.5×10^{10}
15) Polychlorostyrenes (cont.)													
Poly-3,4-dichloro-styrene[a]	25	ϵ'/ϵ_0	2.94	2.93	2.91	2.88	2.85	2.80	2.77	----	2.72	2.70	
		tan δ	85	70	57	50	42	35	27	----	20	20	
	82	ϵ'/ϵ_0	3.0	2.90	2.85	2.83	2.78	2.73	2.68	----	2.65		
		tan δ	203	157	115	77	63	60	55	----	32		
	103	ϵ'/ϵ_0	3.08	2.92	2.85	2.81	2.8	2.8	2.75	----	2.7	2.7	
		tan δ	355	275	200	120	82	78	70	----	60	60	
	153	ϵ'/ϵ_0	6.0	4.9	3.7	3.3							
		tan δ	850	1700	1400	750							
Exp. Plastic Q-409[b]	24	ϵ'/ϵ_0	2.60	2.60	2.60	2.60	2.60	2.60	2.60	----	2.60	2.60	2.60
		tan δ	10	7	5	5	4	4	5	----	8.7	12	16
	90	ϵ'/ϵ_0	2.69	2.69	2.68	2.68	2.68	2.67	2.65	----	2.64	2.60	
		tan δ	50	25	14	8	5.5	5.5	7	----	11	18	
Polyglas D⁺(experimental)[c]	24	ϵ'/ϵ_0	3.22	3.22	3.22	3.22	3.22	3.22	3.22	----	3.22	3.22	
		tan δ	7	6	5	5	6	6	7	----	8	8.5	
	79	ϵ'/ϵ_0	3.2	3.2	3.2	3.2	3.2	3.2	3.2	----	3.2	3.2	
		tan δ	34	17	11	8	7	7	8	----	13	16	
Polyglas D⁺(technical)[d]	24	ϵ'/ϵ_0	3.25	3.25	3.25	3.25	3.25	3.25	3.24	----	3.23	3.22	3.22
		tan δ	3	3	4	4	5	6	7	----	12	13	14
16) Poly-2,5-dichloro-styrene plus fillers													
Poly-2,5-dichlorostyrene (58.1%), TiO_2 (41.9%)[e]	23	ϵ'/ϵ_0	5.30	5.30	5.30	5.30	5.30	5.30	5.30	----	5.30	5.30	
		tan δ	26	14	8	5	3	3	3	----	6	8.5	
Poly-2,5-dichlorostyrene (34.7%), TiO_2 (65.3%)[e]	24	ϵ'/ϵ_0	10.2	10.2	10.2	10.2	10.2	10.2	10.2	----	10.2	10.2	
		tan δ	16	8	6	4	3	3	3	----	7.5	13	
Poly-2,5-dichlorostyrene (18.6%), TiO_2 (81.4%)[e]	23	ϵ'/ϵ_0	23.6	23.4	23.2	23.1	23.0	23.0	23.0	----	----	23.0	
		tan δ	60	41	30	20	12	9	8	----	----	15.6	

a. ca. 95% (Monsanto). b. Copolymer of o- and p-chlorostyrenes (Dow). c. 34.9% Monsanto's poly-2,5-dichlorostyrene; 64.9% Corning's 790 glass powder, 0.1% paraffin wax, 0.1% Dow Corning's Ignition Sealing Compound No. 4 (Lab. Ins. Res.). d. Id. (Monsanto). e. Monsanto's Styramic HT; Titanium Alloy's Tam Ticon T, heavy grade (Lab. Ins. Res.).

Dielectric Data

I. Solids B. Organic 3. Plastics j. Polyvinyl Resins (cont.)

16) Poly-2,5-dichlorostyrene plus fillers (cont.)

Values for tan δ are multiplied by 10^4; frequency given in c/s.

	T°C		1×10^2	1×10^3	1×10^4	1×10^5	1×10^6	1×10^7	1×10^8	3×10^8	3×10^9	1×10^{10}
Poly-2,5-dichlorostyrene (38.2%), MgTiO$_3$ (61.8%)[a]	23	ϵ'/ϵ_0	6.10	6.10	6.10	6.09	6.09	6.08	6.05	6.01	5.91	4.90
		tan δ	35	22	14	10	7	6	6	7	8.9	
Poly-2,5-dichlorostyrene (63.0%), SrTiO$_3$ (37.0%)[b]	25	ϵ'/ϵ_0	5.20	5.20	5.20	5.19	5.18	5.17	5.15	----	4.97	14.1
		tan δ	20	10	6	4	3	4	5	----	11.7	
Poly-2,5-dichlorostyrene (40.5%), SrTiO$_3$ (59.5%)[b]	24	ϵ'/ϵ_0	9.65	9.65	9.63	9.62	9.61	9.60	9.57	----	----	9.36
		tan δ	41	37	16	12	10	9	10	----	----	23
Poly-2,5-dichlorostyrene (25.2%), SrTiO$_3$ (74.8%)[b]	23	ϵ'/ϵ_0	17.9	17.4	17.4	17.3	17.3	17.2	17.0	----	----	15.2
		tan δ	220	130	78	42	30	22	23	----	----	64
Poly-2,5-dichlorostyrene (14.4%), SrTiO$_3$ (80.6%)[b]	23	ϵ'/ϵ_0	24.5	22.8	21.6	21.2	21.2	21.0	20.8	----	----	19.5
		tan δ	780	380	180	100	60	40	33	----	----	50
Poly-2,5-dichlorostyrene (66.6%), BaTiO$_3$ (33.4%)[c]	23	ϵ'/ϵ_0	4.10	4.08	4.07	4.06	4.04	4.03	4.03	----	4.02	4.02
		tan δ	24	14	8	6	4	3	3	----	17	38
Poly-2,5-dichlorostyrene (32.8%), BaTiO$_3$ (67.2%)[c]	25	ϵ'/ϵ_0	9.7	9.7	9.7	9.7	9.7	9.7	9.7	----	----	9.7
		tan δ	60	42	25	14	7	6	6	----	----	143
Poly-2,5-dichlorostyrene (23.5%), BaTiO$_3$ (76.5%)[c]	23	ϵ'/ϵ_0	15.8	15.8	15.8	15.8	15.7	15.7	15.7	----	----	15.5
		tan δ	30	30	25	12	7	7	13	----	----	214
Poly-2,5-dichlorostyrene (21.0%), BaTiO$_3$ (79.0%)[c]	23	ϵ'/ϵ_0	19.4	19.3	19.1	18.9	18.9	18.9	18.9	----	----	18.4
		tan δ	52	35	25	16	11	7	8	----	----	314
Poly-2,5-dichlorostyrene (94%), Fe (6%)[d]	25	μ'/μ_0	----	----	----	----	----	----	----	----	1.02	.98
		tan δ$_m$	----	----	----	----	----	----	----	----	250	270
		ϵ'/ϵ_0	----	----	----	----	----	----	----	----	2.78	2.78
		tan δ$_d$	----	----	----	----	----	----	----	----	7	15
	83	μ'/μ_0	----	----	----	----	----	----	----	----	1.04	
		tan δ$_m$	----	----	----	----	----	----	----	----	300	
		ϵ'/ϵ_0	----	----	----	----	----	----	----	----	2.7	
		tan δ$_d$	----	----	----	----	----	----	----	----	40	

a. Monsanto's Styramic HT; Titanium Alloy's Tam Ticon MC (Lab. Ins. Res.). b. Monsanto's Styramic HT; Titanium Alloy's Tam Ticon S (Lab. Ins. Res.). c. Monsanto's Styramic HT; Titanium Alloy's Tam Ticon B (Lab. Ins. Res.). d. Monsanto's D-1795; Plastic Metals' Plast-Iron, minus 200 mesh, annealed (Lab. Ins. Res.).

I. Solids B. Organic 3. Plastics J. Polyvinyl Resins (cont.) Values for tan δ are multiplied by 10^4; frequency given in c/s.

16) Poly-2,5-dichlorostyrene plus fillers (cont.)

Material	T°C		1×10^2	1×10^3	1×10^4	1×10^5	1×10^6	1×10^7	1×10^8	3×10^8	3×10^9	1×10^{10}
Poly-2,5-dichlorostyrene (60%), Fe (40%)[a]	25	ϵ'/ϵ_0	----	----	----	----	----	----	----	----	5.3	5.2
		μ'/μ_0	----	----	----	----	----	----	----	----	1.16	.96
		tan δ_d	----	----	----	----	----	----	----	----	270	190
		tan δ_m	----	----	----	----	----	----	----	----	2300	2620
	85	ϵ'/ϵ_0	----	----	----	----	----	----	----	----	5.1	
		μ'/μ_0	----	----	----	----	----	----	----	----	1.27	
		tan δ_d	----	----	----	----	----	----	----	----	----	
		tan δ_m	----	----	----	----	----	----	----	----	3000	
Poly-2,5-dichlorostyrene (49.3%), Fe (50.7%)[a]	25	ϵ'/ϵ_0	9.8	9.7	9.5	9.2	9.0	8.7	----	8.08	6.93	2.64
		μ'/μ_0	----	3.00	----	----	----	2.70	----	2.00	1.24	1.03
		tan δ_d	110	130	160	190	220	320	----	601	644	420
		tan δ_m	----	----	----	----	----	340	----	2254	3170	
Poly-2,5-dichlorostyrene (40.7%) Fe (59.3%)[a]	25	ϵ'/ϵ_0	----	----	----	----	----	----	----	13.8	10	
		μ'/μ_0	----	----	----	----	----	----	----	2.6	1.37	
		tan δ_d	----	----	----	----	----	----	----	1950	1400	
		tan δ_m	----	----	----	----	----	----	----	2810	4500	
Poly-2,5-dichlorostyrene (92%), Fe_3O_4 (8%)[b]	25	ϵ'/ϵ_0	----	----	----	----	----	----	----	2.78	2.76	2.64
		μ'/μ_0	----	----	----	----	----	----	----	1.05	1.03	1.03
		tan δ_d	----	----	----	----	----	----	----	13	----	
		tan δ_m	----	----	----	----	----	----	----	77	200	420
Poly-2,5-dichlorostyrene (83.7%) Fe_3O_4 (16.3%)[b]	25	ϵ'/ϵ_0	----	----	----	----	----	----	----	3.2	3.2	3.2
		μ'/μ_0	----	----	----	----	----	----	----	1.12	1.07	.99
		tan δ_d	----	----	----	----	----	----	----	18	----	
		tan δ_m	----	----	----	----	----	----	----	200	420	580
Poly-2,5-dichlorostyrene (75%) Fe_3O_4 (25%)[b]	25	ϵ'/ϵ_0	----	----	----	----	----	----	----	3.45	3.44	
		μ'/μ_0	----	----	----	----	----	----	----	1.19	1.12	
		tan δ_d	----	----	----	----	----	----	----	26	----	
		tan δ_m	----	----	----	----	----	----	----	250	590	

a. Monsanto's D-1795; Plastic Metals' Plast-Iron, minus 200 mesh, annealed (Lab. Ins. Res.). b. Monsanto's D-1795; Mallinckrodt's magnetite (Lab. Ins. Res.).

I. Solids B. Organic 3. Plastics j. Polyvinyl Resins (cont.) Values for tan δ are multiplied by 10^4; frequency given in c/s.

16) Poly-2,5-dichlorostyrene plus fillers (cont.)

		T°C	1×10^2	1×10^3	1×10^4	1×10^5	1×10^6	1×10^7	1×10^8	3×10^8	3×10^9	1×10^{10}	3×10^{10}
Poly-2,5-dichlorostyrene (65.9%) Fe_3O_4 (34.1%)[a]	ϵ'/ϵ_o	25	----	----	----	----	----	----	----	4.08	4.07		
	μ'/μ_o		----	----	----	----	----	----	----	1.30	1.17		
	tan δ_d		----	----	----	----	----	----	----	42			
	tan δ_m		----	----	----	----	----	----	----	390	1010		
Poly-2,5-dichlorostyrene (56.2%) Fe_3O_4 (43.8%)[a]	ϵ'/ϵ_o	25	----	----	----	----	----	----	----	5.00	4.98		
	μ'/μ_o		----	----	----	----	----	----	----	1.45	1.23		
	tan δ_d		----	----	----	----	----	----	----	75			
	tan δ_m		----	----	----	----	----	----	----	560	1720		
Poly-2,5-dichlorostyrene (46.0%) Fe_3O_4 (54.0%)[a]	ϵ'/ϵ_o	25	----	----	----	----	----	----	----	7.4	7.3	7.3	
	μ'/μ_o		----	----	----	----	----	----	----	1.65	1.32	.91	
	tan δ_d		----	----	----	----	----	----	----	150			
	tan δ_m		----	----	----	----	----	----	----	780	2400	3500	
Poly-2,5-dichlorostyrene (35.7%) Fe_3O_4 (64.3%)[a]	ϵ'/ϵ_o	25	----	----	----	----	----	----	----	9.8	9.5		
	μ'/μ_o		----	----	----	----	----	----	----	1.94	1.41		
	tan δ_d		----	----	----	----	----	----	----	450	180		
	tan δ_m		----	----	----	----	----	----	----	980	3000		
Poly-2,5-dichlorostyrene (24.5%) Fe_3O_4 (75.5%)[a]	ϵ'/ϵ_o	25	----	----	----	----	----	----	----	20	18		
	μ'/μ_o		----	----	----	----	----	----	----	2.38	1.60		
	tan δ_d		----	----	----	----	----	----	----	2000	300		
	tan δ_m		----	----	----	----	----	----	----	1430	3700		
Poly-2,5-dichlorostyrene (38.55%) $(Mn,Fe)_3O_4$ (61.45%)[b]	ϵ'/ϵ_o	25	24.8	16.9	11.9	10.0	7.6	5.0	----	3.76			
	μ'/μ_o		----	----	----	----	----	----	----	2.56			
	tan δ_d		2500	2500	2400	2300	1900	900	----				
	tan δ_m		----	----	----	----	----	----	----	4000			
Poly-2,5-dichlorostyrene (21.3%) $(Mn,Fe)_3O_4$ (78.7%)[b]	ϵ'/ϵ_o	25	----	99	39	29	21	7.4	----	6.9	6.5		
	μ'/μ_o		----	----	----	----	----	----	----	4.3	1.32		
	tan δ_d		----	27700	10000	6200	4800	930	----	90	<50		
	tan δ_m		----	----	----	----	----	----	----	4900	13200		

a. Monsanto's D-1795; Mallinckrodt's magnetite (Lab. Ins. Res.).   b. Monsanto's D-1795; iron-manganese oxide (Lab. Ins. Res.).

I. Solids B. Organic 3. Plastics j. Polyvinyl Resins (cont.) Values for tan δ are multiplied by 10^4; frequency given in c/s.

	T°C		1×10^2	1×10^3	1×10^4	1×10^5	1×10^6	1×10^7	1×10^8	3×10^8	3×10^9	1×10^{10}
17) Polyvinyl cyclohexane												
Polyvinylcyclohexane[a]	24	ϵ'/ϵ_o	2.25	2.25	2.25	2.25	2.25	2.25	2.25	----	2.25	2.25
		tan δ	15	2	<2	<2	<2	<2	<2	----	1.8	3.9
18) Poly-α-vinylnaphthalene												
Poly-α-vinylnaphthalene[b]	24	ϵ'/ϵ_o	----	2.6	2.6	2.6	2.6	2.6	2.6			
		tan δ	----	9.5	6.7	7.5	8.7	13	22			
19) Poly-2-vinylpyridine												
Poly-2-vinylpyridine[b]	22	ϵ'/ϵ_o	4.91	4.64	4.26	3.77	3.56	3.33	3.1	3.06	2.98	
		tan δ	360	460	560	600	560	420	280	240	135	
2-Vinylpyridine-styrene copolymer[b]	23	ϵ'/ϵ_o	3.44	3.38	3.26	3.13	3.00	2.90	2.82	----	2.76	
		tan δ	168	230	270	290	260	185	120	----	65	
20) Poly-N-vinylcarbazole												
Polectron #24[c]	25	ϵ'/ϵ_o	2.95	2.95	2.95	2.95	2.95	2.95	2.95	----	2.88	
		tan δ	13	9	6	5	4	5	6	----	9.3	
	80	ϵ'/ϵ_o	2.95	2.95	2.95	2.95	2.95	2.95				
		tan δ	19	10	8	9	9	7				

K. Polyesters

	T°C		1×10^2	1×10^3	1×10^4	1×10^5	1×10^6	1×10^7	1×10^8	3×10^8	3×10^9	1×10^{10}
Laminac 4115[d]	25	ϵ'/ϵ_o	3.24	3.22	3.20	3.17	3.12	3.07	2.94	2.87*	2.83	2.82
		tan δ	37.5	43.2	68.3	113	135	147	141	107*	93	88
Experimental Resin PDL7-669[d]	25	ϵ'/ϵ_o	4.15	4.05	3.93	3.77	3.58	3.44	3.20	----	3.00	2.95
		tan δ	155	192	247	306	313	282	240	----	155	142
Laminac PDL7-627[d]	25	ϵ'/ϵ_o	3.26	3.25	3.23	3.18	3.08	3.06	3.03	2.96*	2.92	2.86
		tan δ	50.5	42.8	67.8	115	140	148	135	107*	95	91.4
Laminac PDL7-650[d]	25	ϵ'/ϵ_o	3.02	3.00	2.98	2.95	2.89	2.82	2.78	----	2.74	2.72
		tan δ	39.1	45.0	60.2	100	119	114	108	----	92	96
Laminac 4-205[d]	25	ϵ'/ϵ_o	4.45	4.36	4.24	4.03	3.86	3.69	3.49	----	3.21	3.15
		tan δ	147	162	219	301	360	345	340	----	240	243
Formica Z65[e]	25	ϵ'/ϵ_o	5.21	5.10	5.00	4.89	4.70	4.35	----	3.83*	3.69	3.49
(field ∥ laminate)		tan δ	191	162	212	317	434	570	----	485*	445	425
(field ⊥ laminate)	25	ϵ'/ϵ_o	4.77	4.70	4.68	4.60	4.45	4.17				
		tan δ	125	108	131	185	302	418				

a. Hydrogenated polystyrene (Dow). b. Lab. Ins. Res. c. Poly-N-vinylcarbazole, 1.3% HB-40 oil (Gen. Aniline). d. Amer. Cyanamid.
e. 50% Shell's DAP 85/80, 50% paper (Formica).
*Freq. = 1×10^9.

I. Solids B. Organic 3. Plastics (cont.) Values for tan δ are multiplied by 10^4; frequency given in c/s.

k. Polyesters (cont.)

	T°C		1×10^2	1×10^3	1×10^4	1×10^5	1×10^6	1×10^7	1×10^8	3×10^8	3×10^9	1×10^{10}
Formica Z80[a]	25	ϵ'/ϵ_o	4.76	4.75	4.72	4.65	4.46	4.08	3.77	3.53*	3.41	3.36
(field ∥ laminate)		tan δ	175	160	210	310	420	565	500	390*	310	250
(field ⊥ laminate)	25	ϵ'/ϵ_o	4.52	4.42	4.38	4.33	4.15	3.91				
		tan δ	95	96	130	197	330	440				
Glastic S[b]	26	ϵ'/ϵ_o	3.50	3.39	3.29	3.22	3.16	3.14	3.07			
		tan δ	330	250	160	98	100	105	130			
Glastic MF[b]	26	ϵ'/ϵ_o	3.92	3.82	3.76	3.72	3.65	3.57	3.53			
		tan δ	270	185	135	125	130	130	135			
Plaskon 911[c]	24	ϵ'/ϵ_o	3.88	3.81	3.75	3.66	3.56	3.42	3.25	3.17	3.07	3.02
		tan δ	89	125	165	210	240	240	220	217	175	158
	80	ϵ'/ϵ_o	4.27	4.17	4.06	3.95	3.83	3.70	3.55	----	3.26	3.10
		tan δ	165	150	155	200	270	320	340	----	299	248
Marco Resin MR-21C[d]	25	ϵ'/ϵ_o	3.37	3.35	3.31	3.25	3.16	3.08	----	2.90	2.84	2.82
		tan δ	53	51	65	102	150	170	----	149	106	123
Marco Resin MR-23C[d]	25	ϵ'/ϵ_o	4.56	4.50	4.40	4.28	4.14	3.88	----	3.24	2.92	2.82
		tan δ	130	140	170	220	340	570	----	810	500	410
Marco Resin MR-25C[d]	25	ϵ'/ϵ_o	3.27	3.24	3.20	3.15	3.10	3.06	2.90	2.86	2.77	2.73
		tan δ	76	72	82	108	138	165	190	185	130	135
Laminating Resin MP, Glass reinforced[e]	25	ϵ'/ϵ_o	----	----	----	----	----	----	----	----	----	4.26
		tan δ	----	----	----	----	----	----	----	----	----	110
Laminating Resin MT, Glass reinforced[f]	25	ϵ'/ϵ_o	----	----	----	----	----	----	----	----	----	4.24
		tan δ	----	----	----	----	----	----	----	----	----	88
Laminate BD-44[g]	24	ϵ'/ϵ_o	3.32	3.28	3.22	3.19	3.14	3.08	3.02	2.99	2.93	2.91
		tan δ	95	81	85	105	125	140	150	132	109	110

a. 52% Shell's DAP 25/50, 48% yarn fabric (Formica). b. With Fiberglas (Laminated Plastics). c. Unsaturated polyester (Libby-Owens-Ford). d. Unsaturated polyester (Marco). e. Resin of 50% Vibron x-1039 alkyd and 50% triallyl cyanurate plus catalyst and Fiberglas ECC-181-114 (Naugatuck). f. Resin of 45% polyethylene tetraclophthate-maleate alkyd and 55% triallyl cyanurate plus catalyst and Fiberglas ECC-181-114 (Naugatuck). g. Selectron 5003, Fiberglas (Owens-Corning).

*Freq. = 1×10^9.

348 Tables of Dielectric Materials

I. Solids B. Organic 3. Plastics (cont.)   Values for tan δ are multiplied by 10^4; frequency given in c/s.

k. Polyesters (cont.)	T°C		1×10^2	1×10^3	1×10^4	1×10^5	1×10^6	1×10^7	1×10^8	3×10^8	3×10^9	1×10^{10}	2.5×10^{10}
Laminate BK 164[a]	24	ϵ'/ϵ_o	4.12	4.10	4.09	4.07	4.05	4.03	4.00	3.98	3.94	3.90	
		tan δ	67	58	63	80	96	110	112	108	120	130	
Stypol 16B[b]	25	ϵ'/ϵ_o	----	----	----	----	----	----	----	----	----	2.88	
		tan δ	----	----	----	----	----	----	----	----	----	180	
Stypol 16C[b]	25	ϵ'/ϵ_o	----	----	----	----	----	----	----	----	----	2.75	
		tan δ	----	----	----	----	----	----	----	----	----	160	
Stypol 16D[b]	25	ϵ'/ϵ_o	----	----	----	----	----	----	----	----	----	2.81	
		tan δ	----	----	----	----	----	----	----	----	----	140	
Paraplex P13[c]	25	ϵ'/ϵ_o	4.02	4.00	3.92	3.92	3.65	3.32	3.08	2.89*	2.77	2.77	
		tan δ	73	108	184	310	530	590	600	440*	320	290	
Paraplex P43[c]	25	ϵ'/ϵ_o	3.23	3.22	3.19	3.16	3.11	3.04	2.98	2.89*	2.85	2.85	
		tan δ	33	43	68	98	130	160	160	110*	100	80	
Polydiallyl phthalate[d]	26.8	ϵ'/ϵ_o	3.60	3.57	3.50	3.42	3.35	3.24	3.1	----	2.95	2.95	
		tan δ	104	95	118	150	200	241	195	----	118	118	
Allymer CR-39[e]	24	ϵ'/ϵ_o	4.18	4.14	4.03	3.85	3.52	3.27	3.07	----	2.88	----	2.76
		tan δ	90	120	210	380	520	460	330	----	203	----	165
	84	ϵ'/ϵ_o	4.90	4.84	4.70	4.47	4.18	3.85	3.45	----	3.02	----	
		tan δ	130	140	170	250	440	840	740	----	370	----	
Allymer CR-39 laminate[f]	25	ϵ'/ϵ_o	4.84	4.80	4.73	4.57	4.37	4.17	3.98	----	3.78	----	
		tan δ	73	80	120	208	315	295	205	----	140	----	
Phoresin[g]	25	ϵ'/ϵ_o	3.98	3.84	3.70	3.65	3.49	3.32	3.15	3.10	3.03	3.02	2.96
		tan δ	260	270	320	330	280	210	165	145	171	165	164
m. Alkyd resins													
Chlorinated alkyd diisocyanate, foamed[h]	25	ϵ'/ϵ_o	----	----	----	----	----	----	----	----	----	1.179	
		tan δ	----	----	----	----	----	----	----	----	----	27	
Alkyd, diisocyanate, foamed[k]	25	ϵ'/ϵ_o	1.223	1.223	1.223	1.223	1.218	1.205	1.20	----	1.20	1.19	
		tan δ	19.8	14.7	22.7	33.5	41	42	38	----	34	22	
Red Glyptal #1201[m]	25	ϵ'/ϵ_o	4.9	4.5	4.1	4.0	3.9	3.8					
		tan δ	760	600	500	400	320	290					

a. 38% polyester (Bakelite BRS-16631), 62% Fiberglas (Owens-Corning). b. Robertson. c. Rohm and Haas. d. Shell Dev. e. Allyl resin (Southern Alkali Corp.) f. 40% resin, 60% ECC-11-148 Fiberglas (Southern Alkali Corp.). g. Diallyl phenyl phosphonate resin (Victor). h. Cornell Aeronautical Labs. k. Goodyear Aircraft. m. General Electric.
*Frequency = 1×10^9.

Dielectric Data

I. Solids B. Organic 3. Plastics (cont.)

Values for tan δ are multiplied by 10^4; frequency given in c/s.

	T°C		1×10^2	1×10^3	1×10^4	1×10^5	1×10^6	1×10^7	1×10^8	3×10^8	3×10^9	1×10^{10}
m. Alkyd resins (cont.)												
#51 Permo Potting Compound[a]	25	ϵ'/ϵ_o	3.18	2.95	2.83	2.74	2.70	2.64	2.59	2.55*	2.53	2.52
		tan δ	730	410	260	174	124	101	120	182*	125	122
Glastic Grade GF[b]	25	ϵ'/ϵ_o	----	----	----	----	3.76					
		tan δ	----	----	----	----	95					
Glastic Grade MM[b]	25	ϵ'/ϵ_o	4.17	4.13	4.09	4.02	3.96	3.88	3.78			
		tan δ	102	87	84	99	115	125	136			
Glastic Grade MP[b]	25	ϵ'/ϵ_o	----	----	----	----	3.56					
		tan δ	----	----	----	----	267					
Glastic Grade A-2[b]	25	ϵ'/ϵ_o	----	----	----	----	4.34					
		tan δ	----	----	----	----	518					
Plaskon Alkyd Special Electrical Granular[c]	25	ϵ'/ϵ_o	5.32	5.10	4.96	4.90	4.76	4.65	4.55	4.54*	4.50	4.47
		tan δ	366	236	170	147	149	152	138	121*	108	138
Plaskon Alkyd 411[c]	25	ϵ'/ϵ_o	6.02	5.77	5.60	5.36	5.19	4.95	4.63	----	4.32	4.31
		tan δ	340	240	220	260	310	320	288	----	220	178
Plaskon Alkyd 420[c]	25	ϵ'/ϵ_o	5.50	5.35	5.24	5.10	5.03	4.90	4.82	4.63*	4.62	
		tan δ	270	160	130	147	153	147	145	124*	141	
Plaskon Alkyd 422[c]	25	ϵ'/ϵ_o	5.47	5.26	5.14	5.01	4.92	4.85	4.77	4.75*	4.75	4.72
		tan δ	365	213	151	134	120	113	110	100*	104	126
Plaskon Alkyd 440[c]	25	ϵ'/ϵ_o	5.13	5.04	4.95	4.85	4.73	4.61	4.50	4.42*	4.38	4.33
		tan δ	191	151	154	185	196	188	172	133*	137	146
Plaskon Alkyd 440A[c]	25	ϵ'/ϵ_o	5.28	5.19	5.10	4.97	4.90	4.78	4.65	4.63*	4.62	4.61
		tan δ	165	133	131	143	146	146	134	124*	141	149
Plaskon Alkyd 442[c]	25	ϵ'/ϵ_o	5.02	4.89	4.80	4.71	4.59	4.50	4.44	4.37*	4.34	4.29
		tan δ	212	189	140	141	146	142	122	127*	131	141
n. Epoxy resins												
Araldite Casting Resin Type B[d]	25	ϵ'/ϵ_o	3.67	3.67	3.67	3.65	3.62	3.49	3.35	3.28	3.09	3.01
		tan δ	17	24	50	110	190	270	340	340	270	220
Araldite E-134[d]	25	ϵ'/ϵ_o	7.3	6.1	5.3	4.7	4.4	4.1	3.7	3.5	3.2	3.1
		tan δ	1200	1050	920	760	770	1000	1300	750	460	390
Araldite Casting Resin G[d]	25	ϵ'/ϵ_o	4.05	3.99	3.92	3.81	3.69	3.54	3.39	3.25*	3.15	3.10
		tan δ	105	104	141	210	270	288	300	311*	310	263

a. Hardman. b. Laminated Plastics. c. Libbey-Owens-Ford. d. Ciba.
*Freq. = 1×10^9.

I. Solids B. Organic 3. Plastics (cont.)   Values for tan δ are multiplied by 10^4; frequency given in c/s.

(cont.) n. Epoxy Resins	T°C	ε'/ε₀ tan δ	1×10^2	1×10^3	1×10^4	1×10^5	1×10^6	1×10^7	1×10^8	3×10^8	3×10^9	1×10^{10}	2.5×10^{10}
Araldite Adhesive Type I Natural[a]	25	ε'/ε₀	4.00	3.97	3.91	3.82	3.71	3.46	3.27	----	3.14	----	
		tan δ	29	52	120	200	300	350	340	----	230	----	
Araldite Adhesive Type I Silver[a]	25	ε'/ε₀	11.6	11.6	11.5	11.3	11.0	10.5	10.2	9.9	9.6	9.2	
		tan δ	38	62	120	190	300	370	360	360	520	980	
Hysol 6000[b]	25	ε'/ε₀	3.47	3.43	3.38	3.30	3.26	3.17	3.10	3.07	3.00	2.96	
		tan δ	73	61	70	92	130	170	170	150	120	120	
Hysol 6020[b]	25	ε'/ε₀	3.96	3.90	3.82	3.67	3.54	3.42	3.29	----	3.01	2.99	
		tan δ	68	113	206	260	272	266	299	----	274	252	
Hysol 6030 flexible Potting Compound[b]	25	ε'/ε₀	6.65	6.15	5.75	5.37	4.74	4.15	3.61	----	3.20	3.11	
		tan δ	605	485	469	593	840	1010	900	----	380	324	
Hysol 6000 PR[b]	25	ε'/ε₀	3.85	3.80	3.69	3.64	3.57	3.44	3.33	----	3.18	3.18	
		tan δ	90	99	107	163	227	248	250	----	192	180	
Hy-tuf Laminate Grade GF181[b] (field ⊥ laminate)	25	ε'/ε₀	4.67	4.62	4.52	4.45	4.34	4.21	4.15	----	4.08	4.05	
		tan δ	92	99	125	145	212	239	242	----	177	180	
(field ∥ laminate)	25	ε'/ε₀	----	----	----	----	----	----	----	4.14*	----	----	
		tan δ	----	----	----	----	----	----	----	171*	----	----	
Epon Resin RN-48[c]	25	ε'/ε₀	3.64	3.63	3.61	3.57	3.52	3.44	3.32	3.13*	3.04	2.91	
		tan δ	31	38	68	111	142	191	264	220*	210	184	
o. Miscellaneous Plastics													
M Resin[d]	23	ε'/ε₀	2.49	2.49	2.48	2.48	2.47	2.45	2.45	----	2.43	2.42	
		tan δ	7	4	<3	<5	5	8	8	----	6	5	
	80	ε'/ε₀	2.5	2.5	2.5	2.5	2.5	2.45	2.45	----	2.4	2.4	
		tan δ	25	14	8	5	4	6	7	----	15	17	
Same material after 3 1/2 yrs. at room temp. and humidity	25	ε'/ε₀	2.56	2.56	2.55	2.54	2.54	2.54	2.53	2.5	----	2.44	2.43
		tan δ	8	11	16	19	21	22	19	16	----	7	11
Permafil 3256[e]	24	ε'/ε₀	4.27	4.22	4.12	4.01	3.86	3.70	3.5	----	----	3.1	3.0
		tan δ	98	120	170	230	300	330	340	----	----	276	290
	99	ε'/ε₀	4.95	4.85	4.75	4.61	4.48	4.28	4.05	----	----	3.34	
		tan δ	200	155	160	200	300	440	570	----	----	580	
	125	ε'/ε₀	5.12	5.04	4.92	4.77	4.62	4.4	4.1	----	----	3.38	
		tan δ	280	180	170	185	260	410	550	----	----	560	
Piccopale Resin[f]	25	ε'/ε₀	2.33	2.33	2.33	----	2.33	2.33	2.33	----	2.30		
		tan δ	<3	<3	<3	----	<5	2	8	----	3.5		

a. Cibas. b. Houghton Labs. c. Shell Chem. d. Cross-linked addition hydrocarbon polymer (Esso Lab.). e. Cross-linked addition polymer (Gen. Elec.). f. Linear addition hydrocarbon copolymer made with aliphatic dienes and olefins (Penn. Industrial Chem.).
*Freq. = 1×10^9.

I. Solids, B. Organic 4. Elastomers   Values for tan δ are multiplied by 10^4; frequency given in c/s.

	T°C		1×10^2	1×10^3	1×10^4	1×10^5	1×10^6	1×10^7	1×10^8	3×10^8	3×10^9	1×10^{10}
a. Natural Rubber												
Hevea rubber[a]	-12	ϵ'/ϵ_o	2.4	2.4	2.4	2.4	2.4	2.4	2.4	----	----	
		tan δ	10	14	24	44	55	38	33	----	----	
	25	ϵ'/ϵ_o	2.4	2.4	2.4	2.4	2.4	2.4	2.4	----	2.15	
		tan δ	28	18	14	14	18	32	50	----	30	
	80	ϵ'/ϵ_o	2.4	2.4	2.4	2.4	2.4	2.4	2.4			
		tan δ	132	79	48	27	21	21	30			
Hevea rubber, vulcanized[b]	27	ϵ'/ϵ_o	2.94	2.94	2.93	2.88	2.74	2.52	2.42	----	2.36	
		tan δ	48	24	62	220	446	410	180	----	47	
Hevea rubber compound[c]	27	ϵ'/ϵ_o	4.09	4.01	3.92	3.80	3.64	3.44	3.3	----	3.25	
		tan δ	130	155	182	230	308	405	265	----	148	
Hevea rubber compound[d]	27	ϵ'/ϵ_o	----	2.4	2.4	14	9.0	7.0	6.8	----	6.3	
		tan δ	----	36	27	4000	2500	1600	850	----	234	
Cellular Rubber[e]	25	ϵ'/ϵ_o	----	----	12000	----	----	----	----	----	1.31	1.31
		tan δ	----	----							75	94
#49 Permo Potting Compound[f]	25	ϵ'/ϵ_o	3.46	3.39	3.28	3.18	2.96	2.80	2.72	2.63*	2.60	
		tan δ	169	230	274	320	370	350	215	170*	180	
b. Gutta-percha												
Gutta-percha[g]	25	ϵ'/ϵ_o	2.61	2.60	2.58	2.55	2.53	2.50	2.47	2.45	2.40	2.38
		tan δ	5	4	9	21	42	80	120	110	60	50
c. Balata												
Balata, precipitated[h]	25	ϵ'/ϵ_o	2.50	2.50	2.50	2.50	2.50	2.47	2.42	2.41	2.40	2.39
		tan δ	9	5	4	5	15	33	62	63	37	30
d. Cyclized Rubbers												
Pliolite[i]	27	ϵ'/ϵ_o	2.5	2.5	2.5	2.5	2.5	2.5	2.45	----	2.4	
		tan δ	52	35	31	31	37	46	46	----	31	
Pliolite GR[i]	25	ϵ'/ϵ_o	2.6	2.6	2.6	2.55	2.55	2.55	----	----	----	2.5
		tan δ	92	98	78	60	44	33	----	----	----	33

a. Pale crepe (Rubber Res. Corp.). b. 100 pts. pale crepe, 6 pts. sulfur (Rubber Res. Corp.). c. 100 pts. pale crepe, 1 pt. stearic acid, 10 pts. United Black, 5 pts. Kadox, 0.5 pts. Captax, 3 pts. sulfur (Rubber Res. Corp.). d. 100 pts. pale crepe, 1 pt. stearic acid, 40 pts. United Black, 5 pts. Kadox, 1 pt. Captax, 3 Pts. sulfur (Rubber Res. Corp.). e. Cellular Rubber Prod. f. Depolymerized rubber (Hardman). g. Palaquium Oblongifolium (Hermann Weber). h. Minusops Globosa (Hermann Weber). i. Goodyear.
*Freq. = 1×10^9.

I. Solids B. Organic 4. Elastomers (cont.) Values for tan δ are multiplied by 10^4; frequency given in c/s.

d. Cyclized Rubbers (cont.)	T°C		1×10^2	1×10^3	1×10^4	1×10^5	1×10^6	1×10^7	1×10^8	3×10^8	3×10^9	1×10^{10}
Plioband M-190-C[a]	24	ϵ'/ϵ_o	14	12	9.5	7.9	6.3	4.8	4.0	----	3.76	2.8
		tan δ	7800	1900	850	900	2000	1900	1000	----	740	120
e. Buna Rubbers												
Air seal[b]	24	ϵ'/ϵ_o	15.3	8.9	6.2	4.4	3.6	3.3	3.1	3.0	2.8	2.8
		tan δ	5200	3100	2300	1700	990	740	580	410	180	120
GR-S (Buna S) uncured[c]	-12	ϵ'/ϵ_o	----	2.5	2.5	2.5	2.5	2.45				
		tan δ	----	13	24	62	80	50				
	26	ϵ'/ϵ_o	2.5	2.5	2.5	2.5	2.50	2.45	2.45		2.45	
		tan δ	6	9	10	18	38	69	71		44	
	80	ϵ'/ϵ_o	----	2.5	2.5	2.5	2.50	2.45	2.45			
		tan δ	----	4	4	4	6	22	60			
GR-S (Buna S) compound[d]	26	ϵ'/ϵ_o	2.66	2.66	2.66	2.65	2.56	2.52	2.52		2.49	2.44
		tan δ	7	9	25	60	120	160	95		56	50
GR-S (Buna S) compound BXG-117-G[e]	25	ϵ'/ϵ_o	2.98	2.97	2.95	2.93	2.91	2.88	2.82		2.78	2.77
		tan δ	10	24	54	100	120	128	80		57	48
	80	ϵ'/ϵ_o	2.84	2.83	2.82	2.81	2.78	2.76	2.72	2.70	2.69	
		tan δ	86	31	26	33	65	110	120	114	83	
Hycar OR Cell-tite[f]	25	ϵ'/ϵ_o	1.41	1.40	1.39	1.38	1.38	1.38	1.38		1.38	1.37
		tan δ	115	58	50	51	56	54	47		39	38
Marbon S, Buna S Hardboard[g]	25	ϵ'/ϵ_o	1.31	1.31	1.31	1.30	1.30	1.30	1.29		1.28	1.27
		tan δ	14	12	22	36	37	28	19		17	16
f. Butyl Rubbers												
GR-I (butyl rubber)[h]	25	ϵ'/ϵ_o	2.39	2.38	2.37	2.36	2.35	2.35	2.35		2.35	2.35
		tan δ	34	35	27	13	10	10	10		9	8
GR-I compound[i]	25	ϵ'/ϵ_o	2.43	2.42	2.41	2.40	2.40	2.40	2.39		2.38	2.38
		tan δ	50	60	58	38	22	15	10		9.3	9.9

a. Solution and suspension of thermosetting resins and vulcanizable elastomers (Goodyear). b. Kearney. c. Copolymer of 75% butadiene and 25% styrene (Rubber Res. Corp.). d. 100 pts. GR-S, 1 pt. stearic acid, 5 pts. Kadox, 5 pts. Captax, 3 pts. Sulfur (Rubber Res. Corp.). e. Sulfur, Selanac, Altax, Methyl Tuads, zinc oxide, Akroflex, Heliozone, Cumar MH, stearic acid, limestone, Marbon S-1, etc. (Gen. Cable). f. Based on butadiene polymer (Sponge Rubber Prod.). g. U.S.Rubber. h. Copolymer of 98-99% isobutylene, 1-2% isoprene (Rubber Res. Corp.). i. 100 pts. GR-I 44-7 M_2K, 5 pts. zinc oxide (from $ZnCO_3$), 1 pt. Tuads, 1.5 pts. sulfur (Rubber Res. Corp.).

Dielectric Data

I. Solids, B. Organic 4. Elastomers (cont.) Values for tan δ are multiplied by 10^4; frequency given in c/s.

g. Nitrile Rubbers	T°C		1×10^2	1×10^3	1×10^4	1×10^5	1×10^6	1×10^7	1×10^8	3×10^8	3×10^9	1×10^{10}	2.5×10^{10}
Kralastic BE Natural[a]	25	ϵ'/ϵ_o	4.80	4.75	4.70	4.40	4.00	3.30	3.00	2.85*	2.80	2.70	
		tan δ	280	130	180	530	1100	1050	560	240*	180	130	
Kralastic BM Natural[a]	25	ϵ'/ϵ_o	4.48	4.45	4.35	4.20	3.78	3.15	2.91	2.83*	2.76	2.70	
		tan δ	130	96	160	490	990	940	490	220*	160	110	
Kralastic D Natural[a]	25	ϵ'/ϵ_o	3.54	3.54	3.51	3.44	3.20	2.90	2.78	2.68*	2.66	2.62	
		tan δ	105	52	82	260	530	550	270	130*	93	65	
Kralastic EBMU Natural[a]	25	ϵ'/ϵ_o	4.47	4.42	4.36	4.29	4.05	3.38	2.98	2.88*	2.80	2.72	
		tan δ	102	75	102	230	720	1300	760	300*	210	140	
Kralastic F Natural[a]	25	ϵ'/ϵ_o	4.20	4.20	4.15	4.00	3.61	3.11	2.87	2.80*	2.70	2.63	
		tan δ	140	65	120	440	940	860	460	190*	150	100	
Royalite 149-11[b]	25	ϵ'/ϵ_o	5.41	5.20	5.12	4.87	4.41	3.62	----	3.18*	3.13	3.03	
		tan δ	320	165	250	590	1080	900	----	260*	200	190	
Expanded Royalite M2198-1[b] (field ⊥ plane of sample)	25	ϵ'/ϵ_o	1.28	1.26	1.25	1.20	1.15	----	----	----	----	----	
		tan δ	175	108	150	240	184	----	----	----	----	----	
(field ∥ plane of sample)	25	ϵ'/ϵ_o	----	----	----	----	----	----	----	1.18*	1.15	1.11	
		tan δ	----	----	----	----	----	----	----	54*	56	36	
Expanded Royalite M2219O[b] (field ⊥ plane of sample)	25	ϵ'/ϵ_o	1.23	1.22	1.19	1.16	1.14	----	----	----	----	----	
		tan δ	157	111	141	162	143	----	----	----	----	----	
(field ∥ plane of sample)	25	ϵ'/ϵ_o	----	----	----	----	----	----	----	1.15*	1.15	1.15	
		tan δ	----	----	----	----	----	----	----	40*	32	31	
h. Neoprene													
Neoprene GN[c]	26	ϵ'/ϵ_o	7.5	6.5	6.2	6.1	5.7	4.7	3.4	----	2.84		
		tan δ	6000	860	330	300	950	2000	1600	----	480		
Neoprene compound[d]	24	ϵ'/ϵ_o	6.70	6.60	6.54	6.47	6.26	5.54	4.5	4.24	4.00	4.00	4
		tan δ	160	110	115	150	380	1190	900	636	339	261	250
	80	ϵ'/ϵ_o	6.70	6.20	6.00	5.86	5.75	5.65	5.2	4.73	4.40	4.40	
		tan δ	2700	430	140	92	93	320	900	1200	960	700	

a. Naugatuck. b. Polystyrene-acrylonitrile and polybutadiene-acrylonitrile (U.S.Rubber). c. Poly-2-chlorobutadiene-1,3 stabilized with tetraethylthiuram disulfide (Du Pont). d. 38% GN, 28.4% Catalpo Clay, 14% blanc fixe, 0.4% Gastex (carbon black), 1.9% paraffin, 3.8% Circo (light process oil), 0.2% stearic acid, 0.8% Neozone A, 0.4% Parazone, 7.6% litharge (DuPont).
*Freq. = 1×10^9.

I. Solids B. Organic 4. Elastomers (cont.) Values for tan δ are multiplied by 10^4; frequency given in c/s.

1. Thiokol	$T^\circ C$		1×10^2	1×10^3	1×10^4	1×10^5	1×10^6	1×10^7	1×10^8	3×10^8	3×10^9	1×10^{10}	2.5×10^{10}
Thiokol, Type FA compound[a]	23	ϵ'/ϵ_0	----	2260	515	200	110	70	30	24	16	14	13.6
		tan δ	----	12900	8000	5100	3900	3200	2800	2800	2200	1500	1000
Thiokol PRI compound[b]	25	ϵ'/ϵ_0	----	17900	15200	7550	740	106	----	20.5*	17.1	8.4	
		tan δ	----	36300	56400	88000	40000	27000	----	2600*	2300	2200	
Thiokol ST compound[c]	25	ϵ'/ϵ_0	----	17300	5100	600	215	90	----	17*	14.8	8.5	
		tan δ	----	10000	17000	36000	18000	7600	----	3000*	3600	4000	
J. Silicone Rubbers													
Silastic 120[d]	25	ϵ'/ϵ_0	5.78	5.76	5.75	5.75	5.75	5.75	5.75	5.74	5.73	5.72	
		tan δ	51	30	16	10	8	12	27	52	254	510	
Silastic 125[d]	25	ϵ'/ϵ_0	7.1	7.1	7.0	7.0	6.9	6.9	6.8	----	7		
		tan δ	40	26	18	13	12	14	30	----	430		
Silastic 150[e]	25	ϵ'/ϵ_0	5.87	5.85	5.79	5.61	5.39	5.09	5.0	5.0	5.0	5.0	
		tan δ	43	48	108	200	270	280	230	250	440	800	
Silastic 152[f]	25	ϵ'/ϵ_0	2.96	2.95	2.95	2.95	2.95	2.95	2.95	2.93*	2.90	2.85	
		tan δ	5.8	5.2	4	3.5	5.4	10.4	20	59*	100	167	
Silastic 160[g]	25	ϵ'/ϵ_0	9.32	9.27	9.2	9.0	8.8	8.4	8.0	7.9	7.8	7.6	
		tan δ	47	31	60	150	250	270	310	310	600	1400	
Silastic 167[h]	25	ϵ'/ϵ_0	8.6	8.6	8.6	8.5	8.5	8.5	8.5	----	8		
		tan δ	52	32	19	12	11	13	16	----	360		
Silastic 180[i]	25	ϵ'/ϵ_0	4.66	4.60	4.56	4.53	4.53	4.52	4.5	4.5	4.5	4.5	
		tan δ	67	63	54	40	30	27	35	56	170	320	
Silastic 181[j]	25	ϵ'/ϵ_0	3.36	3.30	3.26	3.23	3.20	3.19	3.18	3.16	3.11	3.09	
		tan δ	62	67	65	58	37	28	29	36	100	174	
Silastic 250[k]	25	ϵ'/ϵ_0	3.19	3.18	3.17	3.16	3.10	3.07	3.05	3.04	3.02	3.00	
		tan δ	55	30	69	106	64	29	28	44	190	200	

a. 100 pts. Thiokol (organic polysulfide), 10 pts. zinc oxide, 60 pts. semi-reinforcing carbon black, 1 pt. stearic acid, 0.5 pt. diphenylguanidine, 0.35 pt. Altax (Thiokol). b. 100 pts. polysulfide copolymer of bis (2-chloroethyl) formal and ethylene dichloride, 60 pts. carbon blacks, compounding ingredients. (Thiokol). c. 100 pts. polysulfide polymer of bis (2-chloroethyl) formal, 60 pts. carbon blacks, compounding ingredients. (Thiokol). d. 50% siloxane elastomer, 50% TiO_2 (Dow Corning). e. 35% siloxane elastomer, 35% ZnO, 30% $CaCO_3$. (Dow Corning). f. Dow Corning. g. 33% siloxane elastomer, 33% ZnO, 33% TiO_2 (Dow Corning). h. 33% siloxane elastomer, 67% TiO_2 (Dow Corning). i. 35% siloxane elastomer, 35% SiO_2, 30% TiO_2 (Dow Corning). j. 45% siloxane elastomer, 55% SiO_2 (Dow Corning). k. 70% siloxane elastomer, 30% SiO_2 (Dow Corning).
*Freq. = 1×10^9.

Dielectric Data

I. Solids, B. Organic 4. Elastomers (cont.)   Values for tan δ are multiplied by 10^4; frequency given in c/s.

J. Silicone Rubbers (cont.)

	T°C		1×10^2	1×10^3	1×10^4	1×10^5	1×10^6	1×10^7	1×10^8	3×10^8	3×10^9	1×10^{10}
Silastic X4342[a]	25	ϵ'/ϵ_0	3.22	3.18	3.16	3.16	3.14	3.12	3.11	3.08	3.08	3.00
		tan δ	51	64	60	48	29	20	23	35	83	110
Silastic 6167[b]	25	ϵ'/ϵ_0	10.1	10.1	10.0	10	10	10	10	10	10	10
		tan δ	41	26	17	13	9.5	10	27	55	450	1000
Silastic 6181[c]	25	ϵ'/ϵ_0	3.66	3.53	3.42	3.35	3.31	3.30	3.27	3.25	3.23	3.20
		tan δ	200	250	190	100	57	34	32	36	88	160
Silastic X-6734[d]	25	ϵ'/ϵ_0	3.21	3.12	3.12	3.12	3.10	3.08	3.08	3.08	3.01	3.00
		tan δ	28	19	31	53	37	23	29	44	144	213
Silastic 7181[c]	25	ϵ'/ϵ_0	3.72	3.55	3.45	3.37	3.33	3.31	3.31	3.28	3.20	3.15
		tan δ	210	250	190	97	53	29	24	29	75	170
SE-450[e]	25	ϵ'/ϵ_0	3.09	3.08	3.08	3.08	3.07	3.06	3.05	3.00*	2.97	2.88
		tan δ	16	7.2	5.3	7	11	17	30	74*	158	183
SE-460[e]	25	ϵ'/ϵ_0	3.14	3.12	3.11	3.10	3.10	3.09	3.07	3.05*	3.02	2.94
		tan δ	56	54	44	25	12	15	23	57*	98	180
SE-550[e]	25	ϵ'/ϵ_0	3.14	3.12	3.10	3.10	3.10	3.08	3.06	3.02*	3.00	2.94
		tan δ	13	7.8	6.5	7	9.5	16	31	84*	143	195
SE-972[e]	25	ϵ'/ϵ_0	3.40	3.35	3.25	3.22	3.20	3.18	3.16	3.15*	3.13	3.08
		tan δ	64	67	58	46	30	27	32	65*	97	146

5. Natural Resins

	T°C		1×10^2	1×10^3	1×10^4	1×10^5	1×10^6	1×10^7	1×10^8	3×10^8	3×10^9	1×10^{10}
Amber[f]	25	ϵ'/ϵ_0	2.7	2.7	2.7	2.7	2.65	2.65	----	2.6	2.6	
		tan δ	12.5	18	31	43	56	68	----	82	90	
	80	ϵ'/ϵ_0	2.85	2.85	2.85	2.8	2.8	2.75	----	2.65	2.60	
		tan δ	24	19	22	40	70	93	----	110	115	
Shellac, natural XL[g]	28	ϵ'/ϵ_0	3.86	3.81	3.75	3.66	3.47	3.26	3.10	----	2.86	
		tan δ	65	74	128	225	310	350	300	----	254	
	70	ϵ'/ϵ_0	6.50	5.65	5.10	4.60	4.33	4.00	3.80	----	3.45	
		tan δ	1050	860	650	460	400	540	700	----	732	

a. 50% siloxane elastomer, 50% SiO_2 (Dow Corning). b. 33% siloxane elastomer, 67% TiO_2 (Dow Corning). c. 45% siloxane elastomer, 55% SiO_2 (Dow Corning). d. 70% siloxane elastomer, 30% SiO_2 (Dow Corning). e. General Electric. f. Fossil resin (Amber Mines). g. Contains ca. 3.5% wax (Zinsser).

*Freq. = 1×10^9.

Tables of Dielectric Materials

I. Solids B. Organic (Cont.) Values for tan δ are multiplied by 10^4; frequency given in c/s.

5. Natural Resins (cont.)

	T°C		1×10^2	1×10^3	1×10^4	1×10^5	1×10^6	1×10^7	1×10^8	3×10^8	3×10^9	1×10^{10}	2.5×10^{10}
Shellac, natural Zinfo[a]	27	ϵ'/ϵ_0	3.84	3.82	3.75	3.66	3.50	3.30	3.1	----	2.86		
		tan δ	58	64	132	250	313	341	345	----	290		
Shellac, pure C garnet[b]	28	ϵ'/ϵ_0	3.99	3.92	3.84	3.69	3.50	3.32	3.2	----	2.94		
		tan δ	137	102	131	205	304	347	340	----	270		
Shellac, garnet dewaxed[c]	26	ϵ'/ϵ_0	3.60	3.56	3.48	3.40	3.30	3.20	3.05	----	2.75		
		tan δ	58	64	118	205	285	317	310	----	267		
Chemlac B-3[d]	23	ϵ'/ϵ_0	3.19	3.12	3.09	3.04	3.00	2.98	2.98	2.91	2.90	2.87	
		tan δ	105	85	86	91	100	110	118	125	135	120	

6. Asphalts and Cements

	T°C		1×10^2	1×10^3	1×10^4	1×10^5	1×10^6	1×10^7	1×10^8	3×10^8	3×10^9	1×10^{10}	2.5×10^{10}
Gilsonite[e]	26	ϵ'/ϵ_0	2.68	2.66	2.63	2.61	2.58	2.57	2.56	----	----	2.55	
		tan δ	58	35	25	19	16	13	11	----	----	9.3	
Millimar[f]	22	ϵ'/ϵ_0	2.65	2.65	2.65	2.64	2.64	2.64	2.64	----	2.5	2.5	
		tan δ	39	35	26	14	11	10	11	----	11	12	
Cenco Sealstix[g]	23	ϵ'/ϵ_0	3.90	3.75	3.55	3.37	3.23	3.13	----	----	2.96		
		tan δ	452	335	275	240	240	272	----	----	210		
Plicene Cement[h]	25	ϵ'/ϵ_0	2.48	2.48	2.48	2.48	2.48	2.48	2.47	----	2.40	2.35	
		tan δ	43.9	35.5	28.9	25.9	25.5	22.2	15	----	7.8	6.8	

7. Waxes

	T°C		1×10^2	1×10^3	1×10^4	1×10^5	1×10^6	1×10^7	1×10^8	3×10^8	3×10^9	1×10^{10}	2.5×10^{10}
Acrawax C[i]	24	ϵ'/ϵ_0	2.60	2.58	2.57	2.55	2.54	2.53	2.52	----	2.48	2.45	2.44
		tan δ	157	68	45	30	20	15	12	----	15	19	21
	85	ϵ'/ϵ_0	----	5.0	3.2	2.63	2.50	2.46	----	----	2.45		
		tan δ	----	7000	3500	560	180	130	----	----	228		
Apiezon Wax "W"[j]	22	ϵ'/ϵ_0	2.75	2.69	2.66	2.64	2.63	2.63	----	----	2.62		
		tan δ	186	120	70	39	25	19	----	----	16		
	70	ϵ'/ϵ_0	3.29	3.14	2.97	2.88	2.82	2.79	----	----			
		tan δ	450	420	340	210	110	61	----	----			

a. Contains ca. 3.5% wax (Zinsser). b. Natural, ca. 2% wax (Zinsser). c. Natural, wax-free (Zinsser). d. From zein and rosin (Filtered Rosin Products). e. 99.9% natural bitumen (U.S.Rubber). f. Asphaltic product (U.S.Rubber). g. De Khotinsky Cement (Central Scientific). h. Central Scientific. i. Cetylacetamide (Glyco). j. Shell Oil.

Dielectric Data

I. Solids B. Organic (cont.)

7. Waxes (cont.)

Values for tan δ are multiplied by 10^4; frequency given in c/s.

Material	T °C		1×10^2	1×10^3	1×10^4	1×10^5	1×10^6	1×10^7	1×10^8	3×10^8	3×10^9	1×10^{10}	2.5×10^{10}
Beeswax, white[a]	23	ϵ'/ϵ_0	2.65	2.63	2.56	2.48	2.43	2.41	2.39	----	2.35	2.35	
		tan δ	140	118	266	190	84	68	60	----	50	48	
	55	ϵ'/ϵ_0	2.52	2.50	2.46	2.43	2.39	2.36	2.34	----	2.31		
		tan δ	82	44	25	25	35	58	125	----	216		
Beeswax, yellow[b]	23	ϵ'/ϵ_0	2.73	2.66	2.59	2.53	2.49	2.45	2.42	----	2.39	2.36	
		tan δ	240	140	240	150	92	84	90	----	75	62	
Cerese Wax AA[c]	24*	ϵ'/ϵ_0	2.35	2.35	2.35	2.34	2.33	2.32	----	----	2.29	2.27	
		tan δ	5	6	7	8	8	7	----	----	4.6	5.0	
	46*	ϵ'/ϵ_0	2.35	2.35	2.35	2.34	2.33	2.32	----	----	2.28	2.27	
		tan δ	<5	<3	3	5	8	10	----	----	6.2	6	
Cerese Wax, brown[c]	22	ϵ'/ϵ_0	----	2.28	2.28	2.28	2.28	2.28	2.27	----	2.25		
		tan δ	----	5	10	13	11	7	2	----	2.8		
Ceresin, white[d]	25	ϵ'/ϵ_0	2.3	2.3	2.3	2.3	2.3	2.3	2.3	----	2.25	2.24	
		tan δ	8	6	5	5	4	4	4	----	4.6	6.5	
Ceresin, yellow[e]	20	ϵ'/ϵ_0	----	2.25	2.25	2.25	2.25	2.25	2.25	----	2.25	2.25	
		tan δ	----	4	5	7	7	5	4	----	4.2	4	
Estawax[f]	25	ϵ'/ϵ_0	2.30	2.30	2.30	2.30	2.30	2.30	2.30	2.30	2.29	2.28	2.28
		tan δ	7	<2	2	4	5	4	2.5	2.3	2.9	3.7	4.6
	80	ϵ'/ϵ_0	2.18	2.18	2.18	2.18	2.18	2.18	2.18	2.18	2.17	2.15	
		tan δ	11	7	3	<2	<2	<4	5	6.4	7.9	8.5	
Halowax #1001[g] cold-molded	26	ϵ'/ϵ_0	5.45	5.45	5.45	5.40	5.40	5.30	4.2	3.46	2.92	2.86	2.84
		tan δ	18	17	7	8	45	450	2700	1620	583	296	200
hot-molded	25	ϵ'/ϵ_0	3.78	3.78	3.76	3.74	3.70	3.62	3.37	3.17	2.57		
		tan δ	16	8	7	8.6	32	400	1200	1210	496		

a. Bromund. b. Candy. c. Mainly petroleum aliphatic hydrocarbons (Socony-Vacuum). d. Vegetable and mineral waxes (Kuhne-Libby).
e. Vegetable and mineral waxes (Mitchell-Rand). f. Long-chain, singly unsaturated (Lovell). g. Tri- and tetrachloronaphthalenes (Bakelite).
*Similar data were obtained after incorporation of 0.5% phenyl mercuric stearate.

I. Solids B. Organic (Cont.)

Values for tan δ are multiplied by 10^4; frequency given in c/s.

7. Waxes (cont.)

	T°C		1×10^2	1×10^3	1×10^4	1×10^5	1×10^6	1×10^7	1×10^8	3×10^8	3.10^9	1×10^{10}	2.5×10^{10}
Halowax #11-314[a]	23	ϵ'/ϵ_0	3.11	3.04	3.01	2.99	2.98	2.96	2.93	2.92	2.89	2.87	
		tan δ	700	110	17	5	3	7	17	38	37	9.5	
	45	ϵ'/ϵ_0	3.11	3.04	3.01	2.99	2.98	2.96	2.93	----	----	2.87	
		tan δ	700	110	17	5	3	3	5	----	----	31	
Kel-F Wax #150[b]	25	ϵ'/ϵ_0	3.01	2.97	2.91	2.75	2.52	2.35	2.25	2.24	2.23	2.22	
		tan δ	83	93	252	510	540	425	270	218	113	90	
Opalwax[c]	24	ϵ'/ϵ_0	13.2	10.3	7.0	4.3	3.2	2.9	2.7	----	2.55	2.52	2.5
		tan δ	1250	2100	2900	2700	1450	580	270	----	167	160	160
	47	ϵ'/ϵ_0	14.2	11.4	8.2	5.4	3.7	3.0	2.7	----	----	2.47	
		tan δ	450	1300	2600	3000	2100	860	430	----	----	236	
Ozokerite[d]	20	ϵ'/ϵ_0	2.26	2.26	2.26	2.26	2.26	2.26	2.26	----	2.26		
		tan δ	6	6	8	10	9	7	6	----	6		
Paraffin Wax 132° ASTM[e]	25	ϵ'/ϵ_0	2.25	2.25	2.25	2.25	2.25	2.25	2.25	----	2.25	2.24	2.2
		tan δ	<2	<2	<2	<2	<2	<2	<2	----	2.0	2.1	<3
	81	ϵ'/ϵ_0	2.02	2.02	2.02	2.02	2.02	2.02	----	----	2.00		
		tan δ	5	1.2	.5	<2	<2	<3	----	----	5.2		
Paraffin Wax 135° AMP[f]	24	ϵ'/ϵ_0	2.25	2.25	2.25	2.25	2.25	2.25	----	----	2.22	2.22	
		tan δ	12.5	5.5	<3	<2	<2	<4	----	----	1	2	
Parowax[g]	20	ϵ'/ϵ_0	----	2.25	2.25	2.25	2.25	2.25	2.25	2.25	2.25	2.25	
		tan δ	----	2	2	2	3	2	1.4	1	2	2.5	
Polinel[h]	25	ϵ'/ϵ_0	2.89	2.87	2.87	2.85	2.84	2.80	2.72	2.62*	2.59	2.58	
		tan δ	67	44	36	43	82	166	240	180*	195	199	
Sealing wax, Red Express[i]	25	ϵ'/ϵ_0	3.68	3.52	3.40	3.32	3.29	3.27	3.2	----	3.09		
		tan δ	249	150	99	79	80	100	120	----	122		
Thermoplastic Composition 1766MX[j]	25	ϵ'/ϵ_0	3.77	3.72	3.68	3.63	3.60	3.58	3.56	----	----	3.58	
		tan δ	122	106	87	60	44	32	28	----	----	20	
Thermoplastic Composition 3738[j]	25	ϵ'/ϵ_0	2.65	2.60	2.57	2.56	2.55	2.54	2.54	2.52*	2.52	2.49	
		tan δ	112	91	67	41	25	16	15	13*	13	13	

a. 80% 1,4-, 10% 1,5-, 10% 1,2-dichloronaphthalenes (Bakelite). b. Polychlorotrifluoroethylene (Kellogg). c. Mainly 12-hydroxystearin (DuPont). d. Natural paraffin (Allison). e. Mainly C_{22} to C_{29} aliphatic, saturated hydrocarbons (Stand. Oil N.J.). f. Gulf. g. Paraffin wax (Socony Vacuum). h. Lovell. i. Dennison. j. Mitchell-Rand.
*Freq. = 1×10^9.

Dielectric Data

I. Solids B. Organic (cont.)

7. Waxes (cont.)

Values for tan δ are multiplied by 10^4; frequency given in c/s.

	T°C		1×10^2	1×10^3	1×10^4	1×10^5	1×10^6	1×10^7	1×10^8	3×10^8	3×10^9	1×10^{10}	2.5×10^{10}
Thermoplastic composition 3767A[a]	25	ϵ'/ϵ_0	2.84	2.79	2.76	2.64	2.59	2.54	2.51	2.48	2.45	2.45	
		tan δ	140	160	170	120	98	78	69	59	32	27	
Vistawax[b]	25	ϵ'/ϵ_0	2.34	2.34	2.34	2.34	2.34	2.32	2.30	----	2.27	2.26	
		tan δ	2	3	9.2	12.1	13.3	13.6	13.3	----	9.0	5	
Wax 3760[c]	25	ϵ'/ϵ_0	2.36	2.36	2.36	2.36	2.36	2.36	2.36	2.31*	2.31	2.31	
		tan δ	21.7	8.8	7.1	5.6	4.3	2.7	2.6	3.6*	4.9	14.2	
Wax S-1167[d]	25	ϵ'/ϵ_0	19.9	10.2	5.9	4.1	3.3	3.1	2.9	2.6	2.6	2.6	
		tan δ	7200	3900	2300	1700	1000	600	470	300	190	92	
Wax S-1184[d]	25	ϵ'/ϵ_0	2.43	2.43	2.43	2.40	2.40	2.40	2.37	2.34*	2.30	2.21	
		tan δ	29.4	21.5	21.5	16.4	11.3	9.3	13.8	20*	24	54	
Wax Compound F-590[e]	26	ϵ'/ϵ_0	2.37	2.34	2.32	2.32	2.32	2.32	2.32	----	2.32		
		tan δ	80	99	78	44	20	8	5	----	6.2		
Wax Compound #1340[e]	26	ϵ'/ϵ_0	2.30	2.30	2.30	2.30	2.30	2.30	2.30	----	2.25		
		tan δ	1	2	2	<5	4	4	4	----	4.2		

8. Woods

	T°C		1×10^2	1×10^3	1×10^4	1×10^5	1×10^6	1×10^7	1×10^8	3×10^8	3×10^9	1×10^{10}	2.5×10^{10}
Balsa	26	ϵ'/ϵ_0	1.4	1.4	1.4	1.4	1.37	1.35	1.30	----	1.22	1.20	1.87
		tan δ	50	40	43	77	120	135	135	----	100	83	260
Yellow Birch	25	ϵ'/ϵ_0	2.91	2.88	2.82	2.78	2.70	2.60	2.47	2.40	2.13	1.95	1.78
		tan δ	72	90	140	220	290	360	400	390	330	280	320
Fir, Douglas (field ⊥ to grain)	25	ϵ'/ϵ_0	2.04	2.00	1.97	1.95	1.93	1.90	1.88	1.86	1.82	1.80	1.5
		tan δ	48	80	130	190	260	310	330	320	270	290	220
Fir, Douglas, plywood (field ⊥ to grain)	25	ϵ'/ϵ_0	2.1	2.1	2.05	1.95	1.90	1.80	----	1.7	----	----	
		tan δ	115	105	130	170	230	320	----	360	----	----	3.2
Fir, Douglas, staypak (field ⊥ to grain)	25	ϵ'/ϵ_0	7.3	7.05	6.9	6.65	6.3	5.6	4.6	4.0	3.5	3.3	460
		tan δ	82	130	200	320	500	750	800	850	700	550	1.6
Mahogany (field ⊥ to grain)	25	ϵ'/ϵ_0	2.42	2.40	2.36	2.30	2.25	2.17	2.07	2.01	1.88	1.7	200
		tan δ	86	120	150	195	250	310	320	300	250	210	1.4
Poplar, yellow (field ⊥ to grain)	25	ϵ'/ϵ_0	1.84	1.79	1.78	1.76	1.75	1.70	----	1.60	1.50	1.42	170
		tan δ	42	54	83	133	190	220	----	200	150	200	

a. Mitchell-Rand. b. Polybutene (Cantol Wax). c. Mitchell-Rand. d. Glyco. e. Zophar Mills.
*Freq. = 1×10^9.

Tables of Dielectric Materials

I. Solids B. Organic (cont.) Values for tan δ are multiplied by 10^4; frequency given in c/s.

9. Miscellaneous

	T°C		1×10^2	1×10^3	1×10^4	1×10^5	1×10^6	1×10^7	1×10^8	3×10^8	3×10^9	1×10^{10}	2.5×10^{10}
Paper, Royalgrey[a]	25	ϵ'/ϵ_o	3.30	3.29	3.22	3.10	2.99	2.86	2.77	2.75	2.70	2.62	
		tan δ	58	77	117	200	380	570	660	660	560	403	
	82	ϵ'/ϵ_o	3.57	3.52	3.49	3.40	3.31	3.14	3.08	3.00	2.94	2.84	
		tan δ	170	74	61	85	230	440	630	720	800	827	
Leather, sole, dried	25	ϵ'/ϵ_o	4.1	3.9	3.6	3.4	3.2	3.1	3.1				
		tan δ	450	350	300	280	280	300	380				
Leather, sole, ca. 15% moisture	25	ϵ'/ϵ_o	38	14.0	9.3	6.9	5.6	4.9	4.5				
		tan δ	14000	7000	3700	2200	1400	1000	1000				
Soap, Ivory[b]	25	ϵ'/ϵ_o	----	----	----	----	----	----	----	----	2.96		
		tan δ	----	----	----	----	----	----	----	----	1765		
Steak (bottom round)	25	ϵ'/ϵ_o	----	----	24400	----	197	50	----	50	40	30	15
		tan δ	----	----	405000	----	610000	260000	----	7800	3000	3700	4000
Suet	25	ϵ'/ϵ_o	----	750	210	----	14	4.5	2.6	2.5	2.5	2.5	2.4
		tan δ	----	30000	25000	----	17000	9300	1500	1200	700	500	500

a. Rogers. b. Procter and Gamble.

II. LIQUIDS

Values for tan δ are multiplied by 10^4; frequency given in c/s.

A. Inorganic

	T°C		1×10^5	1×10^6	1×10^7	1×10^8	3×10^8	3×10^9	1×10^{10}	2.5×10^{10}**
Water, conductivity[a]	1.5	ϵ'/ϵ_o	87.0	87.0	87	87	86.5	80.5	38	15
		tan δ	1900	190	20	70	320	3100	10300	4250
	5	ϵ'/ϵ_o	----	85.5	----	----	85.2	80.2	41	17.5
		tan δ	----	220	----	----	273	2750	9500	3950
	15	ϵ'/ϵ_o	----	81.7	----	----	81.0	78.8	49	25
		tan δ	----	310	----	----	210	2050	7000	3300
	25	ϵ'/ϵ_o	78.2	78.2	78.2	78	77.5	76.7	55	34
		tan δ	4000	400	46	50	160	1570	5400	2650
	35	ϵ'/ϵ_o	----	74.8	----	----	74.0	74.0	58	41
		tan δ	----	485	----	----	125	1270	4400	2150
	45	ϵ'/ϵ_o	----	71.5	----	----	71.0	70.7	59	46
		tan δ	----	590	----	----	105	1060	4000	2750
	55	ϵ'/ϵ_o	----	68.2	----	----	68	67.5	60	49
		tan δ	----	720	----	----	92	890	3600	2450
	65	ϵ'/ϵ_o	----	64.8	----	----	64.5	64.0	59	50.5
		tan δ	----	865	----	----	84	765	3200	1250
	75	ϵ'/ϵ_o	----	61.5	----	----	61	60.5	57	51.5
		tan δ	----	1030	----	----	77	660	2800	1050
	85	ϵ'/ϵ_o	58	58	58	58	57	56.5	54	
		tan δ	12400	1240	125	30	73	547	2600	
	95	ϵ'/ϵ_o	----	55	----	----	52	52		
		tan δ	----	1430	----	----	70	470		

Aqueous sodium chloride[b]

	T°C		1×10^5	1×10^6	1×10^7	1×10^8	3×10^8	3×10^9	1×10^{10}	2.5×10^{10}**
0.1 molal solution	25	ϵ'/ϵ_o	78.2*	----	----	----	76	75.5	54	
		tan δ	24,000,000	----	----	----	7800	2400	5600	
0.3 molal solution	25	ϵ'/ϵ_o	78.2*	----	----	----	71	69.3	52	
		tan δ	63,000,000	----	----	----	24000	4350	6050	
0.5 molal solution	25	ϵ'/ϵ_o	78.2*	----	----	----	69	67.0	51	
		tan δ	99,000,000	----	----	----	39000	6250	6300	
0.7 molal solution	25	ϵ'/ϵ_o	78.2*	----	----	----	----	----	50	
		tan δ	130,000,000	----	----	----	----	----	6600	

a. Research Laboratory of Physical Chemistry, M.I.T.   b. NaCl, Mallinckrodt's Analytical Reagent.
*. ϵ'/ϵ_o of conductivity water assumed for purpose of calculating tan δ from conductivity measurements.
** Data of Collie, Hasted and Ritson, Proc. Phys. Soc. 60, 145 (1948).

II. Liquids B. Organic

Values for tan δ are multiplied by 10^4; frequency given in c/s.

1. Aliphatic		$T°C$	1×10^2	1×10^3	1×10^4	1×10^5	1×10^6	1×10^7	1×10^8	3×10^8	3×10^9	1×10^{10}
Heptane[a]	ϵ'/ϵ_0	25	1.971	1.971	1.971	----	----	----	----	1.97	1.97	1.97
	tan δ		< 3	< 0.4	< 0.4	----	----	----	----	< 2.5	1	16
Methyl alcohol[b]	ϵ'/ϵ_0	25	----	----	----	----	31	31.0	31.0	30.9	23.9	8.9
	tan δ		----	----	----	----	2000	260	380	800	6400	8100
Ethyl alcohol[c]	ϵ'/ϵ_0	25	----	----	----	----	24.5	24.1	23.7	22.3	6.5	1.7
	tan δ		----	----	----	----	900	330	620	2700	2500	680
n-Propyl alcohol[d]	ϵ'/ϵ_0	25	----	----	----	----	20.1	20.1	19.0	16.0	3.7	2.3
	tan δ		----	----	----	----	180	170	2000	4200	6700	900
n-Butyl alcohol[d]	ϵ'/ϵ_0	25	----	----	----	17.4	17.4	17.4	14.8	11.5	3.5	
	tan δ		----	----	12000	1000	95	240	2700	5500	4700	
Ethylene Glycol[e]	ϵ'/ϵ_0	25	----	----	42	41	41	41	41	39	12	7
	tan δ		----	----	30000	3000	300	80	450	1600	10000	7800
Butyraldehyde[e]	ϵ'/ϵ_0	25	----	6.7	6.7	----	6.7					
	tan δ		----	51000	5200	----	50					
Dibutyl sebacate[f]	ϵ'/ϵ_0	26	4.92	4.73	4.63	4.60	4.58	4.56	----	4.55	3.80	
	tan δ		790	110	14	3	3	16	----	383	2120	
Dioctyl sebacate[f]	ϵ'/ϵ_0	26	4.05	4.05	4.03	4.02	4.01	4.00	----	3.77	2.75	
	tan δ		85	12	5	4	7	55	----	1040	1290	
Carbon tetrachloride[g]	ϵ'/ϵ_0	25	2.17	2.17	2.17	2.17	2.17	2.17	2.17	2.17	2.17	2.17
	tan δ		60	8	0.4	< 0.4	< 0.4	< 2	< 2	< 1	4	16
Tetrachloroethylene[h]	ϵ'/ϵ_0	25	2.28	2.28	2.28	2.28	2.28	2.28	----	----	2.28	
	tan δ		15	2	0.7	1	2	2	----	----	10	
Hexachlorobutadiene[i]	ϵ'/ϵ_0	25	2.55	2.55	2.55	2.55	2.55	2.55	2.55	2.55	2.51	2.47
	tan δ		3	1.5	1	< 1	< 1	2	16	55	240	130

a. Practical (Solvents Supply Laboratory, M.I.T.). b. Absolute, analytical grade (Mallinckrodt). c. Absolute (U.S. Industrial Chemicals). d. Eastman Kodak. Dried and refractionated, Lab. Ins. Res. e. Eastman Kodak. f. Resinous Products. g. Purified, Lab. Ins. Res. h. Eastman Kodak, fractionated Lab. Ins. Res. i. Hooker, fractionated Lab. Ins. Res.

Dielectric Data

II. Liquids B. Organic (cont.)

Values for tan δ are multiplied by 10^4; frequency given in c/s.

	T°C		1×10^2	1×10^3	1×10^4	1×10^5	1×10^6	1×10^7	1×10^8	3×10^8	1×10^9	3×10^9	1×10^{10}
1. Aliphatic (Cont.)													
Dichloropentanes #40[a]	25	ϵ'/ϵ_0	----	334	17.1	8.65	----	7.76	----	7.57	----	6.81	
		tan δ	----	5200000	1060000	135000	----	2700	----	840	----	1980	
Dichloropentanes #14[a]	25	ϵ'/ϵ_0	----	----	8.24	8.06	8.05	8.05	----	2.53	----	2.34	2.15
		tan δ	----	----	7500	720	85	16	----	380	----	870	860
Kel-F Oil, Grade #1[b]	25	ϵ'/ϵ_0	2.61	2.61	2.61	2.60	2.61	2.61	2.58	2.48	----	2.22	2.12
		tan δ	22	2.3	.3	1.3	2.0	17	140	380	----	870	420
Kel-F Oil, Grade #3[b]	25	ϵ'/ϵ_0	2.73	2.73	2.73	2.73	2.73	2.73	----	2.48	----	2.22	2.12
		tan δ	7.9	.8	<.3	1.3	7.3	65	----	930	----	696	420
Kel-F Oil, Grade #10[b]	25	ϵ'/ϵ_0	2.83	2.83	2.83	2.83	2.83	2.78	----	2.40	----	2.26	2.13
		tan δ	2.1	.45	1.1	6.1	68	393	----	820	----	300	274
Kel-F Grease #40[b]	25	ϵ'/ϵ_0	2.88	2.88	2.88	2.86	2.78	2.57	----	2.25	----	2.20	2.19
		tan δ	5.4	3.8	18	111	430	570	----	350	----	140	106
Perfluorodihexyl ether (experimental)[c]	25	ϵ'/ϵ_0	1.871	1.871	1.871	----	1.87	1.87	----	1.86	----	1.86	1.85
		tan δ	0.8	0.3	<.4	----	.3	3.5	----	55	----	122	920
Heptacosefluorotributyl amine (exp.)[c]	25	ϵ'/ϵ_0	1.853	1.853	1.853	1.853	1.85	1.85	1.85	1.85	----	1.85	1.85
		tan δ	.15	<.1	<1	<1	<1	1.2	11	25	----	28	20
2. Aromatic													
HB-40 oil[d]	25	ϵ'/ϵ_0	2.59	2.59	2.59	2.59	2.58	2.57	2.54	2.48	----	2.40	2.34
		tan δ	1.3	<.4	<.4	<3	13	76	160	93	----	30	17
Pyranol 1467[e]	25	ϵ'/ϵ_0	4.42	4.40	4.40	4.40	4.40	4.40	4.08	3.19	----	2.84	2.62
		tan δ	36	3	<4	3.6	25	260	1300	1500	----	1200	740
Aroclor 1221[f]	25	ϵ'/ϵ_0	----	----	----	----	4.55	4.53	4.35	3.85	3.10	2.75	2.65
		tan δ	----	----	----	----	9	80	800	2000	1900	1030	590
Aroclor 1232[g]	25	ϵ'/ϵ_0	----	----	----	----	5.88	5.85	4.60	3.65	3.00	2.82	2.75
		tan δ	----	----	----	----	22	220	2600	3240	2000	95	51
Aroclor 1242[h]	25	ϵ'/ϵ_0	----	----	----	----	5.89	5.85	3.50	2.93	2.80	2.72	2.69
		tan δ	----	----	----	----	70	700	3000	1850	780	400	223
Aroclor 1248[i]	25	ϵ'/ϵ_0	----	----	----	5.58	5.57	5.10	2.80	2.76	2.78	2.71	2.68
		tan δ	----	----	----	26	260	1900	1000	410	300	151	118
Aroclor 1254[j]	25	ϵ'/ϵ_0	----	----	5.05	5.04	3.70	2.90	2.75	2.72	2.71	2.70	2.69
		tan δ	----	----	42	415	2380	1130	170	78	52	44	40

a. Sharples. b. Polychlorotrifluoroethylene (Kellogg). c. Minn. Mining. d. Monsanto. e. Chlorinated benzenes and diphenyls (Gen. Elec.). f. Monochlorobiphenyl (Monsanto). g. Dichlorobiphenyl (Monsanto). h. Trichlorobiphenyl (Monsanto). i. Tetrachlorobiphenyl (Monsanto). j. Pentachlorobiphenyl (Monsanto).

II. Liquids B. Organic (cont.)

2. Aromatic (cont.)

Values for tan δ are multiplied by 10^4; frequency given in c/s.

	T°C		1×10^2	1×10^3	1×10^4	1×10^5	1×10^6	1×10^7	1×10^8	3×10^8	3×10^9	1×10^{10}
Pyranol 1476[a]	26	ϵ'/ϵ_0	5.04	5.04	5.04	4.91	3.85	2.81	----	2.74	2.70	2.70
		tan δ	.6	6	54	480	2500	1100	----	130	42	38
Aroclor 1260[b]	25	ϵ'/ϵ_0	4.33	4.26	3.46	2.89	2.83	2.79	----	2.73	----	2.72
		tan δ	86	593	1500	645	185	37	----	4.6	----	9.4
Aroclor 1262[c]	25	ϵ'/ϵ_0	4.03	3.44	2.86	2.76	2.75	2.75	----	----	2.75	2.75
		tan δ	392	1400	683	186	46	15.7	----	----	6.0	4.9
Aroclor 5442[d]	25	ϵ'/ϵ_0	----	----	----	----	----	----	----	----	2.78	2.78
		tan δ	----	----	----	----	----	----	----	----	7.2	7.2
Halowax oil 1000[e]	25	ϵ'/ϵ_0	4.77	4.76	4.76	4.75	4.74	----	----	4.67	3.52	2.99
		tan δ	490	50	5	<1	<2	----	----	500	2500	1900
	80	ϵ'/ϵ_0	4.30	4.30	4.30	4.29	4.26	----	----	4.16	3.96	3.30
		tan δ	7800	930	80	15	11	----	----	170	1400	2800
Nitrobenzene[f]	25	ϵ'/ϵ_0	----	----	36	36	35.6	34.4	----	----	31.1	----
		tan δ	----	----	3900	350	80	90	----	----	1660	----
Styrene N-100[g]	22	ϵ'/ϵ_0	2.40	2.40	2.40	2.40	2.40	2.40	----	----	2.40	2.36
		tan δ	38	5	<1	<3	<3	<3	----	----	20	58
Styrene N-100,[g] purified[h]	25	ϵ'/ϵ_0	2.40	2.40	2.40	2.40	----	----	----	----	2.38	2.38
		tan δ	15	1.8	<0.5	<3	----	----	----	----	13	37
Styrene N-100,[g] sat. with water	27	ϵ'/ϵ_0	2.40	2.40	2.40	----	----	----	----	----	2.40	----
		tan δ	47	4.8	1.8	----	----	----	----	----	14	----
Styrene dimer[g]	25	ϵ'/ϵ_0	----	----	----	2.7	2.7	2.7	2.7	----	2.5	----
		tan δ	----	----	----	9	3	2	18	----	110	----
2,5-Dichlorostyrene[i]	24	ϵ'/ϵ_0	2.58	2.58	2.58	2.58	2.58	2.58	----	2.58	2.52	----
		tan δ	85	9	1	<5	<3	<3	----	37	114	----
Pyranol 1478[j]	26	ϵ'/ϵ_0	4.55	4.53	4.53	4.53	4.53	4.53	----	4.50	3.80	----
		tan δ	150	14	2	<5	2	12	----	381	2310	----
β-chloroethyl-2,5-dichlorobenzene[k]	24	ϵ'/ϵ_0	----	6.05	5.45	5.22	5.20	5.20	5.20	5.18	3.31	----
		tan δ	----	5000	500	25	5	30	430	1100	3240	----
Ethylpolychlorobenzene[m] (discontinued)	25	ϵ'/ϵ_0	4.12	4.12	4.12	4.12	4.12	4.12	4.10	3.84	2.70	2.55
		tan δ	13.6	1.3	1	<2	6	55	550	1400	1260	660
	100	ϵ'/ϵ_0	3.62	3.62	3.62	3.62	3.62	3.62	----	3.46	3.12	2.59
		tan δ	380	40	3.7	<1	1	5	----	220	1720	3800

a. Isomeric pentachlorodiphenyls (Gen. Elec.). b. Hexachlorobiphenyl (Monsanto). c. Heptachlorobiphenyl (Monsanto). d. Pentachloroterphenyls (Monsanto). e. 50% mono-, 40% di- and trichloronaphthalenes (Bakelite). f. Purified Lab. Ins. Res. g. Dow Chemical Co. h. Fractionated (Lab. Ins. Res.). i. Monsanto (fractionated Lab. Ins. Res.). j. Isomeric trichlorobenzenes (Gen. Elec.). k. Monsanto. m. DuPont.

II. Liquids, B. Organic (cont.)

Values for tan δ are multiplied by 10^4; frequency given in c/s.

3. Petroleum Oils

		T°C		1×10^2	1×10^3	1×10^4	1×10^5	1×10^6	1×10^7	1×10^8	3×10^8	3×10^9	1×10^{10}
Aviation gasoline 100 octane		25	ϵ'/ϵ_0	----	----	1.94	----	----	----	----	1.94	1.92	
			tan δ	----	----	1	----	----	----	----	.8	14	
Aviation gasoline 91 octane		25	ϵ'/ϵ_0	----	----	1.95	----	----	----	----	1.95	1.94	
			tan δ	----	----	4	----	----	----	----	.4	11.5	
Jet fuel JP-1		25	ϵ'/ϵ_0	----	----	2.12	----	----	----	----	2.12	2.09	
			tan δ	----	----	<1	----	----	----	----	12	68	
Jet fuel JP-3		25	ϵ'/ϵ_0	----	----	2.08	----	----	----	----	2.08	2.04	
			tan δ	----	----	<1	----	----	----	----	7	55	
Kerosene		25	ϵ'/ϵ_0	----	----	----	----	----	----	----	----	2.09	
			tan δ	----	----	----	----	----	----	----	----	45	
Vaseline		25	ϵ'/ϵ_0	2.16	2.16	2.16	2.16	2.16	2.16	2.16	----	2.16	2.16
			tan δ	3	2	<2	<1	<1	<3	<4	----	6.5	10
		80	ϵ'/ϵ_0	2.10	2.10	2.10	2.10	2.10	----	----	----	2.10	2.10
			tan δ	16	3.6	.9	<1	<1	----	----	----	9.2	22
Cable Oil 5314[a]		25	ϵ'/ϵ_0	2.25	2.25	2.25	2.25	----	----	----	2.24	2.22	2.22
			tan δ	3	<0.4	<0.4	<1	----	----	----	39	18	22
		80	ϵ'/ϵ_0	2.18	2.18	2.18	----	----	----	----	----	2.18	
			tan δ	38	4	0.5	----	----	----	----	----	47	
Cable Oil PL101270[b]		26	ϵ'/ϵ_0	2.20	2.20	2.19	2.18	2.18	2.17	2.17	2.17	2.16	2.16
			tan δ	1.3	0.9	0.6	3	17	42	51	43	20	18
Transil Oil 10C[a]		26	ϵ'/ϵ_0	2.22	2.22	2.22	2.22	2.22	2.22	2.20	2.19	2.18	2.10
			tan δ	4	<1	<1	<6	<5	8	48	55	28	20
Bayol-D[c]		24	ϵ'/ϵ_0	2.06	2.06	2.06	2.06	2.06	2.06	----	----	2.06	2.06
			tan δ	1	<1	<1	<2	<3	<3	----	----	13.3	25
Bayol-F[d]		24	ϵ'/ϵ_0	2.14	2.14	2.14	----	----	----	----	----	2.13	2.13
			tan δ	6.7	1.6	<1	----	----	----	----	----	10.5	14
Marcol[e]		24	ϵ'/ϵ_0	2.14	2.14	2.14	2.14	2.14	2.14	----	----	2.14	2.14
			tan δ	1	<1	<1	<2	<2	<3	----	----	9.7	11

a. Aliphatic and aromatic hydrocarbons (Gen. Elec.). b. California Res. Corp. c. 77.6% paraffins, 22.4% naphthenes (Stanco). d. 74.5% paraffins, 25.5% naphthenes (Stanco). e. 72.4% paraffins, 27.6% naphthenes (Stanco).

II. Liquids, B. Organic (cont.) 3. Petroleum Oils (cont.) Values for tan δ are multiplied by 10^4; frequency given in c/s.

		T°C		1×10^2	1×10^3	1×10^4	1×10^5	1×10^6	1×10^7	1×10^8	3×10^8	3×10^9	1×10^{10}	2.5×10^{10}
Bayol[a]		24	ϵ'/ϵ_0	2.14	2.14	2.14	----	----	----	----	----	2.14	2.14	
			tan δ	12.6	2	<1	----	----	----	----	----	9.3	18	
Bayol-16[b]		24	ϵ'/ϵ_0	2.15	2.15	2.15	2.15	2.15	2.15	----	----	2.15	2.15	
			tan δ	7.7	1	<1	<2	----	<3	----	----	9.9	13.5	
Fractol A[c]		26	ϵ'/ϵ_0	2.17	2.17	2.17	2.17	----	2.17	----	----	2.17	2.16	2.12
			tan δ	<1	<1	<1	<2	----	<3	----	----	7.2	11.3	19
Primol-D[d]		24	ϵ'/ϵ_0	2.17	2.17	2.17	2.17	2.17	2.17	----	----	2.17	2.16	
			tan δ	<1	<1	<1	<2	<2	<3	----	----	7.7	10.6	
Diala Oil 15[e]		25	ϵ'/ϵ_0	2.16	2.16	2.16	2.16	2.16	2.16	2.16	2.16	2.16	2.16	2.12
			tan δ	<0.8	<0.8	<0.4	<5	<3	3	10	16.3	13.4	18.6	20
4. Silicones														
DC500 0.65 cs. at 25°C[f]		-15	ϵ'/ϵ_0	2.20	2.20	2.20	2.20	2.20	2.20	----	----	2.20		
			tan δ	<5	<3	<3	<5	<3	<2	----	----	18.6		
		22	ϵ'/ϵ_0	2.20	2.20	2.20	2.20	2.20	2.20	----	2.20	2.20	2.19	2.13
			tan δ	1	<0.4	<0.4	<3	<3	<2	----	1.4	14.5	30	60
DC500 10 cs. at 25°C[f]		23	ϵ'/ϵ_0	2.66	2.66	2.66	2.66	2.66	2.66	----	----	2.65	2.63	2.48
			tan δ	12	1.5	<0.4	<3	<3	3	----	----	68	270	410
DC500 100 cs. at 25°C[f]		22	ϵ'/ϵ_0	2.76	2.76	2.76	----	----	----	----	----	2.75	2.72	2.61
			tan δ	0.4	<0.4	<0.4	----	----	----	----	----	94	240	390
DC200 100 cs. at 25°C[f]		23	ϵ'/ϵ_0	2.76	2.76	2.76	2.78	2.78	2.78	----	----	2.76	2.70	2.55
			tan δ	0.8	0.4	<0.4	<3	<3	<2	----	----	96	320	550
DC200 1000 cs. at 25°C[f]		22	ϵ'/ϵ_0	2.78	2.78	2.78	2.78	2.78	2.78	----	----	2.74	2.74	
			tan δ	0.8	0.8	0.6	<5	<3	<2	----	----	96	330	
DC200 7600 cs. at 130°C[f]		-17	ϵ'/ϵ_0	2.90	2.90	2.90	2.90	2.90	2.90	----	----	2.87		
			tan δ	<5	<3	<3	<5	<3	2	----	----	130		
		23	ϵ'/ϵ_0	2.75	2.75	2.75	2.75	2.75	2.75	----	2.74	2.72	2.69	
			tan δ	0.5	0.2	<0.4	<5	<3	2	----	11	103	225	
		83	ϵ'/ϵ_0	2.56	2.56	2.56	2.56	2.56	2.56	----	----	2.55		
			tan δ	<5	<3	<3	<5	<3	2	----	----	60		

a. 72.0% paraffins, 28.0% naphthenes (Stanco). b. 68.9% paraffins, 31.1% naphthenes (Stanco). c. 57.4% paraffins, 42.6% naphthenes (Stanco). d. 49.4% paraffins, 50.6% naphthenes (Stanco). e. Petroleum hydrocarbons, mainly naphthenes (Shell). f. Methyl or ethyl siloxane polymer (Dow Corning).

Dielectric Data

II. Liquids, B. Organic (cont.)   Values for tan δ are multiplied by 10^4; frequency given in c/s.

4. Silicones (cont.)

	T°C		1×10^2	1×10^3	1×10^4	1×10^5	1×10^6	1×10^7	1×10^8	3×10^8	3×10^9	1×10^{10}
DC550[a]	25	ϵ'/ϵ_0	2.91	2.90	2.90	2.90	2.90	2.88	----	2.88	2.77	2.60
		tan δ	1630	170	18	3.7	3.8	12	----	130	210	220
DC710[a]	25	ϵ'/ϵ_0	2.98	2.98	2.98	2.98	2.98	2.97	----	2.93	2.79	2.60
		tan δ	13	1.6	.7	3	10	50	----	200	140	170
Ignition Sealing Compound #4[b]	25	ϵ'/ϵ_0	2.75	2.75	2.75	2.75	2.75	2.75	2.74	2.72	2.65	2.49
		tan δ	15	6	5	4	4	6	15	28	92	270
	80	ϵ'/ϵ_0	2.6	2.6	2.6	2.6	2.6	----	----	----	2.54	
		tan δ	64	18	6	4	<4	----	----	----	47	
SF96-40[c]	25	ϵ'/ϵ_0	2.71	2.71	2.71	2.71	2.71	2.71	----	2.71	2.70	2.67
		tan δ	<.3	<.03	<.03	<.3	<1	<1	----	11	95	186
SF96-100[c]	25	ϵ'/ϵ_0	2.73	2.73	2.73	2.73	2.73	2.73	----	2.73	2.71	2.69
		tan δ	<.6	<.06	<.03	<1	<1	<1	----	11	107	200
SF96-1000[c]	25	ϵ'/ϵ_0	2.73	2.73	2.73	2.73	2.73	2.73	----	2.73	2.71	2.695
		tan δ	<.3	<.03	<.03	<.3	<1	.8	----	13	106	203

a. Methyl and methyl phenyl polysiloxane (Dow Corning). b. Organosiloxane polymer (Dow Corning). c. Gen. Elec.

Tables of Dielectric Materials

Supplementary High-Temperature Data on Plastics

Frequency given in c/s.

Material	T°C		10^2	10^3	10^4	4×10^4	10^5	10^6	3×10^8	3×10^9
Formica FF-41 (after 5 yrs. storage, various samples)	25	ϵ'/ϵ_o	6.27	6.19	6.12	6.08	5.97	----	3.87	4.2
		tan δ	.018	.0085	.0081	.0090	.012	----	.0143	.021
	100	ϵ'/ϵ_o	----	11.2	6.73	6.23	6.07	----	4.00	4.3
		tan δ	----	.50	.18	.055	.040	----	.0144	.020
Formica LE (after 5 yrs. storage)	25	ϵ'/ϵ_o	5.92	5.32	5.01	4.87	4.86	4.85	----	3.63
		tan δ	.112	.057	.038	.036	.040	.042	----	.037
	100	ϵ'/ϵ_o	----	16.0	8.16	6.69	6.22	5.84	----	4.10
		tan δ	----	.42	.29	.173	.151	.080	----	.067
Formica MF-66	200	ϵ'/ϵ_o	----	12.8	7.17	6.4	----	----	----	4.94
		tan δ	----	.847	.245	.134	----	----	----	.059
Formica XX	100	ϵ'/ϵ_o	9.77	6.66	5.87	5.66	5.61	5.44	----	4.09
		tan δ	.42	.195	.073	.046	.035	.029	----	.102
Micarta 259	100	ϵ'/ϵ_o	9.39	6.75	6.14	6.03	----	----	5.2	
		tan δ	.44	.16	.043	.022	----	----	.025	
Micarta 299	200	ϵ'/ϵ_o	12.8	6.7	5.0	4.8	4.75	4.6	4.4	
		tan δ	.59	.36	.158	.081	.046	.015	.016	
Micarta 496 (after 5 yrs. storage)	25	ϵ'/ϵ_o	7.31	6.35	5.89	5.73	7.26	6.31		
		tan δ	.16	.074	.045	.041	.116	.064		
	100	ϵ'/ϵ_o	----	15.5	8.53	7.42				
		tan δ	----	.68	.305	.178				

Dielectric Data

Supplementary High-Humidity Data

	T°C		10^2	10^3	10^4	10^5	10^6	10^7	4×10^7	3×10^8	10^{10}
Mycalex 400, dry	25	ϵ'/ϵ_o	7.45	7.45	7.42	7.40	7.39	7.38	7.36	----	7.12
		$\tan\delta$	.0029	.0019	.0016	.0014	.0013	.0013	.0012	----	.0033
after 48 hrs. in H_2O	25	ϵ'/ϵ_o	7.45	7.45	7.42	----	7.39	7.38	7.36	----	7.18
		$\tan\delta$	.012	.0097	.0068	----	.0029	.0021	.0019	----	.010
Bakelite BM 120, dry	25	ϵ'/ϵ_o	4.87	4.74	4.62	4.50	4.36				
		$\tan\delta$	.030	.022	.020	.021	.028				
after 19 days, 90% rel. hum.	25	ϵ'/ϵ_o	7.4	5.4	4.4						
		$\tan\delta$	.32	.19	.11						
after 18 mos., 90% rel. hum.	25	ϵ'/ϵ_o	11.1	7.8	6.3	5.4	4.54				
		$\tan\delta$	.37	.20	.12	.078	.056				
Bakelite BM 262, dry	25	ϵ'/ϵ_o	4.85	4.80	4.74	4.72	4.67			----	
		$\tan\delta$	.0095	.0082	.0075	.0060	.0055				
after 20 days, 90% rel. hum.	25	ϵ'/ϵ_o	3.87	3.78	3.70						
		$\tan\delta$	.0184	.015	.012						
after 7 mos., 90% rel. hum.	25	ϵ'/ϵ_o	6.72	5.78	5.15						
		$\tan\delta$	.147	.09	.057						
after 19 mos., 90% rel. hum.	25	ϵ'/ϵ_o	9.10	6.75	5.51	----	4.18			----	
		$\tan\delta$	.244	.169	.111	----	.047				
Bakelite BM 1895, dry	25	ϵ'/ϵ_o	4.80	4.72	4.70	4.67	4.64				
		$\tan\delta$	.0087	.0077	.0080	.0065	.0052				
after 20 days, 90% rel. hum.	25	ϵ'/ϵ_o	5.1	5.0	4.9						
		$\tan\delta$	.023	.018	.016						
after 19 mos., 90% rel. hum.	25	ϵ'/ϵ_o	----	61.4	24	11.7	8.0				
		$\tan\delta$	----	1.04	.75	.41	.22				
Formica FF-55, dry	25	ϵ'/ϵ_o	----	----	----	----	----	----	----	5.55	
		$\tan\delta$	----	----	----	----	----	----	----	.027	
after 20 hrs. in H_2O	25	ϵ'/ϵ_o	----	----	----	----	----	----	----	6.5	
		$\tan\delta$	----	----	----	----	----	----	----	.064	

Frequency given in c/s.

Supplementary High-Humidity Data (cont.)

	T°C		Frequency given in c/s.				
			10^2	10^3	10^4	10^5	10^6	3×10^9
GMG Melamine, dry	25	ϵ'/ϵ_o	8.2	7.0	6.7	6.6	6.4	
		tan δ	.19	.069	.019	.010	.011	
after 6 or 8 mos., 90% rel. hum.	25	ϵ'/ϵ_o	42.5	16.8	10.4	7.65	6.57	
		tan δ	.75	.54	.27	.10	.030	
Hysol 6030, dry and after 4 days in H₂O, no change at 1×10⁹.								
Lucite HM-119, dry and after 18 months at 90% relative humidity, 25°C, no change in the range 10² to 10⁶ c/s.								
Lumarith 22361, dry	25	ϵ'/ϵ_o	----	----	----	----	----	2.74
		tan δ	----	----	----	----	----	.0196
after 20 days or more at 90% rel. hum.	25	ϵ'/ϵ_o	----	----	----	----	----	3.14
		tan δ	----	----	----	----	----	.047
Micarta 496, dry	25	ϵ'/ϵ_o	----	----	----	----	----	3.78
		tan δ	----	----	----	----	----	.059
after 24 hrs., 90% rel. hum.	25	ϵ'/ϵ_o	----	----	----	----	----	3.94
		tan δ	----	----	----	----	----	.071
Nylon, See page 23								
Poly-2,5-dichlorostyrene, dry	25	ϵ'/ϵ_o	----	----	----	----	----	2.59
		tan δ	----	----	----	----	----	.00026
after 20 days, 90% rel. hum.	25	ϵ'/ϵ_o	----	----	----	----	----	2.59
		tan δ	----	----	----	----	----	.00055
Polystyrene, See pages 36 and 37								
Polythene, dry	25	ϵ'/ϵ_o	----	----	----	----	----	2.26
		tan δ	----	----	----	----	----	.00067
after 10 days, 90% rel. hum.	25	ϵ'/ϵ_o	----	----	----	----	----	2.26
		tan δ	----	----	----	----	----	.00085
Teflon, dry and after 18 months at 90% relative humidity, 25°C, no change in the range 10² to 10⁶ c/s.								
Vinylite VU-1900, dry and after 18 months at 90% relative humidity, 25°C, no change in the range 10² to 10⁶ c/s.								

Data at Fixed Frequencies f as a Function of Temperature

Inorganic Crystals

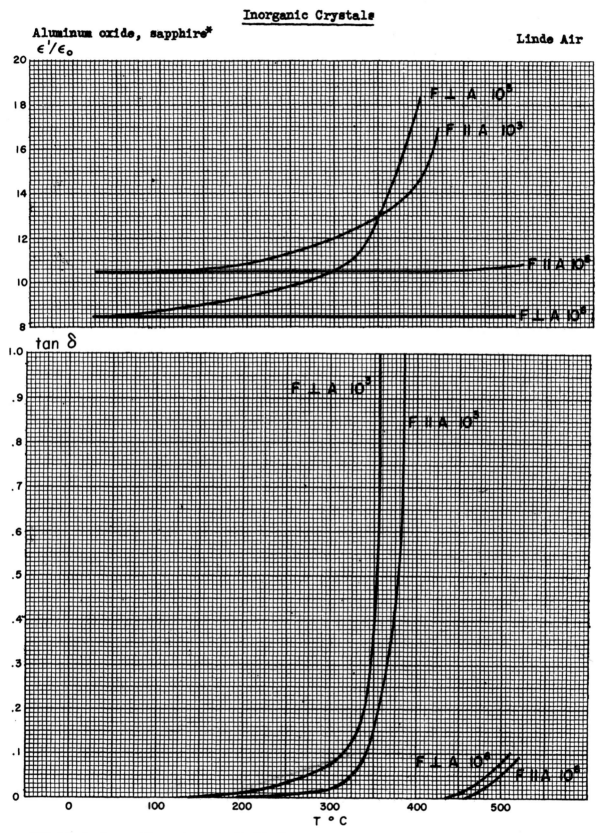

*Field ⊥ to optic axis and field ∥ to optic axis.

Inorganic Crystals (cont.)

Potassium dihydrogen phosphate*   Brush

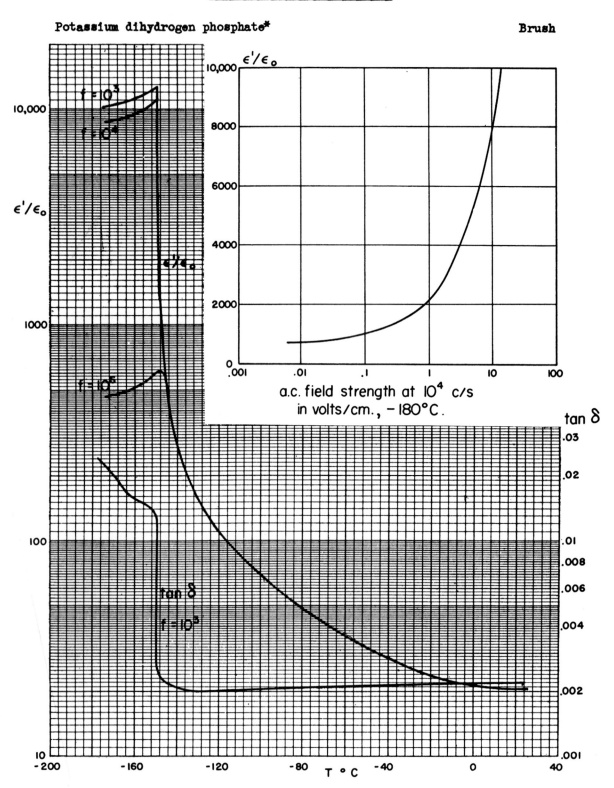

*Field ‖ to optic axis, 15 volts/cm.

Inorganic Crystals (cont.)

Thallium iodide (pressed sample) — Eng. Res. and Dev. Lab.

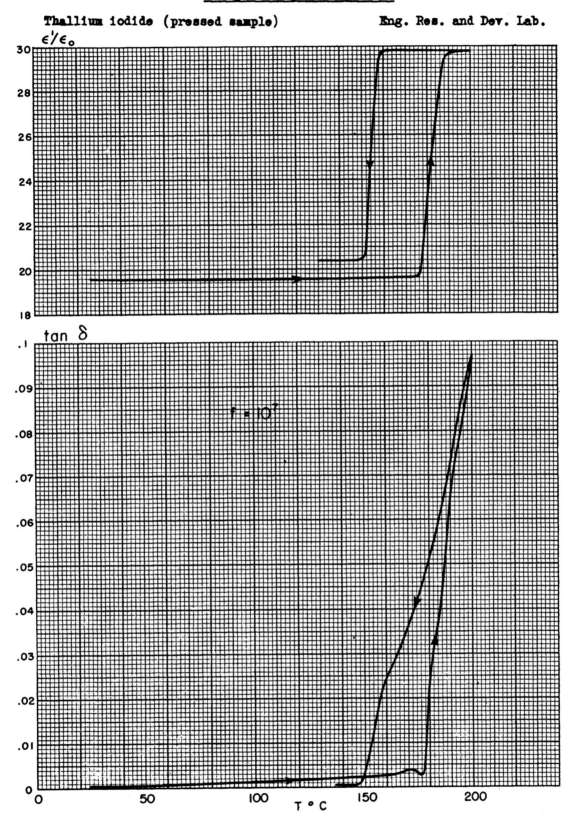

Inorganic Crystals (cont.)

KRS-6 (TlCl, 40%, TlBr, 60%)
Thallium bromide

Eng. Res. and Dev. Lab.

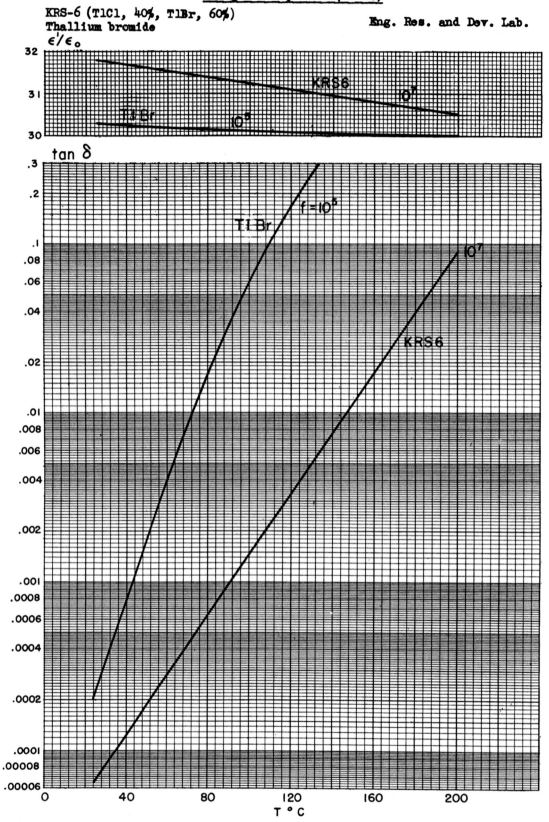

Inorganic Crystals (cont.)

KRS-5 (TlBr, 42%, TlI, 58%)   Eng. Res. and Dev. Lab.

Inorganic Crystals (cont.)

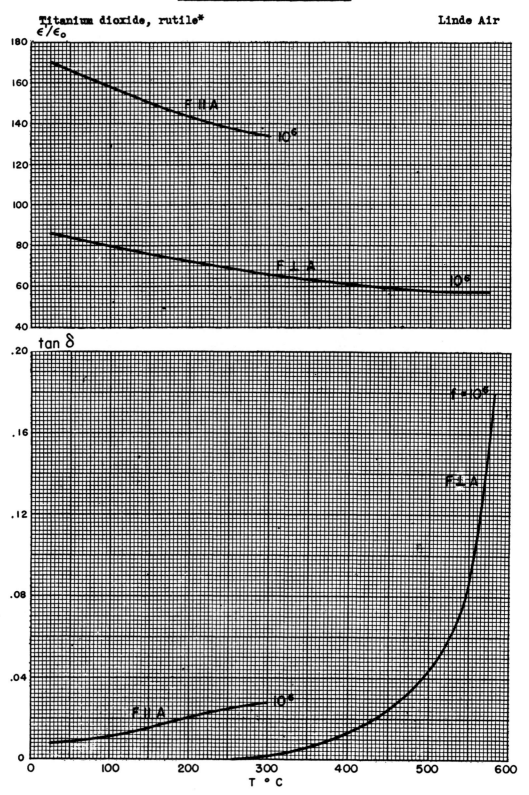

*Field ⊥ to optic axis and field ∥ to optic axis.

Steatite Bodies

AlSiMag A-196 — Amer. Lava

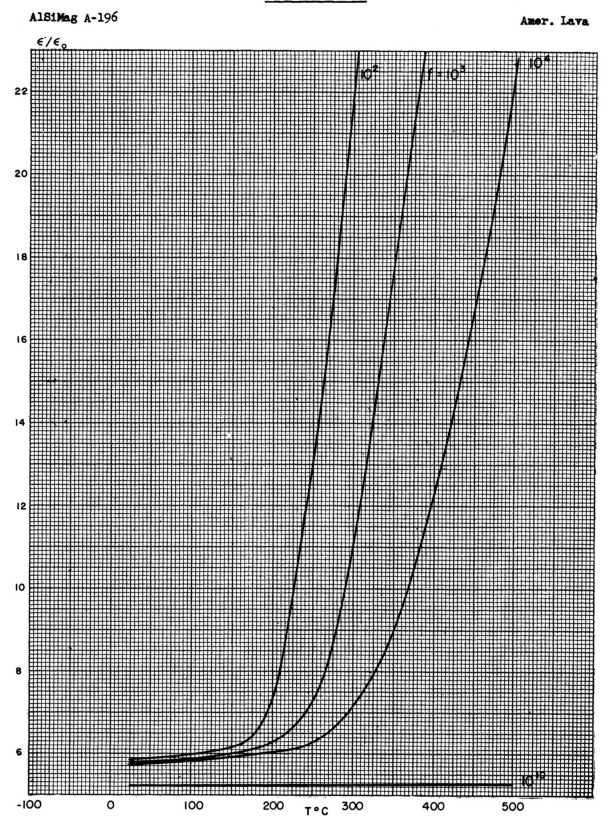

Steatite Bodies (cont.)

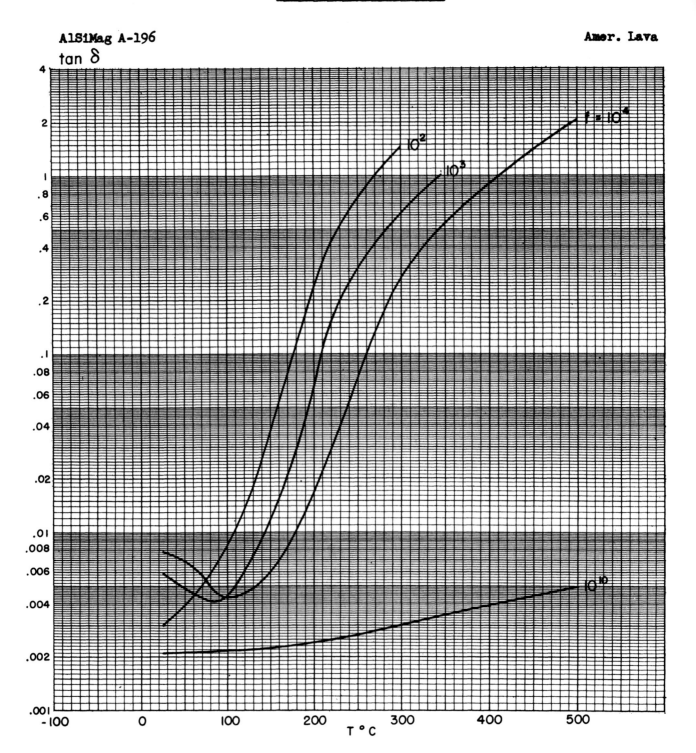

AlSiMag A-196 — Amer. Lava

Steatite Bodies (cont.)

AlSiMag A-35 — Amer. Lava

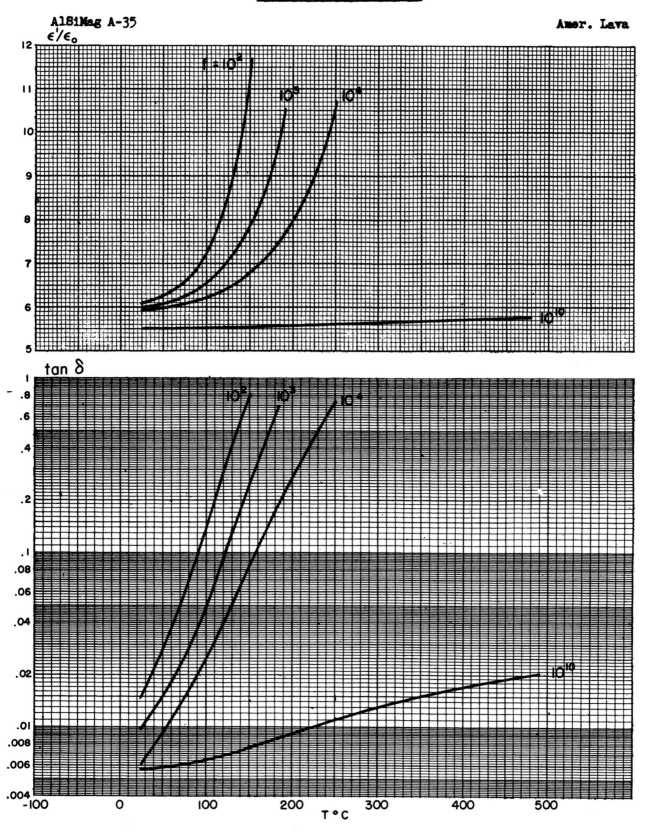

Steatite Bodies (cont.)

AlSiMag 228 — Amer. Lava

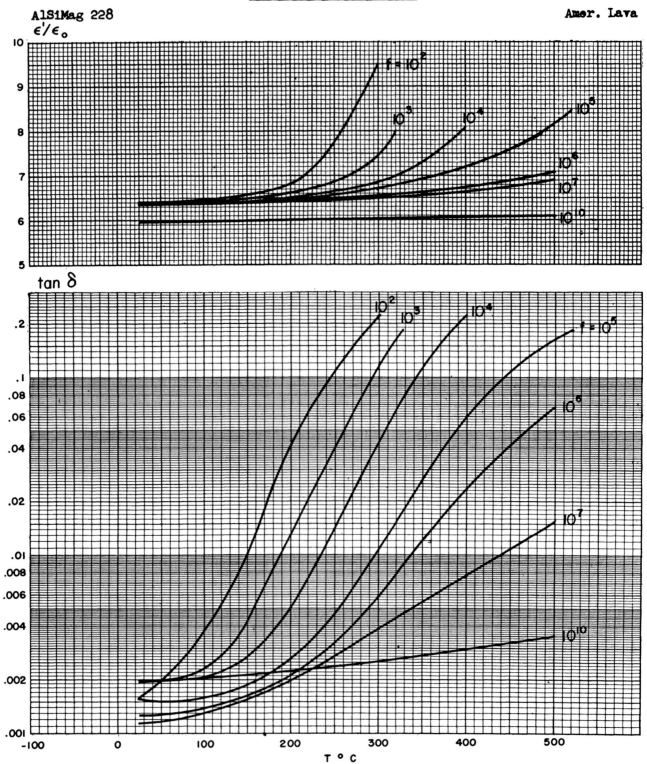

Steatite Bodies (cont.)

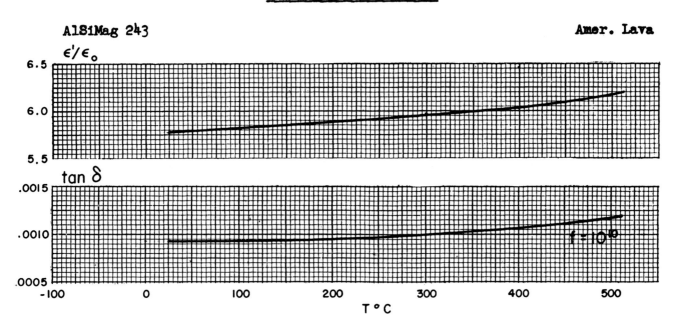

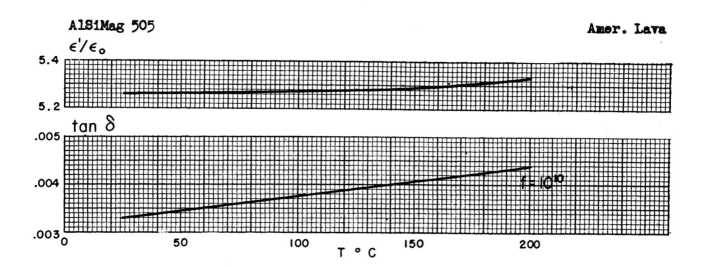

Steatite Bodies (cont.)

Ceramic F-66 — Bell

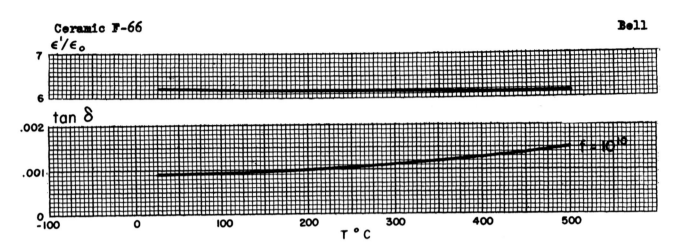

Steatite Body 7292 — Gen. Ceramics and Steatite

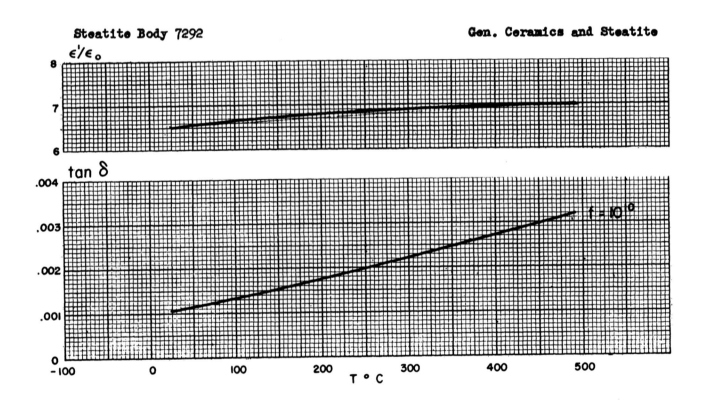

Steatite Bodies (cont.)

Steatite Type 302 — Centralab

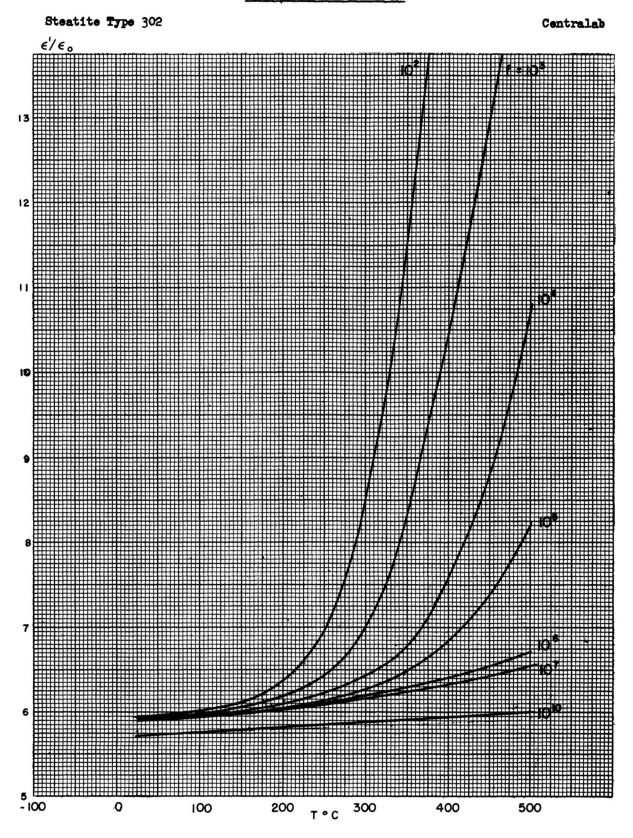

Steatite Bodies (cont.)

Steatite Type 302 — Centralab

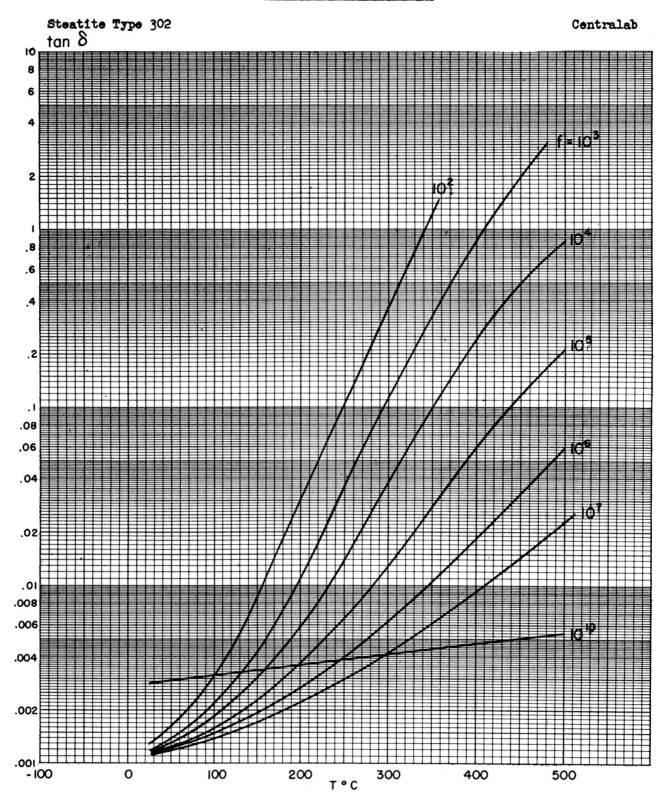

Steatite Bodies (cont.)

Steatite Type 400, 452 — Centralab

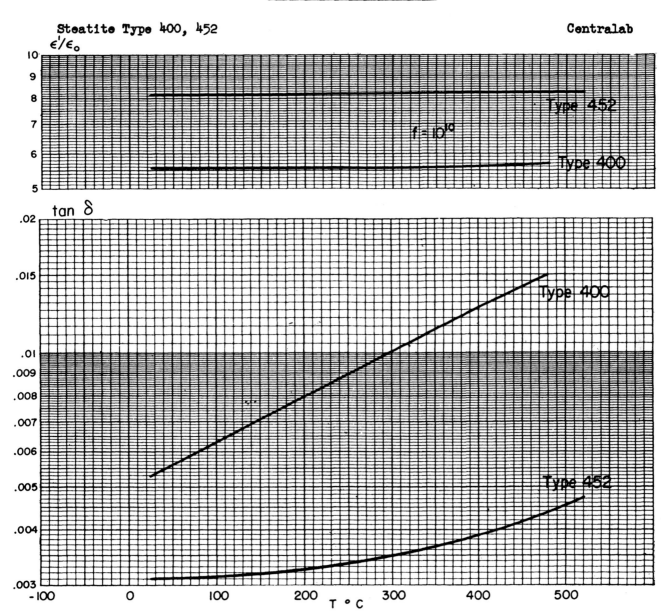

Steatite Bodies (cont.)

Steatite Type 410 — Centralab

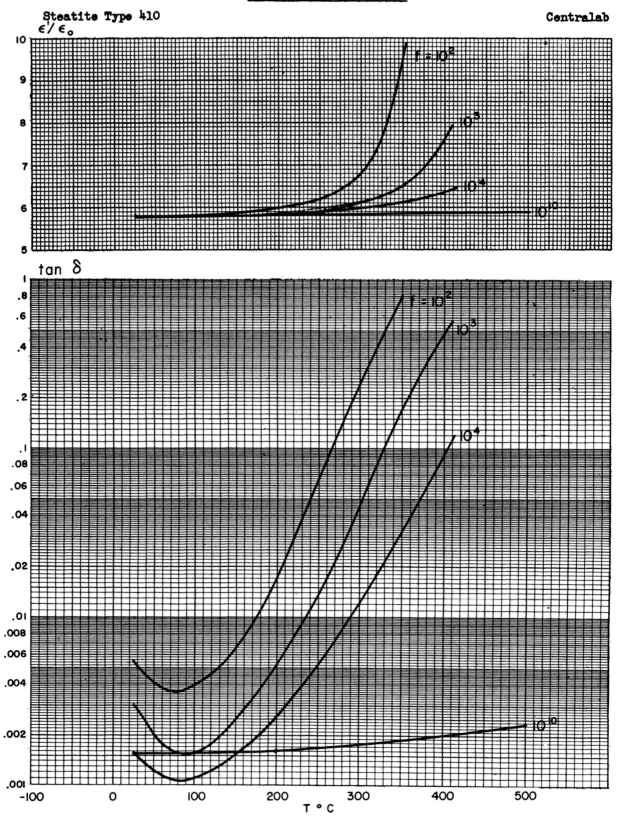

Titania and Titanate Bodies

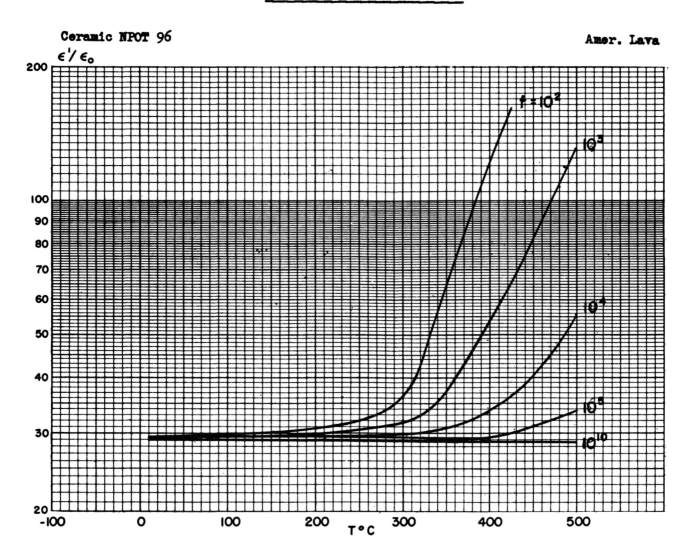

Titania and Titanate Bodies (cont.)

Ceramic NPOT 96 — Amer. Lava

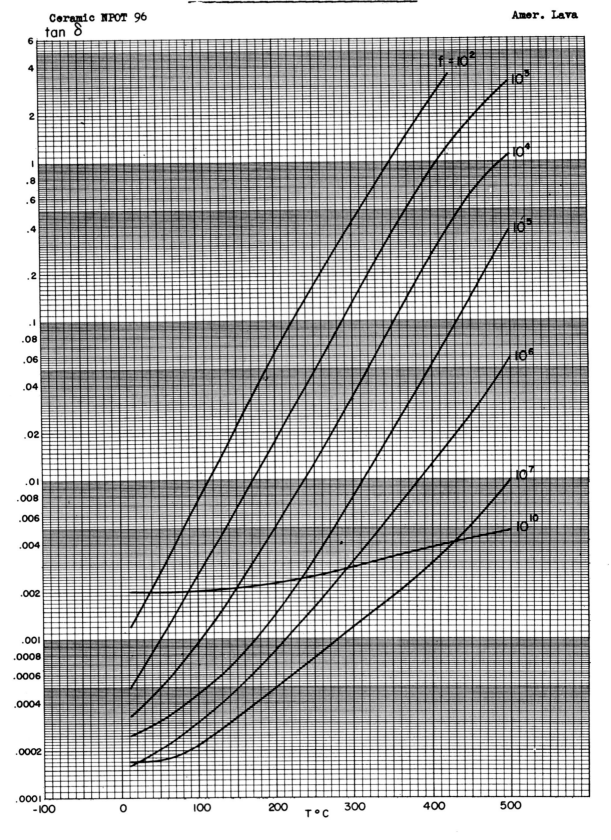

Titania and Titanate Bodies (cont.)

Ceramic N750T96   Amer. Lava

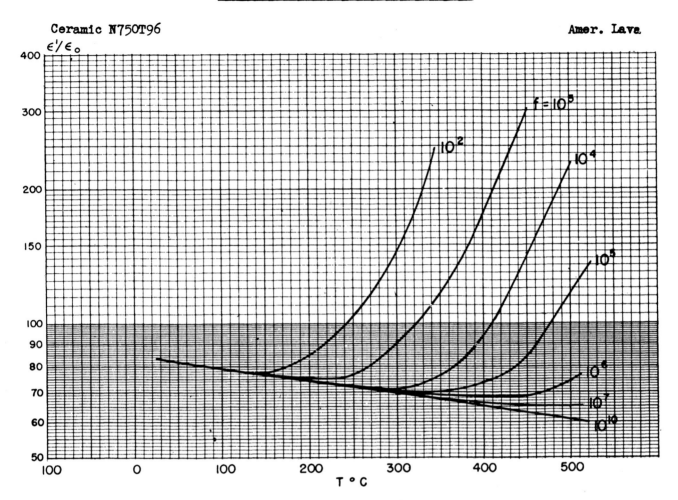

Tables of Dielectric Materials

Titania and Titanate Bodies (cont.)

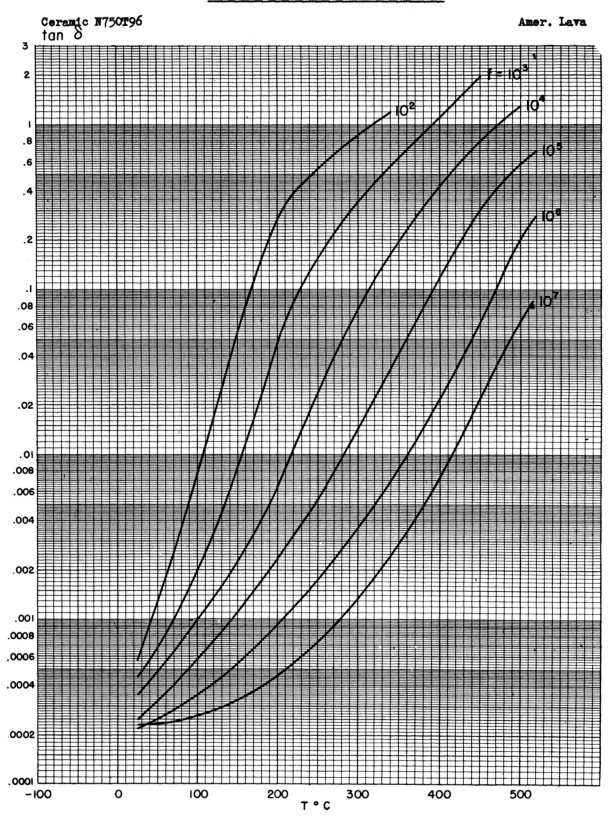

Ceramic N750T96, Amer. Lava

Titania and Titanate Bodies (cont.)

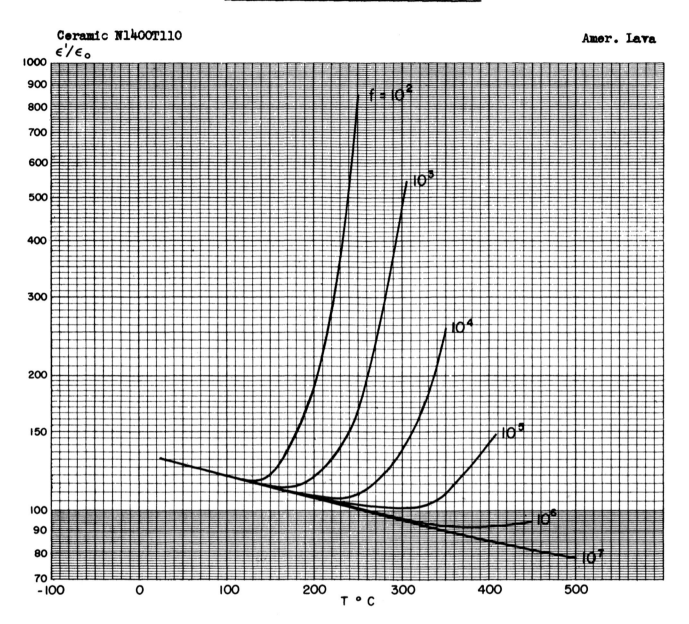

Ceramic N1400T110
Amer. Lava

Titania and Titanate Bodies (cont.)

Ceramic N1400T110 — Amer. Lava

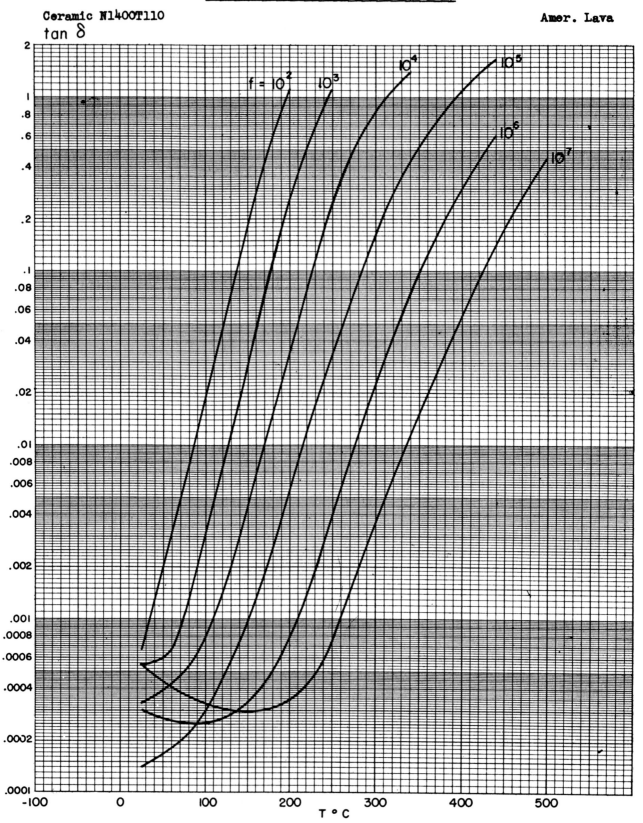

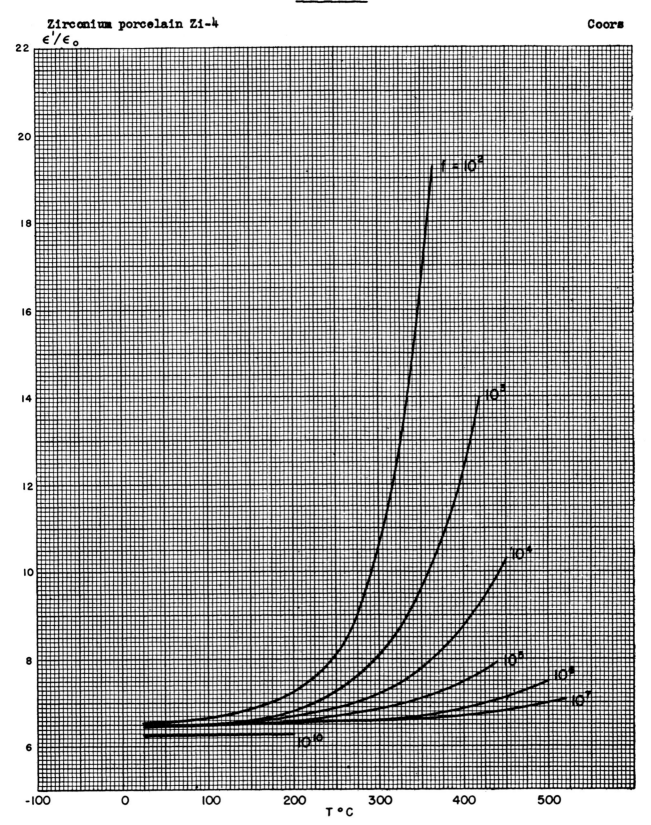

Porcelains (cont.)

Zirconium porcelain Zi-4 — Coors

tan δ

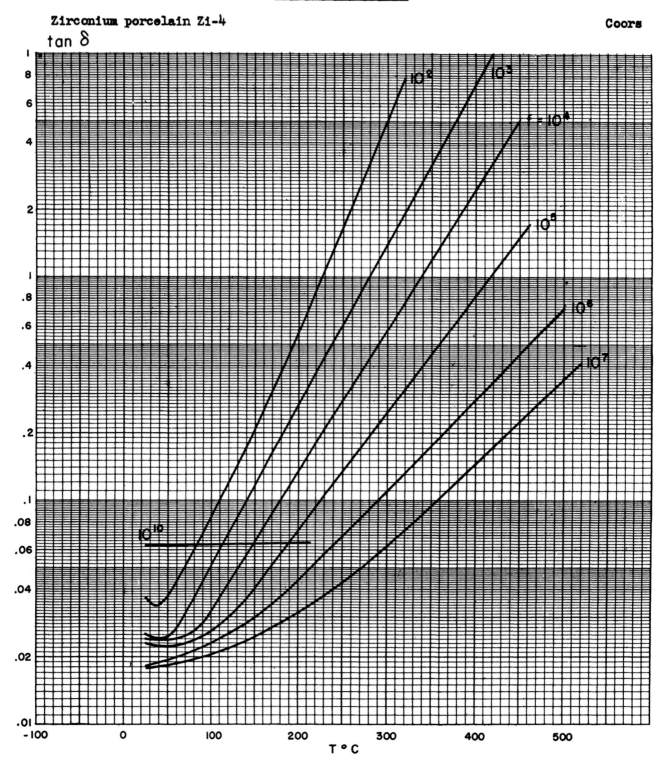

Porcelains (cont.)

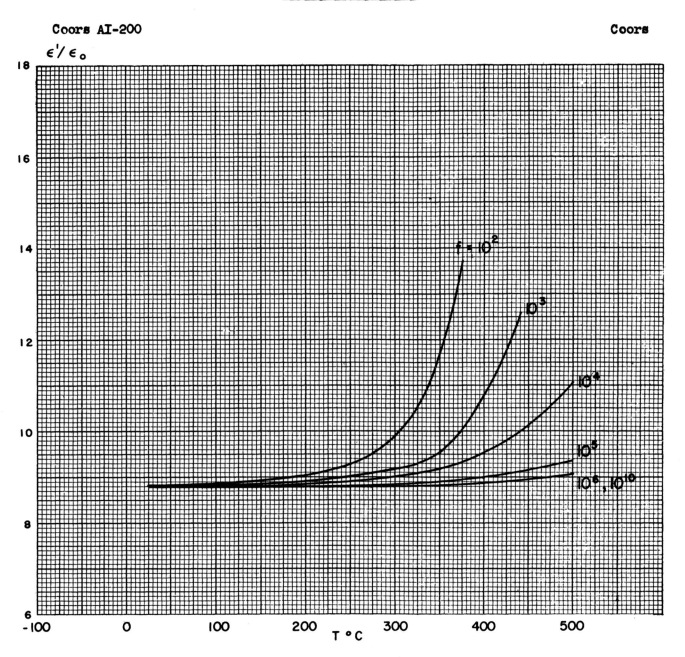

Porcelains (cont.)

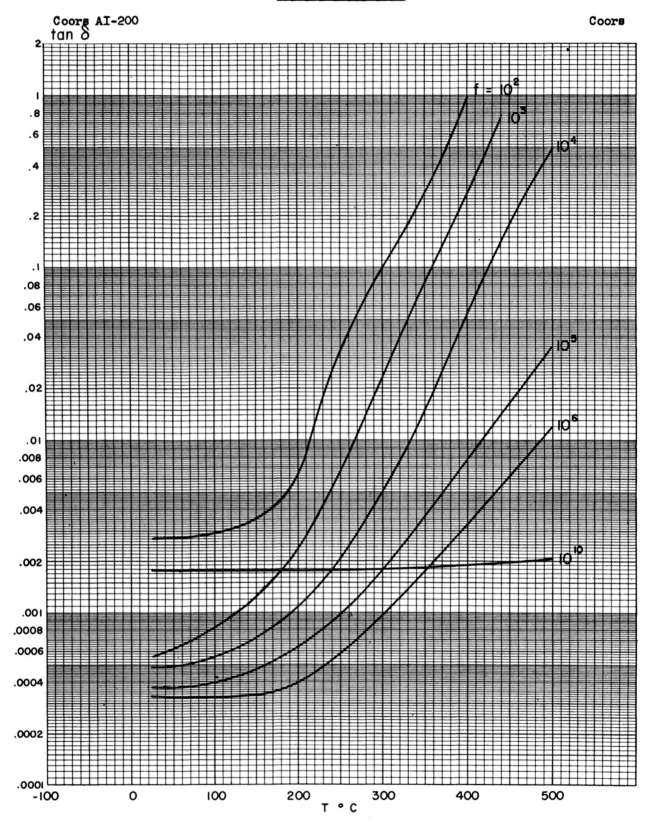

Porcelains (cont.)

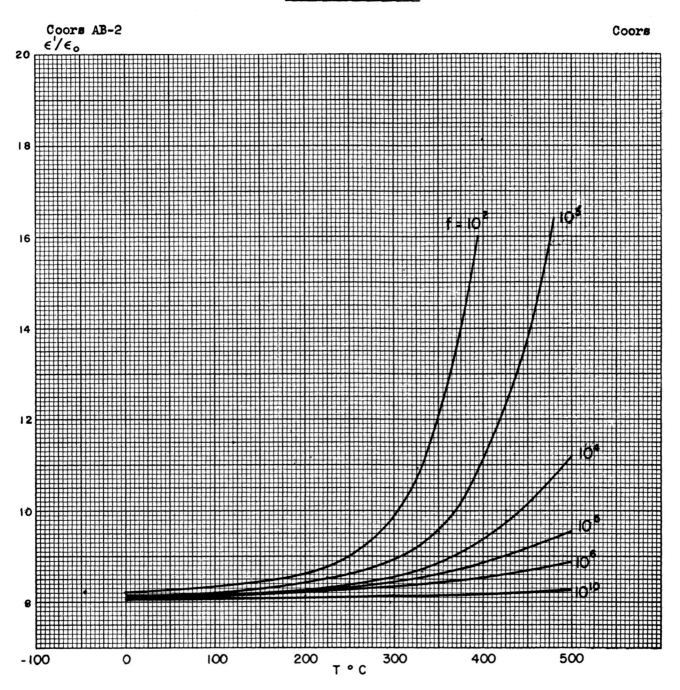

Porcelains (cont.)

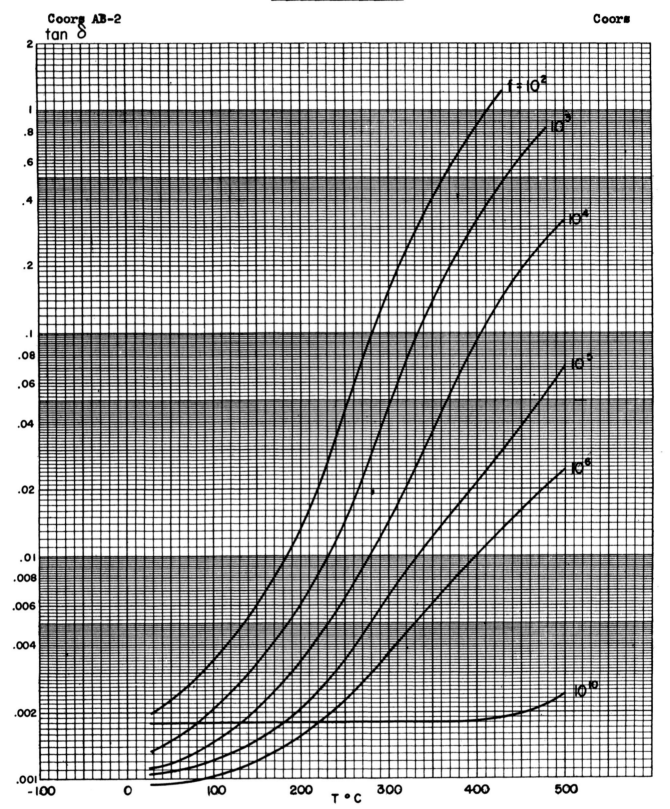

Coors AB-2 — Coors

Porcelains (cont.)

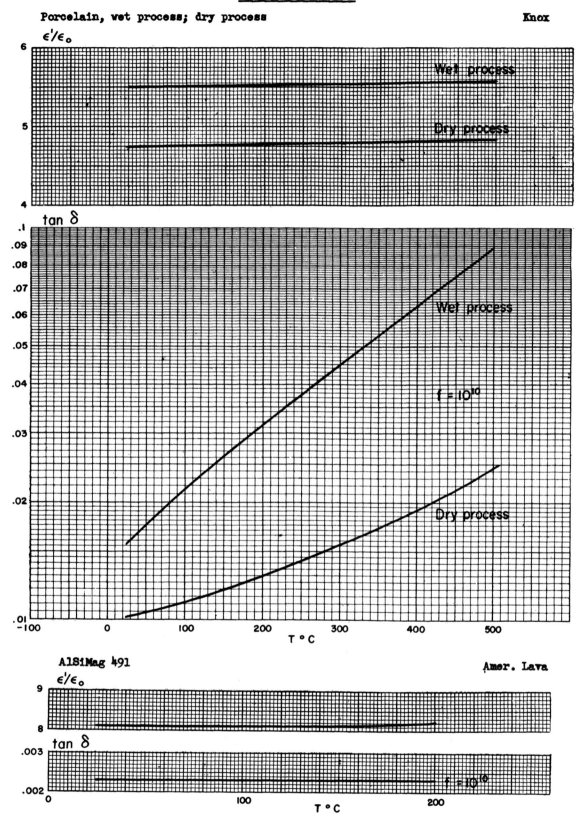

Tables of Dielectric Materials

Miscellaneous Ceramics

Porous Ceramic AF-497 — Stupakoff

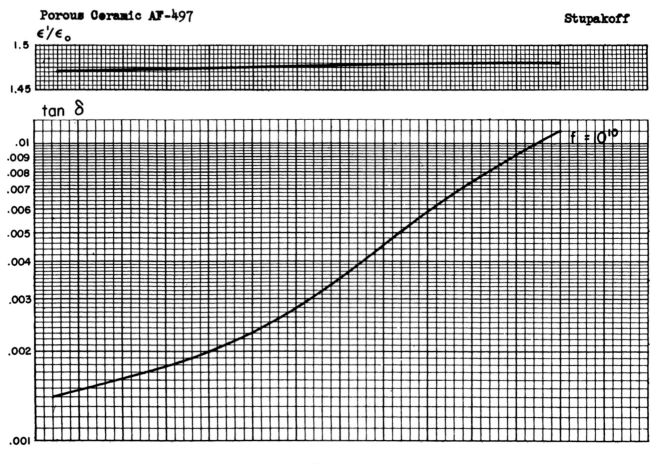

Glasses

Borosilicate Glass — Components and Systems Lab.

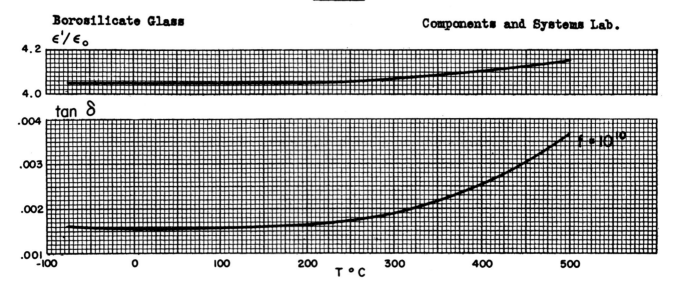

Glasses (cont.)

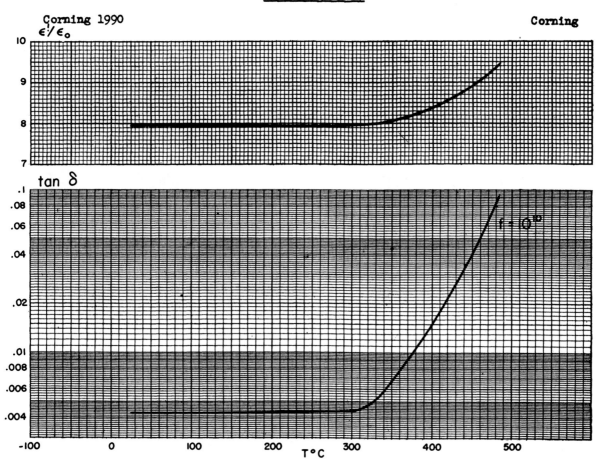

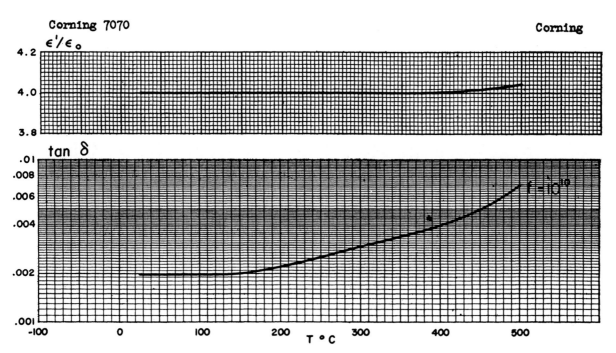

Glasses (cont.)

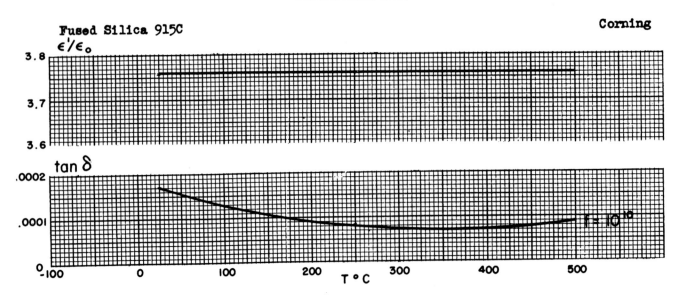

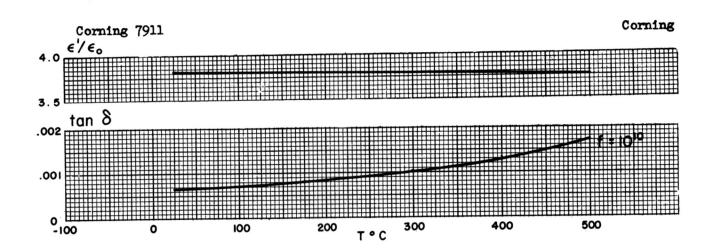

Glasses (cont.)

Fused Quartz — General Electric

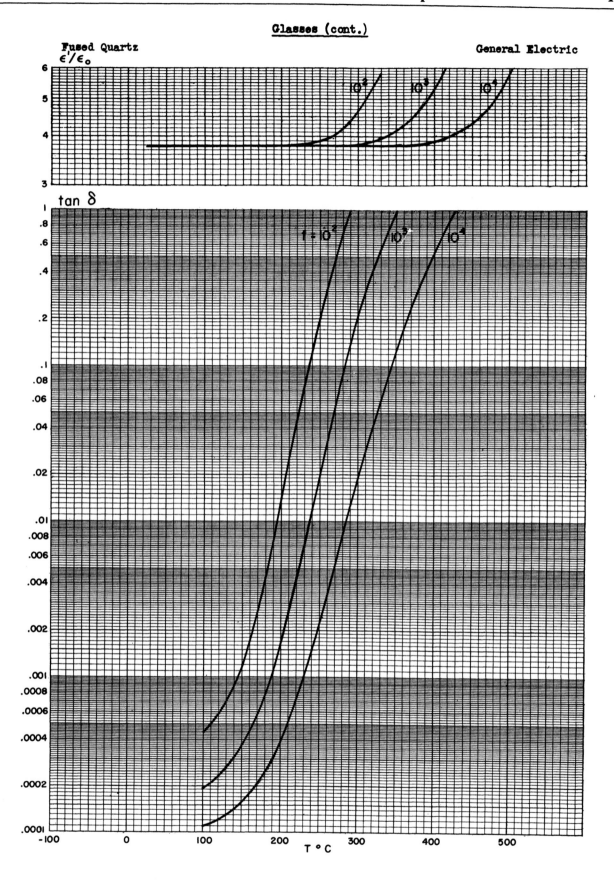

Glasses (cont.)

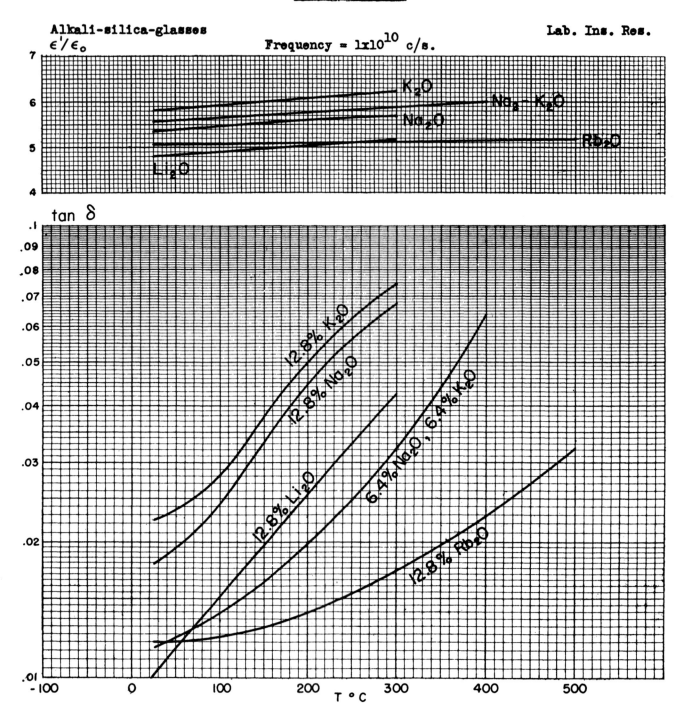

Mica and Glass

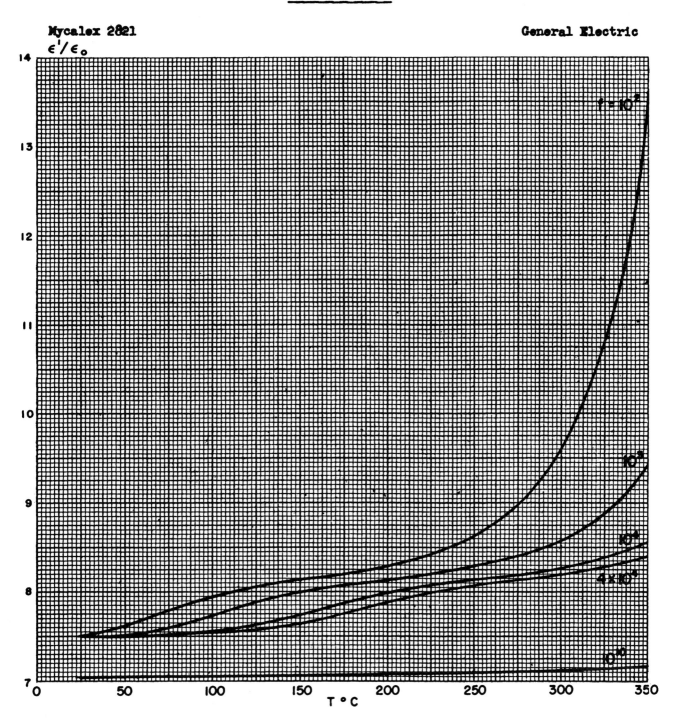

Mica and Glass (cont.)

Mycalex 2821 — tan δ — General Electric

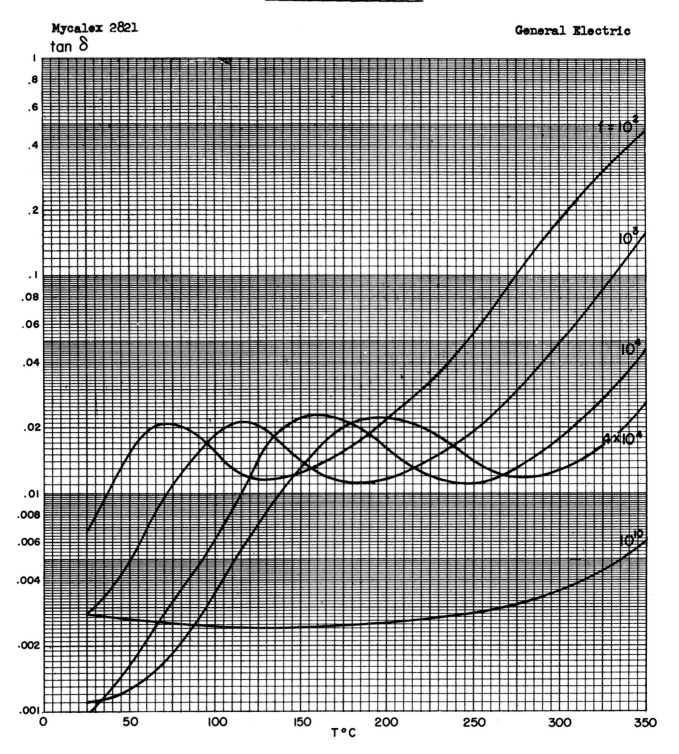

Mica and Glass (cont.)

Mykroy Grade 8, Grade 38 — Electronic Mech.

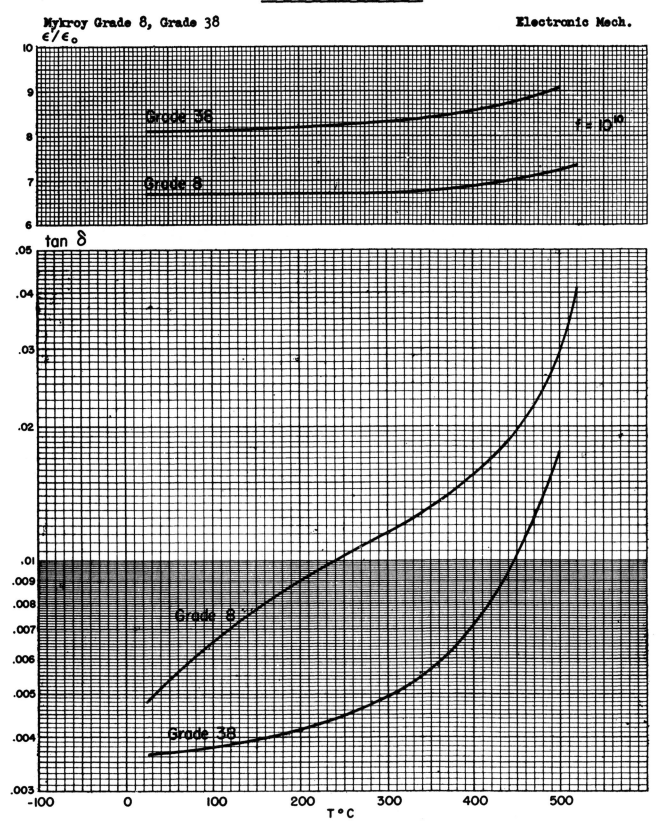

Phenol-formaldehyde Resins

Resinox 10231 — Monsanto

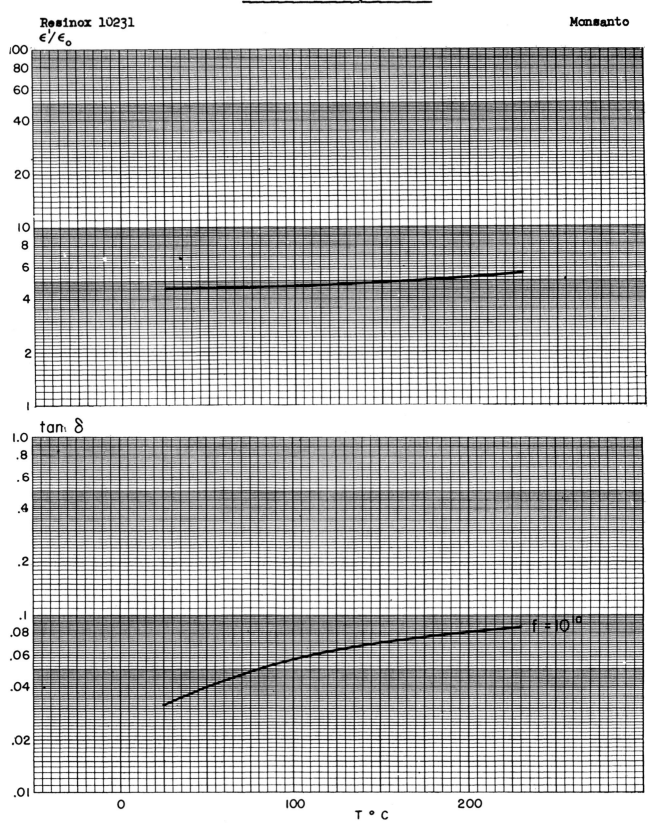

Phenol-formaldehyde Resins (cont.)

Resinox 10900 — Monsanto

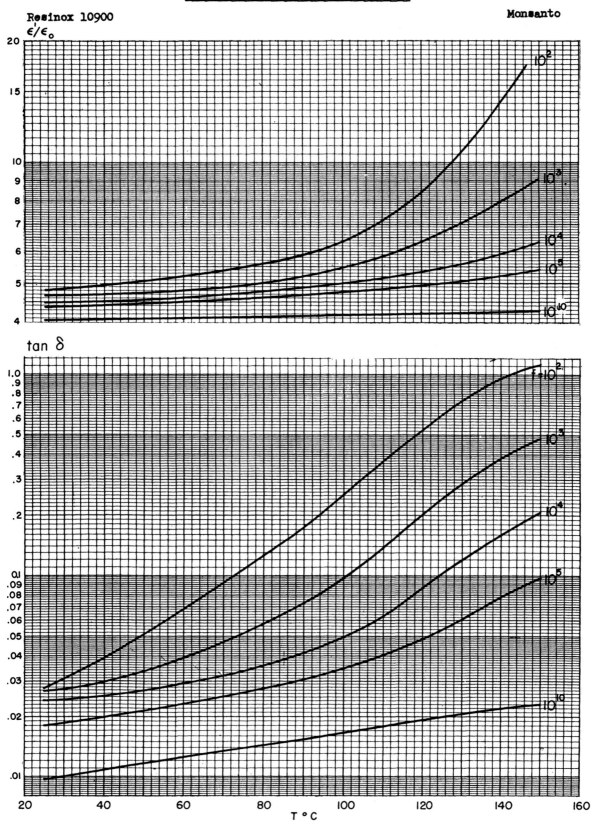

Phenol-formaldehyde Resins (cont.)

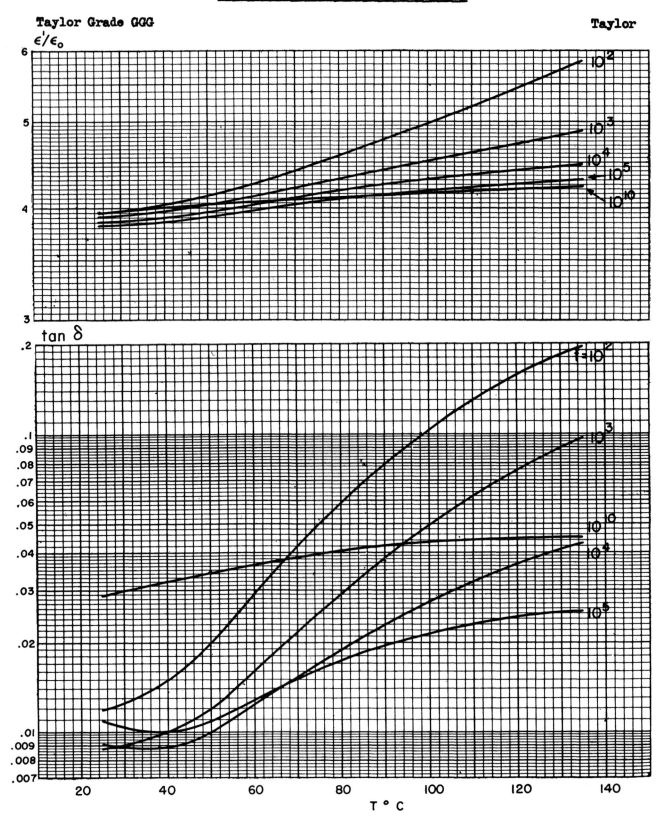

Melamine-formaldehyde Resin

Melmac Molding Compound 1502 — Amer. Cyanamid

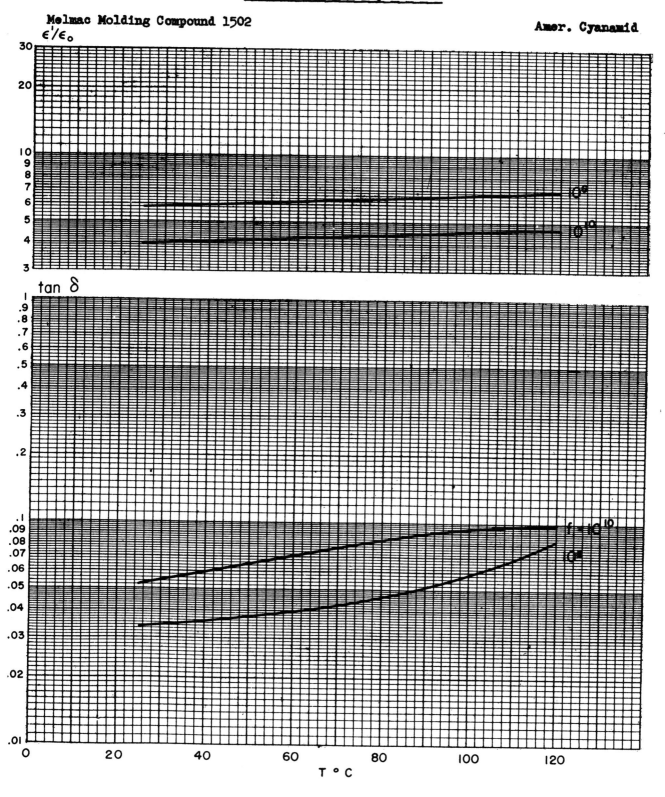

Polyamide Resin

Nylon FM 10,001 — DuPont

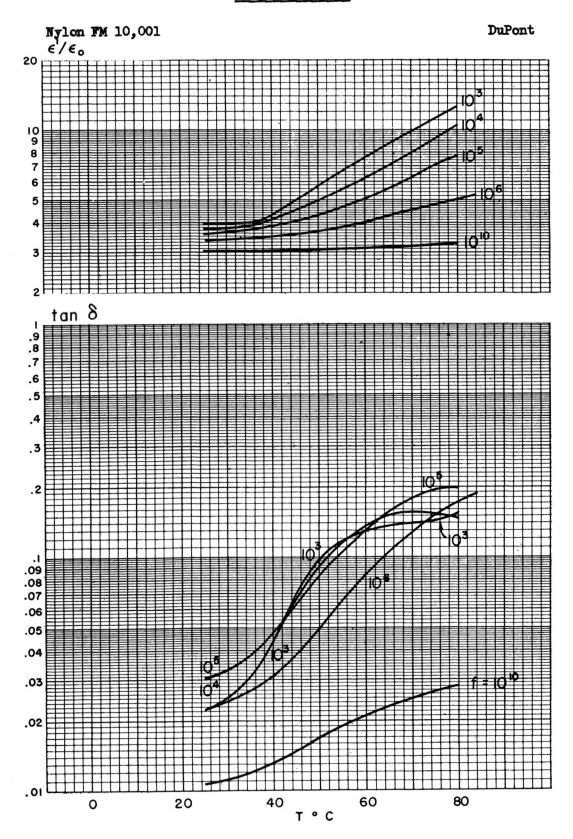

Cellulose Propionate

Forticel — Celanese

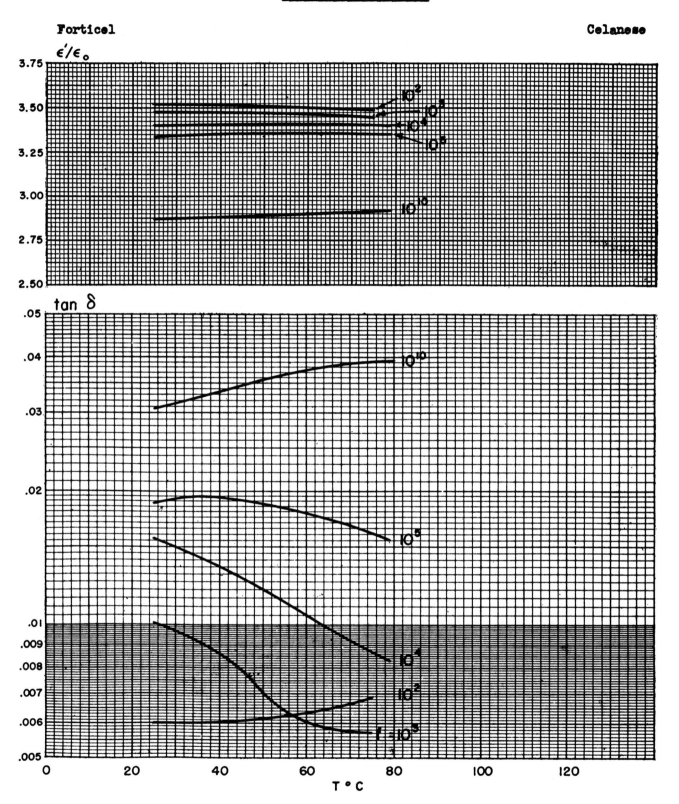

Ethyl Cellulose

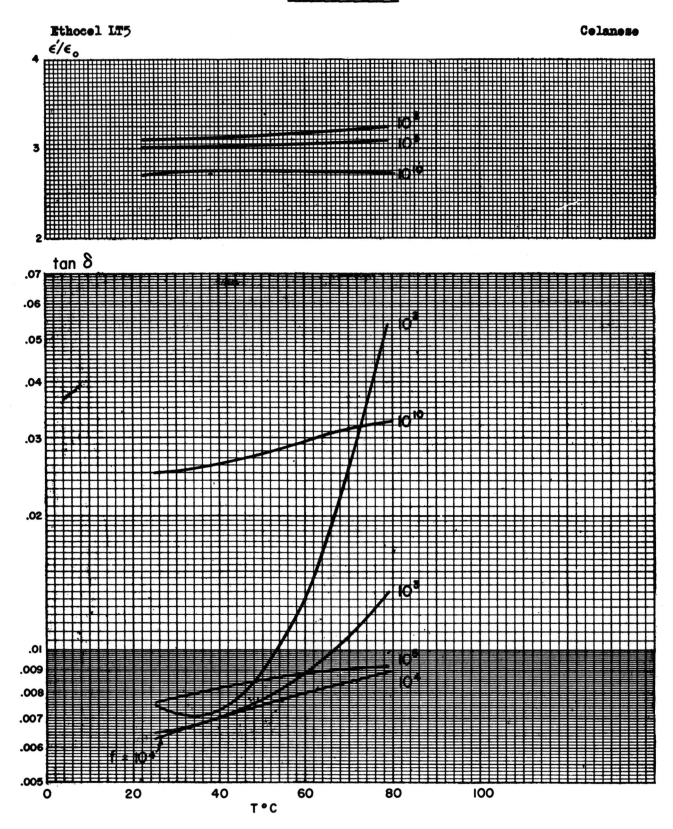

Polychlorotrifluoroethylene

Kel-F Grade 300 — Kellogg

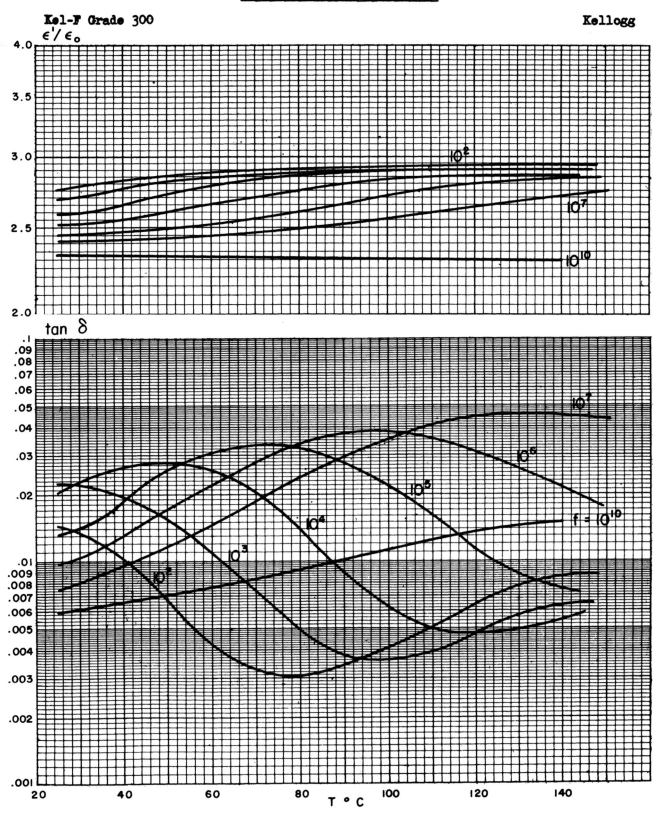

Miscellaneous Vinyls

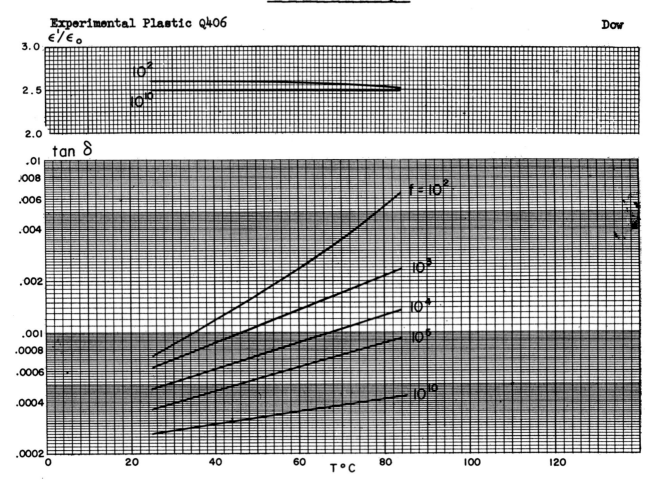

Experimental Plastic Q406 — Dow

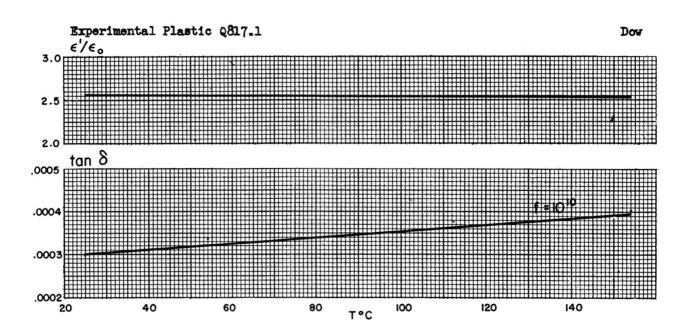

Experimental Plastic Q817.1 — Dow

Styrene Copolymers, Linear

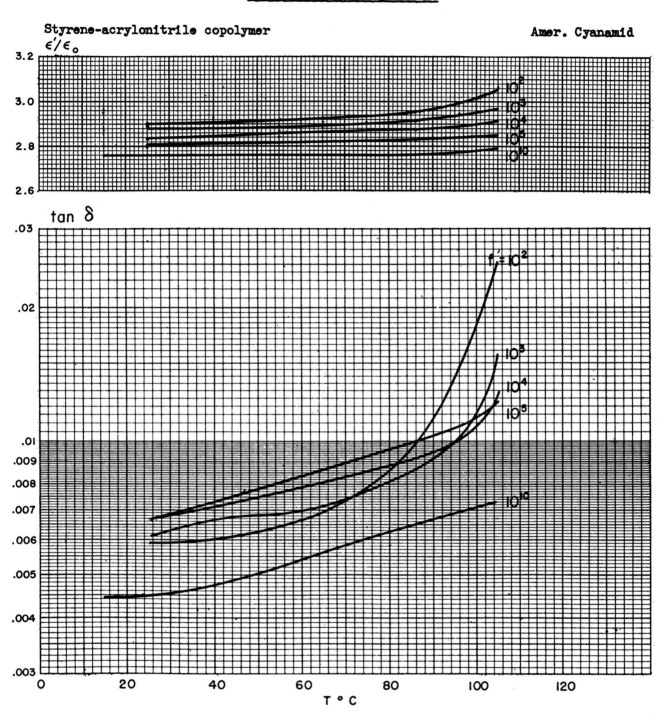

Styrene Copolymers, Linear (cont.)

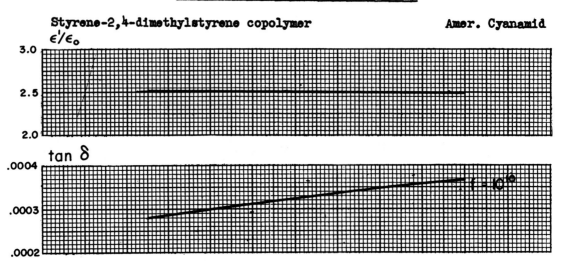

Styrene Copolymers, Cross-linked

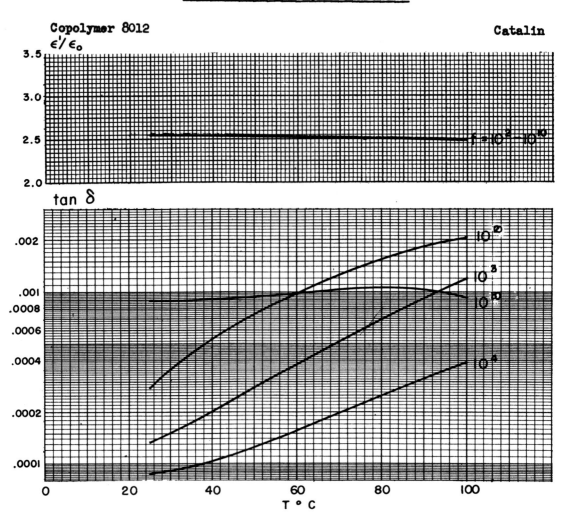

Polyesters

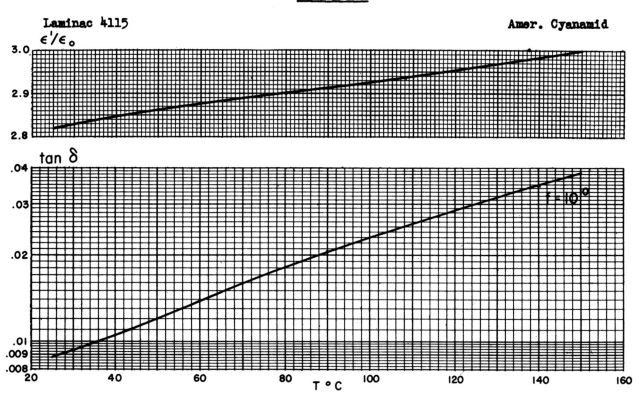

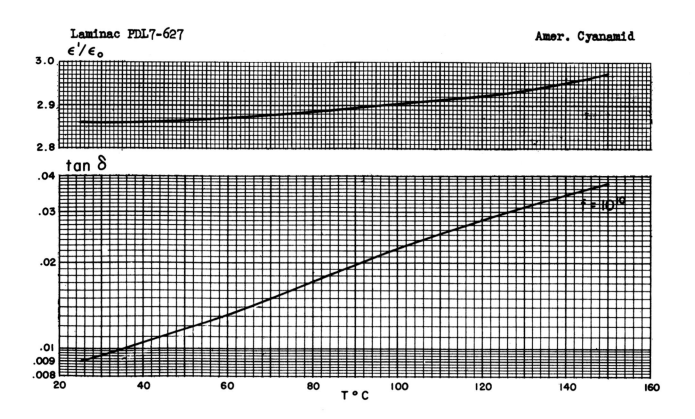

Polyesters (cont.)

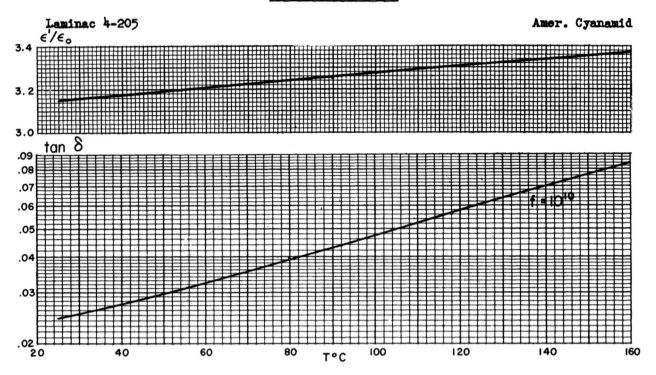

Laminac 4-205 — Amer. Cyanamid

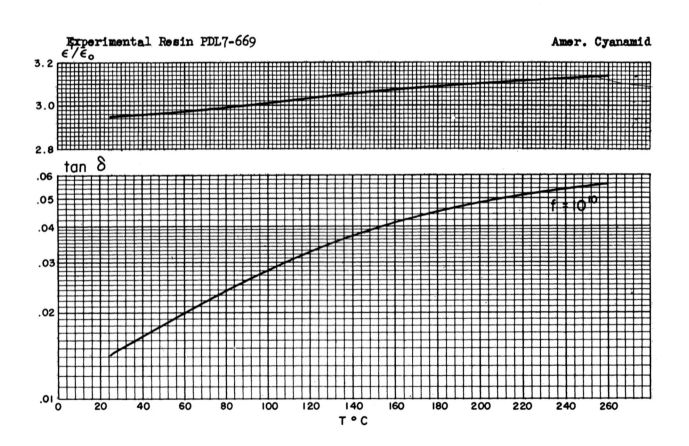

Experimental Resin PDL7-669 — Amer. Cyanamid

Polyesters (cont.)

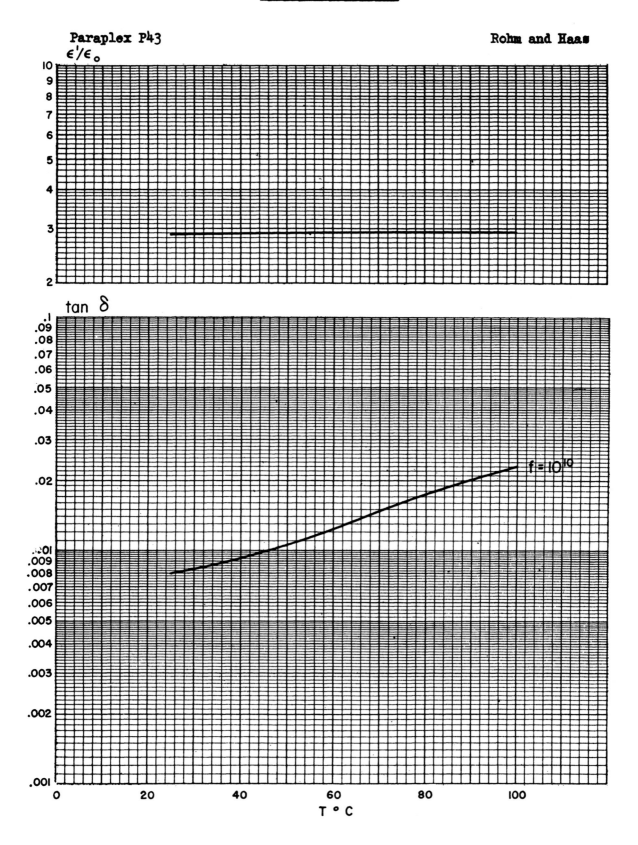

Polyesters (cont.)

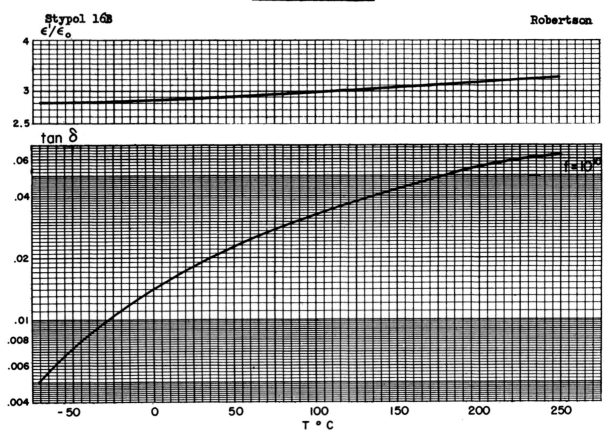

Alkyd Resins

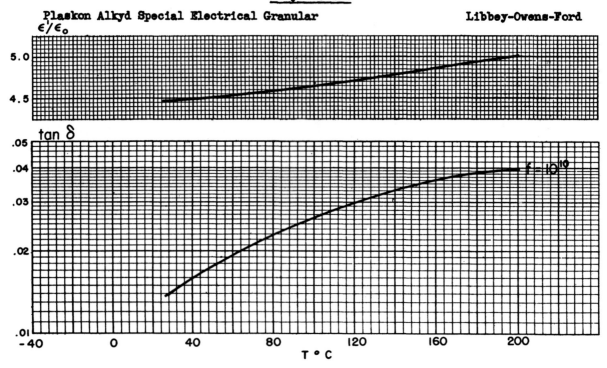

Alkyd Resins (cont.)

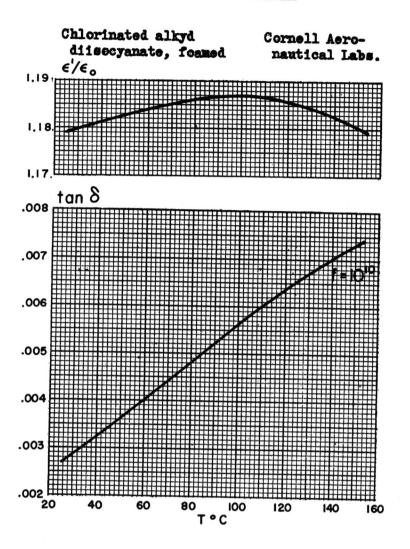

Chlorinated alkyd diisocyanate, foamed

Cornell Aeronautical Labs.

Epoxy Resin

Araldite Casting Resin Type B — Ciba

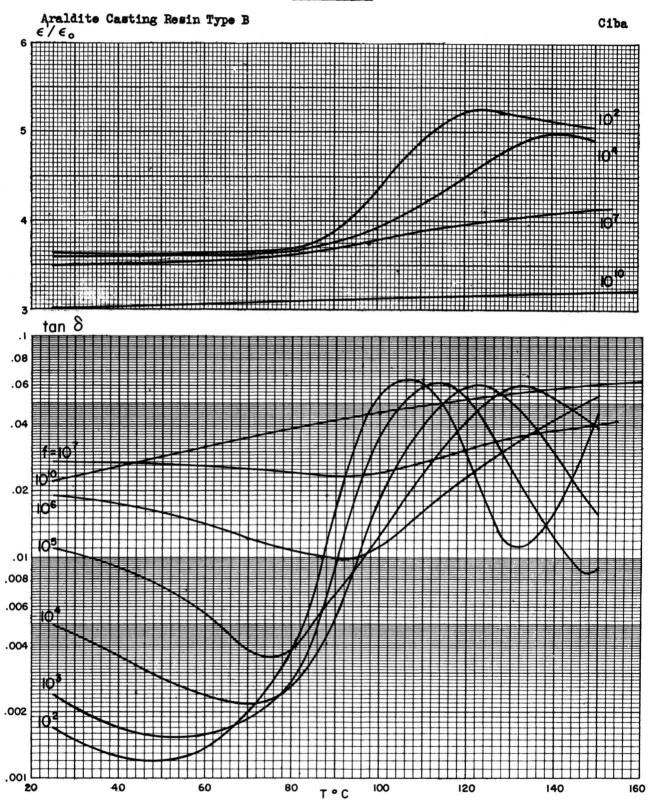

Silicone Rubbers

Silastic 181, 250, 6167 — Dow Corning

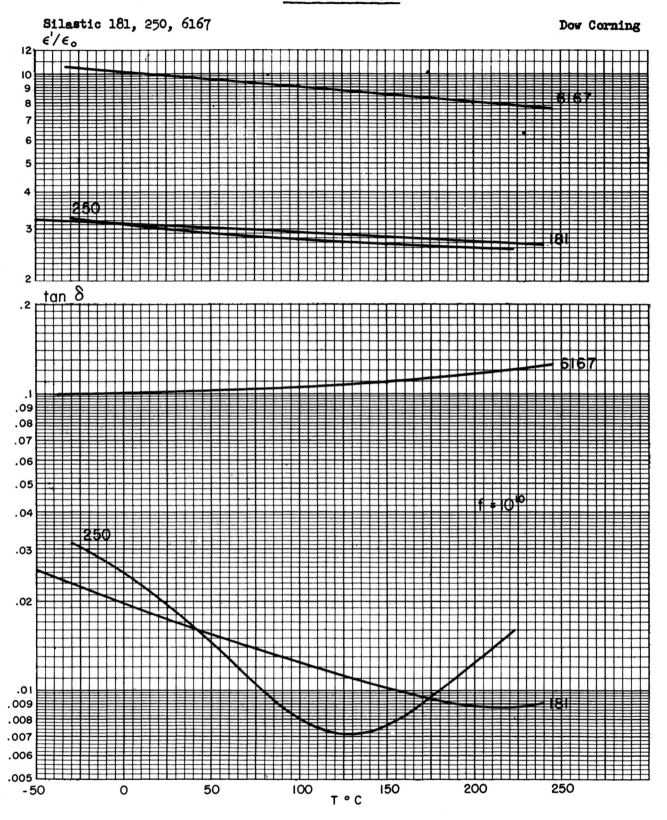

Company Index

Aircraft-Marine Products, Inc., 314
    1523 N. Fourth St., Harrisburg, Pa.
Allison and Co., W. M., 358
    162 Water St., New York, N. Y.
Amber Mines, Inc., 355
    353 Fifth Ave., New York, N. Y.
American Cyanamid Co., 321–323, 338, 346, 411, 417–420
    Plastics Development Laboratories, 1937 West Main St., Stamford, Conn.
American Lava Corp., 303, 304, 306, 377–381, 387–392
    Chattanooga 5, Tenn.
American Optical Co., 309
    Southbridge, Mass.
American Phenolic Corp., 328
    1830 South 54th Ave., Chicago 50, Ill.
American Smelting and Refining Co., 313
    120 Broadway, New York 5, N. Y.

Bakelite Corp., 315, 316, 320, 327–329, 336, 348, 357, 358, 364
    30 East 42nd St., New York 17, N. Y.
Bell Telephone Laboratories, 303, 382
    463 West St., New York, N. Y.
Bromund and Co., E. A., 357
    258 Broadway, New York, N. Y.
Brush Development Co., 301, 372
    3405 Perkins Ave., Cleveland 14, Ohio

California Research Corp., 365
    Box 1627, Richmond, Calif.
Candy and Co., Inc., 357
    35th St. at Maplewood Ave., Chicago, Ill.
Cantol Wax Co., 359
    211 N. Washington St., Bloomington, Ind.
Catalin Corporation of America, 316–318, 336, 339, 418
    Plant and Laboratories, Fords, N. J.
Celanese Corporation of America, 324, 325, 413, 414
    Plastics Division, 290 Ferry St., Newark 5, N. J.
Cellular Rubber Products, Inc., 351
    P.O. Box 126, Willimantic, Conn.
Central Scientific Co., 356
    79 Amherst St., Cambridge, Mass.
Centralab Globe-Union, Inc., 303, 383–386
    932 E. Keefe Ave., Milwaukee 1, Wis.
Ciba Co., Inc., 321, 349, 350, 424
    Plastics Division, 627 Greenwich St., New York, N. Y.
Components and Systems Laboratory, see U. S., Components and Systems Laboratory
Continental-Diamond Fibre Co., 318, 319, 321, 326, 327, 331, 332
    Bridgeport, Pa.
Coors Porcelain Co., 306, 393–398
    Golden, Colo.
Cornell Aeronautical Laboratories, 348, 423
    Cornell University, Ithaca, N. Y.
Corning Glass Works, 309–311, 326, 341, 342, 401, 402
    Physical Laboratory, Research and Development Dept., Corning, N. Y.

Crowley and Co., Henry L., 304, 307
    1 Central Ave., West Orange, N. J.

Dennison Manufacturing Co., 358
    45 Ford Ave., Framingham, Mass.
Dewey and Almy Chemical Co., 338
    Organic Chemicals Division, Cambridge 40, Mass.
Dow Chemical Co., 336–342, 346, 364, 416
    Plastics Division, Midland, Mich.
Dow Corning Corp., 326, 327, 341, 342, 354, 355, 366, 367, 425
    Midland, Mich.
Du Pont de Nemours and Co., Inc., E. I.:
    Electrochemicals Dept., 333, 364
        Niagara Falls, N. Y.
    Organic Chemicals Dept., 353
        P.O. Box 525, Wilmington, Del.
    Pigments Dept., 304
        Newport, Del.
    Polychemicals Dept., 323, 325, 327, 332, 334, 335, 358, 412
        Wilmington, Del.
Durez Plastics and Chemicals, Inc., 317, 320
    North Tonawanda, N. Y.
Durite Plastics, Inc.,* 317, 319
    5000 Summerdale Ave., Philadelphia, Pa.

Eastman Kodak Co., 315, 362
    Rochester 4, N. Y.
Electronic Mechanics, Inc., 313, 407
    70 Clifton Blvd., Clifton, N. J.
Engineer Research and Development Laboratory, see U. S., Engineer Research and Development Laboratory
Enjay Co., Inc., 328, 339, 350
    15 West 51st St., New York 19, N. Y.
Esso Laboratories, see Enjay Co., Inc.

Filtered Rosin Products, Inc., 356
    Brunswick, Ga.
Formica Co., The, 317, 320, 321, 325, 346, 347
    4614 Spring Grove Ave., Cincinnati 32, Ohio
Frenchtown Porcelain Co., 306
    Frenchtown, N. J.

General Aniline and Film Corp., 334, 346
    Coal and Lincoln Sts., Easton, Pa.
General Cable Corp., 352
    Bayonne, N. J.
General Ceramics and Steatite Corp., 304, 308, 382
    Keasbey, N. J.
General Electric Co., 363–365
    100 Woodlawn Ave., Pittsfield, Mass.
    Chemical Div., 348
        Pittsfield, Mass.
    Plastics Dept., 350
        1 Plastics Ave., Pittsfield, Mass.
    1 River Rd., 313, 405, 406
        Schenectady, N. Y.

* Now The Borden Co., Chemical Div., 350 Madison Ave., New York 17, N. Y.

Company Index

General Electric Co. (*continued*)
  Chemicals Dept., 355, 367
    Waterford, N. Y.
  Lamp Dept., 311, 403
    113 East 152nd St., Cleveland 10, Ohio
General Mills, Inc., 323
  2010 East Hennepin Ave., Minneapolis 13, Minn.
Glastic Corp., The, 347, 349
  1823 East 40th St., Cleveland 3, Ohio
Glenco Corp., 305, 306
  Metuchen, N. J.
Glyco Products Co., 356, 359
  26 Court St., Brooklyn 2, N. Y.
Goodrich Chemical Co., B. F., 329, 330
  Avon Lake Experimental Station, P.O. Box 122, Avon Lake, Ohio
Goodyear Aircraft Corp., 348
  Akron 15, Ohio
Goodyear Tire and Rubber Co., Inc., The:
  Plastics Dept., 340, 351, 352
    Akron 16, Ohio
Gulf Oil Co., 358
  P.O. Box 2038, Pittsburgh, Pa.

Hardman Co., H. V., 349, 351
  571 Cortlandt St., Belleville 9, N. J.
Harshaw Chemical Co., 101, 102
  1945 East 97th St., Cleveland 6, Ohio
Hercules Powder Co., Inc., 323
  Cellulose Products Dept., Parlin, N. J.
Hood Rubber Co., 322
  98 Nichols Ave., Watertown 72, Mass.
Hooker Electrochemical Co., 362
  Niagara Falls, N. Y.
Houghton Laboratories, Inc., 350
  322 Bush St., Olean, N. Y.

Johns-Manville, 313
  22 East 40th St., New York 16, N. Y.

Kearney Corp., James R., 352
  4236 Clayton Ave., St. Louis 10, Mo.
Kellogg Co., The M. W., 331, 358, 363, 415
  P.O. Box 469, Jersey City 3, N. J.
Knox Porcelain Corp., 306, 399
  Knoxville 11, Tenn.
Kuhne-Libby Co., 357
  54 Front St., New York 4, N. Y.

Laminated Plastics, Inc., *see* Glastic Corp.
Libbey-Owens-Ford Glass Co., 322, 323, 347, 349, 422
  Plaskon Div., 2112-24 Sylvan Ave., Toledo 6, Ohio
Linde Air Products Co., The, 301, 302, 371, 376
  East Park Dr. and Woodward Ave., Tonawanda, N. Y.
Lovell Chemical Co., 357, 358
  Watertown 72, Mass.
Lucoflex Plastic Fabricating, Inc., 330
  242 Village St., Medway, Mass.

Mallinckrodt Chemical Works, 344, 345, 361, 362
  3600 North Second St., St. Louis, Mo.
Marbon Corporation, 338
  1926 West 10th Ave., Gary, Ind.

Marco Chemicals, Inc., 347 (now Celanese Corporation of America)
  Sewaren, N. J.
Massachusetts Institute of Technology, 301, 302, 305, 311, 312, 326, 337, 341-346, 362, 364, 404
  Laboratory for Insulation Research, 77 Massachusetts Ave., Cambridge 39, Mass.
Mathieson Chemical Corp., 341
  P.O. Box 480, Niagara Falls, N. Y.
Mica Insulator Co., 321
  801 Broadway, Schenectady, N. Y.
Minnesota Mining and Manufacturing Co., 363
  900 Fauquier Ave., St. Paul 6, Minn.
Mitchell-Rand Insulation Co., 357-359
  51 Murray St., New York, N. Y.
Monsanto Chemical Co., 315, 318, 320, 322, 324, 330, 336, 337, 341-345, 363, 364
  Plastics Division, Springfield 2, Mass.
Mycalex Corporation of America, 313
  60 Clifton Blvd., Clifton, N. J.

Naugatuck Chemical, 347, 353
  Naugatuck, Conn.
Norton Co., 301, 306
  Worcester 6, Mass.

Owens-Corning Fiberglas Corp., 311, 316, 322, 347, 348
  Newark, Ohio

Pennsylvania Industrial Chemical Corp., 339, 350
  Clairton, Pa. (Products distributed by Standard Chemical Co., Akron 8, Ohio)
Pittsburgh Corning Corp., 311
  632 Duquesne Way, Pittsburgh, Pa.
Plaskon Division, 322, 323, 347, 349
  Libbey-Owens-Ford Glass Co., 2112-24 Sylvan Ave., Toledo 6, Ohio
Plastic Metals, Inc., 343, 344
  Johnstown, Pa.
Plax Corp., 335
  133 Walnut St., Hartford 5, Conn.
Polaroid Corp., 335, 337, 340
  Cambridge, Mass.
Procter and Gamble Manufacturing Co., 360
  Cincinnati 1, Ohio

Resinous Products and Chemical Co., *see* Rohm and Haas Co.
Rex Corp., 340
  Hayward Rd., West Acton, Mass.
Rezolin, Inc., 319
  5736 West 96th St., Los Angeles 45, Calif.
Robertson Co., H. H., 348, 422
  2400 Farmers Bank Bldg., Pittsburgh 22, Pa.
Rogers Paper Manufacturing Co., 360
  Manchester, Conn.
Rohm and Haas Co., 330, 334, 348, 362, 421
  Washington Square, Philadelphia 5, Pa.
Rubber Reserve Corp., 351, 352
  Washington 1, D. C.

St. Regis Paper Co., 317, 322
  Panelyte Division, Enterprise Ave., Trenton, N. J.
Sharples Chemicals, Inc., 363
  123 South Broad St., Philadelphia 9, Pa.

Shawinigan Products Corp., 334
    Empire State Bldg., New York 1, N. Y.
Shell Chemical Corp., 350
    Resins and Plastics Dept., 50 West 50th St., New York 20, N. Y.
Shell Development Co., 346–348
    4560 Horton St., Emeryville 8, Calif.
Shell Oil Co., 356, 366
    Special Products Dept., 50 West 50th St., New York 20, N. Y.
Socony-Vacuum Oil Co., Inc., 357, 358
    Technical Service Division, 412 Greenpoint Ave., New York, N. Y.
Southern Alkali Corp., 348
    Barberton, Ohio
Sponge Rubber Products Co., 319, 331, 352
    Shelton, Conn.
Sprague Electric Co., 339
    North Adams, Mass.
Stanco Distributors, Inc., 365, 366
    Chemical Products Dept., 26 Broadway, New York 4, N. Y.
Standard Oil Co. of N. J., 358
    26 Broadway, New York, N. Y.
Standard Oil Development Co., *see* Enjay Co., Inc.
Stupakoff Ceramic and Manufacturing Co., 306, 400
    Hillview Ave., Latrobe, Pa.

Taylor Fibre Co., 318, 327, 410
    Norristown, Pa.
Tennessee Eastman Corp., 324
    Kingsport, Tenn.
Tennessee Marble, Inc., 313
    Knoxville, Tenn.
Thiokol Corp., 354
    780 North Clinton Ave., Trenton 7, N. J.
Titanium Alloy Manufacturing Division, 304, 305, 342, 343
    National Lead Co., Hyde Park Blvd., Niagara Falls, N. Y.

U. S. Gasket Co., 332, 333
    P.O. Box 93, Camden, N. J.
United States (Government):
    Components and Systems Laboratory, 309, 400
        Air Matériel Command, Wright-Patterson Air Force Base, Dayton, Ohio
    Engineer Research and Development Laboratory, 302, 373–375
        Fort Belvoir, Va.
    National Bureau of Standards, 340
        Washington, D. C.
    War Department, 315
        Picatinny Arsenal, Dover, N. J.
U. S. Industrial Chemicals, Inc., 362
    60 East 42nd St., New York 17, N. Y.
U. S. Rubber Co., 331, 340, 352, 353, 356
    General Laboratories, Passaic, N. J.

Victor Chemical Works, 348
    Board of Trade Building, 141 West Jackson Blvd., Chicago 4, Ill.

War Department, Picatinny Arsenal, *see* U. S. War Department, Picatinny Arsenal
Weber and Co., Hermann, 351
    76 Beaver St., New York 5, N. Y.
Westinghouse Electric Corp., 318, 319, 322
    Research Laboratories, East Pittsburgh, Pa.

Zinsser and Co., Wm., 355, 356
    516 West 59th St., New York, N. Y.
Zophar Mills, Inc., 359
    112–130 26th St., Brooklyn, N. Y.

Materials Index

Acrawax C, p. 356
Acrylate resins, pp. 334, 335
Acrylonitrile-butadiene copolymer, p. 353
Air seal, p. 352
Alathon, pp. 327, 370
Alcohols, p. 362
Alkyd resins, pp. 347–349, 422, 423
Allyl resins, pp. 347, 348
Allymer CR-39, CR-39 + glass, p. 348
AlSiMag A-35, pp. 303, 379
AlSiMag A-196, pp. 303, 377, 378
AlSiMag 211, p. 303
AlSiMag 228, pp. 303, 380
AlSiMag 243, pp. 303, 381
AlSiMag 393, p. 303
AlSiMag 491, pp. 306, 399
AlSiMag 505, p. 382
Aluminum oxide, pp. 301, 303, 306, 372, 395–399
Alvar 11/90, p. 334
Amber, p. 355
Ammonium dihydrogen phosphate, p. 301
Amplifilm, p. 314
Aniline-formaldehyde resins, p. 321
Apiezon Wax "W," p. 356
Araldite Adhesive, Type I, natural and silver, p. 350
Araldite Casting Resin, Type B, pp. 349, 424
Araldite Casting Resin G, p. 349
Araldite E-134, p. 349
Aroclor 1221, 1232, 1242, 1248, 1254, p. 363
Aroclor 1260, 1262, 5442, p. 364
Aroclor 1268, 4465, 5460, p. 315
Asbestos, p. 313
Asphalts and cements, p. 356

Bakelite BM-120, pp. 315, 316, 369
Bakelite BM-250, BT-48-306, BM-16981, and BM-16981 powder, p. 316
Bakelite BM-262, BM-1895, pp. 320, 369
Bakelite BRS-16631 + glass, p. 348
Bakelite BV-17085 + glass, p. 316
Balata, precipitated, p. 351
Barium-strontium titanate, pp. 305, 306
Barium titanate, p. 305
Barium titanate and plastic mixtures, p. 343
Bayol, Bayol-16, p. 366
Bayol-D, Bayol-F, p. 365
Beeswax, white, yellow, p. 357
Beetle resin, p. 323
Bentonite, p. 314
Benzenes, chloro-, p. 364
Benzenes and diphenyls, chlorinated, p. 363
Benzoguanamine-formaldehyde resin, p. 323
Beryllium oxide, p. 306
Biphenyls, chlorinated, pp. 315, 363, 364
Bitumen, natural, p. 356
Buna S (GR-S) and compounds, p. 352
Bureau of Standards Casting Resin, p. 340
Butadiene, chloro-, pp. 353, 362

Butadiene-acrylonitrile copolymer, p. 353
Butadiene-styrene copolymer, pp. 338, 352
Butvar, Low OH and 55/98, p. 334
n-Butyl Alcohol, p. 362
Butyl rubbers, p. 352
Butyraldehyde, p. 362

Cable Oil 5314 and PL101270, p. 365
Calcium titanate, p. 305
Carbon and plastic mixtures, p. 341
Carbon tetrachloride, p. 362
Catalin 200, 500, and 700 base, pp. 316, 317
Catalin 8012, pp. 339, 418
Catalin EK 2784, p. 339
Cellulose Acetate LL-1, p. 323
Cellulose acetate + plasticizer, p. 324
Cellulose acetates, pp. 323, 324
Cellulose derivatives, pp. 323–325
Cellulose nitrate and camphor, p. 325
Cellulose propionate, pp. 325, 413
Cements and asphalts, p. 356
Cenco Sealstix, p. 356
Ceramic F-66, pp. 303, 382
Ceramic NPOT 96, pp. 304, 387, 388
Ceramic N750T96, pp. 304, 389, 390
Ceramic N1400T110, pp. 304, 391, 392
Ceramic T106, p. 304
Ceramics, pp. 303–308, 378, 400
Cerese Wax AA and brown, p. 357
Ceresin, white and yellow, p. 357
Cetylacetamide, p. 356
Chemelac M1405, M1406, M1407, M1411, M1412, and M1414, p. 332
Chemelac M1418-2, M1418-5, M1422, and M1423, p. 333
Chemlac B-3, p. 356
Chlorinated benzenes and biphenyls, pp. 315, 363, 364
β-Chloroethyl-2,5-dichlorobenzene, p. 364
Chlorostyrenes, ortho and para, copolymer, p. 342
Cibanite, p. 321
Coors AB-2, pp. 306, 397, 398
Coors AI-200, pp. 306, 395, 396
Copolene B, p. 328
Corfoam 114, p. 319
Corning Glass Nos. 0010, 0014, 0080, 0090, 0100, 0120, 1770, p. 309
Corning Glass No. 1990, pp. 309, 401
Corning Glass Nos. 1991, 3320, 7040, 7050, p. 309
Corning Glass Nos. 7052, 7055, 7060, p. 310
Corning Glass No. 7070, pp. 310, 401
Corning Glass Nos. 7230, 7570, 7720, 7740, 7750, 7900, p. 310
Corning Glass No. 7911, pp. 310, 402
Corning Glass Nos. 8460, 8830, p. 310
Corning Glass Nos. 8871, 9010, Lab. No. 189CS, p. 311
Crepe, pale and compounds, p. 351
Cresylic-acid, formaldehyde resins, pp. 318, 319, 368, 370
Crolite No. 29, p. 304
Crowloy 20, 70, BX113, BX114, p. 307
Crystals, inorganic, pp. 301, 302, 371–376
Crystals, organic, p. 315

Darex No. 3, 43E, X-34, X-43, p. 338
DeKhotinsky Cement, p. 356
Diala Oil, p. 366
Diallyl phenyl phosphonate resin, p. 348
Diatomaceous-earth ceramic, pp. 306, 400
Dibutyl sebacate, p. 362
Dichloronaphthalenes, mixture of the 1,2-, 1,4-, and 1,5-isomers, p. 358
Dichloropentanes #14, #40, p. 363
2,5-Dichlorostyrene, p. 364
Dilectene 100, p. 321
Dilecto GB-112S, p. 327
Dilecto GB-112T, pp. 331, 332
Dilecto GB-261S, p. 326
Dilecto (hot punching), p. 319
Dilecto (mecoboard), p. 318
Dioctyl sebacate, p. 362
Diphenyl, see Biphenyls
Dow C-244, p. 336
Dow Experimental Plastic Q-166, Q-166 + Fiberglas, Q-200.5, p. 339
Dow Experimental Plastic Q-247.1, p. 336
Dow Experimental Plastic Q-344, p. 340
Dow Experimental Plastic Q-406, pp. 337, 416
Dow Experimental Plastic Q-409, p. 342
Dow Experimental Plastic Q-475.5, p. 340
Dow Experimental Plastic Q-764.6, Q-767.2, p. 337
Dow Experimental Plastic Q-817.1, pp. 337, 416
Durez 1601, natural, p. 317
Durez 11863, p. 320
Durite #500, p. 317
Durite #221X, p. 319

E Resin, p. 350
Elastomers, pp. 351–355
Elvacet 42A-900, p. 333
Elvanol 51A-05, 50A-42, 70A-05, 72A-05, 72A-51, p. 333
Ensolite M22240, M22239, 3036, p. 331
Epon Resin RN-48, p. 350
Epoxy resins, pp. 349, 350, 424
Estawax, p. 357
Ethocel LT5, pp. 325, 414
Ethyl alcohol, p. 362
Ethyl cellulose, pp. 325, 414
Ethylene Glycol, p. 362
Ethylpolychlorobenzene, p. 364

Ferramic A, B, C, D, G, H, I, J, p. 308
Ferrites, pp. 307, 308
Fiberglas, pp. 322, 368
Fiberglas BK-174, laminated, p. 316
Fibestos 2050TVA C-1686, p. 324
Foamglas, p. 311
Formaldehyde resins, aniline, p. 321
Formaldehyde resins, benzoguanamine, p. 323
Formaldehyde resins, cresol, pp. 318, 319, 368, 370
Formaldehyde resins, melamine, pp. 321, 322, 411
Formaldehyde resins, phenol, pp. 315–319, 408–410
Formaldehyde resins, phenol-aniline, p. 320
Formaldehyde resins, urea, p. 323
Formica FF-41 (sheet, rod stock), pp. 321, 368
Formica FF-55, pp. 321, 369
Formica G7, G6, p. 325
Formica Grade MF-66, pp. 320, 368

Formica XX, LE, pp. 317, 368
Formica YN-25, p. 317
Formica Z65, p. 346
Formica Z80, p. 347
Formvar, Type E, p. 334
Forticel, pp. 325, 413
Fractol A, p. 366
Furfuraldehyde resin, phenol, p. 319

Gafite cast polymer, p. 334
Gasoline, aviation, 100 and 91 octane, p. 365
Geon 2046, 80365, 80384, p. 329
Gilsonite, p. 356
Glass, alkali-silica, pp. 312, 404
Glass, alkaline lead silicate, p. 311
Glass, aluminum borosilicate, p. 310
Glass, aluminum zinc-phosphate, p. 309
Glass, barium borosilicate, p. 310
Glass, borosilicate, pp. 309, 400
Glass, "E," p. 311
Glass, iron-sealing, pp. 309, 401
Glass, lead-barium, p. 309
Glass, lime-alumina-silicate, p. 309
Glass, low alkali, potash-, lithia-borosilicate, pp. 310, 401
Glass, Phosphate 2043x, 2279x, p. 309
Glass, potash-lead-silicate, p. 309
Glass, potash-soda-barium-silicate, p. 309
Glass, silica, pp. 311, 401
Glass, soda-borosilicate, pp. 309, 310
Glass, soda-lead-borosilicate, p. 310
Glass, soda-lime-silicate, pp. 309, 311
Glass, soda-potash-borosilicate, p. 309
Glass, soda-potash-lead-silicate, p. 309
Glass, soda-potash-lithia-borosilicate, p. 310
Glass, soda-silica, p. 311
Glass Lamicoid #6038, pp. 321, 370
Glass and mica, pp. 313, 369, 405–407
Glass and plastic mixtures, pp. 341, 342
Glasses, pp. 309–312, 400–404
Glastic GF, MM, MP, and A-2, p. 348
Glastic S and MF, p. 347
Glyptol #1201 (red), p. 348
GR-I (butyl rubber) and compound, p. 352
GR-S (Buna S) and compounds, p. 352
Gutta-percha, p. 351

Halowax #1001, p. 357
Halowax #11-314, p. 358
Halowax Oil 1000, p. 364
Heptacosafluorotributyl amine, p. 363
Heptane, p. 362
Hexachlorobutadiene, p. 362
Hexamethylene-adipamide polymer, pp. 323, 412
Hycar OR Cell-tite, p. 352
Hydrocarbon polymer, cross-linked, p. 350
Hydrocarbons, petroleum, pp. 365, 366
12-Hydroxystearin, p. 358
Hysol 6000 and 6020, p. 350
Hysol 6030, pp. 350, 370
Hy-tuf Laminate Grade GF181, p. 350

Ice, p. 301
Ignition Sealing Compound #4, p. 367
Iron and plastic mixtures, pp. 343, 344

Materials Index

Iron-manganese oxide and plastic mixtures, p. 345
Isobutylene-isoprene copolymer, p. 352

Jet fuel JP-1 and JP-3, p. 365

Kel-F, p. 331
Kel-F Grade 300 and 300-P25, pp. 331, 415
Kel-F Grease #40, p. 363
Kel-F Oil, Grade #1, #3, #10, p. 363
Kel-F Wax #150, p. 358
Kerosene, p. 365
Koroseal 5CS-243, p. 330
Kralastic BE, BM, D, EBMU, F, p. 353
KRS-5, pp. 302, 375
KRS-6, pp. 302, 374

Laminac 4115, pp. 346, 419
Laminac 4-205, pp. 346, 420
Laminac PDL7-627 and PDL7-650, pp. 346, 419
Laminate BD-44 and BK 164, pp. 347, 348
Leather, sole, p. 360
Liquids, aliphatic, pp. 362, 363
Liquids, aromatic, pp. 363, 364
Liquids, inorganic, p. 361
Liquids, organic, pp. 362–367
Liquids, petroleum, pp. 365, 366
Liquids, silicone, pp. 366, 367
Lithium fluoride, p. 301
Loalin, p. 336
Lucite, sintered, p. 334
Lucite HM-119 and HM-140, pp. 334, 370
Lucoflex, p. 330
Lumarith XFA-H4 and 22361, pp. 324, 325, 370
Lustrex loaded glass mat, p. 341

Magnesium oxide, p. 301
Magnesium silicate, pp. 303, 377–381
Magnesium titanate, p. 304
Magnesium titanate and plastic mixture, p. 343
Magnetite and plastic mixtures, pp. 344, 345
Marble S-3030, p. 313
Marbon S, Buna S Hardboard, p. 352
Marbon S (Code 7206), S-1 (Code 7254), 8000 and 9200, p. 338
Marco Resin MR-21C, MR-23C and MR-25C, p. 347
Marcol, p. 365
Mathieson Plastic CY-8 and CQ-10DM, p. 341
Meat, p. 360
Melamine-formaldehyde resins, pp. 321, 322, 411
Melamine GMG, pp. 321, 370
Melmac 7278 + "E" glass, p. 322
Melmac Molding Comp. 1500, 1502, pp. 322, 411
Melmac Resin 592, p. 321
Melmac Type 1077 (Ivory WB 48), p. 321
Methacrylate resins, pp. 334, 335
Methocel, p. 325
Methyl alcohol, p. 362
Methyl cellulose, p. 325
Methylstyrene-styrene copolymer, p. 339
Mica, Canadian, p. 313
Mica, ruby, p. 313
Mica and glass, pp. 313, 369, 405–407
Micarta #254, p. 318
Micarta #259, pp. 322, 368
Micarta #299, pp. 319, 368

Micarta #496, pp. 318, 368, 370
Millimar, p. 356
Muscovite, p. 313
Mycalex K10, p. 313
Mycalex 400, pp. 313, 369
Mycalex 2821, pp. 313, 405, 406
Mykroy Grade 8 and 38, pp. 313, 407

Naphthalene, p. 315
Naphthalene, chloro, p. 364
Naugatuck Laminating Resin MP and MT, p. 347
Neoprene GN and compounds, p. 353
Nitrobenzene, p. 364
Nylon 66 and 610, p. 323
Nylon FM 10,001, pp. 323, 412

Oil, HB-40, p. 363
Oils, petroleum, pp. 365, 366
Opalwax, p. 358
Ozokerite, p. 358

Panelyte Grade 140 and 776, pp. 317, 322
Paper, Royalgrey, p. 360
Paraffin, natural, p. 358
Paraffin Wax 132° ASTM and 135° AMP, p. 358
Paraplex P13 and P43, pp. 348, 421
Parowax, p. 358
Perfluorodihexyl ether, p. 363
Permafil 3256, p. 350
Permo Potting Compound #49 and #51, pp. 349, 351
Petroleum oils, pp. 365, 366
Phenol-aniline-formaldehyde resins, p. 320
Phenol-formaldehyde resins, pp. 315–319, 408–410
Phenol-furfuraldehyde resin, p. 319
Phenolic, Expanded, p. 319
Phenolic paper laminate JH-1410, p. 318
Phoresin, p. 348
Piccolastic D-125, p. 339
Piccopale Resin, p. 350
Plaskon 911, p. 347
Plaskon Alkyd 411, 420, 422, 440, 440A and 442, p. 349
Plaskon Alkyd Special Electrical Granular, pp. 349, 422
Plaskon melamine, p. 322
Plaskon Urea, natural and brown, p. 323
Plasticell, p. 331
Plast-Iron and plastic mixtures, pp. 343, 344
Plexiglas, p. 334
Plicene Cement, p. 356
Pliobond M-190-C, p. 352
Pliolite S5, S3, S6B and S6, p. 340
Pliolite and Pliolite GR, p. 351
Polaroid Resin C, p. 340
Polectron #24, p. 346
Polinel, p. 358
Polyamide resins, pp. 323, 412
Polybutene, p. 359
Polybutyl methacrylate, p. 335
Poly-2-chlorobutadiene-1,3, p. 353
Polychlorostyrenes, pp. 341, 342
Polychlorotrifluoroethylene, pp. 331, 358, 363, 415
Polycyclohexyl methacrylate, p. 335
Polydiallyl phthalate, p. 348
Poly-2,5-dichlorostyrene, pp. 340–342, 370
Poly-2,5-dichlorostyrene + fillers, pp. 342–345

Poly-3,4-dichlorostyrene, p. 342
Polyesters, pp. 346–348, 419–422
Polyethyl methacrylate, p. 335
Polyethylene, pp. 327, 328, 370
Polyethylene (effect of milling), p. 328
Polyethylene DE-3401, p. 327
Polyglas D$^+$, p. 342
Polyglas M, p. 322
Polyglas P$^+$, p. 341
Polyglas S, p. 326
Polyisobutyl methacrylate, p. 335
Polyisobutylene B-100 + Marbon B, p. 328
Polyisobutylene, Run 5047-2, p. 328
Polymethyl methacrylate, pp. 334, 370
Polystyrene, pp. 335–337, 416
Polystyrene, cast in vacuo and cast in air, p. 337
Polystyrene, cross-linked, pp. 339, 340, 418
Polystyrene, hydrogenated, p. 346
Polystyrene, α-methylstyrene, p. 337, 416
Polystyrene + chlorinated diphenyl, p. 337
Polystyrene Fibers Q-107, p. 337
Polystyrene + fillers, p. 341
Polystyrene Foam Q-103, p. 337
Polytetrafluoroethylene, pp. 331–333, 370
Polythene A-3305, pp. 327, 370
Polyvinyl acetal, p. 334
Polyvinyl acetate, p. 333
Polyvinyl alcohol-acetates, p. 333
Polyvinyl butyral, p. 334
Poly-N-vinylcarbazole, p. 346
Polyvinyl chloride 1006, 1018, 1216, 1406, W-174, W-175 and W-176, p. 330
Polyvinyl chloride-acetate, pp. 328–331
Polyvinyl chloride-acetate + plasticizer, pp. 328, 329
Polyvinyl chlorides, pp. 328–331
Polyvinylcyclohexane, p. 346
Polyvinyl formal, p. 334
Polyvinylidene and vinyl chlorides, p. 331
Poly-α-vinylnaphthalene, p. 346
Poly-2-vinylpyridine, p. 346
Polyvinyl resins, pp. 327–346
Polyvinyltoluene, pp. 337, 416
Poly-p-xylylene, p. 337
Porcelain, wet and dry process, pp. 306, 399
Porcelain #4462, p. 306
Porcelains, pp. 306, 393–399
Porous Ceramic AF-497, pp. 306, 400
Potassium bromide, p. 301
Potassium dihydrogen phosphate, pp. 301, 372
Primol-D, p. 366
n-Propyl alcohol, p. 362
Prystal, p. 317
Pyralin, p. 325
Pyranol 1467, 1476 and 1478, pp. 363, 364
Pyrex, p. 310

Quartz, fused, pp. 311, 403
Quinorgo #3000, p. 313
Quinterra, p. 313

Resimene 803-A, p. 322
Resinox 7934, p. 320
Resinox 10231, pp. 318, 408
Resinox 10900, pp. 318, 409

Resins, natural, pp. 355, 356
Rexolite 1422, p. 340
Royalite 149-11, M21982-1 and M22190, p. 353
Rubber, butyl (GR-I), p. 352
Rubber, cellular, p. 351
Rubber, cyclized, pp. 351, 352
Rubber, GR-S (Buna S) and compounds, p. 352
Rubber, Hevea and compounds, p. 351
Rubber, natural, p. 351
Rubber, nitrile, p. 353
Rubber, silicone, pp. 354, 425
Rutile, pp. 302, 304, 376

S-40 and S-60 resins, p. 339
Santicizer 9, p. 315
Sapphire, pp. 301, 371
Saran B-115, p. 331
Sealing Wax, Red Express, p. 358
Selectron 5003 + glass, p. 347
Selenium, amorphous, p. 313
Selenium, multi-crystalline, p. 301
Shellac, natural XL, p. 355
Shellac, natural Zinfo, pure C garnet and garnet dewaxed, p. 356
Silastic 120, 125, 150, 152, 160, 167 and 180, p. 354
Silastic 181 and 250, pp. 354, 425
Silastic X4342, p. 355
Silastic 6167, pp. 355, 425
Silastic 6181, X-6734 and 7181, p. 355
Silica, fused 915c, pp. 311, 402
Silicon dioxide, fused, pp. 311, 402
Silicone fluids DC200 and DC500, p. 366
Silicone fluids DC550 and DC710, p. 367
Silicone fluids SF96-40, SF96-100 and SF96-1000, p. 367
Silicone glass laminates, pp. 326, 327
Silicone Molding Compound XM-3, p. 326
Silicone resin DC996 and DC2101, p. 326
Silicone resins, pp. 325–327
Silicone rubber SE-450, SE-460, SE-550 and SE-972, p. 355
Silicone rubbers, pp. 354, 355, 425
Snow, p. 301
Soap, p. 360
Sodium chloride, p. 302
Sodium chloride, aqueous solutions, p. 361
Soils, p. 314
Steak, p. 360
Steatite bodies, pp. 303, 304, 377–386
Steatite Body 7292, pp. 304, 382
Steatite Type 302, 400, 410 and 452, pp. 303, 383–386
Strontium titanate, p. 305
Strontium titanate and plastic mixtures, p. 343
Stypol 16B, 16C and 16D, pp. 348, 422
Styraloy 22, p. 338
Styramic #18, p. 337
Styramic HT, pp. 342, 343
Styrene copolymers, cross-linked, pp. 339, 340, 418
Styrene copolymers, linear, pp. 338, 339, 417, 418
Styrene dimer, p. 364
Styrene N-100, dry and saturated with water, p. 364
Styrene-acrylonitrile copolymer, pp. 338, 353, 417
Styrofoam 103.7, p. 337
Styron C-176, 411-A, 475, 666 and 671, p. 336
Suet, p. 360
Sulfur, crystalline, p. 302
Sulfur, sublimed, p. 302

Materials Index

Tam Ticon B, BS, C, MC and S, pp. 304, 305, 343
Tam Ticon T-S, T-L and T-M, p. 304
Taylor Grade GGG, pp. 318, 410
Taylor Grade GSC and GSS, p. 327
Teflon, pp. 331–333, 370
Teflon Laminate GB-112T, pp. 331, 332
Tenite I 008A H_2, H_4, M, MH, S and S_4, p. 324
Tenite II 205A H_2, H_4, MH, MS, S_2, S_4, p. 324
Terphenyl, meta-, nona-, ortho-, and para-, p. 315
Terphenyls, chlorinated, p. 364
Tetrachloroethylene, p. 362
Thallium bromide, pp. 302, 374
Thallium bromide-chloride, pp. 302, 374
Thallium bromide-iodide, pp. 302, 375
Thallium iodide, pp. 302, 373
Thermoplastic Composition 1766EX and 3738, p. 358
Thermoplastic Composition 3767A, p. 359
Thiokol, Type FA, PRI and ST, p. 354
TI Pure R-200, p. 304
Titanate ceramics, pp. 305, 306
Titania and titanate bodies, pp. 304–306, 387–392
Titanium dioxide ceramics, pp. 304, 305
Titanium dioxide, rutile, pp. 302, 304, 376
Titanium dioxide + plastic mixtures, p. 342

Toluene sulfonamides, mixture of ortho- and para-isomers, p. 315
Transil Oil 10C, p. 365
Trichlorobenzenes, mixture of isomeric, p. 364
Trichloronaphthalenes, mixture of isomeric, p. 364
α-Trinitrotoluene, p. 315

Ultron Wire Compound UL300, UL1004 and UL2 4001, p. 330
Urea-formaldehyde resins, p. 323

Vaseline, p. 365
Vibron 140 and 141, p. 340
Vinylite QYNA, VG-5544 and VG-5901, p. 328
Vinylite VG-5904, VYHH, VYNS and VYNW, p. 329
Vinylite VU-1900, pp. 329, 370
2-Vinylpyridine-styrene copolymer, p. 346
Vistawax, p. 359

Water, conductivity, p. 361
Wax 3760, p. 359
Wax Compound F-590 and #1340, p. 359
Wax S-1167 and S-1184, p. 359
Waxes, pp. 356–359
Wood, p. 359

Zirconium porcelain Zi-4, pp. 306, 393, 394

INDEX

Absorption, index of, 11
Accelerators, 175
Accelerometer, 188
Acceptor, 245
Accuracy of input impedance measurements, 74–85
Acetylene, 35
Acoustic resonator, 254
Aging of materials, 187, 209
Alkyds, 205
Alumina ceramics, 185
Aluminum conductors, 234
Amines in insulating oils, 166
Ammonia molecule, 35, 135
Ampère's circuital law, 8
Amplifiers, comparison of, 263
  magnetic and dielectric, 261–274
Anisotropic dielectrics, 41, 252
Antinodes, 15
Antioxidants, 174, 219, 225
Arc gases, 203
Arc horns, 202
Arc locus, circular, 39
Armature insulation, 205
Askarels, 157 ff., 195 ff., 223
Atomic structure, 21 ff.
Attenuation factor, 9
Avalanche, streamer theory, 148
Avogadro's number, 21

B-H tracer (after Cioffi), 127
Bacteria cultures from rubber, 229
Band structure of solids, 245
Barium titanate, 42 ff., 187, 188, 260
Benzene, dichloro-, 35
Blocking layer, 244 ff.
Bohr's frequency condition, 25
Bohr magneton, 26, 140
Bohr's quantum theory, 25, 26
Bohr-Stoner arrangement of periodic table, 28, 29
Bonds, carbon, 35
  covalent, 31, 169
  hybrid, 35
  ionic, 31
  van der Waals, 31
Boundary condition, 13
Breakdown, corona, 162, 178, 210, 211, 237
  thermal, 203
  vacuum, 155, 156
Breakdown gradients in gases, 150 ff.

Bridges, 55–58, 71, 127–131
Brittle point of polyethylene, 174, 227
Bushings, 204

Cable, calculations for current-carrying capacity, 239–241
  conductor construction, 234
  microbiological deterioration of, 228, 229
  oils, 157 ff.
  rubber and plastics in, 225–232
  specifications, 227, 241
  wax, 161, 162, 199
Cable engineering, 233 ff.
Cable insulation survey, 236
Capacitance, 3, 51
Capacitors, 186, 187, 197–201, 221–225, 272
Carbon black, addition to polyethylene, 174, 218, 231
Cavity, re-entrant, 61
  resonators, 71 ff.
Cellulose, 171, 193
Ceramics, 179–189
  in circuit breakers, 202
  classification for capacitors, 186
  ferroelectric, 42 ff., 187, 188, 261, 270 ff., 279
  ferromagnetic, 188, 189, 270, 281, 282
  typical properties of, 181
Charge, bound, 5
  elementary, 21
  free, 5
Charts of complex functions $\dfrac{\tanh x}{x}$ and $\dfrac{\coth x}{x}$, 74, 86–104
Circuit breakers, 201 ff.
Classification of dielectrics, 179, 190
Clausius-Mosotti equation, 21, 40
Clay-flint-feldspar diagram, 180
Coax instrument, M.I.T., 72
Coaxial line, 63 ff., 132 ff.
Coil insulation, 206–208
Cole-Cole diagram, 39, 52
Colloidal suspensions, 161, 165
Columbates, sodium and cadmium, 187
Complex functions $\dfrac{\tanh x}{x}$ and $\dfrac{\coth x}{x}$, 74, 86–104
Complex permittivity and permeability, 4, 5
Compounding ingredients for rubber, 226

Conducting plastics, 178, 179, 285
Conduction in liquids, 158 ff.
Conductivity, dielectric, 9, 10
  by irradiation, 220, 221
Conductors of cables, 233 ff.
Conversion formulas for ϵ^* and μ^*, 12, 13, 294
Cordierite, 185
Cores, magnetic, 128, 191, 277, 281
Corona, effects of, 162, 178, 210, 211, 231, 237
Correspondence principle, 25
Corrugations, of insulators, 154
Coulomb force, 22, 31
Coverings, protective, of cables, 227 ff., 238, 239
Cracking temperature, 227
Crystal filter, 260
Crystal resonators, 259 ff.
Curie temperature, 40, 43, 188
Curie-Weiss law, 40
Current, Ampère, 8
  charging, 3
  loss, 3
  pre-breakdown, 149, 156
  time curves, 52, 53

Dacron, 171
De Broglie equations, 26
Debye relaxation equation, 38
Debye unit, 31
Decibel loss, 12
Decomposition products, 160, 161
Deformation of rubber, 172, 173
Degradation of polyethylene, 218
Demagnetizing factor, 126
Detector, square law, 67
Deterioration of insulation, 190 ff., 209, 211, 228, 289
Dielectric constant, field strength, dependence of, 43
  relative, 3
  static, of polar gases, 32
  temperature dependence of, for glycerol, 168
  see also Tables of Dielectric Materials
Dielectric displacement, 5
Dielectric field of E_{010} cavity (derivation), 74, 121, 122
Dielectric parameters, 12, 13, 12, 294
Dielectric polarization, 5–7

Dielectric quality factor Q, 10
Dielectrics, in cables, 225 ff.
  in capacitors, 186, 197, 221
  classification of, 179, 183, 190, 227, 241
  in electronic equipments, 211 ff.
  in power and distribution equipment, 189 ff.
  in rotating machinery, 205 ff.
  in switchgear, 201 ff.
  in transformers, 191 ff.
Dipole moment (electric), 6
  average of polar gases, 32
  of hydrogen atom, 22
  induced, 19, 22
  from microwave spectra, 137
  per mole, 21
  permanent, 19, 31
  per unit volume, 6
  unit of, 31
  vector addition of, 35
Dipole moment (magnetic), of circular current, 25, 139
  of electron, 26
  elementary, 26, 140
  per unit volume, 7
Direct-current amplifier for resistivity measurements, 53
Direct-current breakdown strength, 148 ff., 219
Discharge (electric), in compressed gases, 147 ff.
  flashover, 154
  formation of condensation products, 160
  in liquids, 161
  in vacuum, 154–156
  see also Corona, effects of
Dispersion, normal and anomalous, 24
Dispersion formula of classical physics, 23, 24
Dissipation factor, loss tangent, 4
  maximum, 176, 215
Distribution of relaxation times, 40, 52
Domain structure in $BaTiO_3$, 44
Donor, 245
Double-tee bridge, 71

E_{min}/E_{max}, correction term for, 74, 86
E_{010} mode, 71, 121
Eddy-current shielding, influence on permeability, 125
Electric dipole moment, see Dipole moment (electric)
Electric flux density, 5
Electric susceptibility, 6, 21
Electrode, glass, 166
  influence on breakdown, 150 ff.
Electrode arrangements, 48 ff.
Electrode avalanche, 147, 148
Electrode materials, 48 ff., 150–152
Electron diffraction, 26
Electron emission, 155
Electronegative gases, 149
Electrostatic force, 7
Electrostriction, 41

Elements, periodic system of, 28, 30
Energy, of dissociation, 34
  zero-point, 34
Energy-band picture, 245
Ethane, 35
Ethylene, 35
Expansion coefficients of metal-ceramic seals, 184
Extensional resonance, 256

Faraday's induction law, 4, 8
Ferrites, 188, 189, 270, 281, 282
Ferroelectricity, 42 ff., 270
Ferromagnetic resonance, 146
Field, of E_{010} cavity, 121
  local, 18, 39, 40
  Mosotti, 20
Field equations, Maxwell's, 9
Film material, 177, 197, 219, 221, 224
Flashover strength, 154, 204
Flex temperature, 216, 227
Flexural resonance, 256
Fluorocarbon gases, 149 ff.
Fluorocarbon liquids, 157, 196, 200, 205
Fluorocarbon resins, 177, 210, 217, 232
Forsterite, 183
Freon, 149
Frequency modulation, 73
Frequency standards, 259
Fresnel's equations, 13
Friction factor for dipole rotation, 38

g-factor, 26
GR-S rubber, 171, 226, 228
Gas evolution, 160, 161, 209
Generator, ultrasonic, 260
Germanium rectifier, 243
Glass-ribbon capacitors, 225
Glathart cavity method, 133
Glazes on insulators, 180, 182
Ground insulation of transformers, 193
Guard circuits, 48, 50, 57
Gyromagnetic ratio, 25, 139 ff.

Hardness of polymers, 175, 227
Heisenberg uncertainty relation, 27
Heterogeneity and dielectric loss, 164
Hexafluoride, 149
High-frequency vulcanization, 237
Holes, 245
Horn transducer (Mason), 261
Hydride process, 184
Hydrogen atom, 22, 23
Hydrogen chloride molecule, 34
Hydrogen evolution, 161
Hydrogen-ion concentration, 166
Hypalon, 174
Hysteresis loops, 42, 125, 264, 270, 275

Ice, 169
Ice-phobic surface treatment, 285
Impedance, bridges, for high frequencies, 71 ff.
  input, 64 ff.

Impedance, intrinsic, 11, 15
  measurements, 17, 18, 64 ff.
  of metal, 15
  residual, 56
  terminating, 16
Impregnants for paper capacitors, 223
Indentation hardness, 176
Index, of absorption, 11
  of refraction, 11
Inductance, 4, 128
Induction (magnetic), 7, 122 ff.
Inductive capacity, see Permeability
Inductor (saturable), 264
Inhibitors, 159, 195
Insulation, asphalt bonded, 206 ff.
  classification, 183, 190
  hook-up wire, 212
  migration, 208
Intrinsic impedance of dielectric, 10
Inverse standing-wave ratio, 66
Invisibility condition, 14
Ionizable impurities, 159 ff., 213
Irradiation effects, 162, 220, 286

Jointing of conductors, 235

K shell, 30
k value of plasticizers, 215
Kaolin, 179, 180
Kel-F, 177, 217, 232
Kirkwood theory, 40

Langevin function, 32
Larmor precession, 140
Life time, of insulation, 191, 288, 289
  mean, 142
Light, degradation of polyethylene by, 218
  panalescent, 178
  velocity of, 11, 13
Light quanta, 25
Lithia porcelains, 185
Local field, 18 ff., 39 ff.
Lorentz-Lorenz equation, 21
Loschmidt number, 21
Loss current, 3
Loss factor, 4
  of polyethylene after milling, 174
Loss tangent, 4, 10; see also Tables of Dielectric Materials

Magnetic induction (magnetic flux density), 7
Magnetic metals for amplifiers, 269
Magnetic resonance, 139–146
Magnetite, 125, 188
Magnetization, 7
Magneton, 140
Magneto-resistance of Bi, 273
Magnetostriction, 189
Markites, 178
Matrices for piezoelectric materials, 252, 253
Maxwell's field equation, 9
Maxwell relation, 11

Index

Maxwell-Wien bridge, 129
Measuring cells, 49 ff.
Measuring methods, conductance-variation, 59
  frequency-modulation, 73
  frequency-variation, 59, 70
  line-length-variation, 70
  open-circuit, 18, 66, 133
  resistance-variation, 59
  resonance-rise, 59
  short-circuit, 17, 66, 133
  susceptance-variation, 59, 131
  transmission-line, 63 ff., 72
  two-position, 74
Mechanisms of ionization, 149 ff.
Memory systems, 274 ff.
Metallic shielding of cables, 237
Methacrylates, 171
Mica (flakes, paper, tape), 205, 207, 225
Micrometer-electrode system, 49, 50
Microwave spectroscopy, 134 ff.
Mineral compositions, 179
Miniaturization, 283, 287
Moisture in insulation, 195, 199, 204, 227
Molar polarization, 22, 33
Molar refraction, 21
Mosotti catastrophe, 40, 45
Mosotti field, 20
Mylar, 171, 177

n-p type junction, 245
Neoprene, 226, 232, 239
Nitrogen fixation, 162
Nodes, 15
Nuclear moment, 140, 144
Nuclear resonance, 141 ff.
Nylon, 171, 231

O=C=Se microwave spectrum, 135, 136
Oil deterioration, 159 ff., 195 ff., 209
Onsager theory, 40
Open-circuit method, 18, 66, 133
Orbitals, 30, 34
Orlon, 209
Oscillator, harmonic, 23, 33
Oxidation, 160, 195, 209, 218
Oxygen content of polymers, 220
Ozone, stability to, 175, 210

Paper, 197 ff., 205, 222
Paramagnetic resonance, 144
Paschen's law, 149
Patents, 241
Pauli exclusion principle, 30
Periodic system of elements, 28-30
Permeability, 5, 7, 123 ff.
  notations of, 124
Permeability measurements, 122-134
Permittivity (dielectric constant), 3 ff.
Permittivity measurements, distributed circuits, 63-122
  lumped circuits, 47-62
  see also Tables of Dielectric Materials

Perovskite structure, 42
Phase factor, 9
Phase transitions of $BaTiO_3$, 44
Phase velocity, 10
Phenol-formaldehyde polymer, 169
Photon, 26
Piezoelectricity, 40 ff., 250 ff.
Planck constant, 25
Plasticizer, 175, 213 ff., 231
Plating out of ions, 213
Point-to-plane breakdown, 152, 153
Polar axis, 42
Polarizability, 18 ff.
Polyamides, 171
Polyesters, 170
Polyethylene, 169, 174, 175, 218, 227, 231, 237
Polymer types, 169
Polymerization, 230
Polystyrene, 170, 219, 224
Polytetrafluoroethylene (Teflon), 170, 177, 219, 232
Polyvinylcarbazole, 170
Polyvinylchloride, 170, 175, 176, 215, 231
Porcelains, 180 ff.
Power factor, 3, 10
Pressboard, 194
Probability wave, 26
Propagation factor, 9
Protective covering, for cables, 238
Pyroelectricity, 42

Q (dielectric quality factor), 10
Quantization of rotator, 25, 26
Quantum numbers, 27, 30, 140
Quartz, 180, 188, 254 ff.
Quartz resonators, 260
Quinones, 166

Radii (covalent, ionic, van der Waals), 31
Radome, 284
Rate process, 190, 209
Rectangular hysteresis loops, 125, 275 ff.
Rectifiers, 242 ff.
  germanium, copper oxide, copper sulfide, selenium, 243
Reflection coefficient, 13, 66
Refraction law, 14
Register selection, 276
Relaxation spectrum, 36-40
Relaxation time, 37, 146, 168
Requirements of the Armed Services, 283-290
Resistivity of ceramics, 182
Resonance absorption, 24, 143
Resonance experiment (Bitter), double, 145
Resonance states, 33 ff.
Resonant circuits, 58 ff., 259
Resonators, piezoelectric, 250 ff.
Rochelle salt, 42, 188
Rotation (free or hindered), 35, 40, 169
Rotator, 33 ff.

Rubber, 171-173, 225 ff.
Rutile, 186
Rydberg-Ritz combination principle, 25

Scavenger, 196
Schering bridge, 55 ff.
Schrödinger equation, 27
Sealed conductors of cables, 234
Seals (metal-ceramic), 184
Second-order transition, 173, 220
Selenium rectifiers, 244 ff.
Semiconductors, 125, 245
Shattering temperature, 227
Shells of atoms (K, L, M, etc.), 30
Shielding, 56, 237
Short-circuit methods, 17, 66, 133
Silicone oils, 157
Silicone resins, 171, 178, 210, 231
Similarity of mechanical and electrical relaxation, 176
Simplified calculations for input impedance method, 74, 119-121
Snell's laws, 13
Soluble oxidation products, influence on power factor, 163
Space charge, 19, 147
Sparkplug insulators, 182
Specifications, 183, 190, 227, 241, 289
Stability of polymers, 175, 195, 227
Stabilizers, 159, 196, 200, 219
Standing-wave ratio, 16, 67, 75-80
Standing waves, 15, 65 ff.
Stark effect, 36, 136
Stark modulator, 138
Statistical weights, 36
Steatite, 183
Stokes's law, 38
Stresses (polar and axial), 41, 250 ff.
Sulfur hexafluoride, 149
Susceptance-variation methods, 59 ff., 131 ff.
Susceptibility, 6, 7, 21
Switchgear, 201 ff.

Tables of Dielectric Materials, 291 ff.
Tables of $\dfrac{\tan x}{x}$ and $\dfrac{\cot x}{x}$, 104-117
Talc, 185
Teflon, 170, 177, 203, 217, 232
Temperature characteristics of capacitors, 223
Temperature stability of materials in contact, 185
TEM wave, 10
Testing, 190, 227, 241, 279
Thermal shock resistance of ceramics, 182
Thermalastic, 207
Three-electrode system, 48
Titanates, 42, 186, 271
Torque (electric or magnetic), 6, 7, 31
Total reflection, 15
Townsend process, 147
Transducers, piezoelectric, 250 ff.

Transformer cooling by fluorocarbon liquid, 196
Transformers, 56, 191 ff., 265 ff.
Transition probabilities, 36
Transmission coefficient, 13
Transmission-line methods, 63 ff.
Transmission measurements, 64 ff.
Traveling-probe methods, 63 ff.
Two-position method, 74, 121

Ultrasonic generator, 260
Urea plastics, 205

Valence electrons, 30
Valence molecules, 34
Van der Waals attraction, 31
Vector addition of dipole moments, 36
Vibrator quantization, 34
Vinyl and vinylidene polymers, 171
Viscosity, 38, 157, 159
Voltage rating, of cables, 236
of capacitors, 221
Vulcanization of polyethylene, 231

Wall losses in wave guides, 74, 86

Water absorption, 180, 219, 227
Water molecule, 33–35
Wave equations, 9
Wave forms for saturable inductor, 265
Wave-guide instrument, 72
Wave mechanics, 26
Wheatstone bridge, 54
Wollastonite, 185

Zeeman effect, 36, 137
Zinc ferrite, 189
Zircon, 185

The Artech House Microwave Library

Acoustic Charge Transport: Device Technology and Applications, R. Miller, C. Nothnick, and D. Bailey

Advanced Automated Smith Chart Software and User's Manual, Version 2.0, Leonard M. Schwab

Algorithms for Computer-Aided Design of Linear Microwave Circuits, Stanislaw Rosloniec

Analysis, Design, and Applications of Fin Lines, Bharathi Bhat and Shiban K. Koul

Analysis Methods for Electromagnetic Wave Problems, Eikichi Yamashita, editor

Automated Smith Chart Software and User's Manual, Leonard M. Schwab

C/NL2 for Windows: Linear and Nonlinear Microwave Circuit Analysis and Optimization, Software and User's Manual, Stephen A. Maas and Arthur Nichols

Capacitance, Inductance, and Crosstalk Analysis, Charles S. Walker

Design of Impedance-Matching Networks for RF and Microwave Amplifiers, Pieter L. D. Abrie

Dielectric Materials and Applications, Arthur von Hippel, editor

Dielectrics and Waves, Arthur von Hippel

Digital Microwave Receivers, James B. Tsui

Electric Filters, Martin Hasler and Jacques Neirynck

E-Plane Integrated Circuits, P. Bhartia and P. Pramanick, editors

Feedback Maximization, Boris J. Lurie

Filters with Helical and Folded Helical Resonators, Peter Vizmuller

Fundamentals of Distributed Amplification, Thomas T. Y. Wong

GaAs FET Principles and Technology, J. V. DiLorenzo and D. D. Khandelwal, editors

GaAs MESFET Circuit Design, Robert A. Soares, editor

GASMAP: Gallium Arsenide Model Analysis Program, J. Michael Golio et al.

Handbook of Microwave Integrated Circuits, Reinmut K. Hoffmann

Handbook for the Mechanical Tolerancing of Waveguide Components, W. B.W. Alison

HEMTs and HBTs: Devices, Fabrication, and Circuits, Fazal Ali, Aditya Gupta, and Inder Bahl, editors

High-Power Microwave Sources, Victor Granatstein and Igor Alexeff, edtiors

High-Power GaAs FET Amplifiers, John Walker, editor

High-Power Microwaves, James Benford and John Swegle

Introduction to Microwaves, Fred E. Gardiol

Introduction to Computer Methods for Microwave Circuit Analysis and Design, Janusz A. Dobrowolski

Introduction to the Uniform Geometrical Theory of Diffraction, D. A. McNamara, C. W. I. Pistorius and J. A. G. Malherbe

LOSLIN: Lossy Line Calculation Software and User's Manual, Fred E. Gardiol

Lossy Transmission Lines, Fred E. Gardiol

Low-Angle Microwave Propagation: Physics and Modeling, Adolf Giger

Low Phase Noise Microwave Oscillator Design, Robert G. Rogers

MATCHNET: Microwave Matching Networks Synthesis, Stephen V. Sussman-Fort

Matrix Parameters for Multiconductor Transmission Lines: Software and User's Manual, A. R. Djordjevic et al.

MIC and MMIC Amplifier and Oscillator Circuit Design, Allen Sweet

Microelectronic Reliability, Volume I: Reliability, Test, and Diagnostics, Edward B. Hakim, editor

Microelectronic Reliability, Volume II: Integrity Assessment and Assurance, Emiliano Pollino, editor

Microwave and RF Circuits: Analysis, Synthesis, and Design, Max Medley

Microwave and RF Component and Subsystem Manufacturing Technology, Heriot-Watt University

Microwave Circulator Design, Douglas K. Linkhart

Microwave Engineers' Handbook, 2 Volumes, Theodore Saad, editor

Microwave Materials and Fabrication Techniques, Second Edition, Thomas S. Laverghetta

Microwave MESFETs and HEMTs, J. Michael Golio *et al.*

Microwave and Millimeter Wave Heterostructure Transistors and Applicatons, F. Ali, editor

Microwave and Millimeter Wave Phase Shifters, Volume I: Dielectric and Ferrite Phase Shifters, S. Koul and B. Bhat

Microwave and Millimeter Wave Phase Shifters, Volume II: Semiconductor and Delay Line Phase Shifters, S. Koul and B. Bhat

Microwave Mixers, Second Edition, Stephen Maas

Microwave Transmission Design Data, Theodore Moreno

Microwave Transition Design, Jamal S. Izadian and Shahin M. Izadian

Microwave Transmission Line Couplers, J. A. G. Malherbe

Microwave Tubes, A. S. Gilmour, Jr.

Microwaves: Industrial, Scientific, and Medical Applications, J. Thuery

Microwaves Made Simple: Principles and Applicatons, Stephen W. Cheung, Frederick H. Levien *et al.*

MMIC Design: GaAs FETs and HEMTs, Peter H. Ladbrooke

Modern GaAs Processing Techniques, Ralph Williams

Modern Microwave Measurements and Techniques, Thomas S. Laverghetta

Monolithic Microwave Integrated Circuits: Technology and Design, Ravender Goyal *et al.*

Nonlinear Microwave Circuits, Stephen A. Maas

Nonuniform Line Microstrip Directional Couplers, Sener Uysal

Optical Control of Microwave Devices, Rainee N. Simons

PC Filter: Electronic Filter Design Software and User's Guide, Michael G. Ellis, Sr.

PLL: Linear Phase-Locked Loop Control Systems Analysis Software and User's Manual, Eric L. Unruh

Scattering Parameters of Microwave Networks with Multiconductor Transmission Lines: Software & User's Manual, A. R. Djordjevic et al.

Solid-State Microwave Power Oscillator Design, Eric Holzman and Ralston Robertson

Terrestrial Digital Microwave Communications, Ferdo Ivanek et al.

Time-Domain Response of Multiconductor Transmission Lines: Software and User's Manual, A. R. Djordjevic et al.

Transmission Line Design Handbook, Brian C. Waddell

Yield and Reliability in Microwave Circuit and System Design, Michael Meehan and John Purviance

For further information on these and other Artech House titles, contact:

Artech House	Artech House
685 Canton Street	Portland House, Stag Place
Norwood, MA 02062	London SW1E 5XA England
617-769-9750	+44 (0) 71-973-8077
Fax: 617-769-6334	Fax: +44 (0) 71-630-0166
Telex: 951-659	Telex: 951-659
email: artech@world.std.com	bookco@artech.demon.co.uk